Causes of Photooxidative Stress and Amelioration of Defense Systems in Plants

Causes of Photooxidative Stress and Amelioration of Defense Systems in Plants

Edited by

Christine H. Foyer, Ph.D.
Director of Research
Laboratory of Metabolism
Institut National de la Recherche Agronomique
Versailles, France

Philip M. Mullineaux, Ph.D.
Principal Scientific Officer
Department of Applied Genetics
John Innes Institute
Norwich, England

CRC Press
Taylor & Francis Group
Boca Raton London New York

CRC Press is an imprint of the
Taylor & Francis Group, an **informa** business

First published 1994 by CRC Press
Taylor & Francis Group
6000 Broken Sound Parkway NW, Suite 300
Boca Raton, FL 33487-2742

Reissued 2018 by CRC Press

Library of Congress Cataloging-in-Publication Data

Causes of photooxidative stress and amelioration of defense
systems in plants/edited by Christine H. Foyer and Philip M. Mullineaux.
p. cm.
Includes bibliographical references and index.
ISBN 0-8493-5443-9
1. Plants, Effect of photooxidative stress on. 2. Plant defenses.
I. Foyer, Christine H. II. Mullineaux, Philip M.
QK757.C38 1994
581.2'4 — dc20 93-25101

A Library of Congress record exists under LC control number: 93025101

ISBN 13: 978-1-315-89135-4 (hbk)
ISBN 13: 978-1-351-07045-4 (ebk)

Visit the Taylor & Francis Web site at http://www.taylorandfrancis.com and the
CRC Press Web site at http://www.crcpress.com

PREFACE

It may not be apparent to the casual reader that photooxidative stress is a predominant destructive process in the life of plants. This is indeed the case, even in the areas where the contributory role of light is not immediately apparent. The reader will note that conflicting views are expressed in some topics; this reflects the activity in the field. It is not a moribund area of research.

For many years, oxidative stress in plants has largely been the domain of physiologists and biochemists. However, the advent of plant transformation and molecular genetic techniques has opened this field to new investigators. The experience of the editors testifies to this. The production of this volume presented an intellectual challenge that arose, in part, from a molecular biologist being forced to understand the fundamentals of photochemistry and free radical biology!

In an attempt to improve communication between disciplines in this field, we have aimed to cover what we perceive to be all relevant aspects of photooxidative stress: from primary reactions to molecular genetics and the devising of strategies for engineering stress tolerance in plants. We hope to achieve a forum for new ideas, concepts, and approaches. The intellectual challenge also arose because we wished to produce a work that was accessible to both specialist and nonspecialist. We have encouraged our authors to provide personal perspectives of their topics while discussing them in depth. To this end, the nonspecialist will find that some chapters include relatively simple introductions and conclusions, e.g., Foyer and Harbinson (Chapter 1); Gressel and Galun (Chapter 10).

We hope that molecular biologists will find that the information on biochemistry and physiology is accessible. Similarly, the biochemists and physiologists may appreciate that a genetic component in their experiments could be of immense value. In some respects this book arose out of a perception on our part that laboratories now have to be conversant in more than one discipline or involved in interdisciplinary collaborations. Indeed, we hope that this book will foster such aspirations. The rapid response and clear enthusiasm of our contributors suggest that our views are widely felt. We wish to acknowledge the patience of our contributors in the fulfilment of these exacting guidelines. The editors hope that as a result of this book the authors will not be excluded by the contributors from future collaborations!

There are many opportunities to do original research in this area, and we would encourage workers in other fields to regard this subject as worthy of their attention; for example, it is especially noteworthy that there is a lack of application of classical genetics to the problems of photooxidative stress in plants.

Christine H. Foyer
Philip M. Mullineaux

THE EDITORS

Christine H. Foyer, Ph.D. is Director of Research at the metabolism laboratory of Institut National de la Recherche Agronomique (INRA), Versailles, France.

Dr. Foyer graduated in 1974 from Portsmouth Polytechnic with a B.Sc. degree in Biology (Hons) and obtained her Ph.D. in 1977 from Kings College, University of London, U.K. She remains deeply indebted to Professor Barry Halliwell, her thesis supervisor, for introducing her to the topic of the function of ascorbic acid and glutathione in leaves. From 1977 to 1979 Dr. Foyer was a post-doctoral fellow at the Department of Plant Sciences, Kings College, London and then moved to the Research Institute for Photosynthesis at the University of Sheffield, U.K. In 1988 she moved to her present position in Versailles.

Dr. Foyer has published a book on photosynthesis and has more than 80 published research papers. She is frequently asked to participate in research conferences and seminars. She is a member of the Institute of Biology and the American Society of Plant Physiology. Her research work is funded by INRA and the European Economic Community and she maintains valuable contacts with industry. Her current major research interests include the exploration of the nature of relationships between photosynthesis and the other metabolic pathways in the leaf, notably oxygen, nitrogen, and carbohydrate metabolism, and the use of transformed plants to further understand the mechanisms of integration of these pathways.

Philip Mullineaux, Ph.D. is Principal Scientific Officer in the Department of Applied Genetics in the John Innes Institute at the John Innes Centre for Plant Science Research, Norwich, U.K.

Dr. Mullineaux received his B.Sc. in Biochemistry in 1978 and his Ph.D. in 1981 for research into the relationship between dinitrogen fixation and photosynthesis in unicellular cyanobacteria. Both his degrees are from the University of Wales.

From 1981 to 1983, Dr. Mullineaux was a Medical Research Council Research Fellow in the Department of Molecular Biology at the University of Edinburgh, working on the molecular genetics of conjugation in *Escherichia coli*. From 1983 to 1986 he was an Agrigenetics-funded Senior Scientific Officer in the Department of Virus Research at the John Innes Institute, investigating the molecular biology of plant DNA viruses which infect the Gramineae. Dr. Mullineaux has been in his current post as Principal Scientific Officer since 1986. He is a member of the International Society of Plant Molecular Biology and the Society for Experimental Biology.

Dr. Mullineaux has been recipient of grants from the Agricultural and Food Research Council, the Commission of European Communities, the Department of Industry, and private industry. He has published more than 40 papers. His current research interests include the study and manipulation of antioxidant metabolism in plants (especially in relation to oxidative stress), the development of plant transformation technology, and the interaction of transcription and DNA replication in cereal-infecting plant DNA geminiviruses.

CONTRIBUTORS

William W. Adams III, Ph.D.
Department of Environmental, Population, and Organismic Biology
University of Colorado
Boulder, Colorado

Kozi Asada, Ph.D.
The Research Institute for Food Science
Kyoto University
Uji, Kyoto, Japan

Neil R. Baker, Ph.D.
Department of Biology
University of Essex
Colchester, Essex, England

Gary Creissen, Ph.D.
Department of Applied Genetics
John Innes Institute
Norwich, England

Barbara Demmig-Adams, Dr. rer. nat.
Department of Environmental, Population, and Organismic Biology
University of Colorado
Boulder, Colorado

Alan D. Dodge, Ph.D.
Department of Biological Sciences
Bath University
Bath, England

Noriyuki Doke, Dr. Agr.
Plant Pathology Laboratory
School of Agricultural Sciences
Nagoya University
Nagoya, Japan

Anne Edwards, Ph.D.
Department of Applied Genetics
John Innes Institute
Norwich, England

Christine H. Foyer, Ph.D.
Laboratoire du Métabolisme
INRA
Versailles, France

Esra Galun, Ph.D.
Department of Plant Genetics
Weizmann Institute of Science
Rehovot, Israel

Jonathan Gressel, Ph.D.
Department of Plant Genetics
Weizmann Institute of Science
Rehovot, Israel

Jeremy Harbinson, Ph.D.
ATO Agrotechnologie
Wageningen, The Netherlands

Dirk Inzé, Ph.D.
Laboratoire associé de l'INRA
University of Gent
Gent, Belgium

Kazuhito Kawakita, Dr. Agr.
Plant Pathology Laboratory
School of Agricultural Sciences
Nagoya University
Nagoya, Japan

G. Heinrich Krause, Ph.D.
Institute of Plant Biochemistry
Heinrich Heine University Düsseldorf
Düsseldorf, Germany

Horst Mehlhorn, Ph.D.
Institut für Angewandte Botanik (FB 9.4)
University of Essen
Essen, Germany

Yoshio Miura, M. Agr.
Plant Pathology Laboratory
School of Agricultural Sciences
Nagoya University
Nagoya, Japan

Philip Mullineaux, Ph.D.
Department of Applied Genetics
John Innes Institute
Norwich, England

Andrea Polle, Ph.D.
Institut für Forstbotanik und Baumphysiologie
Freiburg, Germany

Heinz Rennenberg, Ph.D.
Institut für Forstbotanik und Baumphysiologie
Freiburg, Germany

Leandro M. Sanchez, Dr. Agr.
Plant Pathology Laboratory
School of Agricultural Sciences
Nagoya University
Nagoya, Japan

John G. Scandalios, Ph.D.
Department of Genetics
North Carolina State University
Raleigh, North Carolina

Kiyoshi Tanaka, Dr. Agr.
Division of Environmental Biology
The National Institute for Environmental Studies
Tsukuba, Ibaraki, Japan

Wim Van Camp, M.Sc.
Laboratorium voor Genetica
University of Gent
Gent, Belgium

Mark Van Montagu, Ph.D.
Laboratorium voor Genetica
University of Gent
Gent, Belgium

Alan R. Wellburn, Ph.D.
Division of Biological Sciences
Institute of Environmental and Biological Sciences
Lancaster University
Bailrigg, Lancaster, England

TABLE OF CONTENTS

Chapter 1

OXYGEN METABOLISM AND THE REGULATION OF PHOTOSYNTHETIC ELECTRON TRANSPORT

Christine H. Foyer and Jeremy Harbinson

TABLE OF CONTENTS

0-8493-5443-9/94/$0.00+$.50

I. INTRODUCTION

In this chapter, we consider the ways in which oxygen metabolism interacts with the photosynthetic processes. First of all, it is important to consider what is meant by the term "photooxidative stress". Photooxidative stress may be defined as the generation of toxic derivatives of oxygen by light-dependent processes, i.e., the initiation or exacerbation of oxidant production as a result of illumination. Active oxygen species and radicals may be produced directly in response to illumination, e.g., via oxygen reduction by the photosynthetic electron transport chain (as described below), or indirectly, e.g., during the signal transduction response to UV irradiation (see Chapter 7 of this volume). Furthermore, we consider that such processes are a major source of damage in plants exposed to unfavorable environments, and the contributory role of light is often overlooked.

The photosensitizing pigments required for the interception of light energy have the potential to pass that energy to oxygen to create one of the most destructive species in biology, singlet oxygen. Similarly, the photosynthetic requirement for electron carriers with negative electrochemical potentials, which allows electron transport to oxygen, is potentially extremely hazardous. In this chapter, we explore the mechanisms whereby these interactions take place, are controlled, and are limited. In order to do this effectively, we first describe some of the essential features that comprise photosynthetic regulation. This is essential if the nature of avoidance mechanisms and protective devices are to be understood.[1] However, the system is much more complex than simply one that is trying to avoid oxidative stress. Photosynthetic systems have embraced the potential of interaction with oxygen such that oxygen metabolism is now intimately involved with the regulation of photosynthesis.[2,3] The conclusions and perspectives at the end of this chapter may be used as a summary by the non-specialist.

A. AN INTRODUCTION TO THE REGULATION OF PHOTOSYNTHETIC ELECTRON TRANSPORT

Oxygen is one of the products of photosynthetic electron transport, while on the other hand, it is also reduced by the photosynthetic electron transport chain during pseudocyclic electron flow and assimilated via photorespiration.[4] These processes contribute to the overall regulation of the photosynthetic electron transport system and fulfill essential, but rather different roles by providing necessary protective mechanisms.

The coordination of the control of the electron transport processes with the assimilatory processes of photosynthesis, of which CO_2 fixation by the Calvin cycle is the most conspicuous, is an essential feature of photosynthetic regulation.[1-3] It ensures that the rates of ATP and NADPH synthesis match the rate of utilization of these products in metabolism. This coordinate regulation, termed photosynthetic control,[2] serves to avoid potentially damaging overexcitation and overreduction of components within the electron transport system that would lead to damage, loss of function, and decreases in fitness or yield. Precise coordination of reaction rates prevents continuous oscillations in metabolite flux, allows optimization of resources, and yet confers a degree of flexibility that is essential for the avoidance of the detrimental effects of light in a constantly changing environment.

The maintenance of constant ratios of (NADPH)/(NADP) and (ATP)/(ADP) in situations of excess and suboptimal irradiance requires active feedback control of the thylakoid processes and coordination with the feed-forward modulation of carbon assimilation.[3] The feed-forward and feedback mechanisms that act to control the photosynthetic machinery are imperfectly understood. However, they achieve a coordination of the varied reactions that comprise photosynthesis that is good enough to allow the stable, efficient function of photosynthesis without compromising the flexibility of response required of any process operating in a continually changing environment. An inevitable consequence of the operation of the photosynthetic system with its populations of excited and reducing components, in an oxygen-rich environment, is the risk of producing reactive derivates of oxygen, either free radicals or neutral excited states, which can then attack and damage cellular structures.

It appears that the regulation of the light-harvesting components and electron transport activity also serves to reduce the risk of generating these dangerous oxygen derivates. The regulation of photosynthesis is complicated by the necessity to reconcile the conflicting requirements of the thylakoid reactions and the stromal enzymes.[3] In this regard, several regulatory mechanisms are involved in the modulation of the activation states of enzymes of the carbon reduction cycle so as to match their activities to the availability of the products of electron flow.[2,3] Electron transport is restrained when ADP and NADP are in short supply, but the mechanisms operating *in vivo* are not fully understood.[2] Under these conditions, the quantum efficiencies of both photosystems (PSII and PSI) are downregulated. The rapidly reversible mechanisms that serve to decrease the quantum efficiencies of PSII and PSI also facilitate the harmless conversion of light energy directly to heat.

Under conditions of high irradiance, coupled to a limited capacity for electron transport, a slowly reversible downregulation of PSII efficiency develops. This may act as a mechanism of last resort, to control the electron transport rate. The quantum efficiencies of PSII and PSI are generally decreased in unison with respect to increasing irradiance.[2] The availability of excitation energy for photochemistry, and thus electron transport through PSII, is determined to a large degree by the action of imperfectly understood competitive quenching processes within the PSII pigment bed. These quenching processes are clearly stimulated *in vitro* by the establishment of a transthylakoid ΔpH and its associated processes.[1-3] The more excitation energy that is quenched nonradiatively (i.e., via thermal energy dissipation) by these processes, the less energy is available for photochemistry. Just as photochemistry and

TABLE 1
Glossary of Terms Used in This Chapter

AP	ascorbate peroxidase
ϕ_{CO_2}	index of the quantum efficiency of CO_2 fixation
ϕ_F	fluorescence yield
ϕ_{PSI}	index of the relative quantum efficiency for electron transport by PSI
ϕ_{PSII}	index of the relative quantum efficiency for electron transport by PSII
ϕ_T	yield of chlorophyll triplet states
Fd	ferredoxin
Fd_{red}	reduced ferredoxin
F_o	level of modulated Chl *a* fluorescence during a dark interval
F_o'	level of modulated Chl *a* fluorescence observed upon exposure to the measuring modulated light source following a period of light adaptation
F_m	level of modulated Chl *a* fluorescence during a saturating pulse of irradiance following a period of light adaptation
F_m'	level of modulated Chl *a* fluorescence during a saturating light pulse imposed following a period of light adaptation
F_s	the steady-state Chl *a* fluorescence yield under given environmental conditions
F_v	level of variable Chl *a* fluorescence
GSH	reduced glutathione
GSSG	oxidized glutathione
J_{PSI}	product of ϕPSI and incident irradiance
J_{PSII}	product of ϕPSII and incident irradiance
NADPH-MDH	NADP-malate dehydrogenase
O_2^-	superoxide
P_{700^o}	the fraction of the P_{700} pool that remains unoxidized
P_{700^+}	the fraction of the P_{700} pool that is oxidized
PQ	plastoquinone
PSI	photosystem I
PSII	photosystem II
Q_A	the primary stable electron acceptor of PSII
q_E	coefficient for nonphotochemical but energy-dependent quenching of chlorophyll *a* fluorescence associated with the establishment of the transthylakoid ΔpH; a component of q_{NP}
q_P (or q_Q)	coefficient for the photochemical quenching of chlorophyll *a* fluorescence = $(F_m' - F_s)/F_v'$
q_{NP}	coefficient for nonphotochemical quenching of chlorophyll *a* fluorescence = $(F_m' - F_m)/F_m'$
RuBisCo	ribulose-1,5-bisphosphate carboxylase-oxygenase
RuBP	ribulose-bisphosphate
VD	violaxanthin de-epoxidase

electron transport produce a decrease in the yield of chlorophyll *a* fluorescence of PSII, so does the action of these alternative quenchers. However, since they do not produce photochemistry leading to electron transport, this type of quenching is termed nonphotochemical quenching (q_{NP}). Since the majority of q_{NP} is also associated with the establishment of the transthylakoid ΔpH, this energy quenching component is also referred to as energy-dependent quenching or q_E (see Table 1 for glossary of terms). The role of q_{NP} in photoprotection is fundamental, since it acts first to limit the rate of reduction of the first stable electron acceptor to PSII, Q_A. In the event that electron transport is limited, q_{NP} will serve to constrain the steady-state pool of Q_A^- and therefore minimize photoinhibition (see Chapter 2 of this volume). Second, q_{NP} acts to reduce the lifetime of the excited state of singlet chlorophyll in the PSII pigment bed, and this will decrease the probability of the formation of chlorophyll triplet states. The latter can interact directly with ground-state triplet oxygen, leading to the formation of singlet oxygen, $^1O_2$. A major function of the membrane carotenoids and α-tocopherol is to eliminate these highly destructive species.

q_{NP} allows the harmless and controlled thermal dissipation of excess excitation energy within the photochemical apparatus. The molecular mechanisms underlying q_{NP} (or q_E) are still a matter of debate,[2] but they are believed to involve the carotenoid pigments of the xanthophyll cycle and particularly the de-epoxidized states, antheraxanthin and zeaxanthin (see the review in Chapter 4 of this volume). Carotenoid pigments are of two types: (1) the oxygen-free carotenes and (2) the xanthophylls, which contain oxygen in different forms such as hydroxy or epoxy groups. Both of these will quench the triplet state of chlorophyll and the singlet state of oxygen, $^1O_2$. The photoprotective action of the carotenoids in photosynthesis has generally been considered in relation to this function alone. However, the role of zeaxanthin in thermal energy dissipation associated with q_{NP} is rather unique (see Chapter 4 of this volume).

It is evident that the supply of reducing equivalents to the stroma provided by the thylakoid electron transport chain is strictly regulated. Fluxes of reducing equivalents pass through many pathways in the stroma, e.g., through NADPH to the Calvin cycle and the thioredoxin system to the thiol-modulated enzymes. Electrons are donated principally to NADP, but they can also be donated to O_2 or returned to the intersystem electron carriers by the cyclic pathway of electron flow.[5-7] The role of oxygen in the regulation of the rate of cyclic electron flow is only poorly understood. Cyclic electron flow has been demonstrated *in vivo*,[6] but no studies of its regulation *in vivo* have, thus far, been carried out. *In vitro* cyclic electron flow is known to be regulated by the redox states of the intersystem electron carriers[5] and the redox state of the PSI acceptor pool. In the absence of oxygen, cyclic electron flow is inhibited by overreduction of the electron carriers.[5] When oxygen is added, or PSI activity alone is stimulated by far-red illumination, the inhibition is temporarily relieved. When PSII is not excited, oxygen causes inhibition of cyclic electron flow around PSI by oxidizing the electron transport components; thus, cyclic electron flow must be linked to the PSII electron donor system *in vivo* to be functional in the presence of oxygen.[5] One possible function of oxygen in the regulation of electron transport is that of "poising" the electron carriers of the cyclic pathway.

The intrathylakoid pH, that is closely related to the transthylakoid ΔpH, is a major controlling factor regulating electron transport.[1-3] It does this principally by limiting the rate of the reaction between plastoquinone and the cytochrome *b/f* complex in the electron transport chain and via modulation of light harvesting and photochemistry in PSII.[1-3] The roles of carbon and oxygen metabolism in the processes of photosynthetic control need further elucidation, but they are undoubtedly linked, for example, via the process of photorespiration,[4] which is considered to be the dominant O_2-consuming process in leaves.

II. OXYGEN EFFECTS ON PHOTOSYNTHESIS

Oxygen can have both inhibitory and stimulatory effects on photosynthesis, depending on environmental limitations and metabolic constraints. This apparent paradox is a consequence of the intimate association between oxygen metabolism, photosynthetic electron transport, and carbon assimilation. For simplicity, four types of oxygen-consuming processes may be considered to be associated with photosynthesis. These are (1) the oxygenase reaction of ribulose-1,5-bisphosphate carboxylase-oxygenase (RuBisCo) that initiates the pathway of photorespiration. The leaves of C_3 plants may divert a large proportion of recently fixed carbon through this pathway, which leads to the release of CO_2. (2) Direct reduction of molecular oxygen by the PSI electron transport chain. (3) PSII is also capable of reducing O_2, probably from the quinone electron acceptor site termed Q_A, when it becomes reduced and the herbicide-binding Q_B site is empty. (4) Reduction of oxygen resulting from the presence of a respiratory electron transport pathway, with a cyanide-sensitive component, that competes with the photosynthetic electron transport chain for reducing equivalents. This process, called chlororespiration, has been largely studied in microalgae.[8-10] Since there is little evidence to support the operation of a chlororespiratory pathway in higher plants,[11] it is not considered further here. However, it is worthy of note that the accumulation of inorganic carbon by the CO_2-concentrating mechanisms of cyanobacteria appears to stimulate oxygen reduction by direct effects on the potential for oxygen photoreduction at PSI. Oxygen reduction could, thus, provide the energy required by the inorganic-carbon-concentrating mechanisms via pseudocyclic photophosphorylation.

A. THE MEHLER REACTION AND PHOTORESPIRATION

Photosynthesis is decreased when the CO_2 supply is severely restricted. In these circumstances, the quantum efficiency of PSII (ϕ_{PSII}) is determined largely by the intracellular CO_2 concentration, Ci.[12-14] However, increasing O_2 concentrations can further increase ϕ_{PSII} when CO_2 becomes saturating.[13-15] The effect of photorespiration on net CO_2 assimilation is often demonstrated by comparing photosynthesis in air with that observed at low O_2. Decreasing the oxygen concentration does not, however, always lead to an increase in CO_2 assimilation, particularly when metabolic limitations, imposed by the rate of triose phosphate utilization and, hence the rate of recycling of Pi, constrain the photosynthetic rate.[16,17] This occurs at saturating irradiances or at low temperatures. Paradoxically, when O_2 sensitivity is lost, photosynthesis fails to respond to increasing CO_2 concentrations. When the CO_2 concentration is saturating, increasing the O_2 concentration can **enhance** the rate of net CO_2 assimilation by up to 20%. It is in these circumstances that elevated oxygen concentrations can lead to an enhancement of the photochemical yield (ϕ_{PSII} and ϕ_{PSI}).[14,15] Oxygen has direct effects on the electron transport system. Mehler[18] was the first to provide evidence that molecular O_2 was a Hill oxidant in isolated chloroplasts. Rates of photosynthetic electron transport to O_2, termed "pseudocyclic electron flow", are difficult to measure *in vivo,* but have been reported to vary from 3 to 27% of the total electron transport rate according to conditions employed and species used.[19-21] Much controversy still surrounds the *in vivo* relevance of O_2 photoreduction[22] and the contribution of pseudocyclic electron flow to the regulation of photosynthesis. At atmospheric CO_2 levels, the rates of O_2 reduction are considered to be low, but they are greater in certain circumstances, such as the induction phase of photosynthesis, in saturating CO_2, and near the compensation point.[19,21] Furthermore, the coupling of pseudocyclic electron flow to H_2O_2 production and destruction via the ascorbate peroxidase system provides a putative mechanism for the regulation of electron transport.[23-25]

Photorespiration is considered to be more effective than the Mehler reaction in protecting the photosynthetic processes against photoinactivation.[26] Indeed, this protective action could be the main function of photorespiration.[27] In principle, and making certain assumptions,[28] the proportion of electron transport diverted through this process may be calculated from fluorescence and gas exchange data as follows: the ratio of the rates of carboxylation (V_c) to oxygenation (V_o) of ribulose-bisphosphate (RuBP) can be determined as in Equation 1

$$V_c/V_o = (V_cK_o/V_oK_c) \cdot C/O \qquad (1)$$

where C/O is the molar ratio of dissolved CO_2 to O_2 at the enzyme active site, V_c and V_o represent the maximum velocities for carboxylation and oxygenation, respectively, and K_c and K_o are the Michaelis constants for carboxylation and oxygenation respectively. V_cK_o/V_oK_c is called the specificity factor (K_{sp}). This determines the changes in the relative rates of the activities at given C/O values.[29-31]

According to the model of Farquhar et al.,[31] the velocities of carboxylation (V_c) and oxygenation (V_o) can be determined as in Equations 2 and 3.

$$V_c = \left(\frac{V_cC}{c + K_c[1 + O/K_o]}\right) \gamma \qquad (2)$$

$$V_o = \left(\frac{V_oO}{O + K_o[1 + c/K_c]}\right)\gamma$$

(3)

where γ = (RuBP)/(RuBP) + K_r. K_r is the Michaelis constant for RuBP. The rate of net CO_2 assimilation (A) may be derived from $V_c - 0.5\ V_o$ and the total noncyclic electron flow (J) is $(4 + 4\phi)\ V_c$, where $\phi = V_oK_cO/V_cKOC$. The proportion of electron transport used to drive photorespiration (P_{diss}) is given simply by (J − 4A)/J.

The regulatory functions of photorespiration and pseudocyclic electron flow may be quite distinct. In contrast to photorespiration, pseudocyclic electron flow consumes reducing equivalents but not ATP. It has the capacity to generate a large transthylakoid proton gradient, especially when it is coupled to the ascorbate peroxidase reaction and the rereduction of oxidized ascorbate.[23,25] This may have direct repercussions for the regulation of the quantum efficiency of PSII, which is decreased when the proton gradient is increased. It may, thus, provide fine regulation of the electron transport system as opposed to the coarse, but global, protection afforded by photorespiration, which consumes both ATP and reducing power.

B. THE CELLULAR PERSPECTIVE

Implicit to the regulation of photosynthesis involving O_2 is the production and subsequent destruction of active oxygen species. Free radicals and other active derivatives are minor, but not insignificant, products of normal photosynthesis and respiration. In the present discussion, we consider photosynthesis in relation to oxygen metabolism, but we are mindful that chloroplast metabolism is intimately integrated with the metabolism of the whole photosynthesizing cell, where numerous reactions involving oxygen also occur, as illustrated in Figure 1. The chloroplast is potentially a powerful source of oxidants, and chloroplast metabolism must achieve a regulated balance between oxygen radical production and destruction if metabolic efficiency and function are to be maintained in both optimal and stress conditions.

In addition to metabolic effects, photooxidation also has profound effects on chloroplast and nuclear gene expression.[32] These effects are not due to a general loss in mRNA as a result of photooxidation,[32] rather, mRNA accumulation is regulated both transcriptionally and posttranscriptionally in response to photooxidative stress.[33] Light intensity-dependent photooxidation decreases the accumulation of the mRNAs for both the small subunit of RuBisCo and the light-harvesting chlorophyll *a*/*b* binding protein, but not the 28-kDa chloroplast RNA-binding regulatory protein.[32,33] The photodestruction of the photosynthetic complexes and the photooxidation of chlorophyll is closely correlated with the differential regulation of chloroplast and nuclear gene expression during photooxidation.[32] Chloroplast enzymes and proteins

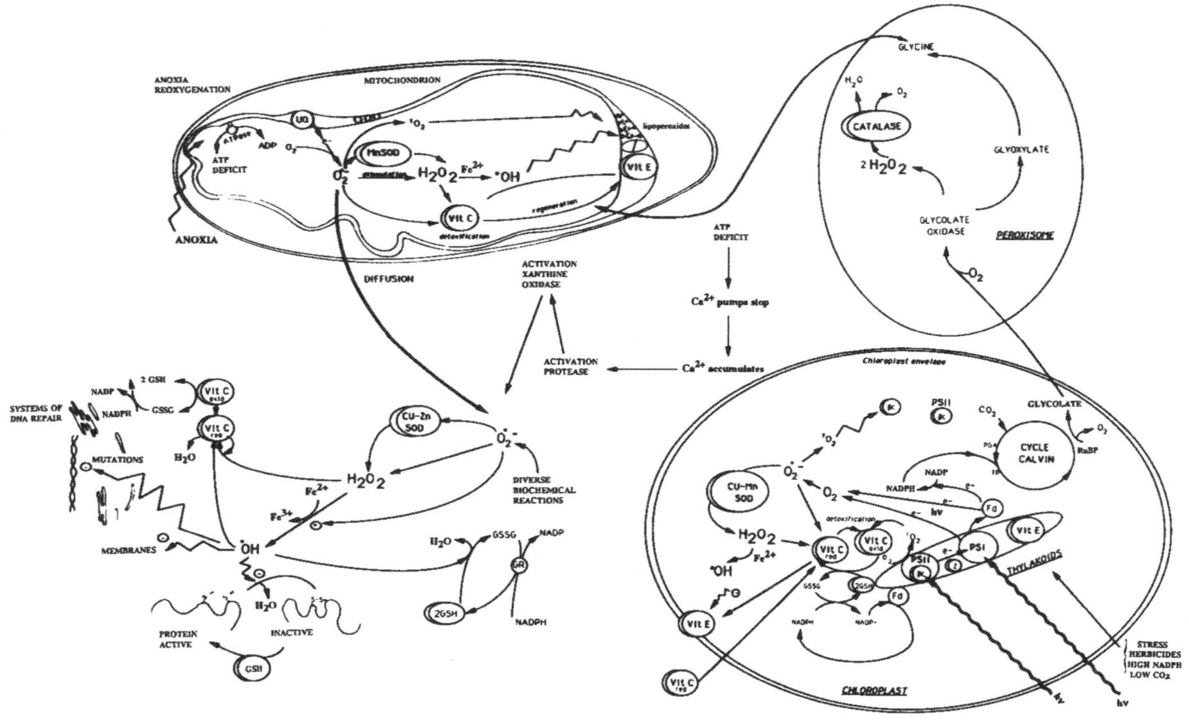

FIGURE 1. Diagram of some of the known reactions involving molecular oxygen and its derivatives in photosynthetic cells.

are greatly decreased as a result of photooxidation, but cytoplasmic and mitochondrial enzymes are maintained at almost normal levels.[32]

III. TRIPLET CHLOROPHYLL AND SINGLET OXYGEN FORMATION

Most molecules exist normally in a singlet ground state; all their electrons have paired spins. Even when excited, they remain in a singlet configuration from which relaxation to the ground state is normally rapid. In addition to the singlet state, molecules can also exist in the triplet configuration in which electrons exist with unpaired spins. In most cases, the triplet state is an excited state which is achieved by relaxation from a higher-energy singlet state. This conversion between singlet and triplet is forbidden and requires spin decoupling to occur, e.g., by interaction with the atomic nucleus or by collisions with other molecules.

Normal, ground-state, molecular oxygen is unusual in having a triplet configuration, with two unpaired electrons each in a π^* orbital and with parallel spins.[34] As most molecules have a singlet ground state with all electrons paired, reactions with molecular oxygen are forbidden. Nonetheless, O_2 can exist in singlet forms which are very reactive because they lack restriction on their reactivity due to spin conservation. Two such forms of O_2 exist, $^1\Sigma_g^+$ and $^1\Delta_g$,[34,35] and both can be formed following the interaction of a triplet O_2 with a triplet chlorophyll molecule. The former is very short-lived in solution and relaxes via collisional processes to the $^1\Delta_g$ state (rate constant for relaxation in water, 10^{11} s^{-1}) before it can react; the latter is more long-lived, lasting for nearly 4 μs in water and up to 100 μs in nonpolar environments. This is long enough for chemical reactions to take place. Therefore, only the $^1\Delta_g$ form is of relevance as a cause of damage in biological systems. It attacks electron-rich sites such as amino acid residues derived from histidine, methionine, and tryptophan.[36] The primary source of singlet oxygen appears to be the photosynthetic pigment bed.[34] A considerable array of defenses within the photosynthetic membranes are present to minimize this type of oxidative stress. The first defense is prevention; the system is regulated to minimize the formation of the chlorophyll triplet states that initiate the process. This avoidance mechanism is considered below.

A. REGULATION OF THE PRODUCTION OF CHLOROPHYLL TRIPLET STATES

Chlorophyll (Chl) has a singlet ground state, and excitation of chlorophyll results in a transition of the molecule to one of two excited singlet states. The higher-energy $S^b(\pi,\pi^*)$ state corresponds to the Soret absorption band in the blue part of the spectrum, and the $S^a(\pi,\pi^*)$ state corresponds to the absorption band in the red. The $S^b(\pi,\pi^*)$ state relaxes in about 10^{-12} s to the lower energy state via internal conversion, and the extra energy is liberated as heat.[37] The $S^a(\pi,\pi^*)$ state is more long-lived, with a lifetime of ~5 ns in ethanolic solution.[38] Decay from this solvated excited state occurs principally via two routes: (1) by radiative decay as fluorescence with a rate constant of $6.5 \times 10^7\ s^{-1}$ (and therefore a lifetime of ~15 ns) and a yield (ϕF) of ~0.32, and (2) by conversion to the triplet state with a rate constant of $1.3 \times 10^8\ s^{-1}$ and a yield (ϕ_T) of ~0.68.[38] In ether, the yields are similar, $\phi_F = 0.32$ and $\phi_T = 0.64$.[39] Shipman[40] calculated a rate constant for $^3Chl^*$ formation from $^1Chl^*$ of $1.9 \times 10^8\ s^{-1}$. Clearly, in solution, the principal route for $^1Chl^*$ relaxation is via the $^3Chl^*$ state. Chlorophylls in solution are vulnerable to light-induced bleaching in the presence of oxygen; whereas, in benzene solution $^3Chl^*$ has a half-life of 400 μs, the addition of 100 μ*M* carotene reduces this to 5 ns[41] and also protects it from damage.[42] *In vitro* chlorophylls can clearly form triplet states, and these triplet states can convert ground-state oxygen to the more reactive singlet form. Carotenes can protect chlorophylls in solution by quenching the chlorophyll triplets.[38-42]

The existence of triplet chlorophyll states in thylakoids was first described by Witt and co-workers,[43-45] who suggested that, as occurs in solution, carotenes could act to quench surplus excitation energy present within the chlorophyll of the thylakoids. This interaction between the triplet states of chlorophyll and carotenes was named the "Triplet Valve". In this model, summarized by Witt,[45] $^3Chl^*$ acts in parallel with fluorescence and nonradiative thermal deexcitation to quench $^1Chl^*$ molecules that have not been quenched by stable charge separation in the reaction center of PSII. $^3Chl^*$ is then rapidly quenched by carotenoids (Car) which forms $^3Car^*$, preventing or minimizing the reaction of the triplet chlorophyll with O_2. $^3Car^*$ can then decay to the ground state by various routes. Mathis and collaborators[46,47] made further quantitative investigations of $^3Chl^*$ and $^3Car^*$ formation in thylakoids following laser-flash excitation, and were able to demonstrate that both were formed following the flash as the $^1Chl^*$ decayed.

Triplet formation has also been detected indirectly: Mauzerall[48] identified a phase of relaxation in the fluorescence quenching in *Chlorella* of $\sim 10^{-6}$ s following a laser flash which he attributed to the decay of quenching $^3Car^*$. Nitsch et al.,[49] using laser-induced optoacoustic calorimetry of PSII particles, found that 5% of the input energy was stored for longer than 1.4 μs, and this they attributed to triplet formation. Clearly, triplet states of both chlorophyll *a* and carotenoids can form *in vitro* under conditions that closely reflect the state of the photosynthetic apparatus in the chloroplast within a leaf. There do not appear, however, to be any measurements of triplet formation *in vivo* under steady-state conditions and, therefore, there is no direct evidence of regulation of the photosynthetic machinery so as to reduce the yield of triplets. Nonetheless, using rate constants for $^3Chl^*$ and $^3Car^*$ formation following laser flashes,[46,47] it is possible to calculate likely yields of triplet formation and to show that the regulation of the photosynthetic machinery *in vivo* serves to reduce triplet formation. Whether the avoidance of triplet formation is the primary function of this regulation is impossible to say.

Within the thylakoids, the chlorophylls (*a* and *b*) and the accessory pigments (principally carotenoids and xanthophylls) are bound in pigment-protein complexes which are structured into two photosystems, each with their own discrete array of complexes. Within these aggregates of pigments and proteins, the lifetime of Chl *a* is much less than in solution and, in the case of PSII, it is variable, depending on physiological conditions.[50] The decrease in lifetime is due to two main factors: (1) photochemistry, the conversion of excitation energy into chemical energy via charge separation and electron transport in the reaction centers associated with each photosystem, and (2) nonradiative quenching of the $^1Chl^*$ to the ground state. The former is controlled by the processes that regulate electron transport and the latter by q_{NP}, photoinhibition, and reaction center oxidation. The effect of a change in the lifetime of chlorophyll singlet states is reflected in the fluorescence yields of the photosystems. If the intrinsic lifetime of fluorescence decay is τ_0, then the yield for chlorophyll fluorescence, when the lifetime of $^1Chl^*$ is reduced to any shorter lifetime, τ, is given by τ/τ_0. If the relationship between the rate constants *in vivo* for the fluorescence decay and singlet/triplet conversion of chlorophyll is known, it becomes possible to predict changes in the yield of $^3Chl^*$ from changes in the yield of chlorophyll fluorescence. The yield of chlorophyll fluorescence *in vivo* is variable, due to changes in PSII photochemistry, whereas PSI fluorescence remains low and constant.[47-49] Relative changes in the yield of chlorophyll fluorescence are easily measured using techniques employing modulated excitation systems.[47-49]

From measurements on thylakoids, Kramer and Mathis[47] concluded that the yields of chlorophyll fluorescence and $^3Car^*$ formation paralleled each other and that the yield of $^3Chl^*$ formation (relative to the light absorbed by both photosystems, when fluorescence was at its maximum and Q_A reduced) was 0.15 (they assumed that the yield of $^3Car^*$ from $^3Chl^*$ was 1.0 in their system). This they calculated was twice the yield of chlorophyll fluorescence,[47] and a similar relationship between the yields of fluorescence and triplet formation is found in chlorophyll solutions.[38,39] This value agrees with an estimate of a 20% quenching via triplet conversion at saturating irradiance made by Wolf.[132] The estimates of triplet yield are consistent with the observed lifetimes of $^1Chl^*$ and measured chlorophyll fluorescence yields *in vivo*. Chlorophyll fluorescence in spinach chloroplasts has a mean lifetime of 0.29 ns at the F_o level, rising to 1.96 ns at the F_m level.[50] Using a fluorescence lifetime of 15 ns, this corresponds to yields of 0.02 and 0.13 at F_o and F_m, respectively, which gives an F_v/F_m ratio of 0.846. The latter value is close to the mean value for F_v/F_m of 0.832 measured for a range of plants.[52]

The lifetime measurements are dominated by PSII fluorescence, especially at F_m. Yields calculated on the basis of lifetime measurements have to be adjusted for the relative cross-sections of the various populations contributing to each component of the mean lifetime. At F_m, using the rise-time of triplet chlorophyll formation found in solution, 5.2 nsec,[40] the yield of triplet formation would be 1.96/5.2, or 0.38.[40] This value, however, only applies to PSII, and taken over the entire chlorophyll population, it is decreased to about 0.19, assuming an equal excitation of both photosystems.[53] The triplet yield at F_m calculated from lifetimes (0.38) is similar to that measured by Kramer and Mathis[47] for PSII alone (~0.3), which confirms their suggestion that the relative rates of ^{1}Chl* decay via the triplet and fluorescence pathways in solution also applied in the thylakoid. At F_o, the triplet yield for PSII calculated from lifetime measurements is 0.29/5.2 (0.06). *In vivo,* the PSI reaction centers are either oxidized or nonoxidized.[54] In both cases, the ^{1}Chl* disappears in 40 ps; this implies a steady quantum yield for fluorescence of 0.003 (which is what is found[53]) and a triplet yield of only 0.008, which again concurs with the very low measured rate of triplet formation by PSI.[56] This suggests that even at F_o, the triplet yield of PSII calculated from the chlorophyll fluorescence lifetime is greater, perhaps by a factor of two, than the yield of the pigment bed as a whole because of the low contribution by PSI.

Genty and co-workers[57] have shown that with increasing nonphotochemical quenching, the mean fluorescence lifetime decreases in proportion to the decreasing yield of chlorophyll fluorescence, just as it does during the development of photochemical quenching.[50] Given the relationship between triplet yield and fluorescence lifetime, or yield, the generation of chlorophyll triplets can be expected to be proportional to the fluorescence yield, and, therefore, the yield of triplets will be influenced by those processes that affect the yield of chlorophyll fluorescence.

B. REGULATION OF THE QUANTUM EFFICIENCY OF PSII

The yield of chlorophyll fluorescence from PSII is essentially modulated by photochemical quenching (q_P) and nonphotochemical quenching (q_{NP}).[58-65] Whereas the mechanism of photochemical quenching is relatively well understood,[61] the q_{NP} mechanisms are by no means resolved.[61,63,64] Photochemical quenching involves electron transport. Its effectiveness depends on the balance between the relative rate of Q_A^- formation via photochemistry and Q_A regeneration via electron transport. These changes are readily reversible. Increasing the rate of Q_A formation, e.g., by increasing irradiance, will cause a fall in ϕ_{PSII}, and if there is no change in q_{NP}, an increase in the fluorescence yield. Similarly, decreasing the capacity for electron transport, e.g., by decreasing temperature, will also result in an increase in Q_A^-. q_{NP} is the general term used for any number of processes that reduce the yield of fluorescence, even though there may be no change in q_{NP}.[60] Some of these components are rapidly reversible, some only slowly.[63] The two most important components are the readily reversible "q_E" quenching and the slowly reversible loss of variable fluorescence related to photoinhibition.[58-61] The latter usually occurs as a result of stress.[64] Both processes depress ϕ_{PSII}, acting competitively with photochemistry to quench excitation in the PSII pigment bed. They act to reduce the rate of Q_A^- formation at any irradiance, decrease the fluorescence yield, and, by implication, decrease the average singlet lifetime in PSII. Therefore, the fluorescence yield of PSII will be determined principally by the interaction between the capacity for electron transport and the degree of nonphotochemical quenching.

With increasing irradiance, ϕ_{PSII} *in vivo* declines.[65,66] In general, the rate of this decline depends on the rate of electron flow between the photosystems, and this varies considerably from leaf to leaf. However, even though large changes may occur in PSII efficiency with increasing irradiance (Figure 2), these changes are not paralleled by large changes in the steady-state yield of chlorophyll *a* fluorescence; F_s' can remain close to the F_o' fluorescence yield values even though large changes occur in F_m', indicating that substantial increases in nonphotochemical quenching have occurred.[65-67] There is some variation between species in the extent to which F_s' is maintained close to F_o'. In two examples (Figure 2), *Saintpaulia ionantha* and *Impatiens* "New Guinea" hybrid, the decline of F_m' is neither as great nor as continuous with increasing irradiance as it is in *Pisum sativum* and *Hedera helix*. As a result, the continuing decline in ϕPSII in the former species is increasingly correlated to a loss of q_P, which results in an increase in F_s' to about twice the F_o' level. In the latter species, the decline in ϕPSII is matched more closely by a parallel decline in q_{NP}, and F_s' remains at close to the F_o' level (Figure 2). As F_o can decline with increasing irradiance (due to q_{NP}), F_s' can fall below the dark-adapted F_o level. The relative values of these fluorescence parameters are linearly related to the yields. The steady-state population of $^1Chl^*$ of PSII (and by implication the $^3Chl^*$ population) is a function of both lifetime and the quantum flux into the photosystem. Therefore, even if the steady-state yield remains constant, an increasing irradiance will result in an increase in the formation of $^1Chl^*$, which may partly explain why plants from high-light environments have more of the quencher tocopherol, in their thylakoids.

Other factors can also produce an increase in the fluorescence yield and, thus, an increase in the rate of triplet formation. The organization of the chlorophyll protein complexes facilitates the movement of excitation energy from the site of absorption to the reaction center. In addition, they may serve to screen the pigment molecules from molecular oxygen so any triplets that do form can transfer more effectively to carotene.[68] The $^3Car^*$ thus produced can decay with less chance of producing an oxygen singlet. Any stress that allows easier access by O_2 to the pigment molecules will increase the rate of $^1O_2$ formation as a result of interaction with any triplets formed photochemically. Similarly, any stress that destroys the regulation of q_{NP} or the organization of the pigment-protein complexes, such that excitation transfer to other pigment molecules is restricted resulting in an increase in F_o, would be expected to increase the rate of singlet oxygen production. High-temperature stress results in a sharp increase in the F_o fluorescence level, while chilling temperatures result in a loss of nonphotochemical quenching in some plants.

IV. ELECTRON TRANSPORT TO O_2

Molecular oxygen becomes reduced to the superoxide radical anion on the acceptor side of PS1.[69-72] This contains electron transfer components, with midpoint potentials from −1290 mV for P_{700}* to −420 mV for iron-sulfur center A, that form an electron transfer chain that ultimately reduces ferredoxin (midpoint potential, −420 mV)[73] (see Chapter 3 of this volume). The exact path and kinetics of electron transfer from P_{700}* to ferredoxin is not known with certainty,[73] but within this path, reduced species are formed that can potentially reduce O_2 to O_2^- (midpoint potential, −160 mV at pH 7).[74] The concentration of ferredoxin in the stroma has been calculated to be between 55 and 1500 μmol dm^{-3}, with available concentrations of about 20 to 750 μmol dm^{-3}.[75] Reduced ferredoxin (Fd_{red}) is reoxidized by a number of metabolites — e.g., NADP via the ferredoxin-NADP oxidoreductase, thioredoxin via ferredoxin-thioredoxin reductase, nitrite via nitrite reductase, sulfate via sulfate reductase — and via the cyclic electron transport pathway, electrons from ferredoxin are fed back into the plastoquinone pool.

The K_m values for the reduction of O_2 by chloroplasts and thylakoids point to at least two sites of reduction.[74] In the absence of ferredoxin, the K_m for O_2 reduction was less than 10 μmol dm^{-3} (at 25°C, the concentration of O_2 in water in equilibrium with air is ~250 μmol dm^{-3}).[76] In the presence of 2 μmol dm^{-3} ferredoxin, it was less than 20 μmol dm^{-3}.[77] At 25 μmol dm^{-3} ferredoxin, the K_m for O_2 was 60 μmol dm^{-3}.[77] In intact spinach chloroplasts, the K_m for O_2 was 75 μmol dm^{-3},[78] while in whole cells of *Scenedesmus,* it was 96 μmol dm^{-3}.[79] The high affinity for O_2 is considered to be due to the reaction of O_2 with reduced ferredoxin.[77] Asada and Takahashi[74,80] have shown that the reaction with the low K_m occurs within the thylakoid membrane (see Chapter 3 of this volume). The rate of the latter reaction could be increased by treatments that disorder the membrane;[74] however, it is not clear if the latter reaction continues in the presence of ferredoxin (Fd). The reduction of O_2 by Fd_{red} also occurs with a maximum rate of about 2.8 times that of the reaction with the low K_m.[77] The Mehler reaction in air should be nearly saturated with respect to O_2. In 2% O_2, used in experiments to suppress photorespiration, there might also be a marked decrease in the rate of the Mehler reaction.

The capacity for the Mehler reaction is determined not only by the O_2 concentration, but also by the concentration of reduced Fd. The effect of Fd concentration and NADP supply on the Mehler reactions in thylakoid membranes has been described by Furbank and Badger[77] and Hosler and Yocum.[81] The Mehler reaction is not saturated by Fd concentrations of over 70 μmol dm^{-3}, but the reduction of NADP is saturated by concentrations above

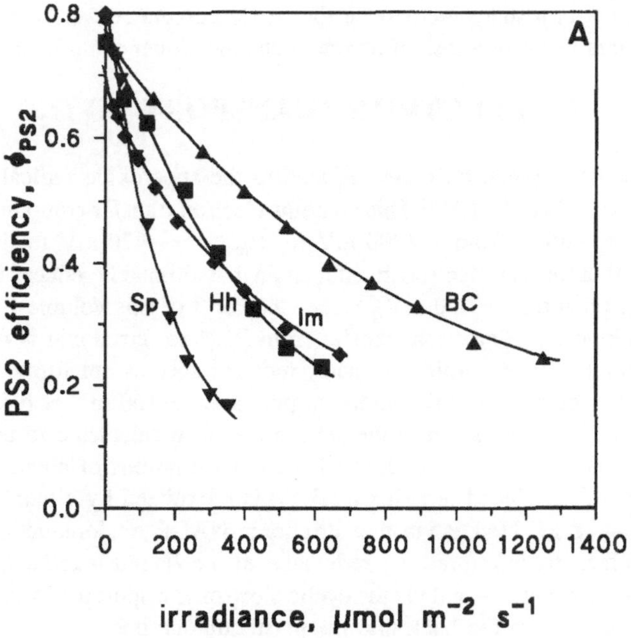

FIGURE 2. Interspecific variations in the changes in the quantum efficiency of PSII (ϕ_{PSII}) and chlorophyll *a* fluorescence parameters with increasing irradiance. The effect of irradiance on ϕ_{PSII} for four plant species is shown in (A). The changes in chlorophyll *a* quenching components, F_o, F_s, and F_m which underly the changes in ϕ_{PSII} are shown for *Pisum sativum* (B), *Hedra helix* (C), *Impatiens*, new Guinea hybrid (D), and *Saintpaulia ionantha* (E). With increasing irradiance, the quantum efficiencies of PSII (A) in leaves decline. A range of values is provided by the following species: *S. ionantha* (Sp), *H. helix* (Hh), *Impatiens*, New Guinea hybrid (Im), and *P. sativum* (Bc). Most of the loss of quantum efficiency is caused by an increase in q_{NP}. As a result of differing degrees of q_{NP} development, with decreasing PSII efficiency, different patterns of change are observed in the yields of chlorophyll fluorescence (B-D) measured at F_o' (■), at steady-state fluorescence, F_s' (▲), and with all Q_A reduced (F_m', ▼). In some cases, the steady-state fluorescence yield scarcely alters with increasing irradiance and is just slightly higher than the F_o level (i.e., B, high-light grown *P. sativum*, and C, *H. helix*, collected from a lightly shaded field site), whereas there are large changes in the yield of chlorophyll *a* fluorescence when Q_A is reduced. In other cases, the fluorescence yield at steady-state increases with increasing irradiance (i.e., D, *Impatiens*, New Guinea hybrid, and E, *S. ionantha*). The latter were grown under shade in a greenhouse. In these instances, the decline in fluorescence yield when Q_A is reduced is not as great.

10 µmol dm^{-3}.[81] NADP reduction, thus, effectively outcompetes the Mehler reaction; in the presence of NADP above 40 µmol dm^{-3}, the Mehler reaction still occurs, but NADP reduction is much greater (at 33 µmol dm^{-3} Fd).[68] Increasing concentrations of Fd will slightly increase the rate of the Mehler reaction at the expense of NADP reduction.[81] In the presence of NADP, the Fd pool is also much more oxidized than it is whenever the Mehler reaction is the only available sink for reducing equivalents.[81] The rate of the Mehler reaction in the absence of NADP is dependent on the Fd concentration, and

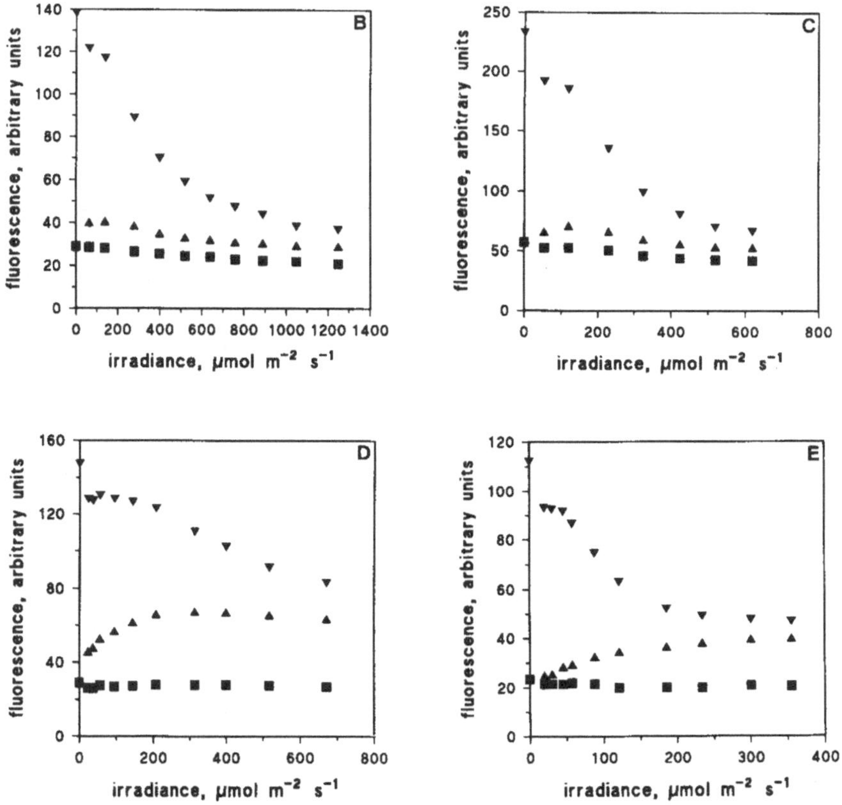

FIGURE 2B–E (continued).

if the latter is high enough (~100 μmol dm^{-3}), the rate of the Mehler reaction can be as great as the rate of NADP reduction.[75,81] The reduction of molecular O_2 is difficult to measure *in vivo* because of the parallel process of photorespiration,[19-21,75,77] but *in vitro* and *in vivo*, at CO_2 concentrations high enough to block photorespiratory O_2 fixation, O_2 uptake still occurs, although the measured rates vary from 1 to 10% of the rate of CO_2 fixation.[75,82-84]

A. RELATIONSHIPS BETWEEN ELECTRON TRANSPORT AND STROMAL ENZYME ACTIVATION STATES

An understanding of how the flux of reducing equivalents to O_2 can be influenced by physiological processes within the thylakoids and stroma ideally requires a detailed kinetic model of the component reactions. Such a model is currently not available. However, a simple qualitative model of thylakoid-stroma interactions can be derived from a basic appreciation of the regulation of flux in cells. Biochemical pathways are never wholly at equilibrium, at least not in living organisms. The description of fluxes in these pathways is best dealt with by reference to disequilibrium thermodynamics.[85] This appears complex, but the basic principles of flux regulation in such systems can be understood quite simply by analogy with other systems such as electrical current flow through a resistance, or gaseous diffusion into, and out of, the leaf. These are also disequilibrium processes and, within certain limits, the analogy with fluxes in biochemical or physiological processes is good. Every flux can be described, in a first approximation, as a function of both a driving force which allows the change in flux and the property of the component or phase through which the flux passes. An example of this would be the conductance of a material for heat flux or electrical flux. The inverse of conductance, resistance, can also be used to describe flux control. A flux (e.g., current) can be increased by either increasing the driving force (e.g., voltage) or by increasing the conductance. This model only holds true in a biochemical pathway until V_{max} is achieved, but, nonetheless, biochemical systems do often obey these relationships,[85] as do physiological processes (e.g., gaseous diffusion in leaves).

Applying a simple model of flux regulation of this kind to the regulation of the redox state of Fd is complicated because it is subject to both feed-forward and feedback processes. The reduction state of ferredoxin determines the reduction state of key enzymes of the Calvin cycle, the thylakoid ATPase, and stromal NADP-malate dehydrogenase via the thioredoxin pathway. Reductive activation of these enzymes serves to increase the metabolic demand for NADPH and to increase the export of reducing equivalents from the chloroplast. By feeding reducing equivalents back into the plastoquinone pool, cyclic electron flow may reduce access to that pool for PSII-derived electrons, thereby causing a decrease in PSII efficiency and a decrease in noncyclic electron flow. Cyclic electron flow will also act by decreasing the intrathylakoid pH. Similarly, other processes allow the metabolism of the stroma to exert feedback control over the resistance for electron flow between the photosystems.[2,3] This is believed to act, in part, via the sensitivity of the reaction between plastoquinone and the cytochrome *b6/f* complex to low pH values in the intrathylakoid space[90] and also to direct effects on PSII.[2] Similarly, it is not yet clear how other means of regulating electron flow, such as the availability of inorganic phosphate, serve to maintain a decreased intrathylakoid pH. Without these effects, mediated by thioredoxin and cyclic electron flow, the control of the redox state of the Fd pool would be easy to understand. If the flux of reducing equivalents into the ferredoxin pool were to increase, then it would become more reduced. This increase in the driving force would increase the fluxes out of the ferredoxin pool until the "out" fluxes balanced the "in" fluxes and vice versa. Ultimately, this would decrease the quantum efficiency of PSI, as reduced electron acceptors accumulated. Increased activation of the thiol-activated stromal enzymes and the effects of cyclic electron flow complicate matters, since their action serves to increase (1) the conductance of the metabolic pathway for reoxidation of

Fd_{red} and (2) the flux of reducing equivalents out of the Fd pool, and decrease (3) the conductance of the pathway for noncyclic electron transport and (4) the flux of reducing equivalents into the Fd pool.

There is currently no means available of measuring the redox state of the Fd pool directly *in vivo,* and measurements of O_2 reduction by the Mehler reaction are complicated by the presence of other O_2-consuming processes such as photorespiration. However, the stromal redox state may be monitored indirectly via the activation state of NADP-malate dehydrogenase.[86] This stromal enzyme is reductively activated by thioredoxin and inhibited by NADP. Such an approach is, however, only semiquantitative[87] and, therefore, it is impossible to use the redox state of this component to calculate accurately that of another. The photosynthetic system is clearly not at redox equilibrium; rather, it is an open system with fluxes of reducing equivalents through pathways of finite and potentially variable conductance. Without quantitative information on these fluxes and conductances, it is not possible to know the degree of redox equilibrium between components such as ferredoxin and NADP malate dehydrogenase. The degree to which P_{700} can be oxidized is another indication of the reduction state of the stroma. Excited P_{700} transfers an electron to its acceptor pool, which is comprised of several components. Back reactions from these to P_{700+} are possible, but these are slower than the rates of electron movement toward ferredoxin. However, in the event that electron transport to Fd and beyond is impossible (due to these pools being reduced), the back reactions will predominate. Exactly which back reactions are important here are not known, but under conditions where no electron transport is possible, no P_{700+} can be detected during irradiation. Using recently developed techniques,[88,89] it has become possible to measure the redox state of P_{700} *in vivo* routinely and to calculate the quantum efficiency of PSI for electron transport. These measurements can be made in parallel with those of the quantum efficiency of PSII and CO_2 fixation.[6,54,88,89] The calculation of the quantum efficiency of PSI electron transport, ϕ_{PSI}, is based on the assumption that any unoxidized P_{700} (P_{700°) that is excited and transformed to P_{700}*, will undergo oxidation to P_{700+}, and produce reduced Fd with a constant yield. The P_{700° pool is, therefore, capable of electron transport and has a quantum efficiency close to one and P_{700+} has a quantum efficiency close to zero.[49] The quantum efficiency of PSI (ϕ_{PSI}) is thus given by the term:

$$\phi_{PSI} = P_{700^\circ}/(P_{700+} + P_{700^\circ}) \qquad (4)$$

If any P_{700° was incapable of undergoing a stable oxidation due to a deficit in stable electron acceptors, P_{700° would become quantum inefficient and no P_{700+} would be detectable. Under these circumstances, the ratio given in Equation 4 will not accurately estimate ϕ_{PSI}. In most leaves, the relationships between ϕ_{PSI} and ϕ_{PSII} are linear. Under nonphotorespiratory conditions, the relationships between ϕ_{PSII} or ϕ_{PSI} are linearly related to the quantum yield of CO_2 fixation, ϕ_{CO_2}.[6,54,88,89] These linear relationships imply that the ratio given in Equation 4 effectively estimates ϕ_{PSI}. Furthermore, they show that no significant restriction of ϕ_{PSI} occurs because of overreduction of the PSI acceptor pool. Rather, these linear relationships show that the regulation of ϕ_{PSI} is achieved on the donor side of PSI and that decreases of ϕ_{PSI} occurring with increasing irradiance, decreasing CO_2 or O_2, or decreasing temperatures are achieved via an increase in the P_{700^+}/P_{700° ratio and not by the accumulation of photochemically inactive P_{700}.[6,54,88,89,91,92]

In the absence of CO_2, when the relationship between ϕ_{PSII} and ϕ_{PSI} is not linear, the oxidation state of P_{700} still does not appear to be restricted as irradiance increases.[6] This may be explained by the presence of significant cyclic electron flow around PSI.[6] Only in exceptional circumstances, such as the induction period of photosynthesis or transitions from low to high irradiance, does any restriction of P_{700} oxidation occur because of a limitation on the acceptor side.[6,87,92] **Taken together, these observations imply that, in most circumstances, control of electron flow between PSII and PSI, occasionally in conjunction with cyclic electron flow around PSI, acts to regulate the reduction state of the acceptor side of PSI such that the redox state of PSI acceptors does not significantly limit electron transport.**

Further evidence for the regulation of the redox state of the electron acceptors on the acceptor side of PSI is provided by the measurement of the activation state of NADP-malate dehydrogenase (NADP-MDH). With increasing irradiance (up to 1300 μmol m^{-2} s^{-1}), the activation state of NADP-MDH of pea leaves in air increased only slightly,[92] while the activation state of the Calvin cycle enzyme, fructose-1,6-bisphosphatase (FBPase), increased substantially. Thus, a combination of increasing demand for reducing equivalents, as estimated by the increasing activity of FBPase (i.e., increasing the conductance of reducing equivalents from ferredoxin) and the limitation of the electron flux through PSI (by restriction of electron flow between the photosystems), serves to control the redox state of the Fd pool, of which the activation state of NADP-MDH is a physiological indicator.[6,87,92] The increase in the activation state of the FBPase, and by analogy the other thiol-modulated enzymes of the Calvin cycle, can be explained in terms of feed-forward regulation from the Fd pool via thioredoxin-dependent activation.[93,94] The regulation of interphotosystem electron transport is more difficult to explain. This may be measured as the half-time for P_{700^+} reduction following a light-to-dark transition, and this remains constant with increasing irradiance,[88,89] although it is altered by other changes in the leaf environment.[6,87]

It is clear that in nonstressed leaves in air, at steady state, electron transport and metabolic activity are coordinated so as to prevent extensive reduction of the electron acceptors (including Fd) on the reducing side of PSI.[6,88,91,92] In contrast, during photosynthetic induction, which occurs when the sun comes up or light intensity increases rapidly, this coordination is not established and considerable reduction of the PSI acceptor side develops, resulting in high

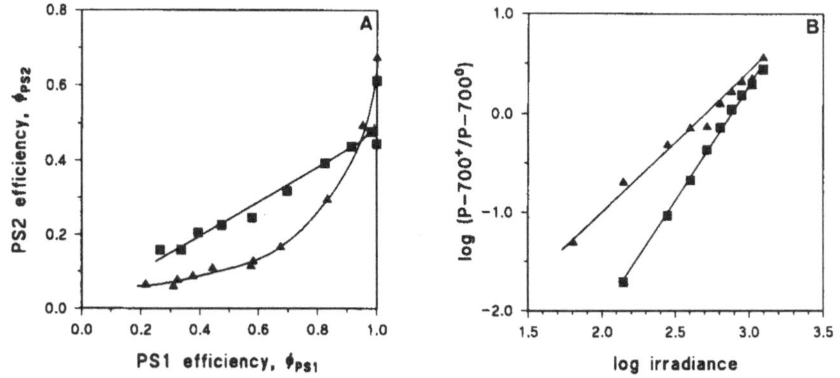

FIGURE 3. Relationship between the quantum efficiencies of the photosystems with changing irradiance. In air, the relationship between ϕ_{PSII} and ϕ_{PSI} in a pea leaf grown in the cold (5°C day, 5°C night) is predominantly linear (A, ■). In an atmosphere consisting of 350 ppm CO_2, 2% O_2, and balance N_2, the linear relationship became curvilinear (▲). Such a curvilinear relationship could be due to the presence of cyclic electron flow around PSI. Alternatively, it could be caused by an overestimation of ϕPSI due to a lack of PSI electron acceptors. However, it is possible to eliminate the latter possibility by analyzing the relationship between irradiance and the P_{700^+}/P_{700^o} ratio. When plotted on a logarithic scale, this relationship is linear in both examples (B), indicating a steady increase in the P_{700^+}/P_{700^o} ratio with increasing irradiance and implying that P_{700} oxidation is not restricted by a lack of electron acceptors.

(NADPH)/(NADP) ratios and a restriction of P_{700} oxidation because of the shortage of electron acceptors.[87]

Even in conditions of stress, leaves are able to control the electron transport carbon metabolism relationships so as to avoid restriction of P_{700} oxidation by reduction of the acceptor pool. In *H. helix* leaves, for example, showing O_2-insensitive photosynthesis, the linear relationships between ϕ_{PSI} and ϕ_{PSII} was maintained even when exposed to an atmosphere (2% O_2, 350 ppm CO_2, N_2 balance) which would largely eliminate photorespiration, implying that no overreduction of the P_{700} acceptor pool occurred. The major restriction of electron transport was still located between the photosystems. Pea plants, grown at low temperatures (5°C, day and night), had very low rates of CO_2 fixation relative to controls grown at 20°C, but they were able to maintain predominantly linear ϕ_{PSI}/ϕ_{PSII} relationships when exposed to increasing irradiances in air (Figure 3). Under nonphotorespiratory conditions (2% O_2, 350 ppm CO_2, balance N_2), however, such leaves developed curvilinear ϕ_{PSI}/ϕ_{PSII} relationships which could be attributed to overreduction of the PSI acceptor pool or to the onset of cyclic electron flow around PSI, or both. The analysis of the pattern of oxidation of P_{700} with increasing irradiance in such circumstances shows that P_{700} oxidation proceeds as though it is not limited on its acceptor side. This suggests that the observed curvilinearity is due to cyclic electron flow.[6] Under conditions where the intracellular CO_2

concentration approaches the compensation point, with unstressed leaves, P_{700} oxidation occasionally appears to be restricted by a shortage of electron acceptors, and the presence of cyclic electron flow is conspicuous. Measurements of NADP-MDH in these circumstances revealed a complex relationship between the enzyme activation state and electron flux, showing that the activation of the enzyme was increased only at low irradiances and then not in all leaves.[6] At high irradiances, very little further activation occurred even though some overreduction of the PSI acceptor pool was detected in some instances. Such contradictory observations could be explained if the presence of cyclic electron flow led to a decreased electron transfer to stromal sinks for reducing equivalents.[83]

At low temperatures, however, no significant overreduction of the PSI acceptor side was evident, and a linear ϕ_{PSI}/ϕ_{PSII} relationship existed in all circumstances. An increase in the activation state of NADPH-MDH with increasing irradiance occurred in these circumstances (Figure 4), implying that, although control over electron flow to stromal sinks is still sufficient to prevent overreduction of the PSI acceptor pool, the stroma becomes significantly more reduced, and by implication, the flux through the Mehler reaction would be increased.

B. PHOTORESPIRATION AS A SINK FOR REDUCING EQUIVALENTS

We have, thus far, considered the reactions of O_2 with the photosynthetic apparatus that produce toxic oxygen species. The Mehler reaction may partly serve to protect the photosynthetic machinery from damage by providing an alternative sink for electrons. Photorespiration can also act as a major sink for reducing equivalents and ATP produced by photosynthetic electron transport and phosphorylation. In pea leaves, the rate of electron flow through PSII in the absence of CO_2 with 20% O_2 and balance N_2 is about 50% of that attained in 20% O_2, 350 ppm CO_2, and balance N_2.[6] In air, photorespiration serves to decrease the quantum efficiency of CO_2 assimilation, leading to a decrease in the measured yield of CO_2 fixation, ϕ_{CO_2}, for a given ϕ_{PSII}.[91] The relationship between ϕ_{CO_2} and ϕ_{PSII} in leaves under nonphotorespiratory conditions (i.e., 2% O_2, 350 ppm CO_2, balance N_2) is linear except at low light (Figure 5). In air, the ϕ_{CO_2}/ϕ_{PSII} relationship is often linear, or nearly so. This implies that the competition between CO_2 fixation and photorespiration appears to progressively favor photorespiration as irradiance is increased.

In sink-limited leaves, the increased use of the photorespiratory pathway relative to the carbon reduction cycle would allow electron transport to continue without much carbohydrate accumulation. However, the nature, regulation, and significance of the conspicuously curvilinear ϕ_{CO_2}/ϕ_{PSII} relationships in these circumstances need to be investigated further. Nevertheless, it is clear that electron transport can continue in air with very low rates of CO_2 fixation. This is achieved by allowing enhanced photochemical quenching and regeneration of PSI electron acceptors.

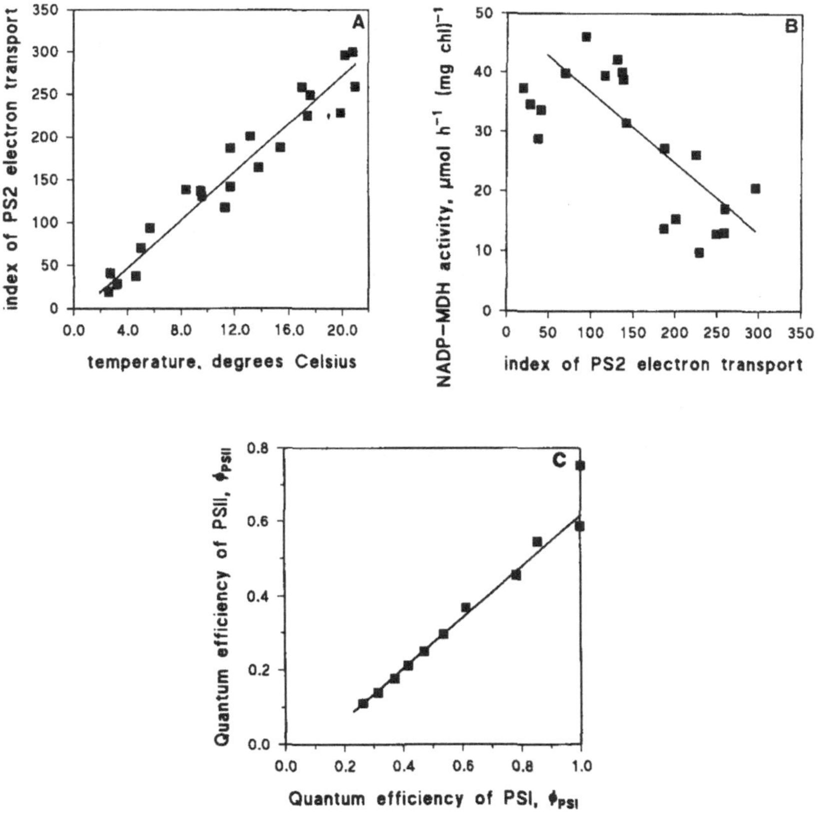

FIGURE 4. Effect of varying temperatures on the index of electron flux through PSII (J_{PSII}, A), the relationship between J_{PSII} and NADP-MDH activity (B), and the relationship between ϕ_{PSII} and ϕ_{PSII} in *P. sativum* leaves (C) Temperature was varied at a fixed irradiance (710 μmol m^{-2} s^{-1}) and chlorophyll *a* fluorescence and NADP-malate dehydrogenase activity were measured in *P. sativum* leaves in air (A-C). The index of electron flux through PSII (J_{PSII}) is given by the product of ϕ_{PSII} and irradiance (A). Maximum NADP-malate dehydrogenase activity was 123 μmol h^{-1} mg^{-1} Chl (B). When measured at a temperature of 5°C, the relationship between the quantum efficiencies of the photosystems is linear (C), implying that noncyclic electron transport still predominates and that the photosystem efficiency is limited on the donor side. In this study, plants were grown at 15 to 20°C. At 5°C in air, the light-saturated rate of CO_2 fixation was approximately half that obtained at the temperature optimum.

V. H_2O_2 GENERATION AND ACTION IN THE CHLOROPLAST

Superoxide is the product of the univalent reduction of oxygen by PSI.[69-72] In the absence of Fd, $O_2^{\overline{\cdot}}$ is generated in the aprotic interior of the thylakoid membranes on the reducing side of PSI.[80] The thylakoid membranes are only poorly permeable to $O_2^{\overline{\cdot}}$[95] and some $O_2^{\overline{\cdot}}$ may be cycled in the membranes to rereduce plastocyanin and cytochrome *f*.[96] However, in the presence of

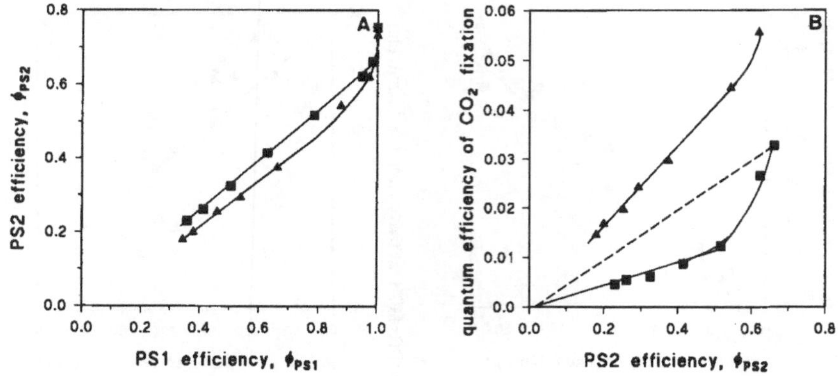

FIGURE 5. The relationships between ϕ_{PSII} and ϕ_{PSI} (A), and ϕ_{PSII} and the quantum efficiency of CO_2 assimilation, ϕ_{CO_2} (B) in *H. helix* leaves with changing irradiances in air and in 2% O_2, 350 ppm CO_2, and balance N_2. Measurements were made on *H. helix* leaves in air (■) or in 2% O_2, 350 ppm CO_2, and balance N_2 (▲) to suppress photorespiration. The relationship between ϕ_{PSI} and ϕ_{PSII} is linear in both cases and shows no significant differences between the two treatments (A). The relationship between ϕ_{PSII} and ϕ_{CO_2} (B) shows quite a different response. Under nonphotorespiratory conditions, the relationship between ϕ_{PSII} and ϕ_{CO_2} is largely linear, implying a simple dependency of ϕ_{CO_2} on electron transport. In air, however, the relationship between ϕ_{PSII} and ϕ_{CO_2} is substantially altered, even though that between ϕ_{PSI} and ϕ_{PSII} is not significantly affected by this change. In addition to the overall decline in ϕ_{CO_2}, caused by the competitive action of photorespiration, the relationship between ϕ_{CO_2} and ϕ_{PSII} is markedly curvilinear. ϕ_{CO_2} is significantly less than expected if the relationship between ϕ_{CO_2} and ϕ_{PSII} has been linear. This trend implies that with increasing irradiance in air, noncyclic electron flow becomes increasingly less efficient at reducing CO_2. This is in marked contrast to the situation in nonphotorespiratory conditions.

ferredoxin, most superoxide will be produced on the thylakoid membrane surface or in the stroma as a result of pseudocyclic electron flow.[75,77] Superoxide is often regarded as the primary agent of oxygen toxicity in cells. This is due (1) to the inactivation of specific enzymes such as ribonucleotide reductase and dihydroxyacid dehydratase and (2) to the cascade sequence of increasingly destructive species to which superoxide gives rise, particularly $^{\cdot}OH$.[97] A multifactorial defense system provides the necessary accommodation to the production of superoxide. Taken alone, the scavenging of superoxide, via superoxide dismutase or by interaction with ascorbate, merely serves to transform one destructive oxygen species to another, H_2O_2, which in terms of photosynthetic metabolism is far more toxic and disruptive. H_2O_2 is produced principally by the dismutation of superoxide, but it may also be generated by other processes in the chloroplast, e.g., by PSII, where it can give rise to localized protein degradation within the membrane.[98] It is possible that the generation of H_2O_2 within PSII may initiate the process of photoinhibition which results in degradation of the PSII core protein D1.[98] The rate of photoinhibition and degradation of D1 are accelerated when the oxygen-evolving complex of PSII is depleted of chloride which is required for the formation of the normal S2 state during the process of oxygen evolution.[99] When the oxygen-evolving complex is deprived of chloride, oxidation of water to H_2O_2 can occur.[99] Furthermore, it is also possible that H_2O_2 can arise from reduction of O_2 at the Q_B site on the reducing site of PSII. H_2O_2 is known to inactivate a number of proteins that contain iron bound to a heme. H_2O_2-induced oxidation of heme-iron to form a ferryl group can result in the oxidation of a nearby tyrosine to form a cation radical.[100,101] The tyrosine radical and the ferryl group can subsequently catalyze the peroxidation of other moieties. This type of reaction can covalently modify proteins and inactivate them. Tyrosine radicals produced by H_2O_2 induce the formation of bityrosine cross-linkages within proteins.[102] This may occur within the PSII reaction center protein called D1, particularly if the tyrosine-161 on D1 becomes cross-linked.[103] This tyrosine is responsible for electron transfer from the oxygen-evolving complex to P_{680}. Another possibility is that H_2O_2 could inactivate PSII via reactions with the histidine that is essential for the function of the oxygen-evolving complex.[104] H_2O_2 generated with the thylakoid membrane complexes or the thylakoid lumen may be relatively inaccessible to antioxidants and detoxifying enzymes.

Many enzymes in the chloroplast and the cytosol are regulated via a process of thiol-disulfide exchange mediated via the thioredoxin system.[93] Specific target enzymes in the chloroplast respond to redox changes in the ferredoxin/thioredoxin system which, in turn, depends on the flux of electrons through the electron transport system, thus linking the supply of electrons, provided by light, to the regulation of enzyme activity and, hence, metabolism.[87,105,106] Thiol-disulfide exchange reactions provide one of the basic mechanisms of regulation of the Calvin cycle, operating to convert several of the component enzymes from inactive forms in the dark to active forms in the light. Reduced thioredoxins recognize and activate fructose-1,6-bisphosphatase, phosphoribulokinase, sedoheptulose-1,7-bisphosphatase, and NADP-glyceraldehyde 3-phosphate dehydrogenase[105,106] through the reduction of specific disulfide bridges.[93] The reduction state of these biosynthetic enzymes reflects the balance between the flux of reducing equivalents through the electron transport system and the oxidizing environment which continuously favors oxidation and, hence, inactivation. The NADP-thioredoxin system is extrachloroplastic and is distinct from the chloroplast system, but will, nevertheless, respond to similar redox influences.[107] In the dark, thioredoxin and the thiol-modulated enzymes become oxidized, and enzymes such as fructose-1,6-bisphosphatase are inactivated (Figure 6). Molecular O_2 is the natural oxidant involved in the inactivation processes.[108] H_2O_2, however, is a powerful oxidant that destroys the balance of this system (Figure 6), since it rapidly oxidizes enzyme thiol groups and prevents thioredoxin-dependent reductive activation.[109-111] H_2O_2 is an extremely potent inhibitor of photosynthetic carbon assimilation even at micromolar concentrations.[109] H_2O_2 will tend to oxidize all exposed thiol groups, and many other enzymes and proteins

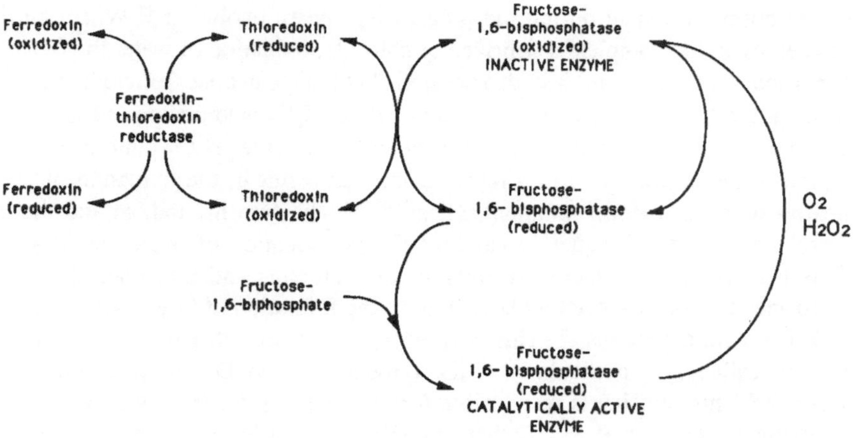

FIGURE 6. Mechanism of activation and inactivation of stromal fructose-1,6-bisphosphatase showing that activation is facilitated by the ferredoxin-thioredoxin system and that inactivation occurs via oxidation by molecular oxygen or hydrogen peroxide.

will be modified and inactivated if H_2O_2 is not removed. The Cu,Zn superoxide dismutase of the chloroplast, for example, is susceptible to H_2O_2-induced damage at alkaline pH values because H_2O_2 can reduce the copper from Cu^{2+} to Cu^{+}, and this is followed by an irreversible inactivation.[112]

Conversely, an oxidizing environment is required for the activation of other pathways. Thiol-dependent regulation also serves to inactivate certain dissimulatory pathways in the light and affords their activation in the dark. Glucose-6-phosphate dehydrogenase activity in the chloroplast, for example, is inactivated by light and by thiols via a decrease in the affinity of the enzyme for its substrate, glucose-6-phosphate.[113] Reactivation is afforded by oxidation.

In addition to these effects, a direct lethal effect of H_2O_2 arises because of the Haber-Weiss reaction in which catalytic levels of metal ions, such as Fe(III), mediate the production of $^{\cdot}OH$, which will react with any cellular material within a few molecular diameters of its cellular generation.[35]

H_2O_2 has, however, important functions in plant cells, particularly in the defenses against pathogen attack (see Chapter 7 of this volume) and in cell wall biosynthesis. Nevertheless, the permitted level of H_2O_2 within the chloroplasts must be very low if photosynthetic carbon assimilation is to function optimally. Catalase is localized in the peroxisomes, has a poor affinity for H_2O_2, and is also subject to inactivation in the light.[114] The mechanism of photoinactivation of catalase is not clear, but requires the presence of O_2. Furthermore, it is distinct from the reversible formation of the inactive catalase compound II observed in the presence of low concentrations of H_2O_2. Rather, photoactivation damages the catalase protein in such a way that it becomes inactive and marked for proteolytic degradation, in a manner similar to the

D1 protein of PSII.[133] Leaf tissues do not normally contain gluthathione peroxidase, but contain instead a remarkably efficient ascorbate-dependent scavenging system with three routes of recycling oxidized ascorbate to maintain the pool of reduced ascorbate (Figure 7). The regeneration of ascorbate requires the participation of the electron transport chain in the chloroplast and has important implications for the regulation of electron transport.[11,114] The production and destruction of H_2O_2 both involve electron transport and are concerted events that amplify the capacity of electron transport to generate the transthylakoid ΔpH that initiates thermal energy dissipation.[12]

VI. THE ROLES OF ASCORBATE IN THE CHLOROPLAST

Ascorbate is an important reductant and antioxidant within the chloroplast, where it serves several functions (Table 2); in particular, H_2O_2 detoxification is pivotal to the efficient functioning of photosynthesis. Plants and certain cyanobacteria have developed a unique ascorbate-dependent peroxidase to remove H_2O_2.[114,115] Ascorbate is an essential factor in preventing undesirable oxidations; it will scavenge superoxide and hydroxyl radicals as well as singlet oxygen and other active species. It is also essential for the regeneration of α-tocopherol, the antioxidant of the membrane environments. The occurrence, metabolism, and functions of ascorbate in plants have recently been reviewed,[114] and hence are not covered here. Instead, we concentrate on the regulatory implications of ascorbate-dependent reactions. When ascorbate functions as an antioxidant in protection of the photosynthesis apparatus, the ensuing cycling of ascorbate between oxidized and reduced forms consumes reducing equivalents and, in consequence, participates in the formation of the transthylakoid ΔpH gradient. With regard to the cycling of ascorbate and its relationships to electron processes, it is first necessary to consider the mechanisms employed to maintain the ascorbate pool of the chloroplast largely in the reduced form. Ascorbic acid has two oxidized forms, monodehydroascorbate and dehydroascorbate (Figure 7). Oxidation of ascorbate occurs in two sequential steps, first producing monodehydroascorbate and, subsequently, dehydroascorbate. The monodehydroascorbate radical is the primary product of the ascorbate peroxidase reaction in chloroplasts and is also produced by the univalent oxidation of ascorbate, e.g., in the reduction of superoxide and hydroxyl radicals, and also in the regeneration of α-tocopherol. If it is not rapidly rereduced to ascorbate, the monodehydroascorbate radical spontaneously disproportionates to ascorbate and dehydroascorbate (Figure 7). Dehydroascorbate is also highly unstable at pH values greater than 6.0 and, therefore, must be rapidly recycled to ascorbate. It is thus not surprising that two enzymes are involved in the regeneration of L-ascorbic acid *in situ*. These are monodehydroascorbate reductase (E.C. 1.6.5.4.), which uses NAD(P)H directly to recycle ascorbate (Equation 5), and dehydroascorbate

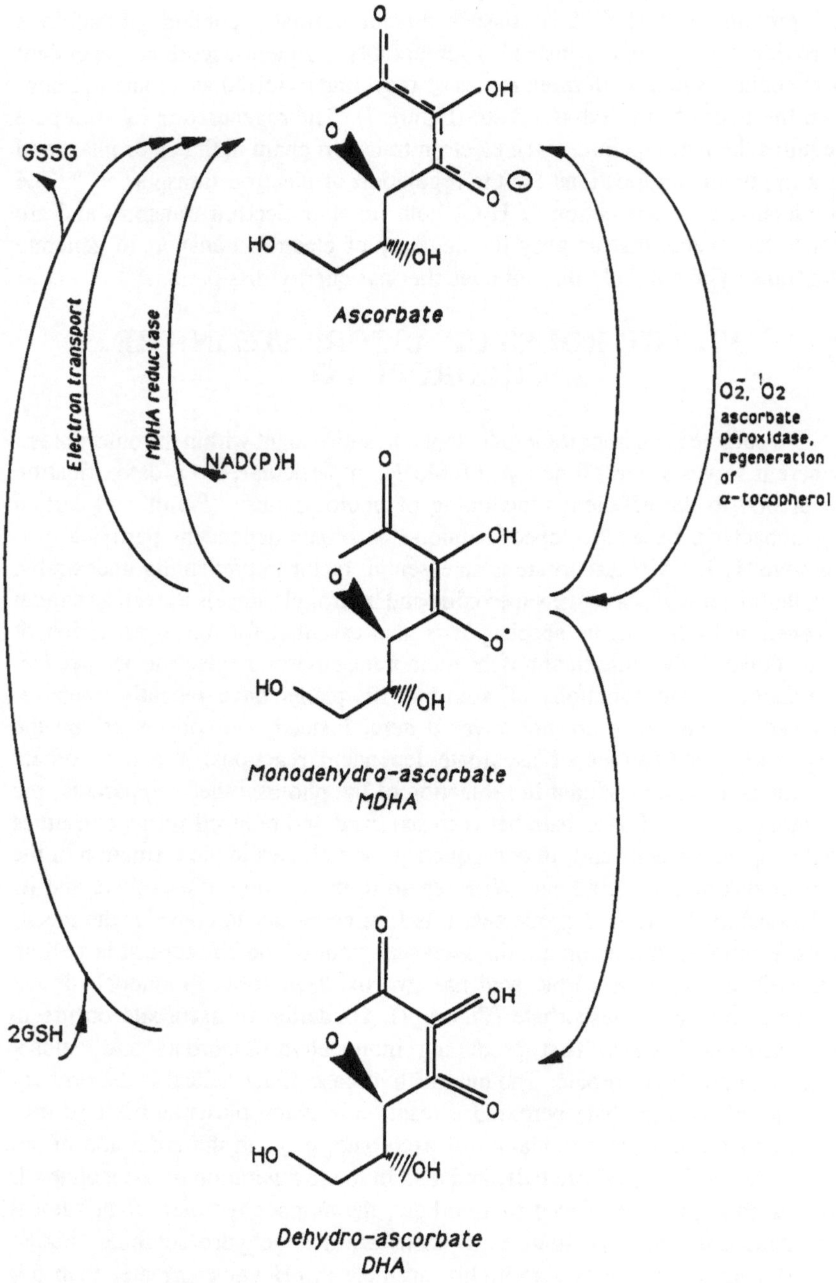

FIGURE 7. Reduced and oxidized forms of L-ascorbic acid.

TABLE 2
The Roles of Ascorbate and the Ascorbate Cycle in the Regulation of Photosynthesis

1. Elimination of active oxygen species such as $O_2^{\cdot-}$, H_2O_2, $^{\cdot}OH$, and $^1O_2$
2. Protection against enzyme inactivation, particularly the thiol-modulated enzymes of the Calvin cycle
3. Protection against the deleterious effects of excess irradiance
   a. All mechanisms of ascorbate regeneration consume reducing equivalents produced by photosynthetic electron transport.
   b. Regeneration of ascorbate participates in the generation of transthylakoid ΔpH and its associated fluorescence quenching components and, therefore, is coupled to ATP synthesis and also to the regulated decreases in ϕPSII
4. Interaction with thylakoid membrane components
   a. Ascorbate is required in the formation of zeaxanthin.
   b. Ascorbate is essential for the recycling of α-tocopherol (vitamin E).
   c. Ascorbate has direct effects on electron transport carriers and other components of the thylakoid membranes.

reductase (E.C. 1.8.5.1), which uses reduced glutathione (GSH) for the regeneration process (Equation 6). The situation is further complicated, however, because monodehydroascorbate itself is an efficient electron acceptor.[111,115] Monodehydroascorbate can be reduced directly to ascorbate using electrons derived from the photosynthetic electron transport chain (Figure 8). In summary, the three routes of recycling the oxidized forms of ascorbate in chloroplasts are as follows.

Monodehydroascorbate reductase

$$\text{Monodehydroascorbate} + \text{NAD(P)H} \longrightarrow \text{Ascorbate} + \text{NADP} \qquad (5)$$

Dehydroascorbate reductase

$$2\text{GSH} + \text{Dehydroascorbate} \longrightarrow \text{GSSG} + \text{Ascorbate} \qquad (6)$$

Electron transport

$$4\ \text{Monodehydroascorbate} + 2H_2O \xrightarrow[\text{Electron transport}]{\text{Light}} 4\ \text{Ascorbate} + O_2 \qquad (7)$$

Chloroplasts contain the majority of the Cu,Zn superoxide dismutase (SOD) activity of the leaf cells,[116,117] and they also contain a large proportion

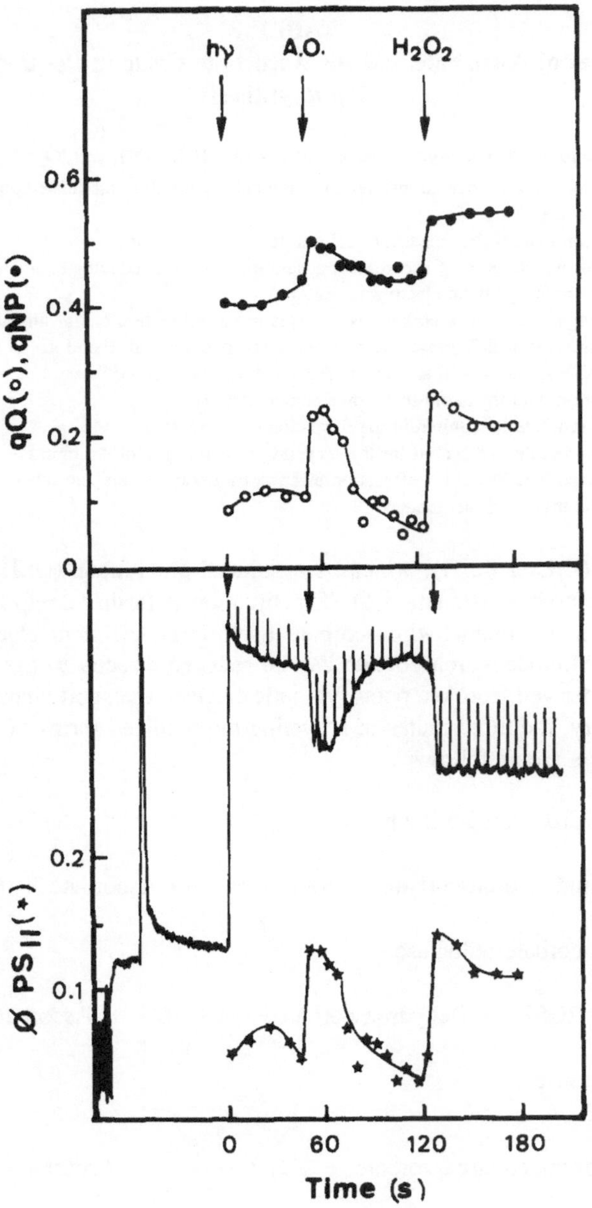

FIGURE 8. Ascorbate oxidase (AO) and H_2O_2-mediated quenching of chlorophyll *a* fluorescence in pea thylakoid membranes. The reaction medium contained 0.33 *M* sorbitol, 5 m*M* L-ascorbate, 10 m*M* KCl, 1 m*M* EDTA, 2 m*M* $MgCl_2$, 150 m*M* Hepes/KOH buffer (pH 7.9), and thylakoids at a chlorophyll concentration of 25 μg ml^{-1}. Irradiance (20 μmol m^{-2} s^{-1}) was applied (hν), and ascorbate oxidase (5 U) and H_2O_2 (500 μ*M*) were added as indicated. Fluorescence quenching components are shown in A, and the original fluorescence trace and ϕ_{PSII} in B.

of the cellular ascorbate peroxidase activity.[118] In the chloroplast, the SOD and ascorbate peroxidase enzymes exist in both soluble[117,118] and thylakoid-bound isoenzymes.[115,117,119] Superoxide generated at the membrane surface can thus be trapped and converted immediately to H_2O_2 to be scavenged by the membrane-bound ascorbate peroxidase.[115,119] This reaction sequence will occur at the interface between the thylakoid membranes and the stroma in the region of PSI. The subsequent production of monodehydroascorbate and the vicinity of the membrane will facilitate its role as an acceptor of electrons from the photosynthetic electron transport chain. Any monodehyroascorbate escaping this means of reduction will spontaneously disproportionate to ascorbate and dehydroascorbate if it is not scavenged by the chloroplast monodehydroascorbate reductase (Figure 7). Dehydroascorbate is reduced to ascorbate via dehydroascorbate reductase via the enzymes of the ascorbate-glutathione cycle.[114] It is important to note that all these cycling reactions directly or indirectly use reducing power provided by the photosynthetic electron transport chain. It has been hypothesized that the rate-limiting reaction of the chloroplastic ascorbate-regenerating system was glutathione reductase[120] (see also Chapter 14 of this volume), however, this does not take into account the mechanisms for recycling monodehydroascorbate. The role of glutathione reductase in maintaining the ascorbate pool is evidenced by studies using transformed plants which express increased glutathione reductase activities.[121,122] Elevated levels of glutathione reductase led to an increase in tolerance of oxidative stress[121] and a better stabilization of the ascorbate pool in these circumstances.[122]

Ascorbate is, without doubt, essential to the anitoxidant defenses of the chloroplast.[114] Any shortage of ascorbate would severely undermine the capacity for oxidant destruction in both the soluble phase of the stroma and in the lipid membrane environment. This is because ascorbate is required for the function of two membrane-bound defense systems. First, water-soluble ascorbate is an antioxidant synergist with the lipid-soluble antioxidant vitamin E (α-tocopherol). In the membranes, tocopherols act as the primary antioxidant, while ascorbate reductively regenerates oxidized tocopherol. Second, ascorbate appears to be the natural reductant required for the synthesis of the xanthophyll, zeaxanthin from violaxanthin in the violaxanthin-depoxidase reaction of the thylakoid membranes (see Chapter 4 of this volume).

The unique position of ascorbate in chloroplast metabolism resides largely in its reducing ability, and ascorbate is implicated in the reduction of many other membrane-bound components.[114] Ascorbate and α-tocopherol are efficient antioxidants because they preferentially reduce oxidizing compounds via a process of hydrogen atom transfer at physiological pH values, rather than separate electron donation and protonation reactions.[123] This minimizes the risk of direct reduction of molecular oxygen while allowing rapid reactivity with free radicals and other oxidizing species. For this reason, ascorbate is a relatively poor electron donor to the photosynthetic electron transport chain in comparison to the rapid reduction of oxidized electron acceptors by PSI.[123]

Ascorbate is, however, often used as a PSI electron donor in *in vitro* measurements of PSI electron transport rates at an effective final concentration of about 5 m*M*, which is close to the lower levels of the measured *in vivo* concentrations.[76] Simple assays of PSI electron transport usually block electron transport from PSII with 10μ*M* 3-(3′,4′-dichlorophenyl)-1,1-dimethyl urea (DCMU) and contain 0.1*M* 2,6-dichlorophenol indophenol (DCPIP) and 5m*M* ascorbate as the electron donor couple, with DCPIP effectively mediating electron transport from ascorbate to P_{700}.[124] This system has been shown to be coupled to photophosphorylation and ATP synthesis *in vitro*. Other PSI electron donors, coupled to ascorbate, can be used in place of DCPIP and give fast electron transport, but are not coupled to ATP synthesis.[124] Ascorbate will reduce several electron transfer components, such as plastoquinone, cytochrome b559, and cytochrome f, on the oxidizing side of PSII[125] and will also reduce superoxide, produced by PSI, to H_2O_2 in competition with SOD (Equation 8).[124]

$$2O_2^{\cdot-} + \text{Ascorbate}(H_2) + 2H^+ \longrightarrow 2H_2O_2 + \text{Ascorbate} \qquad (8)$$

While the ascorbate pool does not effectively compete with PSII in the reduction of the interphotosystem carriers, it will, nevertheless, ensure their reduction in the longer term. Additionally, ascorbate may modify the electrical charge and function of other membrane components by binding to proteins on the membrane surfaces in the stroma and the lumen.[126]

VII. THE REGULATORY SIGNIFICANCE OF THE MEHLER-PEROXIDASE REACTION SEQUENCE

It has long been suspected that pseudocyclic electron flow alone could act as a regulatory device allowing the thylakoid membrane to drain off excess reducing power while providing the ATP that might make up noncyclic electron flow (1.33) and that required by the Calvin cycle (1.5).[127,128] Such speculation has always been overshadowed by the rather poor efficiency of oxygen as an electron acceptor and the resultant low rates of oxygen reduction compared to the rates of reduction of CO_2. Electron flow to O_2 alone may result in substantial q_{NP} formation in certain circumstances. The reduction of O_2 in pseudocyclic electron flow is, however, irrevocably coupled to the dismutation of the product of this reaction, $O_2^{\cdot-}$, to H_2O_2 and the subsequent destruction of H_2O_2 via ascorbate peroxidase (Figure 9).

These reactions all occur on the membrane surface and for simplicity may be envisaged as a concerted reaction sequence that has been termed the "Mehler-peroxidase reaction" sequence. The end product of this reaction, monodehydroascorbate, is a very efficient electron acceptor (V_{max} = 592

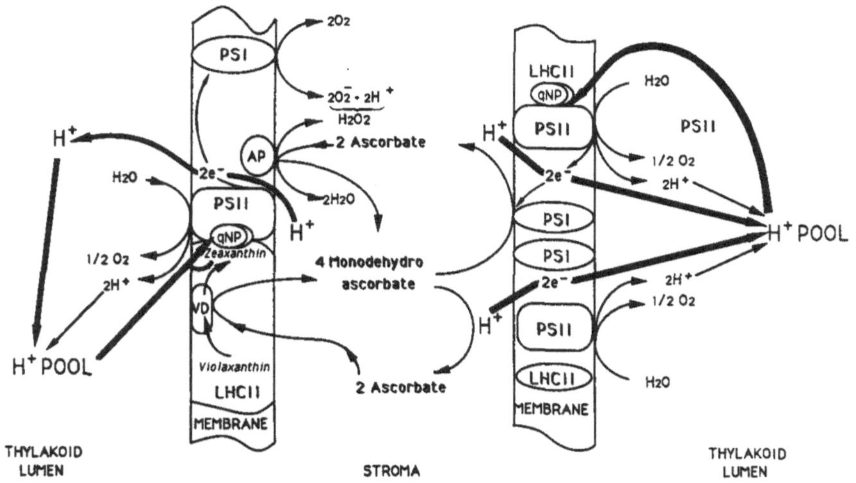

FIGURE 9. Diagram of the central role of ascorbate in membrane-associated reactions. Ascorbate is shown to be the substrate for the membrane-bound ascorbate peroxidase (AP) and violaxanthin depoxidase (VD) reactions. The product of ascorbate oxidation by these reactions, monodehydroascorbate, may then function as an electron acceptor and thus be implicated in the formation of the transthylakoid pH gradient which is required for q_{NP} formation.

μmol h^{-1} mg^{-1} Chl, Figure 8).[115] This reaction has been shown to be capable of producing a rather large amount of q_{NP} (Figure 8). The production of q_{NP} requires the regeneration of a significant transthylakoid ΔpH with the combined action of oxygen and monodehydroascorbate as electron acceptors facilitating the necessary electron flow. In addition, any oxidized ascorbate escaping reduction by this means can still be reduced via monodehydroascorbate reductase or dehydroascorbate reductase. Both enzymes derive the necessary reducing power (NADPH or NADH) from the photosynthetic electron transport chain.

The capacity of the Mehler-peroxidase reaction to generate q_{NP} is further aided by the provision of suitable conditions for the synthesis of zeaxanthin[127] (see Chapter 4 of this volume). If the Mehler-peroxidase sequence is inhibited with cyanide, q_{NP} is inhibited even at high zeaxanthin concentrations.[129] Both the Mehler-peroxidase reaction sequence and the violaxanthin de-epoxidase depend on ascorbate for activity (Figure 9). The affinity of ascorbate peroxidase for ascorbate is significantly lower than that of violaxanthin de-epoxidase for this substrate. The availability of ascorbate will have profound effects on the violaxanthin de-epoxidase reaction, and competition with the ascorbate peroxidase for substrate is possible.

Since q_{NP} is an essential mechanism protecting PSII against the adverse effects of excessive irradiance by decreasing ϕ_{PSII} and increasing thermal energy dissipation, the role of the Mehler-peroxidase reaction in the generation of q_{NP} is potentially very important. While it is difficult to distinguish the contribution made by this reaction sequence to q_{NP} formation from that made by noncyclic electron flow *in vivo,* an indication of its role may be obtained from experiments using the inhibitor of violaxanthin de-epoxidase and ascorbate peroxidase, dithiothreitol. While the application of dithiothreitol to intact leaves must mediate a whole range of other redox changes, the inhibition of violaxanthin de-epoxidase is the only event that has been considered in many experiments.[130,131] When leaves are supplied with dithiothreitol in the transpiration stream, q_{NP} is decreased, indicating that there is a suppression of noncyclic electron flow arising from the inhibition of the violaxanthin de-epoxidase and ascorbate peroxidase reactions.[130,131]

VIII. CONCLUSIONS AND PERSPECTIVES

1. Two major oxygen-consuming reactions are associated with photosynthesis. These are (1) the oxygenase reaction of RuBisCo that initiates the pathway of photorespiration and (2) electron donation to oxygen to form superoxide as a result of pseudocyclic electron flow. These two processes are intrinsically different in their energy requirements. Photorespiration consumes ATP and, therefore, acts as a powerful sink for energy, while pseudocyclic electron flow is a coupled process leading to ATP synthesis. In addition, it appears that pseudocyclic electron flow is only the first step in a reaction sequence that amplifies the capacity for regeneration of the transthylakoid ΔpH. It is associated with the ascorbate peroxidase reaction and the subsequent rereduction of monodehydroascorbate by the electron transport chain. The combined Mehler-peroxidase reaction sequence may generate a large transthylakoid proton gradient *in vivo.* The ascorbate-dependent destruction of H_2O_2 has been shown to be highly efficient, and the monodehydroascorbate thus generated is an electron acceptor equivalent to methyl viologen in its capacity for reduction by the photosynthetic electron transport chain. Photorespiration and the Mehler reaction serve as sinks for the products of electron transport, but their regulatory functions are quite distinct.
2. There is a simple relationship between triplet state chlorophyll formation and the lifetime of the singlet state which, in turn, determines the yield of chlorophyll fluorescence. It is important to emphasize that F'_s is highly regulated and normally only slightly greater than F'_o. This regulation, which is achieved via the combined effects of photochemical and nonphotochemical quenching, serves to limit singlet oxygen production by chlorophylls.
3. The Mehler reaction, *in vivo,* results from the reduction of molecular oxygen by components on the reducing side of PSI, particularly reduced ferredoxin. Regulation of the degree of reduction of the PSI electron acceptors serves to decrease the rate of superoxide production. Steady-state electron transport rates *in vivo* in most conditions are limited on the donor side of PSI

and not on the acceptor side. In addition, the activation state of NADP-MDH is always low in steady-state conditions in leaves. These observations suggest that control processes exist to avoid overreduction of the PSI acceptor side and the stoma, and thus to decrease the rate of O_2 reduction.

4. Both singlet oxygen and superoxide production can be decreased if the metabolic demand for ATP and NADPH can be sustained. Photorespiration, operating in leaves with closed stomata, or under many other constraints, may be acting beneficially to allow continued regeneration of NADP and ADP, and continued turnover of electron transport.

5. Oxygen is an essential electron acceptor for photosynthesis; electron transport can proceed at low O_2 levels, but not in the absence of oxygen. Under anaerobic conditions, the photosynthetic electron transport acceptors become overreduced. Oxygen concentrations as low as $10\mu M$ can prevent this overreduction of the electron carriers and allow poising of cyclic electron flow. Similarly, in the absence of CO_2 ϕ_{PSII} declines more markedly than ϕPSI with increasing irradiance, consistent with the occurrence of a significant cyclic electron flux around PSI. However, a considerable flux of electrons through PSII in these circumstances would support the conclusion that oxygen provides a powerful sink for electrons by either photorespiration or electron transfer to O_2. All the data currently available thus support the hypothesis that oxygen has a necessary regulatory function in electron transport. It provides a sink for excess reducing power and in this way provides the necessary redox poise that facilitates the required equilibrium between noncyclic and cyclic electron flow.

6. In most plants thus far examined, noncyclic electron flow predominates in air. Cyclic electron flow may also occur in air; it appears that under conditions where Calvin cycle activity is depressed (e.g., 100 to 200 ppm CO_2, 2% oxygen), the proportion of electrons passing through the cyclic path increases significantly. Both pseudocyclic and cyclic electron flow may be necessary to supply ATP, in addition to that produced by noncyclic electron flow, and their relative importance may vary according to conditions. However, while pseudocyclic electron flow is slow in most plants and cannot provide much ATP, it functions alongside noncyclic electron flow in leaves constantly, while the occurrence of cyclic electron flow is less evident.

7. The antioxidant defense system of the chloroplast is an integral part of oxygen metabolism. Thiol-disulfide exchange reactions are fundamental to the regulation of photosynthetic carbon metabolism and its responses to changes in irradiance. Effective scavenging of H_2O_2 and other oxidants by the ascorbate-dependent defense systems prevents inhibition of carbon assimilation via uncontrolled oxidation. Ascorbate is essential to this process and also to the membrane-associated production of zeaxanthin, which is required for the thermal energy dissipation that helps to prevent excessive oxidant production.

REFERENCES

1. **Horton, P.,** Interactions between electron transport and carbon assimilation: regulation of light-harvesting and phytochemistry, in *Photosynthesis, Plant Biology Series 8,* Briggs, W. R., Ed., Alan R Liss, New York, 1989, 393.
2. **Foyer, C. H., Furbank, R., Harbinson, J., and Horton, P.,** The mechanisms contributing to photosynthetic control of electron transport by carbon assimilation in leaves, *Photosynth. Res.,* 25, 83, 1990.
3. **Foyer, C. H.,** Interactions between electron transport and carbon assimilation in leaves. Co-ordination of activities and control, in *Photosynthesis Photoreactions to Plant Productivity,* Abrol, Y. P., Mohanty, P., and Govinjee, Eds., Oxford and IBH Publishing, New Dehli, 1992, 199.
4. **Ogren, W. H.,** Photorespiration: pathways regulation and modification, *Annu. Rev. Plant. Physiol.,* 35, 415, 1984.
5. **Ziem-Hanck, U. and Heber, U.,** Oxygen requirement of photosynthetic CO_2 assimilation, *Biochim. Biophys. Acta,* 591, 266, 1980.
6. **Harbinson, J. and Foyer, C. H.,** Relationships between the efficiencies of photosystems I and II and stromal redox state in CO_2-free air, *Plant Physiol.,* 97, 41, 1991.
7. **Heber, U. and Walker, D. A.,** Concerning a dual function of coupled cyclic electron transport in leaves, *Plant Physiol.,* 100, 1621, 1992.
8. **Peltier, G., Ravenel, J., and Vermeglio, A.,** Inhibition of a respiratory activity by short saturating flashes in *Chlamydomonas:* evidence for a chlororespiration, *Biochim. Biophys. Acta,* 893, 83, 1987.
9. **Bennoun, P.,** Evidence for a respiratory chain in the chloroplast, *Proc. Natl. Acad. Sci. U.S.A.,* 79, 4352, 1982.
10. **Garab, G., Lajko, F., Mustardy, L., and Marton, L.,** Respiratory control over photosynthetic electron transport in chloroplasts of higher plant cells: evidence for a chlororespiration, *Planta,* 179, 349, 1989.
11. **Vermeglio, A., Ravenel, J., and Peltier, G.,** Chlororespiration: a respiratory activity in the thylakoid membrane of microalgae and higher plants, in *Experimental Phycologie,* Vol. 1, *Cell Walls and Surfaces, Reproduction, Photosynthesis,* Wiessner, W., Robinson, D. G., and Starr, R. C., Eds., Springer-Verlag, New York, 1990, 188.
12. **Dietz, K. J., Schreiber, U., and Heber, U.,** The relationship between the redox state of Q_A and photosynthesis in leaves at various carbon-dioxide, oxygen and light regimes, *Planta,* 166, 219, 1985.
13. **Peterson, R. B., Sivak, M. N., and Walker, D. A.,** Relationship between steady state fluorescence yield and photosynthetic efficiency in spinach leaf tissue, *Plant Physiol.,* 88, 158, 1988.
14. **Peterson, R. B.,** Effects of O_2 and CO_2 concentrations on quantum yields of photosystems I and II in tobacco leaf tissue, *Plant Physiol.,* 97, 1388, 1991.
15. **Viil, J., Laisk, A., Oja, V., and Pärnik, T.,** Enhancement of photosynthesis caused by oxygen under saturating irradiance and high CO_2 concentrations, *Photosynthetica,* 11, 251, 1977.
16. **Sharkey, T. D., Stitt, M., Heincke, D., Gerhardt, R., Raschke, K., and Heldt, H. W.,** Limitation of photosynthesis by carbon metabolism. II. O_2-insensitive CO_2 uptake results from limitation of triose phosphate utilisation, *Plant Physiol.,* 81, 1123, 1986.
17. **Harris, G. C., Cheesbrough, J. K., and Walker, D. A.,** Measurement of CO_2 and H_2O vapour exchange in spinach leaf discs. Effects of orthophosphate, *Plant Physiol.,* 71, 102, 1983.
18. **Mehler, A. H.,** Studies on reactions of illuminated chloroplasts. I. Mechanisms of the reduction of oxygen and other Hill reagents, *Arch. Biochem. Biophys.,* 33, 65, 1951.

19. **Furbank, R. K., Badger, M. R., and Osmond, C. B.,** Photoreduction of oxygen in mesophyll chloroplasts of C_4 plants. A model system for studying an in vivo Mehler reaction, *Plant Physiol.*, 73, 1038, 1983.
20. **Marsho, T. V., Behrens, P. W., and Radmer, R. J.,** Photosynthetic oxygen reduction in isolated intact chloroplasts and cells from spinach, *Plant Physiol.* 64, 656, 1979.
21. **Canvin, D. T., Berry, J. A., Badger, M. R., Fock, H., and Osmond, C. B.,** Oxygen exchange in leaves in the light, *Plant Physiol.*, 66, 302, 1980.
22. **Robinson, J. M.,** Does O_2 photoreduction occur *in vivo?*, *Physiol. Plant.*, 72, 666, 1988.
23. **Foyer, C. H. and Lelandais, M.,** The roles of ascorbate in the regulation of photosynthesis, *Am. Soc. Plant Physiol.*, in press.
24. **Schreiber, U. and Neubauer, C.,** O_2 dependent electron flow, membrane energisation and the mechanism of non-chemical quenching of chlorophyll fluorescence, *Photosynth. Res.*, 25, 279, 1990.
25. **Neubauer, C. and Schreiber, U.,** Photochemical and non-photochemical quenching of chlorophyll fluorescence induced by hydrogen peroxide, *Z. Naturforsch. Teil C*, 44, 262, 1989.
26. **Wu, J., Weimanis, S., and Heber, U.,** Photorespiration is more effective than the Mehler reaction in protecting the photosynthetic apparatus against photoinhibition, *Bot. Acta*, 104, 283, 1991.
27. **Laing, W. A., Ogren, W. L., and Hageman, R. H.,** Regulation of soybean net photosynthetic CO_2 fixation by the interaction of CO_2, O_2, and ribulose 1,5-diphosphate carboxylase, *Plant Physiol.*, 54, 678, 1974.
28. **Krall, J. P. and Edwards, G. E.,** Relationship between photosystem II activity and CO_2 fixation in leaves. What's new in plant physiology?, *Physiol. Plant.*, 86, 180, 1992.
29. **Jordan, D. B. and Ogren, W. L.,** The CO_2/O_2 specificity or ribulose 1,5-bisphosphate carboxylase/oxygenase, *Planta*, 161, 308, 1984.
30. **Brooks, A. and Farquhar, G. D.,** Effect of temperature on the CO_2/O_2 specificity of ribulose 1,5-bisphosphate carboxylase/oxygenase and the rate of respiration in the light, *Planta*, 165, 397, 1985.
31. **Farquhar, G. H., von Caemmerer, S., and Berry, J. A. A.,** Biochemical model of photosynthetic assimilation in leaves of C_3 species, *Planta*, 149, 78, 1980.
32. **Tonkyn, J. C., Deng, X.-W., and Gruissen, W.,** Regulation of plastid gene expression during photooxidative stress, *Plant Physiol.*, 99, 1406, 1992.
33. **Schuter, G. and Gruissen, W.,** Chloroplast mRNA 3′ end processing requires a nuclear-encoded RNA-binding protein, *EMBO J.*, 10, 1493, 1991.
34. **Cotton, F. A. and Wilkinson, G.,** *Advanced Inorganic Chemistry*, Interscience, New York, 1145.
35. **Cadenas, E.,** Biochemistry of oxygen toxicity, *Annu. Rev. Biochem.*, 58, 79, 1989.
36. **Knox, J. P. and Dodge, A. D.,** Singlet oxygen and plants, *Phytochemistry*, 24, 889, 1985.
37. **Nobel, P. S.,** *Introduction to Biophysical Plant Physiology*, W. S. Freeman, San Francisco, 1974, 488.
38. **Parker, C. A. and Joyce, T. A.,** Delayed fluorescence and some properties of the chlorophyll triplets, *Photochem. Photobiol.*, 6, 395, 1967.
39. **Sauer, K.,** in *Bioenergetics of Photosynthesis*, Govindjee, Ed., Academic Press, New York, 1975, 115.
40. **Shipann, L. L.,** A theoretical study of excitons in chlorophyll *a* photosystems on a picosecond timescale, *Photochem. Photobiol.*, 31, 157, 1980.
41. **Fujimori, E. and Livingston, R.,** Interactions of chlorophyll in its triplet state with oxygen, carotene, etc., *Nature, (London)*, 180, 1036, 1957.
42. **Claes, V. H.,** Energienbertragung von angeregtem Chlorophyll auf Cu_{40}-Polyene mit Verschiedenen Chromophoren Gruppen, *Z. Naturforsch.*, 166, 445, 1961.

43. **Witt, K. and Wolff, C. H.,** Rise time of the absorption changes of chlorophyll *a* and carotenoids in photosynthesis, *Z. Naturforsch. Teil B,* 25, 387, 1970.

44. **Buchwald, H. E. and Wolff, C. H.,** Further evidence for carotenoids engaged in a metastable state in photosynthesis, *Z. Naturforsch. Teil B,* 26, 51, 1971.

45. **Witt, H. T.,** Energy conversion in the functional membrane of photosynthesis. Analysis by light pulse and electric pulse methods. The central role of the electric field, *Biochim. Biophys. Acta,* 505, 355, 1979.

46. **Breton, J. and Mathis, P.,** Photosynthèse. Mise en évidence de l'état triplet de la chlorophylle dans des lamelles chloroplastiques, *C. R. Acad. Sci.* 271, 1094, 1970.

47. **Kramer, H. and Mathis, P.,** Quantum yield and rate of formation of the carotenoid triplet state in photosynthetic structures, *Biochim. Biophys. Acta,* 593, 319, 1980.

48. **Mauzerall, D.,** Light-induced fluorescence changes in *Chlorella* and the primary photoreactions for the production of oxygen, *Proc. Natl. Acad Sci. U.S.A.,* 69, 1358, 1972.

49. **Nitsch, C., Braslavsky, S. E., and Schetz, G. H.,** Lazer-induced optoacoustic calorimetry of primary processes in isolated photosystem I and photosystem II particles, *Biochim. Biophys. Acta,* 934, 201, 1988.

50. **Hodges, M. and Moya, I.,** Modification of room temperature picosecond chlorophyll fluorescence kinetics in photosystem II enriched particles by photochemistry, *Biochim. Biophys. Acta,* 892, 42, 1987.

51. **Wolf, C. H. and Witt, H. T.,** On metastable states of carotenoids in primary events of photosynthesis, *Z. Naturforsch. Teil B,* 24, 1031, 1969.

52. **Demmig, B. and Bjorkamn, O.,** Comparison of the effect of excessive light on chlorophyll fluorescence (77K) and photon yield of O_2 evolution in leaves of higher plants, *Planta,* 171, 171, 1987.

53. **Melis, M., Spangfort, M., and Andersson, B.,** Light-absorption and electron-transport balance between photosystem II and photosystem I in spinach chloroplasts, *Photochem. Photobiol.,* 45, 129, 1987.

54. **Harbinson, J., Genty, B., and Baker, N. R.,** Relationship between the quantum efficiencies of photosystems I and II in pea leaves, *Plant Physiol.,* 90, 1029, 1989.

55. **Thorne, S. W. and Duniec, J. T.,** The physical principles of energy transduction in chloroplast thylakoid membranes, *Q. Rev. Biophys.,* 16, 197, 1983.

56. **Joliot, P., Delosme, R., and Joliot, A.,** 515nm absorption changes in Chlorella at short times (4–100μs) after a flash, *Biochim. Biophys. Acta,* 459, 47, 1977.

57. **Genty, B., Goulas, Y., Dimon, B., Peltier, G., Briantais, J. M., and Moya, I.,** Modulation of efficiency of primary conversion in leaves. Mechanisms involved at PS2, in *Research in Photosynthesis,* Murata, N., Ed., *IV* Kluwer, The Netherlands, 1993, 603.

58. **Briantais, J.-M., Vernotte, C., Krause, G. H., and Weiss, E.,** Chlorophyll a fluorescence of higher plants: chloroplasts and leaves, in *Light Emission by Plants and Bacteria,* Govindgee, Amesz, J., and Fork, D. C., Eds., Academic Press, New York, 1986, 539.

59. **Quick, W. P. and Horton, P.,** Studies on the induction of chlorophyll fluorescence in barley protoplasts. I. Factors affecting the observation of oscillations in the yield of chlorophyll fluorescence and the rate of oxygen evolution, *Proc. R. Soc. London Ser. B,* 220, 361, 1984.

60. **Ogren, E. and Baker, N. R.,** Evaluation of a technique for the measurement of chlorophyll fluorescence from leaves exposed to continuous white light: technical report, *Plant Cell Environ.,* 8, 539, 1985.

61. **Krause, G. H. and Weiss, E.,** Chlorophyll fluorescence and photosynthesis: the basis, *Annu. Rev. Plant Physiol. Plant Mol. Biol.,* 42, 313, 1991.

62. **Horton, P., Ruban, A. V., Pascal, A. A., Noctor, G., and Young, A. J.,** Control of the light harvesting function of chloroplast membranes by aggregation of the LHCII chlorophyll-protein complex, *FEBS Lett.,* 292, 1, 1991.

63. **Horton, P. and Hague, A.,** Studies on the induction of chlorophyll fluorescence in isolated barley protoplasts. IV. Resolution of non-photochemical quenching, *Biochim. Biophys. Acta*, 932, 107, 1988.
64. **Demmig-Adams, B.,** Carotenoids and photoprotection in plants: a role for the xanthophyll zeaxanthin, *Biochim. Biophys. Acta*, 1020, 1, 1990.
65. **Weis, E. and Berry, J. A.,** Quantum efficiency of photosystem II in relation to ''energy''-dependent quenching of chlorophyll fluorescence, *Biochim. Biophys. Acta*, 894, 198, 1987.
66. **Genty, B., Briantais, J.-M., and Baker, N. R.,** The relationship between the quantum yield of photosynthetic electron-transport and photochemical quenching of chlorophyll fluorescence, *Biochem. Biophys. Acta*, 900, 87, 1989.
67. **Van Kooten, O. and Snell, J. F. H.,** The use of chlorophyll fluorescence nomenclature in plant stress physiology, *Photosynth. Res.*, 25, 147, 1990.
68. **Siefermann-Harms, D.,** The xanthophyll cycle in higher plants, in *Lipids and Lipid Polymers in Higher Plants*, Tevini, M. and Lichtenthaler, H. K., Eds., Springer-Verlag, Berlin, 1977, 218.
69. **Epel, B. L. and Neumann, J.,** The mechanism of the oxidation of ascorbate and $Mn2^+$ by chloroplasts. The role of the radical superoxide, *Biochim. Biophys. Acta*, 325, 520, 1973.
70. **Allen, J. F. and Hall, D. O.,** Superoxide reduction as a mechanism of ascorbate-stimulated oxygen uptake by isolated chloroplasts, *Biochem. Biophys. Res. Commun.*, 52, 856, 1973.
71. **Asada, K. and Kiso, K.,** Initiation of aerobic oxidation of sulfite by illuminated spinach chloroplasts, *Eur. J. Biochem.*, 33, 253, 1973.
72. **Asada, K. and Kiso, K.,** The photo-oxidation of epinephrine by spinach chloroplasts and its inhibition by superoxide dismutase. Evidence for the formation of superoxide radicals in chloroplasts, *Agric. Biol. Chem.*, 37, 453, 1973.
73. **Goldbeck, J. H. and Bryant, D. A.,** Photosystem I, *Curr. Topics Bioenerget.*, 16, 83, 1991.
74. **Asada, K. and Takahashi, M.,** Production and scavenging of active oxygen in photosynthesis, in *Photoinhibition*, Kyle, D. J., Osmond, C. B., and Arntzen, C. J., Eds., Elsevier, Amsterdam, 1987, 227.
75. **Badger, M. R.,** Photosynthetic oxygen exchange, *Annu. Rev. Plant Physiol.*, 36, 27, 1985.
76. **Asada, K. and Nakano, Y.,** Affinity for oxygen in photoreduction of molecular oxygen and scavenging of hydrogen peroxide in spinach chloroplasts, *Photochem. Photobiol.*, 28, 917, 1978.
77. **Furbank, R. T. and Badger, M. R.,** Oxygen exchange associated with electron transport and photophosphorylation in spinach thylakoids, *Biochim. Biophys. Acta*, 723, 400, 1983.
78. **Steiger, H. M. and Beck, E.,** Formation of hydrogen peroxide and oxygen dependence of photosynthetic CO_2 assimilation by intact chloroplasts, *Plant Cell Physiol.*, 22, 561, 1981.
79. **Radmer, R., Kok, B., and Ollinger, O.,** Kinetics and apparent Km of the oxygen cycle under conditions of limiting carbon dioxide fixation, *Plant Physiol.*, 61, 915, 1978.
80. **Takahashi, M. A. and Asada, K.,** Superoxide production in aprotic interior of chloroplast thylakoids, *Arch. Biochem. Biophys.*, 267, 714, 1988.
81. **Hosler, J. P. and Yocum, C. F.,** Evidence for two photophosphorylation reactions concurrent with ferredoxin-catalyzed non-cyclic electron transport, *Biochim. Biophys. Acta*, 808, 21, 1985.
82. **Gerbaud, A. and Andre, M.,** Effect of CO_2, O_2 and light on photosynthesis and photorespiration in wheat, *Plant Physiol.*, 66, 1032, 1980.
83. **Behrens, P. W., Marsho, T. V., and Radmer, R. J.,** Photosynthetic O_2 exchange kinetics in isolated soybean cells, *Plant Physiol.*, 70, 179, 1982.

84. **Furbank, R. T., Badger, M. R., and Osmond, C. B.,** Photosynthetic oxygen exchange in isolated cells and chloroplasts of C_3 plants, *Plant Physiol.*, 70, 297, 1982.
85. **Westerhoff, H. V.,** Mosaic Non-Equilibrium Thermodynamics and (the Control of) Biological Free-Energy Transduction, Ph.D. thesis. University of Amsterdam, 368, 1983.
86. **Scheibe, R. and Stitt, M.,** Comparison of NADP-malate dehydrogenase activation, Q_A reduction and O_2 evolution in spinach leaves, *Plant Physiol. Biochem.*, 26, 473, 1988.
87. **Foyer, C. H., Lelandais, M., and Harbinson, J.,** Control of the quantum efficiencies of photosystems I and II, electron flow and enzyme activation following dark to light transitions in pea leaves, *Plant Physiol.*, 99, 979, 1992.
88. **Harbinson, J. and Woodward, F. I.,** The use of light induced absorbance exchanges at 820nm to monitor the oxidation state of P_{700} in leaves, *Plant Cell Environ.*, 9, 131, 1987.
89. **Harbinson, J. and Hedley, C. L.,** The kinetics of P_{700} reduction in leaves: in situ probe of thylakoid functioning, *Plant Cell Environ.*, 12, 357, 1989.
90. **Bendall, D. S.,** Photosynthesis cytochromes of oxygenic organisms, *Biochim. Biophys. Acta*, 683, 119, 1982.
91. **Genty, B., Harbinson, J., and Baker, N. R.,** Relative quantum efficiencies of the two photosystems of leaves in photorespiratory and non-photorespiratory conditions, *Plant Physiol. Biochem.*, 28, 1, 1990.
92. **Harbinson, J., Genty, B., and Foyer, C. H.,** Relationship between photosynthetic electron transport and stromal enzyme activity in pea leaves. Toward an understanding of the nature of photosynthetic control, *Plant Physiol.*, 94, 115, 1990.
93. **Cseke, C. and Buchanan, B. B.,** Regulation of the formation and utilisation of photosynthate in leaves, *Biochim. Biophys. Acta*, 853, 43, 1986.
94. **Heber, U., Neimanis, S., Dietz, K. J., and Vill, J.,** Assimilatory power as a driving force in photosynthesis, *Biochim. Biophys. Acta*, 852, 144, 1986.
95. **Takahashi, M. A. and Asada, K.,** Superoxide anion permeability of phospholipid membranes and chloroplast, *Arch. Biochem. Biophys.*, 226, 558, 1983.
96. **Takahashi, M., Kono, Y., and Asada, K.,** Reduction of plastocyanin with superoxide and superoxide-dismutase-dependent oxidation of plastocyanin by hydrogen peroxide, *Plant Cell Physiol.*, 21, 1431, 1980.
97. **Fridovich, I.,** Molecular oxygen: friend or foe?, in *Active Oxygen/Oxidative Stress and Plant Physiology*, Pell, E. and Steffen, K., Eds., American Society of Plant Physiologists, Rockville, MD, 1991, 1.
98. **Bradley, R. L., Long, K. M., and Frasch, W. D.,** The involvement of photosystem II generated H_2O_2 in photoinhibition, *FEBS Lett.*, 286, 209, 1991.
99. **Wydziynski, T., Angström, J., and Vänngård, T.,** H_20_2 formation by photosystem II, *Biochim. Biophys. Acta*, 973, 23, 1989.
100. **Yonetani, T. and Schleyer, H.,** Studies on cytochrome *c* peroxidase. IX. The reaction of ferrimyoglobin with hydroperoxides and a comparison of peroxide-induced compounds of ferrimyoglobin and cytochrome *c* peroxidase, *J. Biochem.*, 242, 1974, 1967.
101. **Hori, H. and Yonetani, T.,** Powder and single crystal electron paramagnetic resonance studies of yeast cytochrome *c* peroxidase and its peroxide compound, compound ES, *J. Biol. Chem.*, 260, 349, 1985.
102. **Tew, D. and Ortiz de Montellano, P. R.,** The myoglobin protein radical coupling of Tyr-103 to Tyr-151 in the H_20_2-mediated cross-linking of sperm whale myoglobin, *J. Biochem.*, 263, 17880, 1988.
103. **Boussac, A., Zimmermann, J.-L., Rutherford, A. W., and Lavergne, J.,** Histidine oxidation in the oxygen-evolving photosystem II enzyme, *Nature (London)*, 347, 303, 1990.
104. **Miller, A. F. and Brudvig, G. W.,** Manganese and calcium requirements for reconstitution of oxygen evolution activity in manganese-depleted photosystem II membranes, *Biochemistry*, 28, 8181, 1989.

105. **Leegood, R. C., Walker, D. A., and Foyer, C. H.,** Regulation of the Benson-Calvin cycle, in *Photosynthetic Mechanisms and the Environment,* Barber, J. and Baker, N. R., Eds., Elsevier, New York, 1985, 191.
106. **Siebke, K., Laisk, A., Neimanis, S., and Heber, U.,** Regulation of chloroplast metabolism in leaves. Evidence that NADP-dependent glyceraldehyde phosphate dehydrogenase, but not ferredoxin-NADP reductase, controls electron flow to phosphoglycerate in the dark-light transition, *Planta,* 185, 337, 1991.
107. **Florencio, F. J., Yee, B. C., Johnson, T. C., and Buchanan, B. B.,** An NADP/thioredoxin reductase and thioredoxin from spinach, *Arch. Biochem. Biophys.,* 266, 496, 1988.
108. **Leegood, R. C.,** Enzymes of the Calvin cycle, in PJ Lea., ed, *Methods in Plant Biochemistry,* Vol. 3, Lea, P. J., Ed., Academic Press, London, 1990, 15.
109. **Kaiser, W. M.,** Reversible inhibition of the Calvin cycle and activation of oxidative pentose phosphate cycle in isolated chloroplasts by hydrogen peroxide, *Planta,* 145, 377, 1979.
110. **Charles, S. A. and Halliwell, B.,** Effect of hydrogen peroxide on spinach *(Spinacia oleracia)* chloroplast fructose bisphosphatase, *Biochem. J.,* 189, 373, 1980.
111. **Tanaka, K., Otsubo, T., and Kondo, N.,** Participation of hydrogen peroxide in the inactivation of Calvin cycle SH enzymes in SO_2-fumigated spinach leaves, *Plant Cell Physiol.,* 23, 1009, 1982.
112. **Bertini, I., Luchinat, C., Viezzoli, M., and Wang, Y.,** Active site modification of superoxide dismutase by H_2O_2 studied through HNMR of the cobalt derivates, *Arch. Biochem. Biophys.,* 269, 586, 1989.
113. **Scheibe, R., Geissler, A., and Fickenscher, K.,** Chloroplast glucose-6-phosphate dehydrogenase: Km shift upon light modulation and reduction, *Arch. Biochem. Biophys.,* 274, 290, 1989.
114. **Foyer, C. H.,** Ascorbic acid, in *Antioxidants in Higher Plants,* Alscher, R. G. and Hess, J. L., Eds., CRC Press, Boca Raton, FL, 1993, 31.
115. **Miyake, C. and Asada, K.,** Thylakoid-bound ascorbate peroxidase in spinach chloroplasts and photoreduction of its primary oxidation product monodehydroascorbate radicals in thylakoids, *Plant Cell Physiol.,* 33, 541, 1992.
116. **Jackson, C., Dench, J., Moore, A. L., Halliwell, B., Foyer, C. H., and Hall, D. O.,** Subcellular localisation and identification of superoxide dismutase in the leaves of higher plants, *Eur. J. Biochem.,* 91, 339, 1978.
117. **Lumsden, J. and Hall, D. O.,** Soluble and membrane-bound superoxide dismutase in a blue-green alga *(Spirulina)* and spinach, *Biochem. Biophys. Res. Commun.,* 58, 35, 1974.
118. **Chen, G. X. and Asada, K.,** Ascorbate peroxidase in tea leaves: occurrence of two isozymes and the differences in their enzymatic and molecular properties, *Plant Cell Physiol.,* 30, 987, 1989.
119. **Groden, D. and Beck, E.,** H_2O_2 destruction by ascorbate-dependent systems from chloroplasts, *Biochim. Biophys. Acta,* 546, 426, 1979.
120. **Jablonski, P. P. and Anderson, J. W.,** Light dependent reduction of dehydroascorbate by ruptured pea chloroplasts, *Plant Physiol.,* 67, 1239, 1981.
121. **Aono, M., Kubo, A., Saji, H., Natori, T., Tanaka, K., and Kondo, N.,** Resistance to active oxygen toxicity of transgenic *Nicotiana tabacum* that expresses the gene of glutathione reductase in *Escherichia coli., Plant Cell Physiol.,* 32, 691, 1991.
122. **Foyer, C. H., Lelandais, M., Galap, C., and Kunert, K. J.,** Effects of elevated glutathione reductase activity on the cellular glutathione pool and photosynthesis in leaves under normal and stress conditions, *Plant Physiol.,* 97, 863, 1991.
123. **Njus, D. and Kelly, P. M.,** Vitamins C and E donate single atoms *in vivo, FEBS Lett.,* 284, 147, 1991.

124. **Allen, J. F. and Holmes, N. G.,** Electron transport and redox titration, in *Photosynthesis Energy Transduction, A Practical Approach,* Hipkins, M. F. and Baker, N. R., Eds., IRL Press Oxford, 1986, 103.
125. **Aristarkov, A. I., Nikandrov, V. V., and Krasnovikii, A. A.,** Ascorbate permeability of chloroplast thylakoid membrane: reduction of plastoquinones and cytochrome *f* (transl.), *Biochemistry,* 52, 1776, 1987.
126. **Bensch, K. G.,** On a possible mechanism of action of ascorbic acid: formation of ionic bonds with biological molecules, *Biochem. Biophys. Res. Commun.,* 101, 312, 1981.
127. **Egneus, H., Heber, U., Matthiesen, U., and Kirk, M.,** Reduction of oxygen by the electron transport chain of chloroplasts during assimilation of carbon dioxide, *Biochim. Biophys. Acta,* 408, 252, 1975.
128. **Heber, U., Egneus, H., Hanch, U., Jensen, M., and Koster, S.,** Regulation of photosynthetic electron transport and photophosphorylation in intact chloroplasts and leaves of *Spinacia oleracea* L., *Planta,* 143, 41, 1978.
129. **Naubauer, C. and Yamamoto, H. Y.,** Mehler-peroxidase reaction mediates zeaxanthin formation and zeaxanthin-related fluorescence quenching in intact chloroplasts, *Plant Physiol.,* 99, 1354, 1992.
130. **Bailey, K. J. and Walker, D.,** Changes in fluorescence quenching brought about by feeding dithiothreitol to illuminated leaves, *Plant Physiol.,* 99, 124, 1992.
131. **Demmig-Adams, B., Adams, W. W., Heber, U., Neimanis, S., Winter, K., Kruger, A., Czygan, F. C., Bilger, W., and Bjorkamn, O.,** Inhibition of zeaxanthin formation and of rapid changes in radiationless energy dissipation by dithiothreitol in spinach leaves and chloroplasts, *Plant Physiol.,* 92, 293, 1990.
132. **Wolf, C.,** unpublished observation reported by Witt, H. T., in *Excited States of Biological Molecules,* Birk, J. B., Ed., John Wiley & Sons, London, 1975, 250.
133. **Feierabend, J.,** personal communication.

Chapter 2

THE ROLE OF OXYGEN IN PHOTOINHIBITION OF PHOTOSYNTHESIS*

G. Heinrich Krause

TABLE OF CONTENTS

* **Abbreviations:** Chl, chlorophyll; DCMU, 3-(3′,4′-dichlorophenyl)-1,1-dimethyl urea; F_o, F_v, F_m, initial, maximum variable, and maximum total chlorophyll fluorescence yield; Ph, pheophytin; PQ, plastoquinone; PS, photosystem; Q_A, Q_B, primary and secondary quinone-type electron acceptors of photosystem II; SOD, superoxide dismutase; Z, tyrosine 'Z' of the D1 protein, secondary electron donor of photosystem II.

0-8493-5443-9/94/$0.00 + $.50

I. INTRODUCTION

A specific aspect of photooxidative stress in green plants is the inhibition of photosynthesis that results from an inactivation of the photosystem II (PSII) reaction center by excessive absorbed light energy. Numerous studies have suggested that the reactions of dioxygen (O_2) and active oxygen species, such as the superoxide radical ($O_2^{\bar{\cdot}}$), the hydroxyl radical ($^{\cdot}OH$), hydrogen peroxide (H_2O_2), and singlet oxygen ($^1\Delta gO_2$), cause photoinhibition of photosynthesis. However, the emerging picture of the effects of oxygen on photoinhibition is far from complete and is, in part, afflicted with contradictions. One aspect that adds to the difficulty of interpretation is the dual role of oxygen: high levels of O_2 induce photorespiration in C_3 plants. The photorespiratory cycle can diminish photoinhibition, as it dissipates part of the excess energy, and thereby, supposedly relieves the overreduction of the photosynthetic electron transport chain in high light. A further drain of electrons is achieved by the reduction of O_2 at the acceptor side of photosystem I (PSI) in the Mehler reaction (see Reference 1). This forms the potentially deleterious $O_2^{\bar{\cdot}}$ radicals. On the other hand, electron transfer to O_2 may also lower the extent of photoinhibition. This can be demonstrated when the high-light effects on photosynthesis, under anaerobic and aerobic conditions, are compared. However, anaerobic photoinhibition exhibits peculiar characteristics, which are discussed in Section III.B below. The dual role of oxygen in photoinhibition has been reviewed previously.[2]

''Photoinhibition'' in the strict sense should be distinguished from the broader term ''photooxidation'' (see Reference 3), even though oxygen is

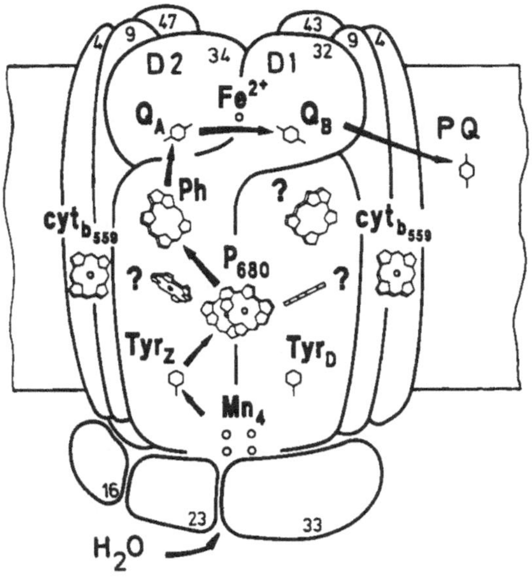

FIGURE 1. Scheme of the reaction sequence of photosynthetic electron transport in the reaction center of PSII. (Modified from Rutherford, A. W., *Trends Biochem. Sci.*, 14, 227, 1989. With permission.)

involved in both phenomena. Photooxidation includes membrane lipid peroxidation and bleaching of antenna pigments in the thylakoids. These processes do not seem to be directly related to photoinhibition. *In vivo,* photoinhibition results predominantly from an inactivation of PSII (see Reference 4). There is general agreement that the site of primary inactivation is located in the reaction center of PSII, which consists essentially of the D1/D2 heterodimer and cytochrome *b*559 (see Reference 5). The intriguing question of which component of the electron transfer chain (Z, P_{680}, Ph, Q_A, or Q_B) of this complex (Figure 1) becomes inactivated initially is very much debated.

There is much evidence that the Q_B-binding D1 protein, with its unusually fast turnover, plays an important role in photoinhibition and the following recovery in low light (see Reference 7). One of the steps in the cycle of photoinhibition and recovery is the degradation of the D1 protein by a specific membrane-bound protease. Oxygen apparently is involved in triggering the degradation. However, degradation, and perhaps even the triggering of degradation, does not seem to be the **initial** step of photoinhibition. A **primary** effect of dioxygen or active oxygen species in photoinhibition has not been clearly proven. On the other hand, the exacerbation of photoinhibition by O_2 and protection by antioxidant systems point to a general role of oxygen in the effects of high-light stress on the photosynthetic apparatus.

This chapter reviews known oxygen effects in photoinhibition and discusses the apparently conflicting evidence and problems of the present debate.

In addition, an outline of the general photooxidative processes occurring in the thylakoid membranes is given in the following section.

II. PHOTOOXIDATION VS. PHOTOINHIBITION

A. DEFINITIONS

Photoinhibition is best defined in the simple meaning of the word as an inhibition of photosynthetic activity (optimal quantum yield and/or capacity of photosynthesis) by visible light (400 to 700 nm).[3] Light energy absorbed in excess of that utilized to drive photosynthetic electron transport can principally cause photoinhibition. However, whether a given excessive photon flux actually causes measurable inhibition in plants depends on many internal and external factors, such as structural state and acclimation of the pigment system, activity of various defense mechanisms, and the impact of environmental stresses. Photoinhibition of photosynthesis occurs independently of the oxidative destruction of bulk pigments.[3] In fact, it has been postulated that photoinhibition contributes to the protection of antenna pigments from gross destruction.[4] It appears that in the photoinhibited state, the inactive PSII reaction center is still able to trap excitation energy, which is dissipated as heat instead of being used in the photochemical reaction.[8] Such an effect would lower the exciton density in the pigment bed and prevent or delay photooxidation.

Usually, substantial bleaching of antenna pigments occurs at a much slower rate than photoinhibition[3] and in the field requires long-lasting extreme conditions (e.g., winter stress[9]). Another oxidative process in the chloroplasts apparently not directly related to photoinhibition is light-induced membrane lipid peroxidation. Oxygen and light-dependent destructive reactions, such as pigment bleaching and lipid peroxidation occurring in the thylakoid membrane, are commonly termed *photooxidation*.[3,10,11] They can be separated from proper *photoinhibition*, which is determined by a specific inactivation of the PSII reaction center. Nevertheless, the same or similar oxygen-mediated reactions that cause photooxidation also might be involved in photoinhibition (see Reference 11a). Therefore, basic evidence regarding photooxidation in chloroplasts is discussed briefly.

B. ROLE OF ACTIVE OXYGEN IN PHOTOINDUCED DESTRUCTION OF THYLAKOID PIGMENTS AND LIPIDS

Photooxidation is enhanced by temperature stress, e.g., by heating[11a] or chilling, but has also been observed at optimum temperatures for photosynthesis. Herbicide action in many cases leads to photooxidation, which often is the basis of herbicide toxicity (see Chapter 9 of this volume). Chilling injury in plants is usually much more pronounced (or exhibits different characteristics) when chilling occurs in the light as compared to darkness. Chilling-sensitive plants exhibited increased rates of O_2^- formation in the light at low temperatures.[12] The involvement of active oxygen species in light-dependent chilling injury has been clearly documented by Wise and Naylor.[10,11] Leaves of cucumber *(Cucumis sativus)*, a chilling-sensitive plant, rapidly lost α-tocopherol during strong irradiation at 5°C. This lipophilic antioxidant located in the thylakoid membrane is supposed to scavenge superoxide radicals as well as lipid peroxy radicals (LOO˙). Furthermore, it is supposed to quench singlet oxygen or the triplet state of chlorophyll (^{3}Chl) (see Reference 11 and, for a review, Reference 12a). Notably, pigment bleaching started only after a lag phase (3 h), when the content of α-tocopherol had been substantially reduced. During these first 3 h, strong photoinhibition developed, as evidenced by the inhibition of photosynthetic CO_2 assimilation.[10] The amounts of ascorbate and glutathione (GSH), soluble antioxidants present at high levels in the chloroplast stroma, strongly decreased at nearly constant rates during the whole time (12 h) of treatment.[11]

1. Superoxide and Hydroxyl Radicals

The bleaching of chlorophyll (Chl) *a* and *b* and of carotenoids (both β-carotene and xanthophylls) is accompanied by substantial lipid peroxidation, as monitored, e.g., by ethane production.[10,11] The latter was largely suppressed in the presence of atrazine. This herbicide blocks the Q_B site in PSII and thereby inhibits electron transfer to PSI and O_2 reduction in the Mehler reaction. From this, it seems that lipid peroxidation is initiated predominantly by $O_2^{\dot{-}}$. The principles of photoinduced lipid peroxidation have recently been reviewed by Girotti.[13] Even though the superoxide radical is rather unreactive and unable to peroxidize unsaturated fatty acyl chains directly, $O_2^{\dot{-}}$ can give rise to the formation of ˙OH radicals in the Haber-Weiss reaction (see Reference 14), which is catalyzed by iron cations:

$$O_2^{\dot{-}} + Fe^{3+} \rightarrow Fe^{2+} + O_2 \qquad (1)$$

$$Fe^{2+} + H_2O_2 \rightarrow Fe^{3+} + OH^- + {}^{\cdot}OH \qquad (2)$$

The hydrogen peroxide in Equation 2 is supplied by the superoxide dismutase (SOD) reaction:

$$2O_2^{\dot{-}} + 2H^+ \rightarrow H_2O_2 + O_2 \qquad (3)$$

The extremely strong oxidant, ˙OH, abstracts a hydrogen atom from an unsaturated acyl residue to form a lipid radical (L·). This starts a radical chain reaction by adding dioxygen (in the ground triplet state, $^3O_2$), yielding the lipid peroxy radical, LOO· (for details, see reviews in References 13 and 15). The generation of ˙OH from $O_2^{\dot{-}}$ has been demonstrated to occur in leaves by means of trapping ˙OH with dimethylsulfoxide. However, this could be observed only in the presence of the herbicide paraquat (methylviologen), which strongly enhances the formation of superoxide radicals.[16] Incubation with paraquat also caused fast lipid peroxidation to occur in chilling-resistant pea (*Pisum sativum* L.).[11]

2. Singlet Oxygen

There is, furthermore, evidence that $^1\Delta gO_2$ (here termed $^1O_2$) is involved in photooxidation. This was shown, e.g., by the increase in ethane formation (indicating lipid peroxidation) in cucumber leaves incubated with D_2O, which prolongs the lifetime of singlet oxygen.[11] The $^1O_2$ generator Rose Bengal was shown to enhance lipid peroxidation in leaves.[17] In high light, excessive excitation might generate triplet chlorophyll (^{3}Chl) in the antennae, which can transfer its excitation energy to ground-state triplet oxygen, forming $^1O_2$, as long as ^{3}Chl is not quenched by carotenoids (see References 18 and 19 for reviews):

$$^1\text{Chl}^* \rightarrow {}^3\text{Chl} \tag{4}$$

$$^3\text{Chl} + {}^3O_2 \rightarrow {}^1\text{Chl} + {}^1O_2 \tag{5}$$

Some $^1O_2$ might also be generated from lipid hydroperoxides (LOOH) formed in the peroxidative radical chain reaction, or from the peroxy radicals (see References 13 and 19 for reviews):

$$2\text{LOO}^{\cdot} \rightarrow \text{LOOL} + {}^1O_2 \tag{6}$$

Singlet oxygen can add to unsaturated acyl residues, directly yielding LOOH in a nonradical reaction. The LOOH may reenter the reaction chain, propagating lipid degradation.

Pigment bleaching is perhaps caused predominantly by the generation of $^1O_2$. Due to its high reactivity, $^1O_2$ might act directly on the pigment molecules next to the site of its formation. Interestingly, Barry et al.[21] have shown that the blocking of electron transport at the Q_B site of PSII, which suppresses lipid peroxidation (see Section II.B.1), enhances pigment bleaching. Chloroplasts isolated from leaves of barley *(Hordeum vulgare)* lost their pigments at a much higher rate when illuminated in the presence of the inhibitor 3-(3′,4′-dichlorophenyl)-1,1-dimethyl urea (DCMU) compared to its absence. Such a block inhibits O_2 reduction via PSI, but also increases excess energy in the antennae, and thereby might enhance ^{3}Chl formation. Carotenoids (Car), particularly β-carotene, were bleached faster than the chlorophylls. Carotenoids are supposed to protect against Chl bleaching by quenching the triplet state (^{3}Chl) and $^1O_2$:

$$^3\text{Chl} + {}^1\text{Car} \rightarrow {}^1\text{Chl} + {}^3\text{Car} \tag{7}$$

$$^1O_2 + {}^1\text{Car} \rightarrow {}^3O_2 + {}^3\text{Car} \tag{8}$$

Triplet carotenoid may return to the singlet ground state by radiationless transition. However, destruction of ^{3}Car occurs with a probability on the order of 10^{-9} (see Reference 18 for a review). Loss of the protective carotenoids may then lead to increased chlorophyll bleaching and lipid peroxidation. Photooxidation has been shown to result in ultrastructural changes of the chloroplast, e.g., partial unstacking of the grana.[10,22]

3. Antioxidant Defense Systems

The lag phase in photooxidation[11] (mentioned in Section II.B.1) demonstrates that under stress conditions, the scavenger systems for active oxygen species present in the chloroplast are efficient only for a limited time, at least in chilling-sensitive plants. The progressive loss of α-tocopherol, glutathione, and ascorbate[11] as well as carotenoids[11,20] supports this view. The same treatment that caused heavy photooxidation in chilling-sensitive cucumber did not induce significant effects in chilling-resistant pea.[11] Thus, faster replenishing of antioxidants and/or more efficient defense systems must be assumed for chilling-resistant plants, whereas the mechanisms of photooxidation seem to be in principle the same in both types of species. As a passive defense, it has been suggested that the structure of the pigment protein complexes limits the diffusion of O_2 to the pigment sites.[23]

III. THE DUAL ROLE OF O_2 IN PHOTOINHIBITION

A. PROTECTIVE EFFECTS OF O_2

1. Photorespiration

A number of studies have shown that the levels of O_2 present in the atmosphere, which promote photorespiration, diminish the extent of photoinhibition (for earlier references, see Reference 2). The effect of O_2 is best seen when leaves are preilluminated at different O_2 levels in a CO_2-free atmosphere. Bean leaves *(Phaseolus vulgaris)* exhibited about 80% photoinhibition of CO_2 assimilation following irradiation in 1% O_2 (in N_2), yet they were not affected when O_2 levels were between 7 and 21%.[24] The protective effect of photorespiration has been attributed to the consumption of photosynthetic energy (NADPH and ATP) occurring through the carbon oxidation cycle initiated by oxygenation of ribulose-1,5-bisphosphate.

It was demonstrated recently that in spinach leaves, photorespiration supports linear electron transport when exogenous CO_2 is absent, and, therefore, net CO_2 assimilation does not occur.[25] In the steady state under high irradiation (600 W m^{-2}), this level of electron flow maintained a substantial fraction of Q_A in the oxidized state, as apparent from the photochemical quenching of Chl fluorescence. The oxidized fraction of Q_A increased with increasing O_2 concentration (Figure 2). Thus, photorespiration may prevent the strong reduction of Q_A that potentially leads to photoinhibition.

Similar results were obtained with severely water-stressed leaves of sunflower *(Helianthus annuus)* and bean[26] as well as of *Digitalis lanata*.[27] Some

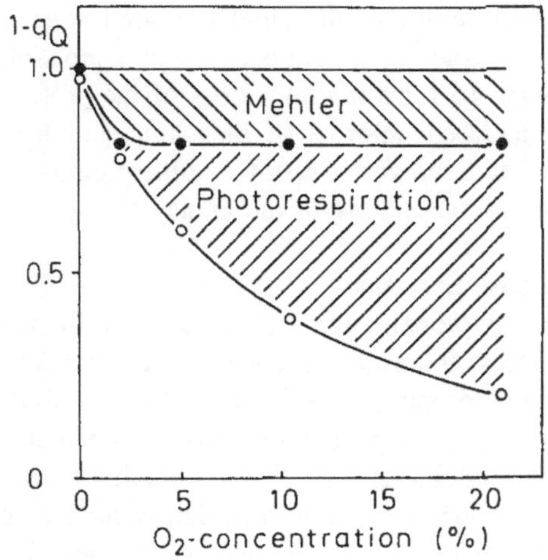

FIGURE 2. The reduction state of Q_A as a function of O_2 concentration in a CO_2-free atmosphere during strong illumination (600 W m^{-2}) of spinach leaves. The fluorescence quenching coefficient q_Q (= q_P, photochemical quenching) is an approximate indication of the redox state of Q_A; $1 - q_Q = 1$ denotes full reduction of Q_A. (●) Mehler reaction only (photorespiration was inhibited by glyceraldehyde) and (○) photorespiration in addition to the Mehler reaction (control leaf without inhibitor). (From Wu, J., Neimanis, S., and Heber, U., *Bot. Acta,* 104, 283, 1991. With permission.)

O_2-dependent energy dissipation in high light related to CO_2 recycling was also evident in water-stressed maize *(Zea mays).*[26] On the other hand, an increase in the O_2 level from 1 to 50% only slightly diminished the photoinhibition in this C_4 species.[28] Photorespiration is largely suppressed in C_4 plants due to CO_2 accumulation in the bundle sheath chloroplasts.

When the stomata are closed under water stress and high irradiation, the integrated carbon reduction and carbon oxidation cycles operate at the CO_2 compensation point. Energy drain by this "futile cycle" of CO_2 release and refixation apparently reduces photoinhibition significantly, even though the rates of electron transport supported by the recycling of CO_2 are not very high.[25,29] Several studies show significant protection against photoinhibition in leaves of C_3 plants under conditions occurring at the CO_2 compensation point.[2,25,30] However, the protective effect cannot be explained simply, in quantitative terms, by energy dissipation through combined photorespiration and CO_2 reassimilation.[2] According to Heber et al.[29] and Wu et al.,[25] the limited rate of coupled electron flow facilitated by photorespiration under conditions of high light and closed stomata exerts protection in two ways: (1) Q_A is maintained in a partly oxidized state (see Figure 2) and (2) a highly energized state of the thylakoid membrane (high proton gradient) is built up, perhaps due to the poising of cyclic electron flow around PSI.

Acidification of the thylakoid interior is supposed to induce a control mechanism in PSII, which leads to increased dissipation of excitation energy manifested by the quenching of Chl fluorescence.[31] This energy-dependent quenching mechanism is known to protect against photoinhibition (see References 2 and 32 for reviews). In conclusion, photorespiration supported by air levels of O_2 appears to be an important factor in stabilizing the photosynthetic apparatus under the conditions of high light and water stress that are frequently encountered by plants.

2. The Mehler Reaction: Reduction of O_2

The action of O_2 as a Hill oxidant on the reducing side of PSI (the Mehler reaction[33]) has been reviewed in detail by Badger[34] and Robinson.[1] O_2 serves as an alternative electron acceptor of PSI when other acceptors such as $NADP^+$ or NO_2^- are insufficient to reoxidize reduced ferredoxin (Fd_{red}) or ferredoxin:NADP oxidoreductase. Reported rates of O_2 reduction are between 10 and 40 μmol O_2 mg^{-1} (Chl) h^{-1} in most cases. The apparent K_m of thylakoids for O_2 is low; values between 2 and 60 μM have been published. (In comparison, the O_2 concentration in water equilibrated with air at 20°C is about 280 μM.) The predominant pathway of O_2 reduction proceeds from Fd_{red} via the superoxide radical (Equation 9). The $O_2^{\cdot-}$ dismutates to hydrogen peroxide (Equation 10). The reaction is catalyzed by superoxide dismutase, but also occurs spontaneously. Probably, most of the H_2O_2 is scavenged within the chloroplasts by the ascorbate peroxidase system (see Chapters 4 and 13 of this volume), which ultimately utilizes NADPH (or Fd_{red}), and thus results in further electron drain from PSI (Equation 11):

$$2O_2 + 2Fd_{red} \rightarrow 2O_2^{\cdot-} + 2Fd_{ox} \tag{9}$$

$$2O_2^{\cdot-} + 2H^+ \rightarrow H_2O_2 + O_2 \tag{10}$$

$$H_2O_2 + NADPH + H^+ \rightarrow 2H_2O + NADP^+ \tag{11}$$

There is evidence that the Mehler reaction operates *in vivo* even under conditions optimal for photosynthesis.[1] In excessive light, O_2 reduction contributes to the electron drain from PSI.[2,25,29,35] The significance of electron flow mediated by H_2O_2 reduction (Equation 11) has also been demonstrated.[36] However, experiments using the inhibitor, glyceraldehyde, showed that the Mehler reaction, in the absence of both assimilatory and photorespiratory carbon metabolism, did not provide efficient protection in high light.[25] As seen from Figure 2, the Mehler reaction had much less effect on the redox state of Q_A than photorespiration. Obviously, the rates of the Mehler reaction are too low for substantial protection.[25,29] This contradicts the reports that high rates of O_2 reduction occur when physiological electron acceptors are exhausted (see References 1 and 34 for reviews). However, Heber et al.[35] found with isolated chloroplasts that acceptors such as 3-phosphoglycerate, oxaloacetate, and nitrite, as well as the artificial acceptor paraquat, are much more efficient in protecting from photoinhibition than O_2. It was postulated by Wu et al.[25] that the Mehler reaction does not efficiently relieve overreduction of the electron transport chain and therefore is incapable of poising cyclic PSI-driven electron flow.

Coupled electron transport is required for protection in high light to support a high transthylakoid ΔpH. This not only downregulates PSII (see above), but also controls electron flow to PSI (for discussion see Reference 32). In fact, in the steady state in excessive light, the thylakoids are known to be highly energized. The reaction center pigment of PSI then is found largely in the oxidized state (P_{700^+}).[37] Thus, PSI is not overreduced, but electron flow is limited by donation to PSI. In this state, P_{700^+} allows radiationless dissipation of excitation energy,[38] and electron flow to PSI acceptors including O_2 is restricted.

Overall, our present knowledge does not finally resolve the question of the physiological role of the Mehler reaction in regulating and stabilizing the photosynthetic electron transport system. One may conclude that the limited rate of O_2 reduction alone is not a decisive factor of protection in high light, but may together with other mechanisms contribute to diminish photoinhibition. The question of whether O_2^- and secondary products of the Mehler reaction promote photoinhibition is discussed in Section IV.B.

B. ENHANCEMENT OF PHOTOINHIBITION BY O_2

There has been some confusion in the literature about the question of whether photoinhibition is enhanced by the level of O_2 in air compared to partial pressures corresponding to 1 to 2% O_2 in the atmosphere. It is now clear that with increasing O_2 levels from 1 or 2 to 21% or higher, photoinhibition potentially is promoted. This O_2 effect can be observed under conditions that overcompensate, diminish, or eliminate the protection provided by photorespiration. This is the case *in vitro* in isolated chloroplasts, as these do not possess the full photorespiratory cycle. The exacerbation of photoinhibition by O_2 can also be evidenced *in vivo* by irradiating leaves at low (chilling) temperatures, when carbon cycling occurs at much reduced rates.

1. Isolated Chloroplasts

Photoinhibition of CO_2 assimilation in intact spinach chloroplasts has been shown to be enhanced with increasing O_2 concentrations.[39] The O_2 effect was observed when irradiation occurred in a largely CO_2-free medium. In the presence of saturating CO_2, photoinhibition was prevented, obviously by carbon metabolism that supported high rates of coupled electron transport.[39,40] Similarly, phosphate-deficient chloroplasts, in which carbon metabolism was severely limited, exhibited a pronounced O_2 dependence of photoinhibition (Table 1), as determined by recording the uncoupled electron transport capacity.[25,35] Release of thylakoids from intact chloroplasts after photoinhibitory

TABLE 1
Effect of O_2 Concentration on Photoinhibition ($600\ W\ m^{-2}$ Red Light) of Intact Spinach Chloroplasts

Time of exposure to light (min)	Oxygen concentration in the chloroplast suspension ($\mu M O_2$)		
	ca. 50	ca. 300	ca. 600
	Loss of electron transport as % of initial activity		
10	23	57	73
20	36	80	86

Note: Light-saturated electron transport to ferricyanide in the presence of 5 m*M* NH_4Cl as uncoupler was measured subsequent to the photoinhibitory treatment.

From Wu, J., Neimanis, S., and Heber, U., *Bot. Acta.* 104, 283, 1991. With permission.

irradiation and subsequent measurement of electron transport in the presence of uncouplers and artificial electron acceptors confirmed that the site of inhibition is within the electron transport system, particularly in PSII (see References 25 and 40). Irradiation of isolated chloroplasts also affects PSI (see Reference 4 for a review), although to a lesser extent than PSII. The quenching of the maximum variable Chl fluorescence (F_v) or lowering of the F_v/F_m ratio (F_m = maximum fluorescence) indicates PSII inactivation. Photoinhibition of PSI appears to be O_2-dependent, i.e., requires the presence of dioxygen, but the O_2 effect has not been studied in detail.[4] In contrast, little or no photoinhibition of PSI is usually observed *in vivo*.

2. Intact Leaf Tissue and Isolated Protoplasts

In intact leaves, the protective effect of photorespiration normally more than compensates for the detrimental O_2 effect that enhances photoinhibition. However, at chilling temperatures, exacerbation of photoinhibition by O_2 can be observed. Presumably, the low rates of photorespiration occurring at temperatures close to 0°C do not relieve overreduction of the electron transport chain. The effects of O_2 during chilling and high-light stress on photosynthesis, measured subsequently, were first reported in 1972 by Rowley and Taylor[41] for the chilling-sensitive C_4 plant, *Sorghum bicolor*. Further support for such an action of O_2 in chilling-sensitive plants was given by Powles et al.[42] Enhancement of photoinhibition by chilling temperatures also occurs (usually to a lesser degree) in chilling-resistant species.[9,43] For instance, in chilling-resistant spinach, preillumination at 4°C that resulted in strong photoinhibition had no such effect at 18°C.[44] Also, O_2 enhancement of photo-

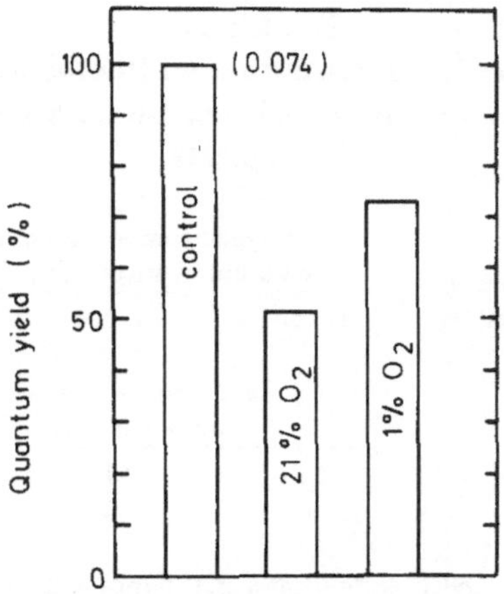

FIGURE 3. Effect of O_2 concentration in the atmosphere (containing 345 μl l^{-1} CO_2) on photoinhibition of spinach leaves at 4°C. After pretreatment for 2 h with 1250 μmol m^{-2} s^{-1} white light in the presence of 21 or 1% O_2, the optimal quantum yield of O_2 evolution was determined. The fluorescence ratios (F_v/F_m) were 0.81 for the control leaves and 0.64 and 0.70 for leaves pretreated in the presence of 21 and 1% O_2, respectively. (From Cornic, G. and Krause, G. H., unpublished experiment.)

inhibition is not confined to chilling-sensitive species.[45] Figure 3 shows that in spinach leaves preilluminated at 4°C in the presence of 21% O_2, the optimal quantum yield of photosynthesis was considerably more photoinhibited than in an atmosphere containing only 1% O_2. The inhibition was related to the inactivation of PSII, as seen from the lowered fluorescence ratio F_v/F_m (legend to Figure 3). The effect of O_2 at chilling temperatures has recently been studied in detail with isolated mesophyll protoplasts of *Valerianella locusta*.[45] Under standard pretreatments, the extent of photoinhibition gradually increased with O_2 concentrations corresponding to 2 to 40% O_2 in the atmosphere. A saturation of the O_2 effect could not be observed. As shown in Figure 4, enhancement of photoinhibition by O_2 was evident from fluorescence data and the rates of PSII-driven electron transport. It appears that extrapolation of the curves of Figure 4, to $[O_2] = 0$, denotes an O_2-independent component of photoinhibition. The slight O_2-independent increase in initial fluorescence (Figure 4a) may also indicate such a component. However, if preillumination is done under anaerobiosis, special effects arise (see below).

In *Elatostema repens*, an extreme shade plant of the tropical rain forest, the enhancing effect of O_2 on photoinhibition was also observed at a non-chilling temperature (24°C).[46] In this case, antioxidant systems might have been insufficiently active. In contrast to intact leaves, spinach mesophyll cells illuminated in a CO_2-depleted medium at 22°C exhibited an O_2-dependent photoinhibition much like that observed in isolated chloroplasts.[39] This could have resulted from depletion of carbohydrate reserves in these cell isolates, which restricted photorespiratory activity.

Current evidence shows that O_2 promotes photoinhibition. The action of O_2 becomes evident when, for various reasons, this is not concealed by the opposing action of photorespiration.

C. ANAEROBIC PHOTOINHIBITION

When spinach leaves are illuminated in pure N_2, photoinhibition is enhanced compared to low (1 to 2%) O_2 levels.[47] This can be explained by extreme overreduction of the electron transport system due to the lack of electron acceptors under anaerobic conditions, and supposedly demonstrates the partially protecting effect of the Mehler reaction (see Section III.A.2 and the review in Reference 4). In recent years, anaerobic photoinhibition, studied with isolated thylakoids and PSII particles, has gained new interest.[48-55] Irradiation in the absence of O_2 and under strongly reducing conditions has been suggested to produce primary or intermediate stages of photoinhibition that lead to full PSII inactivation in the presence of O_2. These stages might also occur under aerobic conditions, but would not accumulate because of interference by oxygen.[53]

The inactivation of PSII under anaerobiosis is, in part, reversible in darkness without a requirement for *de novo* protein synthesis.[50,52,53] In contrast, aerobic photoinhibition needs low light for recovery (see Reference 4 for a review). Anaerobic photoinhibition does not result in degradation of the D1 protein in the PSII reaction center[49,50,53,56] (see Section V.B). However, when O_2 is admitted during irradiation of PSII particles after a period of anaerobiosis, enhanced degradation is observed.[53] In the light, O_2 prevented the recovery of thylakoid membranes from anaerobic photoinhibition.[50] These observations indicate that O_2 acts at a stage of photoinhibition that occurs during anaerobiosis.

Anaerobic photoinhibition is characterized by a pronounced increase in F_o, the initial Chl fluorescence yield[2,47,48,52,53,57] (Figure 5). F_o defines the state when all reaction centers of PSII are "open", i.e., Q_A is fully oxidized. Maximum fluorescence, $F_m = F_o + F_v$, is reached when all Q_A is reduced to the semiquinone, Q_A^- (see Reference 32 for a review). However, as first recognized by Setlik et al.,[48] the increase in F_o seen during anaerobic photoinhibition is only an apparent one and has to be attributed to a state of stabilized Q_A^-. As further steps of anaerobic photoinhibition, Vass et al.[53] suggested double reduction and protonation of Q_A

$$Q_A \xrightarrow{e^-} Q_A^- \rightarrow (Q_A^-)_{stable} \xrightarrow{e^-} Q_A^{2-} \xrightarrow{2H^+} Q_AH_2 \qquad (12)$$

and loss of the hydroquinone, Q_AH_2, from its binding site on the D2 protein. Double reduction of Q_A was also postulated as a step of photoinhibition under aerobic conditions.[58] Such processes have been studied with chemically reduced PSII particles by means of EPR spectroscopy.[59] The double reduction

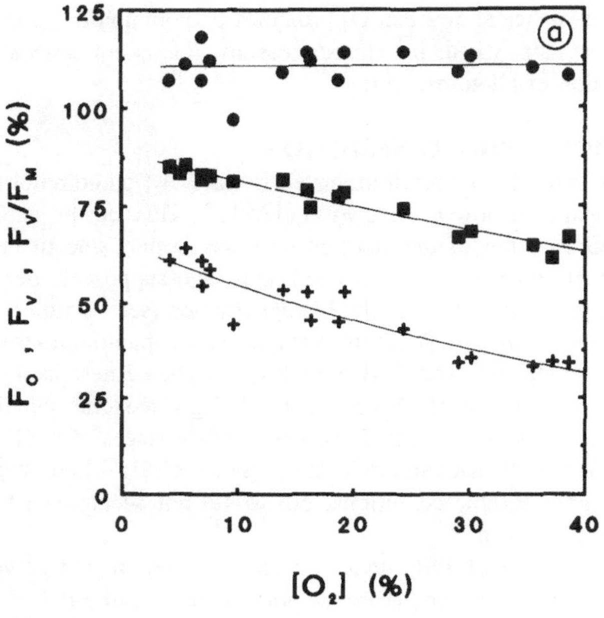

FIGURE 4a. Chlorophyll fluorescence parameters F_o (●), F_v (+), and F_v/F_m (■) of intact mesophyll protoplasts of *Valerianella locusta* after pretreatment with 3 mmol m^{-2} s^{-1} white light in the presence of 2.3 m*M* $KHCO_3$ and different O_2 concentrations (percent of O_2 in the atmosphere equilibrated with the protoplast suspension). Values are expressed as percent of controls kept in the dark.

of Q_A (Equation 12) is characterized by loss of the $Q_A^-Fe^{2+}$ EPR signal. This loss is correlated with an increase in photoinduced Chl triplet formation. Apparently, neutralization of the charge at Q_A increases the yield of primary charge separation between P_{680}, the primary electron donor, and pheophytin (Ph) in the reaction center. Recombination of the radical pair results in a high yield in the triplet state of P_{680}:

$$P_{680}\,Ph \xrightarrow{h\nu} P^{*}_{680}\,Ph \rightarrow P^{+}_{680}\,Ph^{-} \rightarrow {}^{3}P_{680}\,Ph \qquad (13)$$

Under anaerobiosis, the triplet state, ${}^3P_{680}$, probably decays without further consequences. However, in the presence of O_2, it can create singlet oxygen (Equations 4 and 5) that may react close to its site of formation, i.e., on the D1 protein (see also the review in Reference 60). As a surprising and not readily understood result, Vass et al.[53] also observed the Chl triplet EPR signal in the state attributed to $(Q_A^-)_{stable}$ (Equation 12), in which primary charge separation (Equation 13) should be strongly inhibited by the negative charge on Q_A. An explanation could be protonation of $(Q_A^-)_{stable}$, resulting in uncharged Q_AH.[48]

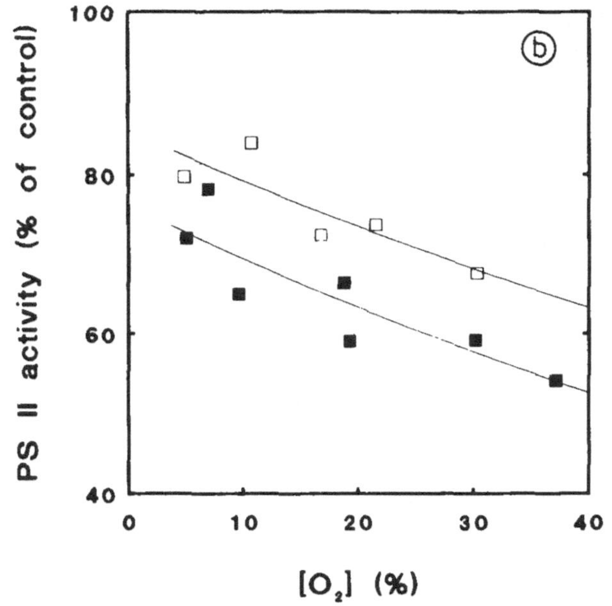

FIGURE 4b. PSII activity of protoplasts from cold-acclimated (□) and nonacclimated (■) leaves pretreated as in (a) in the presence of different O_2 levels. Uncoupled PSII electron transport was determined as O_2 evolution with 1,4-benzoquinone as electron acceptor in saturating red light after osmotic rupture of the protoplasts. (From van Wijk, K. J. and Krause, G. H., *Planta*, 186, 135, 1991. With permission.)

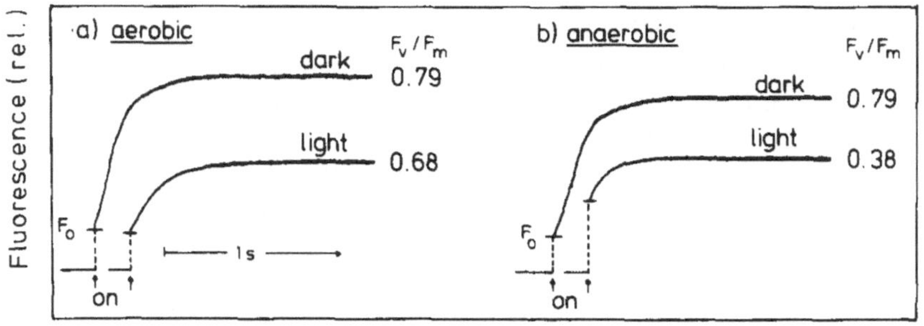

FIGURE 5. Fluorescence induction signals of isolated thylakoids in air-saturated medium recorded at 20°C in the presence of 20 μM DCMU. Samples were pretreated for 5 min in the dark (controls) or in white light (660 W m^{-2}) in air-saturated (a) or anaerobic (b) medium at 20°C. F_v/F_m ratios are noted in the figure. The photoinhibition of uncoupled electron transport (determined with ferricyanide as acceptor) was 13% for (a) and 72% for (b). Anaerobiosis was achieved using the glucose/glucose oxidase/catalase system. (From Krause, G. H. and Laasch, H., in *Progress in Photosynthesis Research*, Vol. 4, Biggins, J., Ed., Martinus Nijhoff, Dordrecht, 1987, 19. With permission.)

The hypothesis that these various stages of a nonfunctional Q_A site observed under anaerobiosis also occur in the presence of normal air appears questionable. Formation of these stages seems to require strongly reducing conditions. In fact, in many cases, no increase in F_O is observed during aerobic photoinhibition *in vivo*. It has been argued that the irreversible state of double-reduced (and dissociated) Q_A does not accumulate because of the immediate attack of $^1O_2$ formed from $^3P_{680}$.[53] This should result in a degradation of the D1 protein simultaneously with photoinhibition. However, it has frequently been shown that inhibition of electron transport occurs much faster, and to a larger extent, than D1 degradation and its triggering (Section V.A).

Interestingly, recovery from anaerobic photoinhibition in the dark requires the presence of O_2.[55] From this, it has been suggested that O_2 reoxidizes stable, reduced Q_A. Low O_2 levels seem to be sufficient for this effect.[55] Thus, O_2 might prevent the accumulation of stable, reduced Q_A, which would be a further protective effect of low O_2 concentrations in addition or as an alternative to the Mehler reaction.

Studies of anaerobic photoinhibition have been complicated by the recently reported effects of bicarbonate.[51,54,55] Replacement of bicarbonate by formate at the site of the nonheme iron in the PSII reaction center (Figure 1) was found to diminish photoinactivation. On the other hand, high concentrations of bicarbonate prevented dark recovery from anaerobic photoinhibition, but protected against photoinactivation of PSII under aerobic conditions.[54]

IV. INVOLVEMENT OF GROUND STATE AND ACTIVE OXYGEN IN PHOTOINHIBITION

A. GROUND STATE DIOXYGEN, $^3O_2$

In the triplet ground state, the peroxidative reactions of dioxygen are spin restricted (see Chapter 1 of this volume). However, electron spin restriction does not exist for reactions of $^3O_2$ with radicals, as is the case in the radical chain of lipid peroxidation (Section II.B.2). In PSII, the semiquinone radicals Q_A^- and Q_B^- participate in photosynthetic electron transport. In excessive light, Q_A^- accumulates, whereas Q_B and the plastoquinone pool are reduced to hydroquinones. The highly reduced state of Q_A promotes photoinhibition.

As an alternative to the double reduction and protonation of Q_A found under anaerobiosis (Equation 12), one could visualize $^3O_2$ reacting directly with Q_A^-. It has indeed been observed that plastosemiquinone in ethanol solution is quenched by oxygen at air levels, but the nature of this reaction was not studied further.[61] *In situ*, a reaction with O_2 may render the quinone functionally inactive and/or the reaction product would dissociate from the Q_A binding site. Like the postulated anaerobic mechanism, this would give rise to formation of the Chl triplet ($^3P_{680}$, Equation 13). Another possibility is that reduction of O_2 via Q_A^- forms active oxygen species, which then exert inhibitory effects (see next section). Such actions of O_2 might be limited by its accessibility to the reaction site, which could explain the gradual increase in photoinhibition with O_2 concentration[25,45] (Table 1, Figure 4). The O_2-independent component of photoinhibition apparent in the study of van Wijk and Krause[45] would be consistent with the assumption that the aforementioned anaerobic mechanism (Equation 12) takes place in the presence of O_2 in parallel with a direct reaction of O_2 with the semiquinone. Both mechanisms, reaction with O_2 or double reduction and protonation of Q_A, could explain the quenching of variable fluorescence (F_v) that always accompanies photoinhibition. However, clear evidence for a direct action of $^3O_2$ on Q_A^- is lacking, and various studies imply the crucial involvement of active oxygen species in photoinhibition.

B. SUPEROXIDE RADICAL, HYDROGEN PEROXIDE, AND HYDROXYL RADICAL

1. Possible Actions in Photosystem II

As discussed above (Sections II.B.1 and III.A.2), the superoxide free radical is created in the Mehler reaction by O_2 reduction via PSI. This gives rise to the formation of H_2O_2 and possibly hydroxyl radicals (Equations 1 to 3). These active oxygen species can promote photoinhibition, as has been demonstrated *in vitro*.[40,61a-65] Dark incubation of thylakoids with xanthine-xanthine oxidase, an $O_2^{\overline{\cdot}}$-generating system (see review in Reference 15), inhibited electron transport.[62] However, $O_2^{\overline{\cdot}}$ itself is not a potent oxidant. Also, H_2O_2 does not readily affect the electron transport system, but rather may inactivate light-regulated enzymes and, in addition, ascorbate peroxidase in the chloroplast stroma (see review in Reference 18). Thus, the $^{\cdot}OH$ radical is often considered to be a most reactive species in photoinhibition, although a direct action of H_2O_2 has also been discussed.[63] The long-lived $O_2^{\overline{\cdot}}$ may diffuse to the site of action where $^{\cdot}OH$ is generated. Involvement of $^{\cdot}OH$ radicals with prolonged lifetimes in a so-called caged form (crypto-$^{\cdot}OH$) has been suggested (see Reference 14). Treating thylakoids with a Fenton reaction mixture, which generates $^{\cdot}OH$ (Equation 2), did not result in effects resembling photoinhibition.[64] Perhaps $^{\cdot}OH$ radicals, when formed in the medium, never reach the shielded reaction centers, and hence must be formed within the membrane for specific photoinhibitory action. Antioxidant systems that scavenge $O_2^{\overline{\cdot}}$ and H_2O_2 and thus prevent $^{\cdot}OH$ generation, such as superoxide dismutase (SOD) plus catalase,[40,64] reduced glutathione plus ascorbate,[64] or catalase alone,[63] substantially diminished the photoinhibition of thylakoids *in vitro*. Full protection of the D1 protein, as determined by means of atrazine binding,[64] but only incomplete protection in terms of PSII activity was obtained[40,64] (Figure 6). The added antioxidants diminished, but did not completely prevent, the loss of α-tocopherol in the membranes.[65] This might indicate the action of another oxygen species, i.e., $^1O_2$ (see below).

One point against the involvement of the Mehler reaction is the gradual increase of photoinhibition with increasing O_2 concentrations to above-air

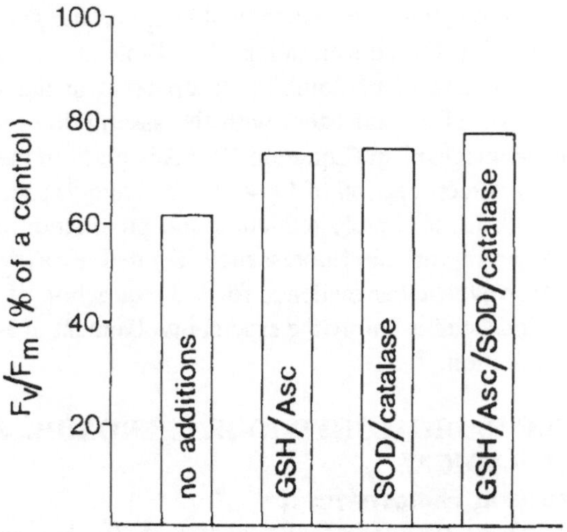

FIGURE 6. Influence of scavengers for active oxygen species on the decline of the F_v/F_m ratio during photoinhibitory treatments (30 min, 100 W m^{-2} white light) of spinach thylakoid membranes. Controls (100%) were kept in the dark (F_v/F_m = 0.79). Additions present during pretreatment are noted in the bars (GSH, 10 m*M* glutathione; Asc, 5 m*M* ascorbate). (From Richter, M., Rühle, W., and Wild, A., *Photosynth. Res.*, 24, 237, 1990. With permission.)

levels, which has been observed both *in vitro* and *in vivo*.[25,41,45] The Mehler reaction should be fully saturated at 21% O_2 or much below (see review in Reference 1). As mentioned above, the effect suggests a diffusion-limited process.

It has been proposed by Kyle[66] that $O_2^{\cdot-}$ and consecutively H_2O_2 and $^{\cdot}OH$ are generated by the reduction of O_2 at the Q_B site of the D1 protein by electron transfer from either Q_A^- or Q_B^-. The $^{\cdot}OH$ radical is then thought to attack the histidine residues responsible for the quinone binding. There seems to be little evidence, from early studies, for a Mehler-type reaction in PSII (see Reference 67). However, some H_2O_2 formation by PSII particles, probably occurring via the reducing side of the reaction center, has been reported.[68,69] Observed rates were low (about 0.6 μmol mg^{-1} [Chl] h^{-1}), but were increased by high O_2 levels.[69] Also, the postulated reoxidation of $Q_A^-{}_{(stable)}$ by O_2 (Section III.C) should lead to H_2O_2 formation in PSII. In Cl^--depleted thylakoids, H_2O_2 can also be formed from H_2O in the water oxidation system (see Reference 63), but this reaction does not appear to be relevant for conditions *in vivo*.

On the other hand, preparations of PSII particles, in contrast to thylakoids,[40,63,64] became readily photoinactivated under aerobic conditions in the absence and presence of added SOD plus catalase (Figure 7a). This result does not support the assumption of a role of $O_2^{\cdot-}$, or H_2O_2, generated in PSII. Alternatively, one has to assume that these species are shielded from exogenous scavenging enzymes.

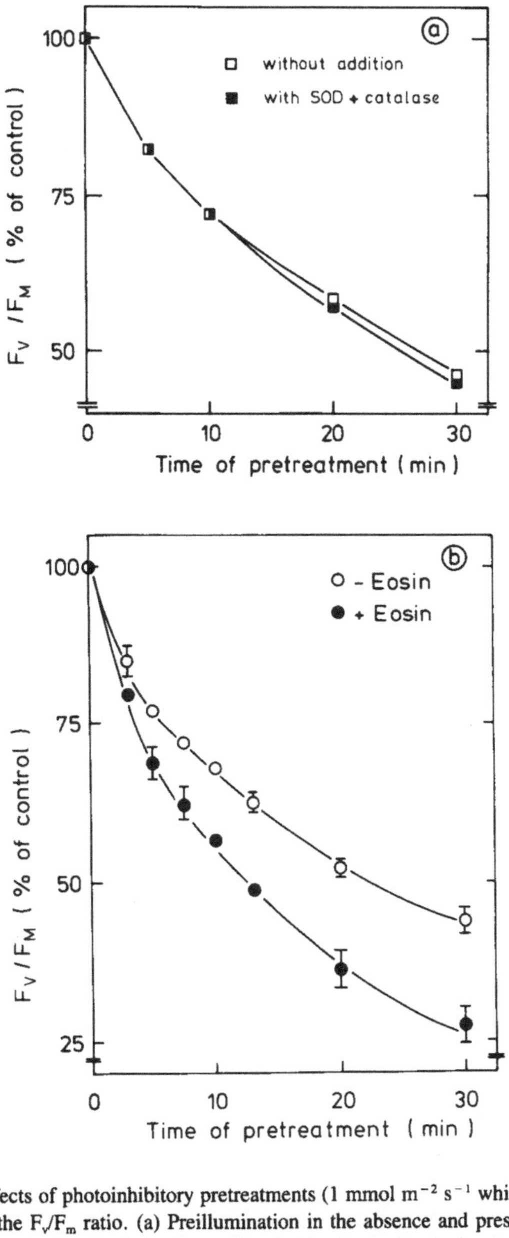

FIGURE 7. Effects of photoinhibitory pretreatments (1 mmol m^{-2} s^{-1} white light at 20°C) of PSII particles on the F_v/F_m ratio. (a) Preillumination in the absence and presence of SOD (100 U ml^{-1}) and catalase (2000 U ml^{-1}). (b) Preillumination in the absence and presence of 25 μM eosin; mean of three to six samples, with standard deviations indicated (when larger than symbols). Values are expressed as percent of controls kept in the dark; control F_v/F_m ratios were 0.75 (without eosin) and 0.71 (with eosin). (From Elsner, C., Graf, M., and Krause, G. H., unpublished experiments.)

2. Scavenger Systems *in Vivo*

The discussed experiments show that active oxygen species formed as a result of the Mehler reaction (or by O_2 reduction in PSII?) *can* induce photoinhibition. However, it is questionable whether they are of great importance to the situation *in vivo*. Defense systems present in the chloroplasts (see Chapter 1 of this volume) should be able to scavenge these oxygen species or prevent their formation, except when the thylakoids are flooded by radicals due to the action of herbicides, e.g., paraquat, or pollutants such as sulfur dioxide, sulfite, or ozone (see Chapter 6 of this volume and review in Reference 18).

Moreover, plants are capable of acclimating to adverse environmental conditions. Acclimation to high-light stress seems to include, among other mechanisms, an increase in the efficiency of the antioxidant systems. Spinach plants cold-acclimated under slightly excess light exhibited diminished susceptibility to photoinhibition at chilling temperatures.[44,70] The acclimated leaves showed increased activities of SOD, ascorbate peroxidase, and monodehydroascorbate reductase (Figure 8).[71] The increased SOD activity seemed to result from induction of an isoenzyme in the chloroplasts.[72] In addition, the ascorbate content was increased.[71] Earlier studies had documented similar acclimations of SOD activity in *Chlorella*[73] and glutathione reductase activity and glutathione levels in spinach.[74,75] Acclimatory changes of scavenging systems (but not SOD) in response to increased irradiation without low-temperature stress were reported for pea leaves.[76] Thus, acclimation leads to both increased activities of enzymes that scavenge products of the Mehler reaction and of enzymes that restore high antioxidant concentrations. The latter are increased as compared to the nonacclimated state. Responses of the antioxidant systems to certain herbicides and pollutants have also been reported (see Chapter 14 of this volume and review in Reference 18).

Thus, the overall evidence does not support the Mehler reaction, including the generation of H_2O_2 and $^{\cdot}OH$ from initially formed $O_2^{\cdot-}$, as a dominant factor in the induction of photoinhibition *in vivo*. Only in extreme cases, when the antioxidant defense systems become overcharged, might these oxygen species have a significant impact.

C. SINGLET OXYGEN

1. Possible Actions in Photosystem II

Chlorophyll is known as an efficient photosensitizer for photodynamic generation of $^1O_2$. In the thylakoid membrane, the dominant source of $^1O_2$ should be energy transfer from ^{3}Chl to $^3O_2$ (Equations 4 and 5). The reactivity of $^1O_2$ is about as high as that of $^{\cdot}OH$ radicals in aqueous media. However, in the lipophilic phase of biomembranes, the lifetime of $^1O_2$ is thought to be prolonged (see reviews in References 15 and 18). This may increase the specificity of its reactions. If, indeed, early stages of photoinhibition give rise to triplet Chl in the PSII reaction center (Section III.C), $^1O_2$ can be

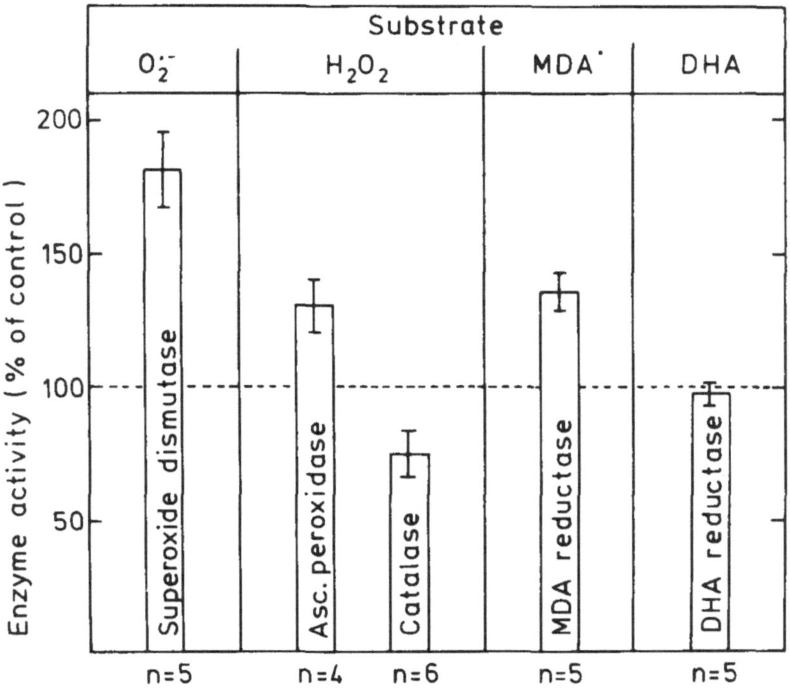

FIGURE 8. Effects of cold acclimation to 1°C in moderate light (260 to 300 μmol m^{-2} s^{-1}) of spinach plants on activities of scavenger enzymes for active oxygen species. Bars denote enzyme activity in cold-acclimated leaves expressed as the percent of activity found in nonacclimated leaves of the same plant batch (= 100%). Calculations were based on Chl content, as the total Chl per leaf area did not change during acclimation. Asc, ascorbate; MDA, monodehydroascorbate; DHA, dehydroascorbate. (From Schöner, S. and Krause, G. H., *Planta*, 180, 383, 1990. With permission.)

generated at a critical site where it may inactivate the center and/or trigger degradation of the D1 protein (see Section V.B). Since $^1O_2$ reacts readily with compounds containing double bonds, e.g., phenols, quinones, and certain amino acids (particularly histidine) (see review in Reference 15), various ways of inactivating the PSII reaction center can be imagined. A tyrosine residue of the D1 protein acts as the secondary electron donor (see review in Reference 5). Two histidine residues on each of the D2 and D1 proteins are responsible for binding the quinones, Q_A and Q_B, and a nonheme iron.[77] The availability of $^3O_2$ in the center could be diffusion limited, which may be another explanation for the dependence of photoinhibition on the O_2 concentration (Section III.B).

In isolated PSII reaction centers, which are devoid of bound quinones, light-induced $^3P_{680}$ formation was monitored by flash-absorption spectroscopy.[78,79] The triplet state resulted in a high yield from the charge recombination in the primary radical pair $P_{680^+}Ph^-$ (Equation 13). The long-lived triplet state (lifetime about 1 ms) was efficiently quenched by O_2. The rapid photoinactivation of these PSII centers occurring under aerobic conditions has been attributed to the action of $^1O_2$ formed by an energy transfer from $^3P_{680}$.[78] Based on photobleaching experiments, Telfer et al.[80] suggested that singlet oxygen directly destroys the primary electron donor, P_{680}.

In PSII particles, generation of $^3P_{680}$ at high rates under anaerobic conditions has been demonstrated (Section III.C). In the presence of O_2, relief of overreduction will probably diminish the yield of $^3P_{680}$. Only a very small steady-state population of $^3P_{680}$ in high light can be expected. However, such a small population might be sufficient to cause inactivation of a substantial fraction of PSII within the timespan of minutes to hours during which photoinhibition becomes manifest.

The assumption that $^1O_2$ is involved in photoinhibition is supported by studies on the effects of the accelerated generation of this oxygen species.[81-83] Pea leaves were incubated with the photosensitizer eosin (2,4,5,7-tetrabromofluorescein). In light, this substance creates $^1O_2$ at high rates by energy transfer from the triplet state:

$$^1\text{Eosin} \xrightarrow{h\nu} {}^{*1}\,\text{Eosin} \rightarrow {}^3\text{Eosin} \tag{14}$$

$$^3\text{Eosin} + {}^3O_2 \rightarrow {}^1\text{Eosin} + {}^1O_2 \tag{15}$$

When pea leaf discs were illuminated in the presence of eosin, a complex pattern of damage, including enzyme inactivation, pigment bleaching, lipid peroxidation, and ultrastructural changes, were observed. However, the photosynthetic electron transport system showed the most sensitive response in the initial stages, before significant Chl destruction took place. Inhibition of electron transport (Figure 9)[81] and quenching of variable fluorescence[82] occurred in a manner apparently identical with "normal" photoinhibition. Specifically, PSII, but not PSI, became inactivated.[81]

To avoid many of the various side effects seen *in vivo,* we illuminated isolated PSII particles in a medium containing eosin. As shown in Figure 7b, the singlet oxygen generated in the medium promoted photoinhibition of PSII. This was accompanied by the characteristic quenching of variable fluorescence. The inhibition took place without detectable degradation of the D1 protein and without Chl destruction.[108] The effect was relatively small, probably because most of the exogenous $^1O_2$ had been quenched or had reacted otherwise before it could reach the reaction center.

The experiments demonstrating the formation of $^3P_{680}$ and its quenching by O_2, as well as the enhancement of specific inactivation of PSII by artificially generated $^1O_2$, make singlet oxygen the most likely oxygen species to have a key role in photoinhibition.

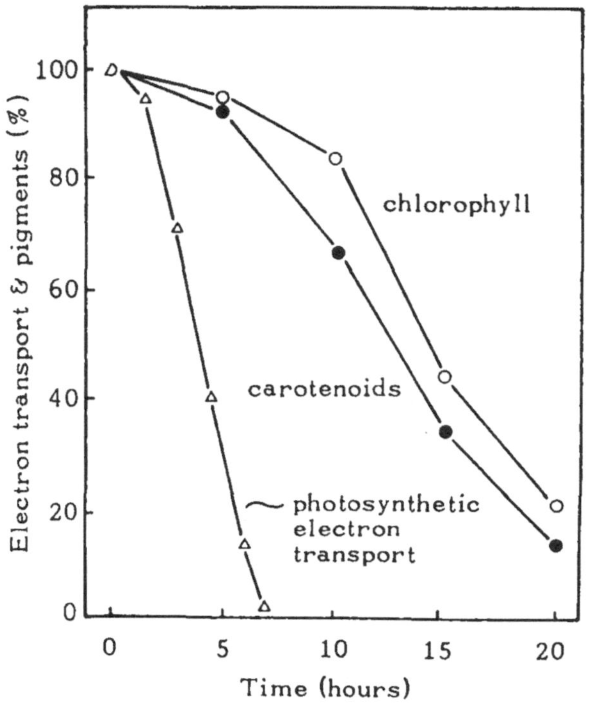

FIGURE 9. Loss of photosynthetic electron transport activity and photosynthetic pigments from pea leaf discs supplied with 1 m*M* eosin and irradiated with white light (350 μmol m^{-2} s^{-1}) for the times indicated. Results are expressed as percent of initial values. Uncoupled electron transport was determined after isolation of thylakoids using ferricyanide as electron acceptor. (From Knox, J. P. and Dodge, A. D., *Planta* 164, 22, 1985. With permission.)

2. Scavengers

β-Carotene has been shown to be a constituent of the PSII reaction center (see Reference 78). One should assume that this, like the carotenoids present in the antennae, exerts a protective function against triplet Chl and $^1O_2$ (Equations 7 and 8). However, present evidence does not unequivocally support this notion. Under conditions that gave high yields of $^3P_{680}$ in isolated PSII reaction centers, very little triplet carotenoid could be detected.[78,84] No quenching of $^3P_{680}$ by the carotene was seen. This lack of energy transfer to β-carotene might be an artifact of the isolation procedure. On the other hand, it is a fact that, *in vivo,* the reaction centers are affected by high light much faster than the antenna pigments (Section II). Protection of the reaction center by carotenoids *in vivo* was demonstrated for *Scenedesmus acutus.*[85] Photoinhibition and light-induced loss of the D1 protein increased with decreasing carotenoid content, which had been modified by culture in the presence of the inhibitor norflurazon. In this case, increased generation in the antennae of $^1O_2$, which attacks the reaction center, might be responsible for the herbicide effect. Whereas the scavenging role of antenna carotenoids is well established, the function of β-carotene in the reaction center awaits experimental proof.

V. OXYGEN EFFECTS ON D1 PROTEIN DEGRADATION

A. PHOTOINHIBITION AND D1 DEGRADATION

The interdependence of turnover of the D1 protein, photoinactivation, and reactivation (recovery) of the PSII reaction center cannot be reviewed here. The reader is referred to a detailed recent review by Prasil et al.[7] Some points of importance to the present article are summarized.

In recent years, it has become clear that D1 degradation is not the cause, but rather a consequence of photoinhibition. There seem to be several steps, which differ in their reversibility, leading to the inhibition of PSII-driven electron transport.[7] Early steps seem to be reversible without the requirement for D1 replacement. For instance, a fast phase of recovery observed in spinach leaves, even at low temperatures, is too rapid to be accounted for by D1 turnover.[44,70,86] Such fast recovery has recently been shown to be insensitive to streptomycin, which inhibits the synthesis of proteins (including D1) in the chloroplasts.[86a] A recovery of electron transport activity without protein synthesis was also described for *Chlamydomonas reinhardtii*.[87] The fast reactivation was sometimes,[44,87] but not always,[70,86] related to reversion of an apparent increase in F_o (indicating stabilized Q_A^-?).

In the inactivated PSII, an alteration of the D1 protein seems to target it for degradation. Once this has occurred, recovery is possible only by replacement of the altered D1 with a newly synthesized peptide (as a precurser D1).[7] Degradation of the targeted D1 is catalyzed by a still-unidentified membrane-bound protease. Degradation has been reported to occur even in isolated reaction centers, and based on this, an endopeptidase activity of the complex was suggested.[88] As an alternative mechanism, induction of proteolysis by H_2O_2 formed in the PSII reaction center has been discussed.[63] At temperatures below 7°C, proteolysis becomes much slower; however, the triggering of degradation still takes place during irradiation of thylakoids.[89] Upon warming, the triggered protein is degraded without a requirement for light.[89] Thus, triggering and degradation are separate events.

Several studies indicate that *in vivo* degradation and resynthesis/replacement of the D1 protein are highly synchronized.[7,86a,90-92] The total amount of D1 scarcely decreases even during extensive photoinactivation of PSII. Thus, photoinhibition, measured as a decline in the rates of PSII-driven electron transport, is not based on net D1 degradation, but rather on an accumulation of inactive reaction centers. Also, a close correlation between D1 degradation and PSII inactivation *in vitro* is usually not found.[90]

The importance of fast D1 turnover lies in the recovery from the stage of photoinhibition that requires replacement of the protein. In *Spirodela oligorrhiza,* D1 turnover at high rates (lifetime around 5 h) was also shown to occur in dim light, far below the irradiance causing photoinhibition.[93,94] Degradation took place faster in nonphotosynthetic ultraviolet than in visible light.[93] Appreciable degradation was induced even by far-red light that preferentially excites PSI.[94] It is still not clear whether the triggering mechanism in these cases is the same as in photoinhibition.

B. ROLE OF OXYGEN

There is general agreement that oxygen is involved in the triggering of D1 degradation. Under anaerobiosis, degradation is usually not observed.[49,50,53,56,95] (but see below). In spinach leaves, the fast phase of recovery from photoinhibition, which was independent of D1 turnover (insensitive to streptomycin), was less pronounced after preillumination in 21% O_2 than in 2% O_2.[86a] This indicates that *in vivo* high O_2 levels enhance the marking of D1 for degradation. Many studies have suggested that the degradation is triggered by active oxygen species (see Reference 96). Indeed, exogenous radical scavengers diminished the degradation of D1 in *Spirodela,* as did incubation with DCMU or atrazine.[96] As these herbicides occupy the Q_B binding site, it seems feasible that this site plays a role in the triggering.

Sopory et al.[96] proposed that the involvement of $^1O_2$ can be excluded, as selenomethionine, a quencher of $^1O_2$, had no effect on degradation. Moreover, exchange of H_2O for D_2O did not increase proteolysis.[96] D_2O is known to prolong the lifetime of $^1O_2$ and thereby promote its action. However, one may argue that the stability of $^1O_2$ created within the membrane and reacting close to the site of generation might be little influenced by D_2O in the aqueous phase. Moreover, not all types of quenchers appear to reach the critical site and act with equal efficiency.[96] In contrast, Vass et al.[53] (Section III.C) and Durrant et al.[78] (Section IV.C) suggested that $^1O_2$ is the species that triggers D1 degradation, based on their observations of $^3P_{680}$ quenching by O_2. This view is supported by the finding that degradation is enhanced *in vivo* when biosynthesis of carotenoids, efficient quenchers of $^1O_2$, has been inhibited.[85]

Another explanation for the discrepancy regarding the triggering species could be that Sopory et al.[96] studied D1 degradation in low, nonphotoinhibitory light, whereas Vass et al.[53] applied high irradiation. However, it appears unlikely that the major oxygen species responsible for the triggering is different under the two conditions. It was suggested that the Q_B^- semiquinone reacts with O_2 and, as a result, leaves its binding site.[96] But this reaction would not require an *active* oxygen species.[61] There are indications that the D1 protein becomes accessible to proteolysis when the Q_B binding site is empty.[97] Contrary to this assumption, it was inferred from recent experiments with a *Chlamydomonas* mutant that D1 degradation can only be initiated when plastoquinone (PQ) is bound to the Q_B site.[7] In low light, high proportions of centers with bound Q_B (PQ) or Q_B^- are present. In contrast, excess light leads to reduction of most of Q_B and PQ to plastoquinol, and a high proportion of Q_B sites will be empty or occupied by Q_BH_2. However, in those reaction centers in which electron transport to Q_A is already blocked (inactive PSII), Q_B or Q_B^- may still be bound to D1.

A study with PSII particles by van Wijk et al.[98] showed that the triggering of D1 degradation coincided with the inhibition of primary charge separation and loss of Mn from the water oxidation system. Inhibition of Q_A reduction and quenching of variable fluorescence occurred much faster than those effects. It was suggested before as a general mechanism that oxidative modification, perhaps via conformational changes, can increase the susceptibility of proteins to enzymatic proteolysis.[99] One could speculate that in the state of blocked electron transport to Q_A (when singlet oxygen is formed via $^3P_{680}$), the triggering action of $^1O_2$ requires a specific conformational state of the D1 protein that is influenced by the Q_B binding site. Firm evidence is needed to clarify these points.

If the water oxidation system has been inactivated before irradiation, D1 degradation in thylakoids also occurs under *anaerobic* conditions.[100] Impairment of water-splitting accelerates photoinhibition and D1 degradation[101,102] (see also the review in Reference 7). In this situation, the lifetimes of strong oxidants (P_{680^+} or Z^+) created by primary charge separation are prolonged. Such oxidants appear to trigger D1 degradation in the absence of O_2. In fact, two sites of photoinhibition in PSII particles have been postulated recently.[103] One site supposedly affects electron transport from P_{680} to Q_A; the second site blocks electron donation from tyrosine 'Z' to P_{680^+}. Inhibition at the second site could trigger D1 degradation independent of O_2 (see also Reference 79). However, in particles active in water splitting, inhibition at the first site proceeded two to three times faster than on the donor side.[103] Recently, it has been suggested that the energy-dependent fluorescence quenching mechanism is based on a limitation of electron donation to tyrosine 'Z', related to a reversible Ca^{2+} release from a specific site in the water-splitting complex.[104,105] This effect is thought to be caused by the low pH in the thylakoid lumen present in high light. Even though the energy-dependent quenching is considered to be a protective mechanism (see Section III.A.1 and review in Reference 32), it might lead (particularly in the presence of efficient electron acceptors) to an O_2-independent photoinactivation of PSII, initiated at the donor side. The significance of such donor-side effects in photoinhibition *in vivo* has yet to be clarified.

An additional effect, seen in high light during heat stress in *Chlamydomonas,* is an aggregation of proteins by cross-linking that includes the D1 protein.[106] The aggregated protein did not degrade, and thus recovery by D1 turnover was made impossible. Such aggregation was also observed, without heat shock, during irradiation of pea thylakoids.[107] Formation of the aggregates was dependent on O_2. Tests with a large array of scavengers pointed to $^1O_2$

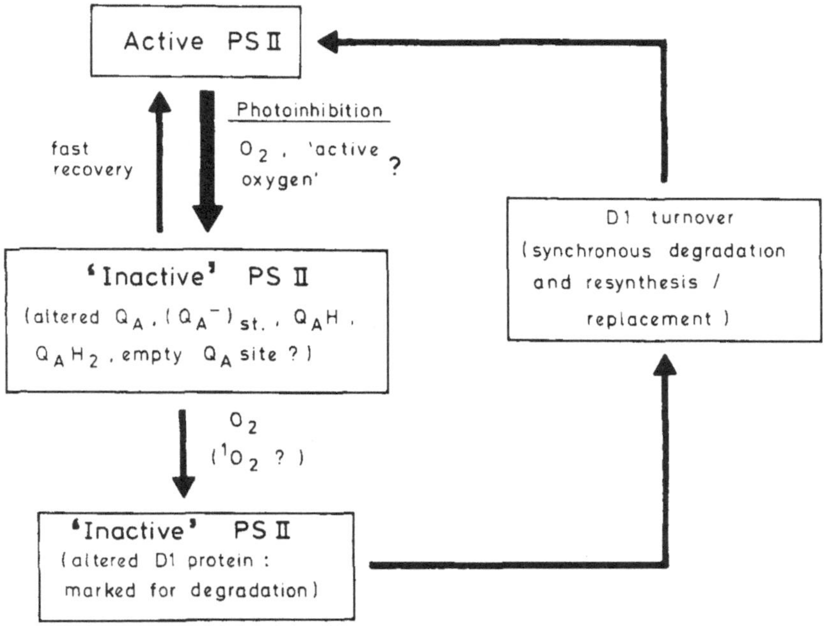

FIGURE 10. Hypothetical scheme of photoinhibition and recovery denoting possible actions of oxygen.

as the major species that caused the cross-linking. Aggregation lagged behind the inactivation of PSII and continued when all activity had been lost. Therefore, this phenomenon can be viewed as a late consequence of photoinhibition.[107]

VI. CONCLUSIONS

Oxygen is involved in the photoinhibition of photosynthesis in various ways. On the one hand, oxygen has protective effects. Photorespiration sustained by the O_2 level in the air affords limited protection against high-light stress, particularly under drought conditions. To a lesser extent, the Mehler reaction (electron transport to O_2) seems to contribute to stabilization of the photosynthetic apparatus by relieving overreduction in strong light. On the other hand, oxygen promotes photoinhibition. However, the nature and sites of oxygen action in PSII are still debated and await further investigations. A hypothetical scheme of photoinhibition and the impact of oxygen is given in Figure 10.

At present, there is no direct evidence — except for the general O_2-dependence of photoinhibition — that oxygen plays a role in the initial steps of PSII inactivation, which are reversible without *de novo* D1 protein synthesis. Singlet oxygen is the most likely species to cause inactivation, but depending on conditions, superoxide radicals, H_2O_2, and hydroxyl radicals, formed from these, may also contribute to inactivation. There may be an O_2-independent component of inhibition in the presence of oxygen. An initial inactivation, without triggering of D1 degradation, also occurs under anaerobiosis. However, it has not been determined whether this is based on the same mechanism as the initial steps of photoinhibition under aerobic conditions.

Recent evidence suggests that the initial effects occur mainly at the Q_A site, but other possibilities (Q_B site and donor side of PSII) not noted in Figure 10 have to be considered.

The triggering of the D1 protein for degradation is a later step in photoinhibition, from which reactivation is only possible via D1 turnover. This slow recovery may be distinguished from the faster reversion of certain primary stages of inhibition. There is wide agreement that oxygen is responsible for the triggering. Again, $^1O_2$, probably formed directly in the reaction center by energy transfer from $^3P_{680}$, appears to be the predominant species involved. However, the contribution of an O_2-independent triggering initiated by strong oxidants at the donor side of PSII cannot be excluded. Such an additional effect appears likely when, due to a high transthylakoid proton gradient and efficient electron acceptance (high rates of photosynthesis), electron donation by the water-splitting system becomes limiting.

In the photoinhibited state, an accumulation of inactivated centers rather than net D1 degradation and disassembly of PSII can be assumed. The inactive centers are supposed to have retained their trap function, dissipating trapped excitation energy as heat. Thus, reversible photoinhibition exerts a protective regulatory function by preventing or diminishing the more severe damaging reactions of O_2 and active oxygen species. Such photooxidations that lead to the destruction of pigments, membrane lipids, and perhaps proteins dominate only under extreme stress. Oxygen-promoted photoinhibition and its reversion, including D1 protein turnover, therefore may be viewed as an adaptive mechanism that has evolved in response to an increasing O_2 concentration in the atmosphere.

ACKNOWLEDGMENTS

I thank Dr. Christa Critchley and Dr. Christine Foyer for critical reading of the manuscript. Our experimental studies concerning the topic of this review have been supported by the Deutsche Forschungsgemeinschaft (SFB 189).

REFERENCES

1. **Robinson, J. M.,** Does O_2 photoreduction occur within chloroplasts in vivo?, *Physiol. Plant.*, 72, 666, 1988.
2. **Krause, G. H. and Cornic, G.,** CO_2 and O_2 interactions in photoinhibition in *Topics in Photosynthesis: Photoinhibition*, Vol. 9, Kyle, D. J., Osmond, C. B., and Arntzen, C. J., Eds., Elsevier, Amsterdam, 1987, 169.
3. **Powles, S. B.,** Photoinhibition of photosynthesis induced by visible light, *Annu. Rev. Plant Physiol.*, 35, 15, 1984.
4. **Krause, G. H.,** Photoinhibition of photosynthesis. An evaluation of damaging and protective mechanisms, *Physiol. Plant.*, 74, 566, 1988.
5. **Hansson, Ö. and Wydrzynski, T.,** Current perceptions of photosystem II, *Photosynth. Res.*, 23, 131, 1990.
6. **Rutherford, A. W.,** Photosystem II, the water-splitting enzyme, *Trends Biochem. Sci.*, 14, 227, 1989.
7. **Prasil, O., Adir, N., and Ohad, I.,** Dynamics of photosystem II: mechanism of photoinhibition and recovery processes in *Topics in Photosynthesis: The Photosystems*, Vol. 11, Barber, J., Ed., Elsevier, Amsterdam, 1992, 295.
8. **Krause, G. H., Somersalo, S., Zumbusch, E., Weyers, B., and Laasch, H.,** On the mechanism of photoinhibition in chloroplasts. Relationship between changes in fluorescence and activity of photosystem II, *J. Plant Physiol.*, 136, 472, 1990.
9. **Öquist, G., Greer, D. H., and Ögren, E.,** Light stress at low temperature, in *Topics in Photosynthesis: Photoinhibition*, Vol. 9, Kyle, D. J., Osmond, C. B., and Arntzen, C. J., Eds., Elsevier, Amsterdam, 1987, 67.
10. **Wise, R. R. and Naylor, A. W.,** Chilling-enhanced photooxidation. The peroxidative destruction of lipids during chilling injury to photosynthesis and ultrastructure, *Plant Physiol.*, 83, 272, 1987.
11. **Wise, R. R. and Naylor, A. W.,** Chilling-enhanced photooxidation. Evidence for the role of singlet oxygen and superoxide in the breakdown of pigments and endogenous antioxidants, *Plant Physiol.*, 83, 278, 1987.

11a. **Mishra, K. R. and Singhal, G. S.,** Function of photosynthetic apparatus of intact wheat leaves under high light and heat stress and its relationship with peroxidation of thylakoid lipids, *Plant Physiol.*, 98, 1, 1992.

12. **Hodgson, R. A. J. and Raison, J. K.,** Superoxide production by thylakoids during chilling and its implication in the susceptibility of plants to chilling-induced photoinhibition, *Planta*, 183, 222, 1991.

12a. **Fryer, M. J.,** The antioxidant effects of thylakoid vitamin E (α-tocopherol), *Plant Cell Environ.*, 15, 381, 1992.

13. **Girotti, A. W.,** Photodynamic lipid peroxidation in biological systems, *Photochem. Photobiol.*, 51, 497, 1990.
14. **Youngman, R. J.,** Oxygen activation: is the hydroxyl radical always biologically relevant?, *Trends Biochem. Sci.*, 9, 280, 1984.
15. **Halliwell, B. and Gutteridge J. M. C.,** *Free Radicals in Biology and Medicine*, Clarendon Press, Oxford, 1989.

16. **Babbs, C. F., Pham, J. A., and Coolbaugh, R. C.,** Lethal hydroxyl radical production in paraquat-treated plants, *Plant Physiol.*, 90, 1267, 1989.
17. **Hodgson, R. A. J. and Raison, J. K.,** Lipid peroxidation and superoxide dismutase activity in relation to photoinhibition induced by chilling in moderate light, *Planta,* 185, 215, 1991.
18. **Asada, K. and Takahashi, M.,** Production and scavenging of active oxygen in photosynthesis, in *Topics in Photosynthesis: Photoinhibition.*, Vol. 9, Kyle, D. J., Osmond, C. B., and Arntzen C. J., Eds., Elsevier, Amsterdam 1987, 227.
19. **Siefermann-Harms, D.,** The light-harvesting and protective functions of carotenoids in photosynthesic membranes, *Physiol. Plant.*, 69, 561, 1987.
20. **Miller, N. and Carpentier, R.,** Energy dissipation and photoprotection mechanisms during chlorophyll photobleaching in thylakoid membranes. *Photochem. Photobiol.*, 54, 465, 1991.
21. **Barr, P., Young, A. J., and Britton, G.,** Photodestruction of pigments in higher plants by herbicide action. I. The effect of DCMU (diuron) on isolated chloroplasts, *J. Exp. Bot.*, 41, 123, 1990.
22. **Musser, R. L., Thomas S. A., Wise, R. R., Peeler, T. C., and Naylor, A. W.,** Chloroplast ultrastructure, chlorophyll fluorescence, and pigment composition in chilling-stressed soybeans, *Plant Physiol.*, 74, 749, 1984.
23. **Siefermann-Harms, D.,** Protective function of the apoprotein of the light-harvesting chlorophyll-*a*/*b*-protein complex in pigment photooxidation, *J. Photochem. Photobiol. B,* 4, 283, 1990.
24. **Powles, S. B. and Osmond, C. B.,** Inhibition of the capacity and efficiency of photosynthesis in bean leaflets illuminated in a CO_2-free atmosphere at low oxygen: a possible role for photorespiration, *Aust. J. Plant Physiol.*, 5, 619, 1978.
25. **Wu, J., Neimanis, S., and Heber, U.,** Photorespiration is more effective than the Mehler reaction in protecting the photosynthetic apparatus against photoinhibition, *Bot. Acta,* 104, 283, 1991.
26. **Scheuermann, R., Biehler, K., Stuhlfauth, T., and Fock, H. P.,** Simultaneous gas exchange and fluorescence measurements indicate differences in the response of sunflower, bean and maize to water stress, *Photosynth. Res.*, 27, 189, 1991.
27. **Stuhlfauth, T., Scheuermann, R., and Fock, H. P.,** Light energy dissipation under water stress conditions. Contributions of reassimilation and evidence for additional processes, *Plant Physiol.*, 92, 1053, 1990.
28. **Powles, S. B., Chapman, K. S. R., and Osmond, C. B.,** Photoinhibition of intact attached leaves of C_4 plants: dependence on CO_2 and O_2 partial pressures, *Aust. J. Plant Physiol.*, 7, 737, 1980.
29. **Heber, U., Neimanis, S., Setlikova, E., and Schreiber U.,** Why is photorespiration a necessity for leaf survival under water stress?, in *Proc. Int. Congr. Plant Physiology, New Delhi 1988,* Sinha S. K., Sane, P. V., Agrawal, P. K., and Bargava, S. C., Eds., Society of Plant Physiology and Biochemistry, New Delhi, 1990, 581.
30. **Sharp, R. E. and Boyer, J. S.,** Photosynthesis at low water potentials in sunflower: lack of photoinhibitory effects, *Plant Physiol.*, 82, 90, 1986.
31. **Krause, G. H., Briantais, J.-M., and Vernotte, C.,** Characterization of chlorophyll fluorescence quenching in chloroplasts by fluorescence spectroscopy at 77K. I. ΔpH-dependent quenching, *Biochim. Biophys. Acta,* 723, 169, 1983.
32. **Krause, G. H. and Weis, E.,** Chlorophyll fluorescence and photosynthesis: the basics, *Annu. Rev. Plant Physiol. Plant Mol. Biol.*, 42, 313, 1991.
33. **Mehler, A. H.,** Studies on the reactions of illuminated chloroplasts. I. Mechanism of the reduction of oxygen and other Hill reagents, *Arch. Biochem. Biophys.*, 33, 65, 1951.
34. **Badger, M. R.,** Photosynthetic oxygen exchange, *Annu. Rev. Plant Physiol.*, 36, 27, 1985.

35. **Heber, U., Viil, J., Neimanis, S., Mimura, T., and Dietz, K.-J.,** Photoinhibitory damage to chloroplasts under phosphate deficiency and alleviation of deficiency and damage by photorespiratory reactions, *Z. Naturforsch. Teil C,* 44, 524, 1989.
36. **Schreiber, U., Reising, H., and Neubauer, C.,** Contrasting pH-optima of light-driven O_2- and H_2O_2-reduction in spinach chloroplasts as measured via chlorophyll fluorescence quenching, *Z. Naturforsch. Teil C,* 46, 635, 1991.
37. **Weis, E. and Lechtenberg, D.,** Steady state photosynthesis in intact plants as analyzed by chlorophyll fluorescence and far-red spectroscopy, in *Applications of Chlorophyll Fluorescence,* Lichtenthaler, H. K., Ed., Kluwer, Dordrecht, 1988, 71.
38. **Nuijs, A. M., Shuvalov, V. A., van Gorkom, H. J., Plijter, J. J., and Duysens, L. M. N.,** Picosecond absorbance-difference spectroscopy on the primary reactions and the antenna-excited states in photosystem I particles, *Biochim. Biophys. Acta,* 850, 310, 1986.
39. **Krause G. H., Kirk, M., Heber, U., and Osmond, C. B.,** O_2-dependent inhibition of photosynthetic capacity in intact isolated chloroplasts and isolated cells from spinach leaves illuminated in the absence of CO_2, *Planta,* 142, 229, 1975.
40. **Barènyi, B. and Krause, G. H.,** Inhibition of photosynthetic reactions by light. A study with isolated spinach chloroplasts, *Planta,* 163, 218, 1985.
41. **Rowley, R. A. and Taylor, A. O.,** Plants under climatic stress. IV. Effects of CO_2 and O_2 on photosynthesis under high light, low-temperature stress, *New Phytol.,* 71, 477, 1972.
42. **Powles, S. B., Berry J. A., and Björkman, O.,** Interaction between light and chilling temperature on the inhibition of photosynthesis in chilling-sensitive plants, *Plant Cell Environ.,* 6, 117, 1983.
43. **Hetherington, S. E., He, J., and Smillie, R. M.,** Photoinhibition at low temperature in chilling-sensitive and -resistant plants, *Plant Physiol.,* 90, 1609, 1989.
44. **Somersalo, S. and Krause, G. H.,** Photoinhibition at chilling temperature. Fluorescence characteristics of unhardened and cold-acclimated spinach leaves, *Planta,* 177, 409, 1989.
45. **van Wijk, K. J. and Krause, G. H.,** Oxygen-dependence of photoinhibition at low temperature in intact protoplasts of *Valerianella locusta* L., *Planta,* 186, 135, 1991.
46. **Le Gouallec, J.-L. and Cornic, G.,** Photoinhibition of photosynthesis in *Elatostema repens, Plant Physiol. Biochem.,* 26, 706, 1988.
47. **Krause, G. H., Köster, S., and Wong, S. C.,** Photoinhibition of photosynthesis under anaerobic conditions studied with leaves and chloroplasts of *Spinacia oleracea* L., *Planta,* 165, 430, 1985.
48. **Setlik, I., Allakhverdiev, S. I., Nedbal, L., Setlikova, E., and Klimov, V. V.,** Three types of photosystem II photoinactivation. I. Damaging processes on the acceptor side, *Photosynth. Res.,* 23, 39, 1990.
49. **Nedbal, L., Masojidek, J., Komenda, J., Prásil, O., and Setlik, I.,** Three types of photosystem II photoinactivation. II. Slow processes, *Photosynth. Res.,* 24, 89, 1990.
50. **Hundal, T., Aro, E.-M., Carlberg, I., and Andersson, B.,** Restoration of light induced photosystem II inhibition without de novo protein synthesis, *FEBS Lett.,* 267, 203, 1990.
51. **Sundby, C.,** Bicarbonate effects on photo-inhibition including an explanation for the sensitivity to photo-inhibition under anaerobic conditions, *FEBS Lett.,* 274, 77, 1990.
52. **Kirilovsky, D. and Etienne, A.-L.,** Protection of reaction center II from photodamage by low temperature and anaerobiosis in spinach chloroplasts. *FEBS Lett.,* 279, 201, 1991.
53. **Vass, I., Styring, S., Hundal T., Koivuniemi, A., Aro, E.-M, and Andersson, B.,** Reversible and irreversible intermediates during photoinhibition of photosystem II: stable reduced Q_A species promote chlorophyll triplet formation, *Proc. Natl. Acad. Sci. U.S.A.,* 89, 1408, 1992.
54. **Sundby, C., Mattsson, M., and Schiött, T.,** Effects of bicarbonate and oxygen concentration on photoinhibition of thylakoid membranes, *Photosynth. Res.,* 34, 263, 1992.

55. **Sundby, C. and Schiött T.,** Characterization of the reversible state of photoinhibition occurring in vitro under anaerobic conditions, *Photosynth. Res.*, 33, 195, 1992.
56. **Arntz, B. and Trebst, A.,** On the role of the Q_B protein of PS II in photoinhibition, *FEBS Lett.*, 194, 43, 1986.
57. **Krause, G. H. and Laasch, H.,** Photoinhibition of photosynthesis. Studies on mechanisms of damage and protection in chloroplasts, in *Progress in Photosynthesis Research*, Vol. 4, Biggins, J., Ed., Martinus Nijhoff, Dordrecht, 1987, 19.
58. **Styring, S., Virgin, I., Ehrenberg, A., and Andersson, B.,** Strong light photoinhibition of electron transport in photosystem II. Impairment of the function of the first quinone acceptor, Q_A, *Biochim. Biophys. Acta*, 1015, 269, 1990.
59. **van Mieghem, F. J. E., Nitschke, W., Mathis, P., and Tutherford, A. W.,** The influence of the quinone-iron electron acceptor complex on the reaction centre photochemistry of photosystem II, *Biochim. Biophys. Acta*, 977, 207, 1989.
60. **Barber, J. and Andersson, B.,** Too much of a good thing: light can be bad for photosynthesis, *Trends Biochem. Sci.*, 17, 61, 1992.
61. **Creed, C., Hales, B. J., and Porter, G.,** Photochemistry of the plastoquinones, *Proc. R. Soc. London Ser. A*, 334, 505, 1973.
61a. **Tschiersch, H., Landgraf, P., and Ohmann, E.,** Photoinhibition in Euglena gracilis: involvement of reactive oxygen species, *Physiol. Plant.*, 85(3) Part 2, A59, 1992.
62. **Takahama, U. and Nishimura, M.,** Formation of singlet molecular oxygen in illuminated chloroplasts. Effects on photoinactivation and lipid peroxidation, *Plant Cell Physiol.*, 16, 737, 1975.
63. **Bradley, R. L., Long, K. M., and Frasch, W. D.,** The involvement of photosystem II-generated H_2O_2 in photoinhibition, *FEBS Lett.*, 286, 209, 1991.
64. **Richter, M., Rühle, W., and Wild, A.,** Studies on the mechanism of photosystem II photoinhibition. II. The involvement of toxic oxygen species, *Photosynth. Res.*, 24, 237, 1990.
65. **Wild, A., Richter, M., and Böthin, B.,** The effect of temperature and radical protection on the photoinhibition of spinach thylakoids, in *Regulation of Chloroplast Biogenensis*, Argyroudi-Akoyunoglou, J. H., Ed., Plenum Press. London, 1992, 445.
66. **Kyle, D. J.,** The biochemical basis for photoinhibition of photosystem II, in *Topics in Photosynthesis: Photoinhibition*, Vol. 9, Kyle, D. J., Osmond, C. B., and Arntzen, C. J., Eds., Elsevier, Amsterdam, 1987, 197.
67. **Salin, M. L.,** Toxic oxygen species and protective systems of the chloroplasts, *Physiol. Plant.*, 72, 681, 1987.
68. **Schröder, W. P. and Akerlund, H.-E.,** H_2O_2 accessibility to the photosystem II donor side in protein-depleted inside-out thylakoids measured as flash-induced oxygen production, *Biochim. Biophys. Acta*, 848, 359, 1986.
69. **Schröder W. P. and Akerlund, H.-E.,** Hydrogen peroxide production in photosystem II preparations, in *Current Research in Photosynthesis*, Vol. 1, Baltscheffsky, M., Ed., Kluwer, Dordrecht, 1990, 901.
70. **Somersalo, S. and Krause, G. H.,** Reversible photoinhibition of unhardened and cold-acclimated spinach leaves at chilling temperatures, *Planta*, 180, 181, 1990.
71. **Schöner, S. and Krause, G. H.,** Protective systems against active oxygen species in spinach: response to cold acclimation in excess light, *Planta*, 180, 383, 1990.
72. **Schöner, S., Foyer, C., Lelandais, M., and Krause, G. H.,** Increase in activities of scavengers for active oxygen in spinach related to cold acclimation in excess light, in *Current Research in Photosynthesis*, Vol. 2, Baltscheffsky, M., Ed., Kluwer, Dordrecht, 1990, 483.
73. **Clare, D. A., Rabinowitch, H. D., and Fridovich, I.,** Superoxide dismutase and chilling injury in *Chlorella ellipsoidea*, *Arch. Biochem. Biophys.*, 231, 158, 1984.
74. **de Kok, L. J. and Oosterhuis, F. A.,** Effect of frost hardening and salinity on glutathione and sulfhydryl levels and on glutathione reductase activity in spinach leaves, *Physiol. Plant.*, 58, 47, 1983.

75. **Guy, C. L. and Carter, J. V.,** Characterization of partially purified glutathione reductase from cold-hardened and non-hardened spinach leaf tissue, *Cryobiology,* 21, 454, 1984.
76. **Gillham, D. J. and Dodge, A. D.,** Chloroplast superoxide and hydrogen peroxide scavenging systems from pea leaves: seasonal variations, *Plant Sci.,* 50, 105, 1987.
77. **Trebst, A.,** The three-dimensional structures of the herbicide binding niche on the reaction center polypeptides of photosystem II, *Z. Naturforsch. Teil C,* 42, 742, 1987.
78. **Durrant, J. R., Giorgi, L. B., Barber, J., Klug, D. R., and Porter, G.,** Characterisation of triplet states in isolated photosystem II reaction centres: oxygen quenching as a mechanism for *photodamage, Biochim. Biophys. Acta,* 1017, 167, 1990.
79. **Mathis, P., Satoh, K., and Hansson, Ö.,** Kinetic evidence for the function of Z in isolated photosystem II reaction centers, *FEBS Lett.,* 251, 241, 1989.
80. **Telfer, A., He, W.-Z., and Barber, J.,** Spectral resolution of more than one chlorophyll electron donor in the isolated photosystem II reaction centre complex, *Biochim. Biophys. Acta,* 1017, 143, 1990.
81. **Knox, J. P. and Dodge, A. D.,** The photodynamic action of eosin, a singlet-oxygen generator. Some effects on leaf tissue of *Pisum sativum* L., *Planta,* 164, 22, 1985.
82. **Knox, J. P. and Dodge, A. D.,** The photodynamic action of eosin, a singlet-oxygen generator. The inhibition of photosynthetic electron transport, *Planta,* 164, 30, 1985.
83. **Knox, J. P., and Dodge, A. D.,** Singlet oxygen and plants, *Phytochemistry,* 24, 889, 1985.
84. **Takahashi, Y., Hansson, Ö., Mathis, P., and Satoh, K.,** Primary radical pair in the photosystem II reaction centre, *Biochim. Biophys. Acta,* 893, 49, 1987.
85. **Sandmann, G., Kuhn, M., and Böger, P.,** Carotenoids in photosynthesis: protection of D1 degradation in the light, *Photosynth. Res.,* 35, 185, 1993.
86. **Somersalo, S. and Krause, G. H.,** Photoinhibition at chilling temperatures and effects of freezing stress on cold acclimated spinach leaves in the field. A fluorescence study, *Physiol. Plant.,* 79, 617, 1990.
86a. **Schnettger, B., Santore, U. J., Critchley, C., and Leitsch, J.,** Light-induced turnover of the D1 protein of the photosystem II reaction centre and photoinhibition *in vivo,* in *Research in Photosynthesis,* Vol. 4, Murata, N., Ed., Kluwer, Dordrecht, 1992, 423.
87. **Kirilovsky, D. L., Vernotte, C., and Etienne, A.-L.,** Protection from photoinhibition by low temperature in *Synechocystis* 6714 and in *Chlamydomonas reinhardtii:* detection of an intermediary state, *Biochemistry,* 29, 8100, 1990.
88. **Shipton, C. A. and Barber, J.,** Photoinduced degradation of the D1 polypeptide in isolated reaction centers from photosystem II: guidance for an autoproteolytic process triggered by the oxidizing side of the photosystem, *Proc. Natl. Acad. Sci. U.S.A.,* 88, 6691, 1991.
89. **Aro, E.-M., Hundal, T., Carlberg, I., and Andersson, B.,** In vitro studies on light-induced inhibition of photosystem II and D1-protein degradation at low temperatures, *Biochim. Biophys. Acta,* 1019, 269, 1990.
90. **Cleland, R. E., Ramage, R. T., and Critchley, C.,** Photoinhibition causes loss of photochemical activity without degradation of D1 protein, *Aust. J. Plant Physiol.,* 17, 641, 1990.
91. **Kettunen, R., Tyystjärvi, E., and Aro, E.-M.,** D1 protein degradation during photoinhibition of intact leaves. A modification of the D1 protein precedes degradation, *FEBS Lett.,* 290, 153, 1991.
92. **Schnettger, B., Leitsch, J., and Krause, G. H.,** Photoinhibition of photosystem 2 *in vivo* occurring without net D1 protein degradation, *Photosynthetica,* 27, 261, 1992.
93. **Greenberg, B. M., Gaba, V., Canaani, O., Malkin, S., Mattoo, A. K., and Edelman, M.,** Separate photosensitizers mediate degradation of the 32-kDa photosystem II reaction center protein in the visible and UV spectral regions, *Proc. Natl. Acad. Sci. U.S.A.,* 86, 6617, 1989.

94. **Jansen, M. A. K., Driesenaar, A. R. J., Kless, H., Malkin, S., Mattoo A. K., and Edelman, M.,** PS II inhibitor binding, Q_B-mediated electron flow and rapid degradation are separable properties of the D1 reaction centre protein, in *Regulation of Chloroplast Biogenesis,* Argyroudi-Akoyunoglou, J. H., Ed., Plenum Press, London, 1992, 303.
95. **Kuhn, M. K. and Böger, P.,** Studies on the light-induced loss of the D1 protein in photosystem-II membrane fragments, *Photosynth. Res.,* 23, 291, 1990.
96. **Sopory, S. K., Greenberg, B. M., Mehta, R. A., Edelman, M., and Mattoo, A. K.,** Free radical scavengers inhibit light-dependent degradation of the 32 kDa photosystem II reaction center protein, *Z. Naturforsch. Teil C,* 45, 412, 1990.
97. **Trebst, A., Depka, B., Kraft, B., and Johanningmeier, U.,** The Q_B site modulates the conformation of the photosystem II reaction center polypeptides, *Photosynth. Res.,* 18, 163, 1988.
98. **van Wijk, K. J., Andersson, B., and Styring, S.,** Spectroscopic characterization of photoinhibited photosystem II and kinetic resolution of the triggering of the D1 reaction center protein for degradation, *Biochim. Biophys. Acta,* 1100, 207, 1992.
99. **Wolff, S. P., Garner, A., and Dean, R. T.,** Free radicals, lipids and protein degradation, *Trends Biochem. Sci.,* 11, 27, 1986.
100. **Jegerschöld, C. and Styring, S.,** Fast oxygen-independent degradation of the D1 reaction center protein in photosystem II, *FEBS Lett.,* 280, 87, 1991.
101. **Wang, W.-Q., Chapman, D. J., and Barber, J.,** Inhibition of water splitting increases the susceptibility of photosystem II to photoinhibition, *Plant Physiol.,* 99, 16, 1992.
102. **Jegerschöld, C., Virgin, I., and Styring, S.,** Light-dependent degradation of the D1 protein in photosystem II is accelerated after inhibition of the water splitting reaction, *Biochemistry,* 29, 6179, 1990.
103. **Eckert, H.-J., Geiken, B., Bernarding, J., Napiwotzki, A., Eichler, H.-J., and Renger, G.,** Two sites of photoinhibition of the electron transfer in oxygen evolving and Tris-treated PS II membrane fragments from spinach, *Photosynth. Res.,* 27, 97, 1991.
104. **Krieger, A. and Weis, E.,** Energy-dependent quenching of chlorophyll-*a*-fluorescence: The involvement of proton-calcium exchange at photosystem II, *Photosynthetica,* 27, 89, 1992.
105. **Krieger, A., Moya, I., and Weis, E.** Energy-dependent quenching of chlorophyll-*a*-fluorescence: effect of pH on stationary fluorescence and picosecond-relaxation kinetics in thylakoid membranes and Photosystem II preparations. *Biochim. Biophys. Acta,* 1102, 167, 1992.
106. **Schuster, G., Even, D., Kloppstech, K., and Ohad, I.,** Evidence for protection by heat shock proteins against photoinhibition during heat shock, *EMBO J.,* 7, 1, 1988.
107. **Roberts, D. R., Kristie, D. N., Thompson, J. E., Dumbroff, E. B., and Gepstein, S.,** In vitro evidence for the involvement of activated oxygen in light-induced aggregation of thylakoid proteins, *Physiol. Plant.,* 82, 389, 1991.
108. **Elsner, C. and Krause, G. H.,** unpublished data.

Chapter 3

PRODUCTION AND ACTION OF ACTIVE OXYGEN SPECIES IN PHOTOSYNTHETIC TISSUES

Kozi Asada

TABLE OF CONTENTS

0-8493-5443-9/94/$0.00+$.50

I. INTRODUCTION

Photosynthetic organisms have a high capacity for conversion of light energy to chemical energy under favorable conditions; eight light quanta are required for the reduction of one molecule of carbon dioxide to the level of carbohydrate; this represents 36% efficiency for 680-nm light. The high efficiency is attained only under conditions where the concentration of carbon dioxide, the availability of mineral nutrients, and other environmental conditions are not limiting, but light intensity is limiting photosynthesis. Natural environments are not always adequate to satisfy these conditions for the high photosynthetic yield. The atmospheric concentration of carbon dioxide often limits photosynthesis. Transport of CO_2 to the chloroplasts is impeded by water and salt stresses. Furthermore, the combination of environmental conditions is not always favorable to photosynthesis. Rather, an unfavorable combination such as high light intensity and water stress is not unusual. In addition to carbon dioxide, the presence of oxygen, varying temperatures, and the physiological state of the leaf, especially in terms of sink capacity, have large effects on photosynthetic efficiency. As a result, the conversion efficiency of sunlight by plants under natural environments is only 3 to 4% at most, compared with the expected maximum efficiency of 12% for solar energy, even when C_4 plants are cultured under optimal conditions.

All conditions that are suboptimal for photosynthesis cause a decrease in the efficiency of energy transfer from photons to carbon dioxide as a whole. High rates of photon flux and low concentrations of carbon dioxide have direct effects. Other environmental conditions such as low or high temperatures, or nitrogen deficiency also interrupt energy transfer from light to carbon dioxide, resulting in excessive irradiation of the chloroplasts. Plants use several mechanisms to dissipate excess light energy as heat so as to lower the rate of excitation of the reaction centers. Failure of these mechanisms to dissipate all of the excess energy allows the generation of reactive molecules: triplet excited chlorophyll and active species of oxygen such as singlet excited dioxygen, superoxide radicals, hydrogen peroxide, and hydroxyl radicals. Plants have several mechanisms to suppress the production of active molecules even when exposed to excess photons (see Chapter 1 of this volume). Most of these active molecules are scavenged to prevent the oxidation of the cellular components that are the target molecules for such active molecules. Failure of the scavenging systems inhibits photosynthetic activity, which further interrupts energy transfer from photons to carbon dioxide.

In this chapter, the production and scavenging mechanism of reactive species of oxygen in chloroplasts are reviewed, including the suppression mechanism of the production. Subsequently, the primary target molecules in chloroplasts and oxidation mechanisms are described.

II. PHOTOREDUCTION OF DIOXYGEN

A. PRODUCTION OF SUPEROXIDE RADICALS

The photoreduction of dioxygen was first described by Mehler in 1951.[1] Chloroplasts were illuminated in the presence of catalase and ethanol, and acetaldehyde was identified as the reaction product. The peroxidatic activity of catalase oxidizes ethanol to acetaldehyde (Reaction 1). Thus, the production of hydrogen peroxide by the photoreduction of dioxygen in chloroplasts was established.

$$C_2H_5OH + H_2O_2 \rightarrow CH_3CHO + 2H_2O \qquad (1)$$

Subsequently, the actual photoreduction of dioxygen by the electrons derived from water was confirmed by measurements of the uptake of $^{18}O_2$ with concomitant evolution of $^{16}O_2$ from water.[2] Until the discovery of the production of superoxide in the 1970s, the primary product of oxygen reduction was believed to be hydrogen peroxide.

Using the enzyme superoxide dismutase, the superoxide anion radical was identified as the primary product of oxygen photoreduction in thylakoids. The superoxide dismutase-sensitive photoreduction of cytochrome *c* provided evidence for the production of superoxide.[3]

$$O_2^{\cdot-} + \text{Cyt } c(Fe^{3+}) \rightarrow O_2 + \text{Cyt } c(Fe^{2+}) \qquad (2)$$

Other probes of superoxide such as ascorbate, sulfite, Mn^{2+}, epinephrine, hydroxylamine, and tiron have also been used to detect the production of superoxide in thylakoids. Inhibition of the reactions of the above probes with superoxide provided further evidence for the production of superoxide, as summarized in a previous review.[4] In addition, superoxide production has been detected via spin-trap reagents using electron spin resonance techniques[4,5] and also by proton consumption during the disproportionation of the superoxide anion radical generated by a flashed light.[4] Pulsed time-resolved photoacoustic spectroscopy has allowed the determination of light-dependent oxygen uptake in intact leaves.[6,7] Detection of superoxide as a photocurrent using a photoelectrochemical cell has been shown to be a useful technique with applications for thylakoids.[8]

Ground-state dioxygen occurs in the triplet state, and this is the reason why the reactivity of dioxygen with the singlet states of cellular components is very low, i.e., due to the difficulty arising from the addition of an electron to the antibonding orbital that is partially occupied. However, ground-state dioxygen can accept electrons from radicals (unpaired electrons) or from singlet molecules, e.g., ligation of dioxygen to transition metals. Among the electron carriers of the thylakoid electron transport system, the semiquinone radicals of plastoquinone, including Q_A and Q_B, are possible sites of interaction with dioxygen, producing superoxide. However, no evidence has been presented thus far for the production of superoxide via the reduction of dioxygen by plastosemiquinone radicals. This is probably due to stabilization of these radicals by interaction with amino acid residues and the *Q*-cycle mechanism, which guarantees a very short radical lifetime within the plastoquinone pool so as to suppress the interaction with dioxygen. Only when the electron transport chain was inhibited by plastoquinone analogs, such as 2,5-dibromo-3-methyl-6-isopropyl-*p*-benzoquinone (DBMIB), has the photoreduction of dioxygen in the plastoquinone region been observed.[9] Similarly, in mitochondria, reduction of dioxygen in the ubiquinone region is only observed when the *Q*-cycle is disturbed by myxothiazol.[10] The reaction center of photosystem II (PSII) is not the major site of dioxygen reduction because superoxide production is inhibited by 3-(3′,4′-dichlorophenyl)-1,1-dimethyl urea (DCMU).[3,5] In isolated PSII membranes, however, photoproduction of hydrogen peroxide, probably derived from superoxide, has been shown.[10a] The oxidation of hydrogen peroxide on the oxidizing side of PSII is another possible reaction by which superoxide can be produced; however, this reaction occurs only when the water-oxidizing system is inactivated.[11]

B. PRODUCTION OF SUPEROXIDE BY MEMBRANE-BOUND REDUCTANTS IN PSI

The major site of superoxide production in chloroplasts under normal conditions is the reducing side of PSI. The photoreduction of oxygen has been observed even in washed thylakoids, depleted in ferredoxin. Therefore, membrane-bound reductants participate in the reduction of oxygen. The photoproduction rate of superoxide in washed thylakoids, as determined by reaction with the superoxide probes, is 15 to 25 μmol (mg Chl)$^{-1}$ h^{-1} (about 4 mol $O_2^{\cdot -}$ (mol P_{700})$^{-1}$ s^{-1}, assuming one mol of P_{700} per 800 mol of Chl) at saturating irradiance.[4,5] This rate is less than 10% that of the maximum electron flow rate observed using other electron acceptors in chloroplasts or in leaves. A low rate of oxygen reduction is also observed when determined in intact chloroplasts using $^{18}O_2$.[12]

Kinetic analysis[13] indicates that the second-order rate constant between oxygen and a reductant in PSI is 10^6 M^{-1} s^{-1}. The apparent K_m value for oxygen is below 10 μ*M*, but it is increased severalfold with an increase in light intensity. Since the oxygen concentration in illuminated chloroplasts may be higher than that in air (250 μ*M* at 25°C), the photoreduction rate of oxygen can always be assumed to be substrate saturated. It should be noted that these kinetic values were determined by changing the oxygen concentration in aqueous media. However, the photoreduction of oxygen can occur in the aprotic environment of the thylakoid membranes. Because the solubility of oxygen in aprotic solvents is approximately one order of magnitude higher than that in water, the actual K_m for dioxygen and the rate constant between oxygen and the membrane-bound reductant would be higher and lower, respectively, than those measured.

When the structure of thylakoid membranes is partially destroyed by detergents, the photoreduction rate of oxygen is dramatically increased. This amounts to a maximum of about 70 mol O_2^- (mol P_{700})$^{-1}$ s^{-1} in PSI particles with a rate constant of 10^8 M^{-1} s^{-1}, two orders of magnitude higher than that of untreated thylakoids.[13,14] These observations implicate the involvement of the supermolecular structure of the PSI reaction center and its electron donors and acceptors in the thylakoid membranes in the suppression of superoxide production.

The tyrosine residues of the P_{700}-containing subunits (*psa*A and *psa*B) of the PSI reaction center are the first to become iodo-labeled when the thylakoids are illuminated in the presence of $^{125}I^-$ and lactoperoxidase.[15] This indicates the generation of superoxide, and hence, hydrogen peroxide in the PSI core subunits, where the hydrogen peroxide activates I^- for the labeling process. In the reaction center of PSI, the electron produced by charge separation of P_{700} is successively transferred to A_o (chlorophyll), A_1 (phylloquinone), center X (Fe-S cluster), and then to the A/B center (an Fe-S cluster).[16] The electron transfer rates up to center X are very fast, and reduced centers A_O and A_1 would have little chance to interact with oxygen. Thus, the rapid labeling of the core subunits suggests that the most probable reductants of oxygen in these conditions are (1) the FeS center, X, which resides in the core subunits and (2) the Fe-S center, A/B, which resides in the 9-kDa polypeptide (*psa*C). Neither of the Fe-S clusters is exposed to the surface of the thylakoid membranes. Center A/B is masked by a 18-kDa polypeptide (*psa*D) and a 10-kDa (*psa*E) peptide[16,17] (Figure 1). It is interesting to note that the 9-kDa center A/B protein is susceptible to autooxidation when isolated.[17]

Further evidence for the univalent reduction of oxygen by the photoreductants in the membrane interior is the very effective inhibition of superoxide-dependent cytochrome *c* photoreduction by hydrophobic catechol derivatives, but not by hydrophilic derivatives,[15] both of which react rapidly with superoxide anions. Thus, superoxide may be produced by the autooxidation of centers X and A/B within the thylakoid membranes.

The spontaneous disproportionation of superoxide radicals requires protons, and the rate of this reaction is dependent on pH. The pK_a of superoxide is 4.7. The disproportionation between the protonated superoxide radical (HO_2, hydroperoxyl radical) and the superoxide anion radical () **has the fastest rate (Reaction 4, 1.0 × 10^8 M^{-1} s^{-1})**, followed by that between HO_2 and HO_2 (Reaction 3, 8.6 × 10^5 M^{-1} s^{-1}). The disproportionation rate between **and is very low (Reaction 5, below 0.35 M^{-1} s^{-1})**.[18] Thus, the disproportionation rate constant is highest at the pK_a of superoxide and decreases by an order of magnitude for each unit increase in pH. The observed disproportionation rate at pH 7 is 5 × 10^5 M^{-1} s^{-1}.

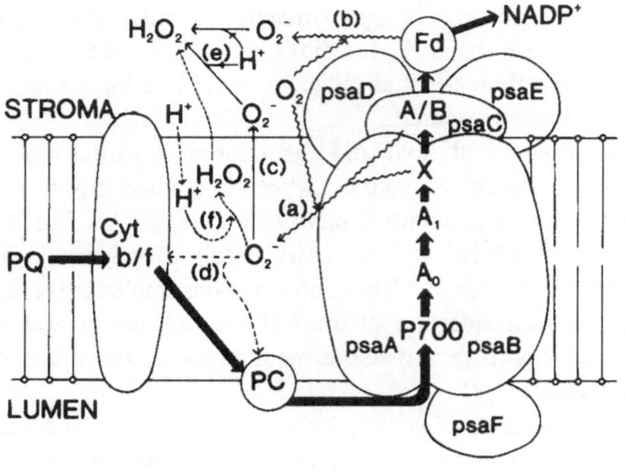

FIGURE 1. Structure of photosystem I reaction center showing the photoreduction of oxygen and fate of the superoxide anion radicals produced in the thylakoid membranes and the stroma. Thick arrows indicate linear electron flow from the plastoquinone pool (PQ) to $NADP^+$ through the cytochrome *b/f* complex (Cyt *b/f*) and the photosystem I complex. The subunit peptides of the photosystem I reaction center are indicated by *psa* A, B, C, D, E, and F, but other subunits of photosystem I (*psa*L, etc.) are not included. A_O, A_1, X, and A/B represent chlorophyll, phylloquinone, Fe-S center X, and Fe-S center A/B, respectively. (a) Univalent reduction of oxygen by centers X and A/B within the thylakoid membranes. (b) Univalent reduction of oxygen by reduced ferredoxin (Fd) in the stroma. (c) Diffusion of the superoxide anion radicals generated within the membranes to the stroma. Diffusion to the lumen side is also likely to occur.[22] (d) Reduction of cytochrome *f* (Cyt *f*) or plastocyanin (PC) by the superoxide anions generated within the thylakoid membranes prior to diffusion to the membrane surfaces. Reactions (a) and (d) comprise the superoxide-mediated pathway of cyclic electron flow around photosystem I. (e) Disproportionation of the ejected superoxide anions outside of the membranes, mostly catalyzed by superoxide dismutase, using protons in the stroma. (f) Spontaneous disproportionation of the superoxide anions within the thylakoid membranes using the protons donated by amines or ammonium ions or from a low-pH medium.

$$HO_2 + HO_2 \rightarrow O_2 + H_2O_2 \quad (H_2O) \tag{3}$$

$$HO_2 + O_2^{\overline{\cdot}} \rightarrow O_2 + HO_2^- \xrightarrow{(H_2O)} O_2 + H_2O_2 + OH^- \tag{4}$$

$$O_2^{\overline{\cdot}} + O_2^{\overline{\cdot}} \rightarrow O_2 + O_2^{2-} \xrightarrow{2H_2O} O_2 + H_2O_2 + 2\,OH^- \tag{5}$$

When oxygen is reduced by Fe-S centers X and A/B, superoxide anion radicals, not hydroperoxyl radicals, are generated as the primary products because of the absence of protons in the membrane interior. In aprotic environments, the lifetime of $O_2^{\overline{\cdot}}$ is determined by the rate of Reaction 5. The following observations indicate that this is the case for superoxide generated within the thylakoid membranes. When protons are donated by ammonium ions or amines to the thylakoid membranes, hydrogen peroxide is produced, even in the presence of cytochrome *c*.[15] This unexpected finding infers that disproportionation of superoxide, produced within the membranes, is supported by protons donated by the ammonium ions or the amines before diffusion of the superoxide to the membrane surface and its interaction with cytochrome *c*, outside the membranes. Thus, the superoxide production rate, as determined by the reduction of cytochrome *c*, represents the rate of superoxide diffusion out of the thylakoid membranes. Further evidence for the production of superoxide within the membranes is provided by the increased reduction of oxygen at pH 5.[7,19,20] At low medium pH values, protons are available near the surface of the membranes, and disproportionation of superoxide to hydrogen peroxide may be expected to be enhanced. Thus, increased light-dependent oxygen uptake at low pH values is caused by an increased rate of disproportionation of the superoxide produced in the aprotic interior of the membranes. This is greater at low pH values than at neutral pH.

C. SUPEROXIDE-MEDIATED CYCLIC ELECTRON FLOW AROUND PSI

As described above, the production rate of superoxide under continuous light, as measured by reaction with superoxide probes, is less than 10% of maximum electron flow at the physiological pH values of the stroma even in the absence of electron acceptors for PSI. Under the same conditions, however, a single excitation of the reaction center by a light flash produces one mole of superoxide per mole of P_{700}.[15] This observation indicates that in the absence of PSI electron acceptors, one molecule of oxygen is reduced to superoxide by each excitation of P_{700}. All of the superoxide produced may not diffuse out from the membranes to interact with the superoxide probes, such as cytochrome *c;* the flash light-induced reduction rate constant of cytochrome *c* is 2 s^{-1}. This is lower than that expected from the reaction between free O_2^- and cytochrome *c* (10 s^{-1}), indicating a limitation of cytochrome *c* reduction by the efflux of O_2^- from its production site into the medium.[15] The permeability of O_2^- through the thylakoid membranes is very slow.[21] When protons are supplied via amines or ammonium ions, the superoxide anions disproportionate to hydrogen peroxide and oxygen prior to diffusion to the membrane surface.[15] Disproportionation is expected to occur near the membrane surface when the pH of the medium is low.[7,19,20] Otherwise, the lifetime of the superoxide anion in the aprotic membrane interior would be long enough to allow interaction with the electron donors to PSI, either plastocyanin or cytochrome *f*.[22] The rate of P_{700^+} reduction by plastocyanin ($t_{1/2}$, 20 to 30 ms) is lower than that of P_{700} photoooxidation. This means that either plastocyanin or cytochrome *f* may accept electrons from superoxide while these electron carriers are respective oxidized forms. It has been shown that both of these electron carriers are reduced by superoxide at appreciable rates (about 10^6 M^{-1} s^{-1}).[23,24] Thus, under physiological conditions, some of the superoxide generated in the membrane interior may donate electrons to the electron donors of PSI, and the superoxide-mediated cyclic electron flow around PSI may operate (Figure 1).

D. PRODUCTION OF SUPEROXIDE BY REDUCED FERREDOXIN

When NADP is not available, reduced ferredoxin can autooxidize to superoxide in illuminated chloroplasts.[25,26] The reduction of oxygen by reduced ferredoxin is a slow reaction.[4,25] When the concentration of oxygen is 250 μM, the pseudo first-order rate constant for the autooxidation of reduced ferredoxin is only 0.081 s^{-1} compared to 250 s^{-1} for the thylakoid-bound reductants. Thus, the lifetime of reduced ferredoxin is much longer than that of the membrane-bound reduced Fe-S centers in chloroplasts in the absence of $NADP^+$. The apparent K_m value for oxygen in this autooxidation (about 60 μM) is higher than that of the reduction of oxygen by thylakoids.[26]

E. RATE OF SUPEROXIDE PRODUCTION IN CHLOROPLASTS

Peripheral ferredoxin-depleted thylakoids produce superoxide at a rate of 15 to 25 μmol (mg Chl)$^{-1}$ h^{-1}.[4,5] This is increased by 1.5-fold at maximum upon the addition of ferredoxin, and is decreased by one half by the addition of $NADP^+$.[5] A similar rate is observed in intact chloroplasts.[12] These responses are attributable to the low reactivity of reduced ferredoxin with dioxygen and to the donation of electrons from reduced ferredoxin to the intersystem electron carriers via ferredoxin-dependent cyclic electron flow. From these observations, the maximum production rate of superoxide, under conditions where reduced ferredoxin donates electrons to cyclic electron flow alone, would be about 30 μmol (mg Chl)$^{-1}$ h^{-1} on average; assuming a chloroplast volume of 35 μl (mg Chl)$^{-1}$, the production rate of superoxide is estimated to be about 240 μM s^{-1}.

F. PRODUCTION OF HYDROGEN PEROXIDE

In leaf peroxisomes, the two-electron oxidase, glycolate oxidase, generates hydrogen peroxide directly, but no such oxidase has been found in chloroplasts. Thus, most of the hydrogen peroxide in chloroplasts arises from the disproportionation of superoxide (Reactions 3 and 4) catalyzed by the thylakoid-bound and stromal superoxide dismutases.

If the superoxide radicals fail to interact with superoxide dismutase, they may interact with other reductants (AH_2) to produce hydrogen peroxide.

$$O_2^{\overline{\cdot}} + AH_2 + H^+ \rightarrow H_2O_2 + AH \qquad (6)$$

Ascorbate, Mn^{2+}, GSH, and reduced ferredoxin can all reduce superoxide, but the reaction rate constants between superoxide and these reductants, together with their stromal concentrations, suggests that Reaction 6 contributes little to the production of hydrogen peroxide.[4]

G. PRODUCTION OF HYDROXYL RADICALS

Direct production of hydroxyl radicals in chloroplasts, as in other organelles, has not been found. The production of hydroxyl radicals, demonstrated in illuminated thylakoids and algal cells, might occur via the transition metal (M^{n+}; Fe^{2+}, and Cu^{+})-catalyzed reduction of hydrogen peroxide.[27,28]

$$H_2O_2 + M^{n+} \rightarrow OH^- + OH + M^{(n+1)+} \qquad (7)$$

The oxidized metal ions (Fe^{3+}, Cu^{2+}) would be re-reduced by superoxide, ascorbate, or GSH. Trace amounts of these metals can catalyze the production of hydroxyl radicals. It is not known what molecular species of chelate complexes of transition metals participate in the production of hydroxyl radicals in chloroplasts. Reduced ferredoxin can reduce hydrogen peroxide, producing hydroxyl radicals.[25]

III. SCAVENGING OF ACTIVE OXYGEN SPECIES

Although the release of the superoxide anion from the thylakoid membranes is suppressed, the production rates of superoxide and its disproportionation product, hydrogen peroxide, are 240 and 120 $\mu M\ s^{-1}$, respectively, in chloroplasts even under conditions favorable to photosynthesis. These amounts inhibit photosynthesis, by inactivation of the stromal enzymes, within several seconds. The operation of the scavenging systems for superoxide and hydrogen peroxide is therefore essential to maintain photosynthetic activity. Lowering the concentrations of superoxide and hydrogen peroxide in chloroplasts protects enzymes from inactivation.

A. DISPROPORTIONATION OF SUPEROXIDE

In chloroplasts, the superoxide anion radical is scavenged via a disproportionation reaction catalyzed by superoxide dismutase (SOD). At stromal pH values, most of the superoxide is present as the anion, and HO_2 is less than 1% of the total superoxide level, since the pK_a of superoxide[18] ($H^+ + O_2^- \rightleftarrows HO_2$) is 4.7.

Three types of SOD have been found; these differ with respect to their prosthetic metals: Cu,Zn, Mn, and Fe. The major stromal SOD is the Cu,Zn enzyme in most angiosperms; the Fe enzyme is present in a few species of plants and most algae.[30-32] Mn SOD has been found in a thylakoid bound form in spinach[32a] and cyanobacteria.[32b,77] Chloroplastic and cytosolic Cu,Zn SODs show amino acid sequences characteristic of each isozyme, but little difference has been found in their enzymatic properties.[32,33] The three-dimensional structure of the chloroplastic Cu,Zn SOD from spinach has been studied.[34] The overall structure is very similar to that of bovine Cu,Zn SOD.

All three types of SOD catalyze the disproportionation of the superoxide anions by the following reduction-oxidation cycle (M^{n+}; Cu^{2+}, Mn^{3+}, and Fe^{3+}):

$$O_2^{\overline{\cdot}} + SOD\text{-}M^{n+} \rightarrow O_2 + SOD\text{-}M^{(n-1)+} \quad (8)$$

$$O_2^{\overline{\cdot}} + {}^2H^+ + SOD\text{-}M^{(n-1)+} \rightarrow H_2O_2 + SOD\text{-}M^{n+} \quad (9)$$

Reactions 8 and 9 proceed at similar rates, about $2 \times 10^9\ M^{-1}\ s^{-1}$ (k_{SOD}).[35] Random collision of the superoxide anion with the prosthetic metal sites of SOD molecules (only 0.1 % of the total surface area of the 30- to 40-kDa molecular size) is not sufficient to explain the observed rates.[29,36] The substrate, the superoxide anion, is attracted to the metal site by positively charged amino acid residues. An arginine residue near the Cu site of Cu,Zn SOD is conserved in the enzymes from various organisms. This guarantees a diffusion-controlled rate of interaction of the superoxide anion with the prosthetic metal of SOD. In contrast to the spontaneous disproportionation of superoxide (Reactions 3 to 5), the SOD-catalyzed disproportionation shows a broad pH optimum (pH 4 to 9 for Cu,Zn SOD; pH 6 to 8 for Mn SOD and Fe SOD)[35] irrespective of the proton requirement of the reduced SOD (Reaction 9). The protonated histidine residue, ligated to both the Cu and the Zn of the reduced Cu,Zn SOD, donates the proton directly to the superoxide anion, which supports a rapid reaction rate over a broad range of values.[35]

This mechanism supports the highest catalytic rate of any SOD and decreases, by at least four orders of magnitude, the steady-state concentration of superoxide in chloroplasts. In the absence of SOD, the steady-state concentration of superoxide is 22 μM from $v = k_{sp}\ (O_2^{\overline{\cdot}})^2$, where v is the production rate of superoxide (240 $\mu M\ s^{-1}$) and k_{sp} ($5 \times 10^5\ M^{-1}\ s^{-1}$)[18] is the spontaneous disproportionation rate of superoxide at pH 7. In the stroma (and the thylakoid lumen also), 50 μM SOD has been found,[4] and this decreases the steady-state concentration of superoxide to 2.4 nM (from $v = k_{SOD}\ [SOD]\ [O_2^{\overline{\cdot}}]$). For this estimation, it is assumed that superoxide is generated uniformly and that SOD is distributed uniformly within the chloroplast. However, superoxide is generated mainly by the PSI complexes on the thylakoid membranes, and SOD may be concentrated at the sites where superoxide is generated.[22] Therefore, the actual steady-state concentration of superoxide would be variable, depending on the localization of superoxide generation and SOD in chloroplasts. Microlocalization of SOD in the sites of superoxide generation would lower the local concentration of superoxide. In *Escherichia coli* cells, the steady-state concentration of superoxide is estimated to be 0.02 to 0.04 nM, as determined by the inactivation of aconitase by superoxide.[37] Photodamage in Fe SOD-defective mutants of cyanobacteria[72] and the protection against photodamage by increased SOD in leaf cells[38] indicate the importance of lowering the superoxide concentration in chloroplasts and other cell compartments.

B. DISPROPORTIONATION OF HYDROGEN PEROXIDE

In leaf tissues, catalase is exclusively localized in peroxisomes, where it is present at high concentrations and sometimes in a crystalline state[39] (about 4 m*M*). It scavenges the hydrogen peroxide produced by glycolate oxidase via a disproportionation reaction:

$$H_2O_2 + H_2O_2 \rightarrow O_2 + 2H_2O \quad (10)$$

The expression of several isozymes of catalase following exposure to oxidative stress has been reviewed previously.[39a] Little catalase has been found in chloroplasts, and almost all of the hydrogen peroxide produced from superoxide is scavenged via a peroxidatic reaction. Hydrogen peroxide is photooxidized, producing oxygen, in a reaction catalyzed by the PSII-water oxidase complex under flashed light, but not under continuous light.[11]

C. REDUCTION OF HYDROGEN PEROXIDE

^{18}O-labeled oxygen is not evolved when $H_2{}^{18}O_2$ is added to chloroplasts in the dark, but $^{16}O_2$ is evolved in the light with a stoichiometry of two moles of hydrogen peroxide reduced to one mole of oxygen evolved.[12] These and other observations[4,40] indicate that in chloroplasts, no scavenging of hydrogen peroxide by disproportionation occurs, and hydrogen peroxide is reduced to water by a peroxidatic reaction (13) using a photoreductant (AH_2) generated by PSI (Reactions 11 and 12).

$$2H_2O \rightarrow 4[e^-] + 4H^+ + O_2 \quad (11)$$

$$4[e^-] + 2A \rightarrow 2AH_2 \quad (12)$$

$$H_2O_2 + 2AH_2 \rightarrow 2H_2O + 2A \quad (13)$$

In angiosperms, AH_2 has been identified as ascorbate. Ascorbate occurs at about 10 m*M* in chloroplasts, together with an ascorbate-specific peroxidase (APX).[4] Even though glutathione occurs at a concentration similar to that of ascorbate and glutathione reductase is found in chloroplasts, little or no glutathione peroxidase has been found in plants.[4] Ascorbate peroxidase has been found also in cell compartments other than chloroplasts and has been referred to as the cytosolic enzyme (cAPX)[41-44] distinct from the chloroplast enzyme.[41,45] In angiosperm chloroplasts, APX occurs in the stroma in a soluble form (sAPX) and also in the thylakoids in a membrane-bound form (tAPX).[46] In spinach chloroplasts, tAPX are localized in the stroma thylakoids at a ratio of one molecule of tAPX per five molecules of P_{700}. The stroma thylakoids are enriched in PSI complexes; thus, tAPX is bound to the thylakoid membranes at the site where superoxide is produced.

APX catalyzes the oxidation of ascorbate by hydrogen peroxide, generating two molecules of the monodehydroascorbate (MDA) radical:[47]

$$2\ \text{Ascorbate} + H_2O_2 \rightarrow 2\text{MDA} + 2H_2O \qquad (14)$$

Similar to other heme peroxidases, APX is oxidized by two equivalents, to the intermediate compound I, by hydrogen peroxide. Reduction to the oxyferryl heme and the occurrence of tryptophane radicals near the heme follow by analogy to the cytochrome *c* peroxidases.[48] Compound I then univalently oxidizes ascorbate, forming compound II and MDA. Subsequently, the interaction of compound II with ascorbate produces a second MDA.

In spinach chloroplasts, the total content of sAPX and tAPX is 70 μM (e_o),[46] assuming a chloroplast volume of 35 μl mg Chl^{-1}. The chloroplast ascorbate content is about 10 m*M*,[49] which is at least one order of magnitude higher than the apparent K_m value for ascorbate of APX (about 0.5 m*M*).[47] Since chloroplasts have a high capacity for photoregeneration of ascorbate from either monohydroascorbate or dehydroascorbate, we can assume that the ascorbate concentration does not limit the reduction of hydrogen peroxide. If we assume a molecular activity of APX of 3.1×10^2 mol s^{-1} (k_o), a K_m for H_2O_2 of 8×10^{-5} *M*, and a production rate of H_2O_2 in chloroplasts ($v(H_2O_2)$) of 120 $\mu M\ s^{-1}$, the steady-state concentration of hydrogen peroxide is estimated to be 4.6×10^{-7} *M*, using the following equation: $v(H_2O_2) = k_oe_o[H_2O_2]/(K_m + [H_2O_2])$. The photoproduction of hydrogen peroxide may not be uniform in whole chloroplasts, but, rather, mainly limited to the PSI complexes of the thylakoid membranes. Also, APX is not uniformly distributed in chloroplasts since tAPX, at least, is bound to the stroma thylakoids.[46] Thus, both the production and scavenging of H_2O_2 seem to be localized in the same region. Therefore, the estimated steady-state concentration of hydrogen peroxide should be regarded as an approximate value.

A comparison of the molecular properties of tAPX, sAPX, and cAPX with those of guaiacol peroxidase (GPX) is presented in Table 1.[47] GPX is a typical plant peroxidase, as represented by horseradish peroxidase (HRP), and is localized in cell walls and vacuoles, but not in chloroplasts. The physiological functions of GPX include the biosynthesis of lignin and ethylene, and the degradation of indoleacetic acid, rather than the scavenging of hydrogen peroxide. The physiological function and molecular properties of APX distinguish it from GPX. Both APX and GPX are protoporphyrin IX-containing enzymes, but the amino acid sequence of APX shows a rather higher degree of homology with the cytochrome *c* peroxidase (CPX) from fungi than with the GPX from plants.[48,50,51] CPX is localized in the mitochondria of yeast and plays a role in H_2O_2 scavenging. Thus, APX and CPX are classified as a superfamily of hemoperoxidases[52] whose physiological function is the scavenging of hydrogen peroxide.

TABLE 1
Comparison of Enzymatic and Molecular Properties of Thylakoid-Bound, Stromal, and Cytosolic Ascorbate Peroxidases (tAPX, sAPX, and cAPX) and Guaiacol Peroxidase (GPX) from Plants[47]

	tAPX	sAPX	cAPX	GPX
Molecular wt. size	40 kDa	30 kDa	30 kDa (57 kDa)	30 kDa
Electron donor to ascorbate specificity	High	High	Low	Very low
Amino acid sequence homology	High with Cyt *c* peroxidase (CPX) from yeast			Low with CPX and APX
	High between tAPX and sAPX		Low with tAPX and sAPX	
Inhibition by				
Thiol reagent	+	+	+	−
Thiols	+	+	+	−
Suicide inhibitors	+	+	+	−
Ascorbate depletion	Rapid	Rapid	Slow	None

tAPX and sAPX show a high degree of homology in the amino acid sequence of the amino terminal, but are not homologous with cAPX[53] in this region. tAPX has a 10-kDa-higher molecular mass than cAPX and sAPX. The additional size of tAPX might contribute to its binding to the thylakoid membranes.[53] One of the characteristic properties of APX is inactivation in the absence of the electron donor.[41,45,46,49] This is especially the case for sAPX and tAPX, whose half-inactivation time is only 15 s. This inactivation is caused by a rapid degradation of the reaction intermediate Compound I of APX, which is formed by a trace of hydrogen peroxide. Under aerobic conditions, the minimum requirement of ascorbate for stabilization of APX is 20 μM. Since the ascorbate content of chloroplasts is approximately 10 mM, it protects APX from inactivation. However, when the ascorbate in chloroplasts is consumed by titration with hydrogen peroxide in the dark, APX is irreversibly inactivated. In contrast, in the light, the continuous addition of hydrogen peroxide does not inactivate APX due to the photoregeneration of ascorbate.[49] These observations indicate the importance of maintaining ascorbate in a reduced form in chloroplasts, not only as an electron donor, but also for stabilization of APX. At night, chloroplast respiration[54] may play a role in the reduction of oxidized ascorbate and, thereby, in the stabilization of APX in chloroplasts.

Another characteristic property of APX is its inhibition by thiols[55] such as GSH, *p*-aminophenol, and hydroxyurea,[56] which are able to be univalently oxidized by hydrogen peroxide in an APX-catalyzed reaction. The enzyme is inactivated by their primary radical products. These compounds are, thus, suicide inhibitors of APX. The thiyl radical of GSH is produced in an APX-catalyzed reaction, but APX in chloroplasts would not be inactivated because ascorbate can protect against inactivation. Dithiothreitol (DTT) is the most effective thiol inhibitor of APX; 5 m*M* DTT inhibits APX in chloroplasts by 50%.[57] When DTT is added to intact chloroplasts either for the stabilization of the thiol enzymes or as an antioxidant, attention should be paid to the inactivation of APX.

D. REGENERATION OF ASCORBATE

For the operation of the peroxidase-dependent scavenging system, the regeneration of ascorbate from monodehydroascorbate (MDA) and its disproportionation product, dehydroascorbate (DHA), is indispensable. Even though the concentration of ascorbate in chloroplasts is high, it can be oxidized within 80 s, as estimated from the production rate of hydrogen peroxide, unless the following regeneration systems operate.

The primary product in an APX reaction, the MDA radical, is directly reduced to ascorbate by two systems. The first is reduction by reduced ferredoxin (Fd_{red})

$$\text{MDA} + \text{Fd}_{red} \rightarrow \text{Ascorbate} + \text{Fd} \tag{15}$$

This is a very rapid reaction, as revealed by the quenching of chlorophyll fluorescence upon the generation of MDA with ascorbate oxidase[46] and also by the competition of the photoreduced ferredoxin between $NADP^+$ and MDA. The ratio of the reactions of MDA and $NADP^+$ with reduced ferredoxin is about 40, indicating preferential photoreduction of MDA in chloroplasts.[58] When hydrogen peroxide is added to intact chloroplasts, the photosynthetic fixation of carbon dioxide is interrupted while hydrogen peroxide is photoreduced,[59] which is inferred from the high reactivity of MDA generated by ascorbate peroxidase with reduced ferredoxin.

Regeneration of ascorbate from MDA is catalyzed by the FAD enzyme, MDA reductase, using NAD(P)H as the electron donor.[60,61]

$$2\text{MDA} + \text{NAD(P)H} \rightarrow 2\ \text{Ascorbate} + \text{NAD(P)}^+ \tag{16}$$

Kinetic analysis indicates that the enzyme is first reduced by NAD(P)H, and the reduced enzyme ($FADH_2$-NAD^+) then donates electrons to MDA through two successive one-electron transfers, with the FAD semiquinone as an intermediate.[61] This enzyme has also been found in nonphotosynthetic tissues.[62] MDA reductase is localized in the stroma in a soluble form at a concentration of 2 to 10 μM,[60] which allows the lowering of the steady-state concentration of MDA from about 50 to 0.1 μM, assuming a generation rate of MDA of 240 $\mu M\ s^{-1}$.

The MDA radicals generated by the thylakoid-bound ascorbate peroxidase on the peripheral space of the thylakoid membranes could be reduced by reduced ferredoxin before its diffusion to the stroma. Even if MDA reductase occurs on the periphery of the thylakoid membranes, it could not catalyze the reduction of MDA because of the decrease in the photoreduction of $NADP^+$ due to a rapid oxidation of reduced ferredoxin by MDA. Furthermore, the K_m value for NADH (4.6 μM) is lower than that for NADPH (23 μM).[61] Thus, it is likely that the MDA generated in the thylakoids is preferentially reduced by photoreduced ferredoxin. MDA reductase in the stroma catalyzes the reduction of the MDA which is not reduced by ferredoxin or is generated in the stroma.

It should be noted that MDA is produced by reactions other than ascorbate peroxidase, such as the reaction of ascorbate with the tocopherol radical, superoxide, the hydroxyl radical, organic radicals such as carbon-centered and phenoxy radicals,[63] and the thiyl radical. When the glutathione thiyl radical (GS·) is generated in chloroplasts by the interaction of GSH with radicals,[64] it would react with ascorbate at a rapid rate ($6 \times 10^8 \ M^{-1} \ s^{-1}$).[65]

$$\text{GS} + \text{Ascorbate} \rightarrow \text{GSH} + \text{MDA} \qquad (17)$$

Glutathione thiyl radicals have been shown to be produced by the interaction of GSH with various radicals.[64]

Furthermore, if ascorbate donates electrons to the intersystem electron transport system, MDA will be generated. Ascorbate serves to de-epoxize violaxanthin to zeaxanthin in the xanthophyll cycle and helps the development of thermal energy dissipation under high light stress.[66] In this reaction MDA radicals would be generated. In cucurbitaceous plants, ascorbate oxidase generates MDA radicals, while oxygen is reduced to water.

$$4 \text{ Ascorbate} + O_2 \rightarrow 4 \text{ MDA} + 2H_2O \qquad (18)$$

If the MDA radical is generated in the stroma by these reactions, the MDA reductase system would operate for the regeneration of ascorbate using NAD(P)H.

The disproportionation rate of MDA to ascorbate and dehydroascorbate (DHA, Reaction 19) is pH dependent;[63] below pH 4, it is $10^8 \ M^{-1} \ s^{-1}$, but decreases to a value similar to that of superoxide anions at neutral pH ($2 \times 10^5 \ M^{-1} \ s^{-1}$ at pH 7):

$$\text{MDA} + \text{MDA} \rightarrow \text{Ascorbate} + \text{DHA} \qquad (19)$$

When MDA is not reduced to ascorbate either by reduced ferredoxin or by NAD(P)H with MDA reductase, DHA will be formed by disproportionation. The thiol enzyme DHA reductase has been found in the chloroplast stroma; it regenerates ascorbate in the following reaction using GSH as the electron donor:[67]

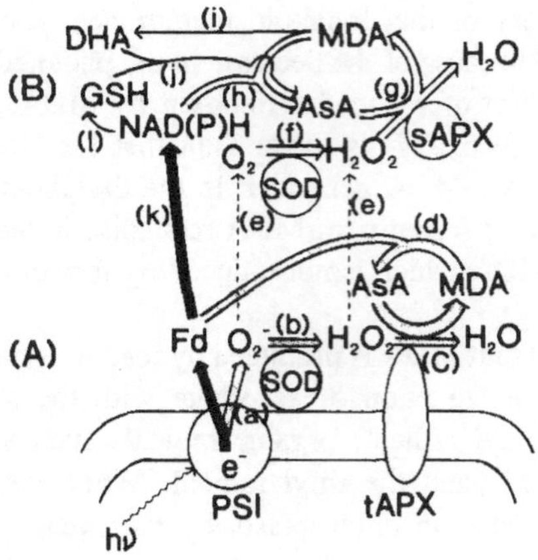

FIGURE 2. Thylakoid and stromal scavenging systems of superoxide and hydrogen peroxide in chloroplasts. (A) Thylakoid scavenging system: (a) diffusion to the stroma of superoxide radicals generated within the membrane (Figure 1), (b) disproportionation of superoxide catalyzed by thylakoid-bound or peripheral SOD (Reactions 8 and 9), (c) reduction of hydrogen peroxide by ascorbate (AsA) catalyzed with thylakoid-bound ascorbate peroxidase (tAPX) localized on the stroma thylakoids (Reaction 14), (d) reduction of monodehydroascorbate radical (MDA) by reduced ferredoxin (Fd) in PSI (Reaction 15). (B) Stromal scavenging system: (e) diffusion of superoxide and hydrogen peroxide to the stroma from the thylakoid periphery, (f) disproportionation of superoxide catalyzed by stromal SOD (Reactions 8 and 9), (g) reduction of hydrogen peroxide catalyzed by stromal ascorbate peroxidase (sAPX), (h) reduction of MDA by NAD(P)H catalyzed by MDA reductase (Reaction 16), (i) spontaneous disproportionation of MDA to dehydroascorbate (DHA) and ascorbate (AsA) (Reaction 19), (j) reduction of DHA to AsA by GSH catalyzed with DHA reductase (Reaction 20), (k) reduction of $NADP^+$ by the reduced Fd catalyzed with Fd-NADP reductase and production of NADH catalyzed by transhydrogenase, (l) reduction of GSSG by NADPH catalyzed by glutathione reductase.

$$\text{DHA} + 2\text{GSH} \rightarrow \text{Ascorbate} + \text{GSSG} \qquad (20)$$

The level of GSH is maintained by glutathione reductase in chloroplasts using NADPH. Since almost all of the MDA is directly reduced to ascorbate, the contribution of DHA reductase and GSH is a minor part of the regeneration of ascorbate. Actually, in some plants and algal cells, DHA reductase could not be detected, although ascorbate peroxidase and MDA reductase are ubiquitously found. It should be noted that in microorganisms, glutaredoxin also catalyzes the GSH-dependent reduction of DHA in addition to disulfides.[68,69] The molecular sizes and properties of spinach DHA reductase (23 kDa) and glutaredoxin (12 kDa) are different. It remains to be demonstrated whether glutaredoxin is bifunctional in cells with respect to thioltransferase and DHA reductase activities.

The proposed scheme for the scavenging systems of superoxide and hydrogen peroxide in chloroplasts is shown in Figure 2. The superoxide anion radicals are produced through the autooxidation of center X or center A/B in the PSI complex within the thylakoid membranes (Figure 1). It is very likely that superoxide dismutase (SOD) is localized at a site where superoxide is ejected from the membranes in the PSI complex, since the hydrogen peroxide produced from superoxide by SOD is immediately scavenged by the stromal thylakoid-bound ascorbate peroxidase (tAPX). Mn SOD has been shown to bind to the thylakoid membranes.[70-72] The thylakoid-bound SOD and the stromal Cu,Zn SOD microlocalized in the peripheral region of the membranes would serve to rapidly disproportionate superoxide. The hydrogen peroxide produced is then reduced to water by tAPX near the PSI complex. Ascorbate is regenerated by ferredoxin, in PSI, from MDA radicals. Thus, SOD, tAPX, and ferredoxin, in or near the PSI complex, would effectively scavenge active species of oxygen produced in PSI. The overall stoichiometry of the thylakoid scavenging system shows no net change of oxygen and electron (the following six reactions).

$$2H_2O \rightarrow O_2 + 4H^+ + 4[e^-] \quad (11)$$

$$2O_2 + 2[e^-] \rightarrow 2O_2^- \quad (21)$$

$$2O_2^- + 2H^+ \rightarrow H_2O_2 + O_2 \quad (8,9)$$

$$H_2O_2 + 2\text{Ascorbate} \rightarrow 2H_2O + 2\text{MDA} \quad (14)$$

$$2\text{Fd} + 2[e^-] \rightarrow 2\text{Fd}_{red} \quad (22)$$

$$2\text{Fd}_{red} + 2\text{MDA} \rightarrow 2\text{Fd} + 2\ \text{Ascorbate} \quad (15)$$

When superoxide and hydrogen peroxide are not scavenged by the thylakoid system on the peripheral region of the thylakoid membranes, they would be scavenged in the stroma. Although little evidence is available for the generation of active species of oxygen in the stroma, it is likely that the autooxidation of stromal components is a source of superoxide which would then be scavenged by the stromal system. In contrast to the thylakoid system, the stromal scavenging system reduces MDA to ascorbate by its reductase using NAD(P)H as the electron donor (Reaction 16) in place of reduced ferredoxin. Furthermore, when MDA is not reduced, DHA is formed by the disproportionation of MDA (Reaction 19) and can be reduced to ascorbate by DHA reductase in the stroma using GSH as the electron donor (Reaction 20). The basic stoichiometry of the stromal scavenging system is the same as that of the thylakoid system since GSH and NAD(P)H are reduced by ferredoxin. The operation of both the thylakoid and stromal scavenging systems shows no net change of dioxygen and is detectable by oxygen exchange only using $^{18}O_2$, but generates the proton gradient in thylakoids.[12]

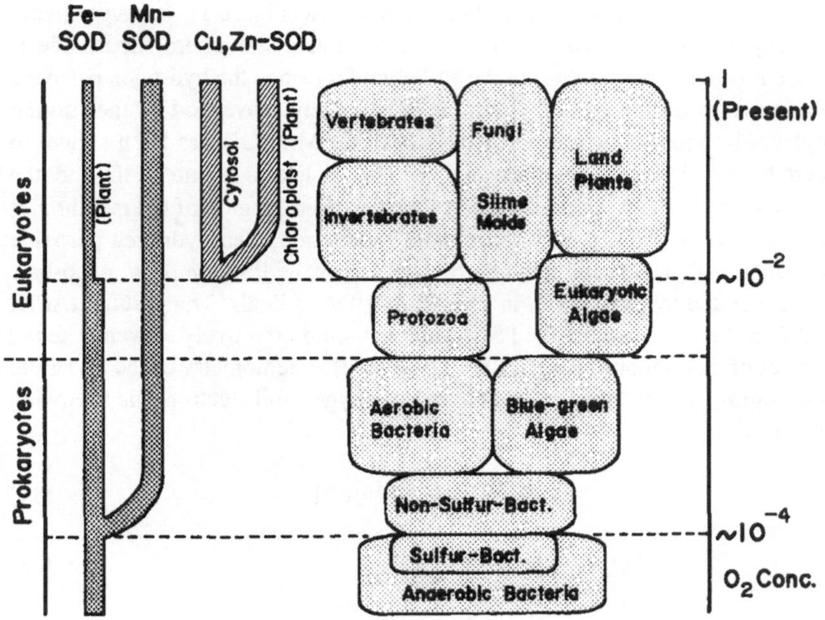

FIGURE 3. Phylogenic distribution of Fe, Mn, and Cu,Zn SODs in organisms at various stages of evolution. The oxygen concentration in the atmosphere relative to that at the present is related to when the indicated organisms appeared on the earth. No Fe SOD has been found in animals or fungi.[30] (From Kanamatsu, S. and Asada, K., *Free Radical Res. Commun.*, 12–13, 383, 1991. With permission.)

IV. EVOLUTION OF PROTECTIVE SYSTEMS IN ALGAE AND PLANTS

The scavenging system for superoxide and hydrogen peroxide in Figure 2 represents that in angiosperm chloroplasts. The operation of a similar scavenging system in other photosynthetic organisms has been deduced by the occurrence of the scavenging enzymes and the light-dependent evolution of either $^{16}O_2$ or $^{18}O_2$ from $H_2{}^{18}O_2$.[12,73]

Superoxide dismutase has even been found in anaerobic photosynthetic bacteria such as *Chromatium*[74] and *Chlorobium*,[75] indicating the necessity of scavenging superoxide radicals generated under extremely low concentrations of oxygen (10^{-4} of the present atmospheric level[30]). In photosynthetic organisms, including the photosynthetic bacteria studied thus far, SOD has been found without exception. There is a characteristic phylogenic and intercellular distribution with respect to Fe-, Mn-, and Cu,Zn-isozymes, as summarized in Figure 3.[30,32] Since a high degree of homology of amino acid sequence exists between Fe and Mn SODs, Mn SOD is supposed to have diverged from Fe SOD before the appearance of cyanobacteria. Cu,Zn SOD was acquired in the most evolved species (*Conjugales* and *Charales*) of the

Chlorophyta[76] and then diverged into the chloroplastic and cytosolic Cu,Zn SODs in land plants,[33] as is also the case for ascorbate peroxidase (see Section III.C). The concentration of SOD is 10 to 50 μM in cyanobacterial cells and angiosperm chloroplasts.

The scavenging system for hydrogen peroxide has been only poorly studied in photosynthetic bacteria. Cyanobacteria are divided into two groups for the scavenging of hydrogen peroxide.[73] One group evolves only $^{18}O_2$ upon the addition of $H_2{}^{18}O_2$ to the illuminated cells. This is not effected by either DCMU or light, indicating that hydrogen peroxide is scavenged by catalase alone. Another group of cyanobacteria evolves $^{16}O_2$ under the same conditions, indicating the reduction of hydrogen peroxide to water via a peroxidase reaction. These observations suggest that the peroxidase-dependent scavenging system of hydrogen peroxide was acquired during the evolution of cyanobacteria to adapt to an increase in the atmospheric concentration of oxygen. Since ascorbate-defective cyanobacteria decompose hydrogen peroxide by a peroxidase reaction and an ascorbate-specific peroxidase has not been found, the electron donor produced and the peroxidase in cyanobacteria remain to be determined.

In *Euglena* cells, ascorbate peroxidase is localized in the cytosol,[77] but the $H_2{}^{18}O_2$ test showed that reduction of hydrogen peroxide occurred via a peroxidase reaction using a photoreductant as the electron donor.[73] *Chlamydomonas* also contains ascorbate peroxidase, but its expression is suppressed upon the addition of selenium, with an increase in glutathione peroxidase activity.[78] In addition to these peroxidases, cytochrome *c* peroxidase activity has been found in *Chlamydomonas*.[101] The $H_2{}^{18}O_2$ test indicates that a photoreductant-peroxidase system also operates in this alga.[73] Thus, in eukaryotic algae, the hydrogen peroxide-scavenging peroxidase seems to occur in several forms, depending on the species and probably on environmental conditions, but the electron donors used for scavenging appear to be photoreductants. In gymnosperms and angiosperms, ascorbate peroxidase is a ubiquitous peroxidase for the scavenging of hydrogen peroxide, and little or no activity of glutathione peroxidase, cytochrome *c* peroxidase, and NAD(P)H peroxidase have been found. In contrast to superoxide dismutase, the hydrogen peroxide-scavenging peroxidases have diverged among organisms, as summarized in Figure 4, with respect to their electron donors and prosthetic groups.

V. INHIBITION OF PHOTOSYNTHESIS BY ACTIVE OXYGEN SPECIES

When superoxide and hydrogen peroxide are not scavenged in chloroplasts, the metal-catalyzed Haber-Weiss reaction (Reaction 7) produces the hydroxyl radical. Because of the high reactivity of the hydroxyl radical with almost any cell components at nearly diffusion-controlled rates (10^8 to 10^9

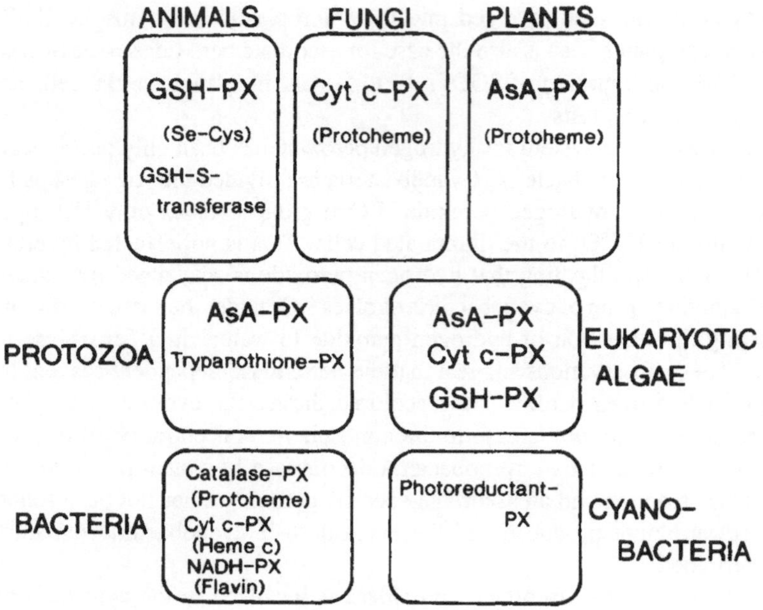

FIGURE 4. Phylogenic distribution of hydrogen peroxide-scavenging peroxidases in organisms at various stages of evolution. AsA, ascorbate; Cyt *c*, cytochrome *c;* PX, peroxidase. Prefixes and parentheses represent the electron donors and prosthetic groups of peroxidases, respectively.

M^{-1} s^{-1}), the hydroxyl radical interacts with chloroplast components at the molecular site where it is generated before it can diffuse. In the case of protein, for example, the hydroxyl radical is oxidized at the specific residue where the radical is generated.[79] A specific scavenging system, therefore, is impossible for the hydroxyl radical.

A. TARGET MOLECULES IN THE STROMA

Hydrogen peroxide rapidly inactivates photosynthesis at low concentrations; a 50% inhibition of CO_2 fixation has been observed with 10 μM H_2O_2 added before illumination.[80] When hydrogen peroxide is generated by thylakoids at a rate which leads to about 10 nM s^{-1} detectable outside intact chloroplasts, CO_2 fixation is inhibited by 90%.[22] Because hydrogen peroxide, at these low concentrations, does not affect electron transport, the inhibition of photosynthesis occurs mainly via the inactivation of the enzymes of the CO_2-fixation cycle. Considering a production rate of hydrogen peroxide (120 μM s^{-1}; see Section II.B) with a steady-state concentration of 500 nM (see Section III.C, estimated assuming a uniform distribution of ascorbate peroxidase in whole chloroplasts), the effective scavenging of hydrogen peroxide is indispensable for photosynthetic activity.[22]

It has been shown that NADP-glyceraldehyde-3-phosphate dehydrogenese,[81] fructose-1,6-bisphosphatase,[82] and ribulose-5-phosphate kinase[81] are very sensitive to hydrogen peroxide because of the oxidation of their functional thiol groups. These enzymes appear to be the target enzymes of hydrogen peroxide in the stroma. The inhibition of any of these enzymes would suppress the operation of the CO_2-fixation cycle in chloroplasts.

Several dehydratases, where the prosthetic group is a nonheme iron, are sensitive to the superoxide anions,[37,83] but this has not yet been shown to be the case for the stromal enzymes. Since superoxide anions have a high reactivity with thiol groups,[84] the stromal thiol-enzymes should be tested as target molecules for superoxide. Catalase, heme peroxidases, and glutathione peroxidase are all inhibited by superoxide anions.[85-87] It has been shown that heme-containing catalase and heme peroxidases are converted to their respective Compound III (an inactive form) by interaction with superoxide. Therefore, it is very likely that ascorbate peroxidase is also converted to an inactive form by superoxide. Furthermore, the superoxide anion is an initiator of the oxidative chain reaction of sulfite[88] and possibly of quinones. Sulfite-induced photooxidative damage is mainly caused by hydroxyl radicals, sulfur trioxide radicals, and sulfate radicals generated by the superoxide-initiated chain reaction of sulfite.[88,89] The scavenging of superoxide is, thus, indispensable for the suppression of photooxidative damage caused by the pollutant, SO_2.

Although catalase is not a stromal protein, its photoinactivation has been shown.[90] The prosthetic heme is considered to be a photosensitizer for the production of singlet oxygen, and hence leaf peroxisomal catalase could suffer this type of inactivation in the light.

B. TARGET MOLECULES IN THYLAKOIDS

In contrast to the stromal enzymes, the thylakoid proteins and electron carriers are comparatively unaffected by superoxide and hydrogen peroxide. At the concentration of hydrogen peroxide which completely inhibits CO_2 fixation, photosynthetic electron transport is not damaged.[4] Under conditions where the photon flux is higher than the capacity of the acceptor system of the chloroplast, a combination of high light and low CO_2 causes inactivation of PSII. The inactivation site has been identified as the D1 protein of the PSII core complex.[91] The photoinactivation of PSII is more rapid under anaerobic conditions than in air,[4,22,92] and active species of oxygen do not participate in the primary step of PSII inactivation.[93]

In conditions of excessive irradiation, the quinone electron acceptor (Q_A) and the pheophytin (Ph) in the PSII reaction center become overreduced. This induces the production of triplet-excited reaction-center chlorophyll ($^3P_{680}$) and reduced pheophytin (Ph^-).[93] In the presence of oxygen, singlet excited oxygen ($^1O_2$) is generated by the rapid reaction between ground-state, triplet oxygen ($^3O_2$) and $^3P_{680}$. The $^1O_2$ generated in the reaction center oxidizes the

D1 protein itself or binding pigments, which triggers its attack by a protease, resulting in its degradation.

The lifetime of singlet oxygen (g^1O_2) in water is very short (mean lifetime, 2 μs; mean diffusion distance, 82 nm), although this is longer in hydrophobic environments, such as the thylakoid membranes. Furthermore, since $^1O_2$ shows a high reactivity with dienes and olefines, these compounds are oxidized at the site of $^1O_2$ generation. In proteins, His, Try, Met, and Cys residues are the targets for $^1O_2$. Although the D1 and D2 proteins of the core complex of the PSII reaction center have a high degree of homology to each other in terms of their amino acid sequences, only the D1 protein shows a high turnover under light stress. This specific targeting must be due to a short diffusion distance from the generation site of $^1O_2$.

Thylakoid membranes contain about one molecule of carotenoid per five molecules of chlorophyll. The isolated PSII core complex also contains carotenoids. Carotenoids can quench ^{3}Chl and $^1O_2$ at near-diffusion-controlled rates. They should, thus, be able to protect the thylakoid components, including the D1 protein, from these active molecules. Another form of protection of the D1 protein in chloroplasts arises from its *de novo* synthesis. When the D1 protein suffers oxidative degradation, translation of the precursor of the D1 protein (encoded in *psb*A of chloroplast DNA) is enhanced. The signal for the degradation of the D1 protein arriving at the translation system is not yet known. The mRNA for *psb*A is very stable and may be able to begin translation immediately after the chloroplasts encounter light stress. The translated precursor is processed on the carboxy terminal in the membranes.[94] The mature D1 protein is incorporated into the D1 protein-defective core complex, forming the functional PSII complex.[91] A similar system for repairing oxidatively damaged enzymes in the stroma is provided by the thioredoxin and GSH systems.

VI. CONCLUDING REMARKS

The production of active oxygen species is inevitable in chloroplasts even under favorable environmental conditions. However, several relaxation mechanisms serve to suppress active oxygen generation under conditions where the photon flux is higher than the capacity of the acceptor system. Light intensity-dependent movements of leaf petioles and chloroplasts in cells tend to decrease effective irradiance, thus providing physiological relaxation mechanisms. Photorespiration is also a dissipation mechanism for excess energy. It can also recycle CO_2 to increase the capacity of the system. Furthermore, ferredoxin- and pyridine nucleotide dehydrogenase-dependent cyclic electron transports[95,96] suppress the production of superoxide via the transfer of electrons, generated in PSI, to the intersystem electron carriers. The putative superoxide-dependent cyclic electron transport cycle around PSI (Section II) decreases the release of superoxide from the thylakoid membranes. In addition, the proton gradient across the thylakoid membrane lowers the apparent quantum yield of PSII by either decreasing the transfer of the excitation energy or decreasing water oxidation via acidification of the thylakoid lumen.[98] These mechanisms can be regarded as means of photochemical relaxation during light stress. Under conditions where the electron acceptors to PSI are not available, the reduction of oxygen, the destruction of hydrogen peroxide (Figures 1 and 2), and the operation of cyclic electron transport[99] would all help to generate the proton gradient protecting against photooxidative degradation of the D1 protein. The high photosensitivity of PSII in the absence of oxygen[4,22,92] indicates that the reduction of oxygen and the scavenging system of superoxide and hydrogen peroxide cooperatively protect PSII from light-induced damage.[100]

Even though the production of reactive oxygen species is suppressed by the above relaxation mechanisms, prompt scavenging is essential for photosynthetic activity. When CO_2 fixation is inhibited by active oxygen species, energy transfer to the acceptor system is blocked even if CO_2 is available. This situation brings about photooxidative damage in PSII. The scavenging system for superoxide and hydrogen peroxide, shown in Figure 2, is indispensable for CO_2 fixation in the stroma and for protection of the D1 protein from degradation in the thylakoids. This system was acquired during the evolution of photosynthetic organisms in order to adapt to the oxidative environment produced when water became the electron donor in photosynthesis.

REFERENCES

1. **Mehler, A. H.,** Studies on reactivities of illuminated chloroplasts. I. Mechanism of the reduction of oxygen and other Hill reagents, *Arch. Biochem. Biophys.*, 33, 65, 1951.
2. **Mehler, A. H. and Brown, A. H.,** Studies on reactivities of illuminated chloroplasts. III. Simultaneous photoproduction and consumption of oxygen studied with oxygen isotopes, *Arch. Biochem. Biophys.*, 38, 365, 1952.
3. **Asada, K., Kiso, K., and Yoshikawa, K.,** Univalent reduction of molecular oxygen by spinach chloroplasts on illumination, *J. Biol. Chem.*, 249, 2175, 1974.
4. **Asada, K. and Takahashi, M.,** Production and scavenging of active oxygen in photosynthesis, in *Photoinhibition*, Kyle, D. J., Osmond, C. B. and Arntzen, C. J., Eds., Elsevier, Amsterdam, 1987, 227.
5. **Hodgson, R. A. J. and Raison, J. K.,** Superoxide production by thylakoids during chilling and its implication in the susceptibility of plants to chilling-induced photoinhibition, *Planta*, 183, 222, 1991.
6. **Mauzerall, D. C.,** Determination of oxygen emission and uptake in leaves by pulsed, time resolved photoacoustics, *Plant Physiol.*, 94, 278, 1990.
7. **Reising, H. and Schreiber, U.,** Pulse-modulated photoacoustic measurements reveal strong gas-uptake component at high CO_2-concentrations, *Photosynth. Res.*, 31, 227, 1992.

8. **Goetze, S. C. and Carpentier, R.**, Monitoring oxygen reduction by photosystem I in whole thylakoid membranes using a photoelectrochemical cell, *J. Photochem. Photobiol. B.*, 8, 17, 1990.
9. **Elstner, E. F. and Frommerer, D.**, Production of hydrogen peroxide by photosystem II of spinach chloroplast lamellae, *FEBS Lett.*, 86, 143, 1978.
10. **Turrens, J. F., Alexandre, A., and Lehninger, A. L.**, Ubisemiquinone is the electron donor for superoxide formation by complex III of heart mitochondria, *Arch. Biochem. Biophys.*, 237, 408, 1985.
10a. **Wydrzynski, T., Ångström, J. and Vänngård, T.**, H_2O_2 formation by photosystem II, *Biochim. Biophys. Acta*, 973, 23, 1989.
11. **Mano, J., Takahashi, M., and Asada, K.**, Oxygen evolution from hydrogen peroxide in photosystem II: flash-induced catalytic activity of water-oxidizing photosystem II membranes, *Biochemistry*, 26, 2495, 1987.
12. **Asada, K. and Badger, M. R.**, Photoreduction of $^{18}O_2$ and $H_2\,^{18}O_2$ with a concomitant evolution of $^{16}O_2$ in intact spinach chloroplasts: evidence for scavenging of hydrogen peroxide by peroxidase, *Plant Cell Physiol.*, 25, 1169, 1984.
13. **Takahashi, M. and Asada, K.**, Dependence of oxygen affinity for Mehler reaction on photochemical activity of chloroplast thylakoids, *Plant Cell Physiol.*, 23, 1457, 1982.
14. **Asada, K. and Nakano, Y.**, Affinity for oxygen in photoreduction of molecular oxygen and scavenging of hydrogen peroxide in spinach chloroplasts, *Photochem. Photobiol.*, 28, 917, 1978.
15. **Takahashi, M. and Asada, K.**, Superoxide production in aprotic interior of chloroplast thylakoids, *Arch. Biochem. Biophys.*, 267, 714, 1988.
16. **Golbeck, J. H.**, Structure and function of photosystem I, *Annu. Rev. Plant Physiol. Plant Mol. Biol.*, 43, 293, 1992.
17. **Ohoka, H., Takahashi, Y., and Matsubara, H.**, Topological considerations of 9-kDa polypeptide which contains centers A and B, associated with the 14- and 19-kDa polypeptides in the photosystem I complex of spinach, *Plant Cell Physiol.*, 30, 869, 1989.
18. **Bielski, B. H. J.**, Reevaluation of the spectral and kinetic properties of HO_2 and O_2^- free radicals, *Photochem. Photobiol.*, 28, 645, 1978.
19. **Schreiber, U., Reising, H., and Neubauer, C.**, Contrasting pH optima of light-driven O_2- and H_2O_2-reduction in spinach chloroplasts as measured via chlorophyll fluorescence quenching, *Z. Naturforsch. Teil C*, 46, 173, 1991.
20. **Hormann, H., Neubauer, C., Asada, K., and Schreiber, U.**, Intact chloroplasts display pH 5 optimum of O_2-reduction in the absence of methyl viologen: evidence for a regulatory role of superoxide protonation, *Photosynth. Res.*, in press.
21. **Takahashi, M. and Asada, K.**, Superoxide permeability of phospholipid membranes and chloroplast thylakoids, *Arch. Biochem. Biophys.*, 226, 558, 1983.
22. **Asada, K.**, Production and scavenging of active oxygen in chloroplasts, in *Molecular Biology of Free Radical Scavenging Systems*, Scandalios, J. G., Ed., Cold Spring Harbor Laboratory Press, Cold Spring Harbor, NY, 1992, 173.
23. **Tanaka, K., Takahashi, M., and Asada, K.**, Isolation of monomeric cytochrome *f* from Japanese radish and a mechanism of autoreduction, *J. Biol. Chem.*, 253, 7397, 1978.
24. **Takahashi, M., Kono, Y., and Asada, K.**, Reduction of plastocyanin with superoxide and superoxide dismutase-dependent oxidation of plastocyanin by hydrogen peroxide, *Plant Cell Physiol.*, 21, 1431, 1980.
25. **Hosein, B. and Palmer, G.**, The kinetics and mechanism of reaction of reduced ferredoxin by molecular oxygen and its reduced products, *Biochem. Biophys. Acta*, 723, 383, 1983.
26. **Furbank, R. T. and Badger, M. R.**, Photoreduction of oxygen in mesophyll chloroplasts of C_4 plants. A model system for studying an *in vitro* Mehler reaction, *Plant Physiol.*, 73, 1038, 1983.

27. **Elstner, E. F., Saran, M., Bors, W., and Legfelder, E.,** Oxygen activation in isolated chloroplasts. Mechanism of ferredoxin-dependent ethylene formation from methionine, *Eur. J. Biochem.*, 89, 61, 1978.
28. **Dykens, J. A., Shick, J. M., Benoit, C., Buettner, G. R., and Winston, G. W.,** Oxygen radical production in the sea anemone *Anthopleura elegantissima* and its endosymbiotic algae, *J. Exp. Bot.*, 168, 9, 1992.
29. **Sharp, K., Fine, R., and Honig, B.,** Computer simulation of the diffusion of a substrate to an active site of an enzyme, *Science*, 236, 1460, 1987.
30. **Asada, K., Kanematsu, S., Okada, S., and Hayakawa, T.,** Phylogenic distribution of three types of superoxide dismutase in organisms and cell organelles, in *Chemical and Biochemical Aspects of Superoxide Dismutase*, Bannister, J. V. and Hill, H. A. O., Eds., Elsevier/North-Holland, Amsterdam, 1980, 136.
31. **de Jesus, M. D., Tabatabai, F., and Chapman, D. J.,** Taxonomic distribution of copper-zinc superoxide dismutase in green algae and its phylogenic importance, *J. Phycol.*, 25, 767, 1989.
32. **Kanamatsu, S. and Asada, K.,** Chloroplast and cytosol isozymes of CuZn-superoxide dismutase: their characteristic amino acid sequences, *Free Radical Res. Comm.*, 12–13, 383, 1991.
32a. **Hayakawa, T., Kanematsu, S., and Asada, K.,** Purification and characterization of thylakoid-bound Mn superoxide dismutase in spinach chloroplasts, *Planta*, 166, 111, 1985.
32b. **Okada, S., Kanematsu, S., and Asada, K.,** Intracellular distribution of manganic and ferric superoxide dismutases in blue-green algae, *FEBS Lett.*, 103, 106, 1979.
33. **Kanematsu, S. and Asada, K.,** Characteristic amino acid sequences of chloroplast and cytosol isozymes of CuZn-superoxide dismutase in spinach, rice, and horsetail, *Plant Cell Physiol.*, 31, 99, 1990.
34. **Kitagawa, Y., Tanaka, N., Hata, Y., Kusunoki, M., Lee, G.-P., Katsube, Y., Asada, K., Aibara, S., and Morita, Y.,** Three dimensional structure of CuZn-superoxide dismutase from spinach at 2.0 A resolution, *J. Biochem.*, 109, 477, 1991.
35. **Asada, K.,** Superoxide dismutase, in *Metalloproteins*, Otsuka, S. and Yamanaka, T., Eds., Elsevier, Amsterdam, 1988, 91.
36. **Getzoff, E. D., Cabelli, D. E., Fisher, C. L., Parge, H. E., Viezzoli, M. S., Banci, L., and Hallewell, R. A.,** Faster superoxide dismutase mutants designed by enhancing electrostatic guidance, *Nature (London)*, 358, 347, 1992.
37. **Gardner, P. and Fridovich, I.,** Inactivation-reactivation of aconitase in *Escherichia coli*. A sensitive measure of superoxide radical, *J. Biol. Chem.*, 267, 8757, 1992.
38. **Bowler, C., Van Montagu, M., and Inze, D.,** Superoxide dismutase and stress tolerance, *Annu. Rev. Plant Physiol. Plant Mol. Biol.*, 43, 83, 1992.
39. **Newcomb, E. H. and Frederick, S. E.,** Distribution and structure of plant microbodies (peroxisomes), in *Photosynthesis and Photorespiration*, Hatch, M. D., Osmond, C. B., and Slatyer, R. O., Eds., Interscience, New York, 1971, 442.
39a. **Scandalios, J. G.,** Regulation of the antioxidant defense genes Cat and Sod of maize, in *Molecular Biology of Free Radical Scavenging Systems*, Scandalios, J. G., Ed., Cold Spring Harbor Laboratory Press, Cold Spring Harbor, NY, 1992, 117.
40. **Nakano, Y. and Asada, K.,** Hydrogen peroxide is scavenged by ascorbate-specific peroxidase in spinach chloroplasts, *Plant Cell Physiol.*, 22, 867, 1981.
41. **Chen, G.-X. and Asada, K.,** Ascorbate peroxidase in tea leaves: occurrence of two isozymes and the differences in their enzymatic properties, *Plant Cell Physiol.*, 30, 987, 1989.
42. **Dalton, D. A., Hanus, F. J., Russel, S. A., and Evans, H. J.,** Purification, properties and distribution of ascorbate peroxidase in legume root nodules, *Plant Physiol.*, 83, 789, 1987.

43. **Mittler, R. and Zilinskas, B. A.,** Purification and characterization of pea cytosolic ascorbate peroxidase, *Plant Physiol.*, 97, 962, 1991.
44. **Elisa, M. R., Borraccino, G., and Dipierro, S.,** Soluble ascorbate peroxidase from potato tubers, *Plant Sci.*, 85, 17, 1992.
45. **Nakano, Y. and Asada, K.,** Purification of ascorbate peroxidase in spinach chloroplasts: its inactivation in ascorbate-depleted medium and reactivation by monodehydroascorbate radical, *Plant Cell Physiol.*, 28, 131, 1987.
46. **Miyake, C. and Asada, K.,** Thylakoid-bound ascorbate peroxidase in spinach chloroplasts and photoreduction of its primary oxidation product, monodehydroascorbate radicals, in thylakoids, *Plant Cell Physiol.*, 33, 541, 1992.
47. **Asada, K.,** Ascorbate peroxidase — hydrogen peroxide-scavenging enzyme in plants, *Physiol. Plant.*, 85, 235, 1992.
48. **Chen, G.-X., Sano, S., and Asada, K.,** The amino acid sequence of ascorbate peroxidase from tea has a high degree of homology to that of cytochrome *c* peroxidase from yeast, *Plant Cell Physiol.*, 33, 109, 1992.
49. **Hossain, M. A. and Asada, K.,** Inactivation of ascorbate peroxidase in spinach chloroplasts on dark addition of hydrogen peroxide: its protection by ascorbate, *Plant Cell Physiol.*, 25, 1285, 1984.
50. **Mittler, R. and Zilinskas, B. A.,** Molecular cloning and characterization of a gene encoding pea cytosolic ascorbate peroxidase, *J. Biol. Chem.*, 267, 21802, 1992.
51. **Kubo, A., Saji, H., Tanaka, K., and Kondo, N.,** Cloning and sequencing of a cDNA encoding ascorbate peroxidase from *Arabidopsis thaliana*, *Plant Mol. Biol.*, 18, 691, 1992.
52. **Welinder, K. G.,** Superfamily of plant, fungal and bacterial peroxidases, *Curr. Opinion Struct. Biol.*, 2, 388, 1992.
53. **Miyake, C., Cao, W.-H., and Asada, K.,** Molecular properties of thylakoid-bound ascorbate peroxidase in spinach chloroplasts, *Plant Cell Physiol.*, 34, 1993.
54. **Kow, Y. W., Erbes, D. L., and Gibbs, M.,** Chloroplast respiration: a means of supplying pyridine nucleotide for dark chloroplastic metabolism, *Plant Physiol.*, 69, 442, 1982.
55. **Chen, G.-X. and Asada, K.,** Inactivation of ascorbate peroxidase by thiols requires hydrogen peroxide, *Plant Cell Physiol.*, 33, 117, 1992.
56. **Chen, G.-X. and Asada, K.,** Hydroxyurea and *p*-aminophenol are the suicide inhibitors of ascorbate peroxidase, *J. Biol. Chem.*, 265, 2775, 1990.
57. **Schreiber, U. and Neubauer, C.,** O_2-dependent electron flow, membrane energization and the mechanism of non-photochemical quenching of chlorophyll fluorescence, *Photosynth. Res.*, 25, 279, 1990.
58. **Miyake, C. and Asada, K.,** Ferredoxin-dependent reduction of monodehydroascorbate radicals in chloroplasts, *Plant Cell Physiol.*, 34, 43(s), 1993.
59. **Nakano, Y. and Asada, K.,** Spinach chloroplasts scavenge hydrogen peroxide on illumination, *Plant Cell Physiol.*, 21, 1295, 1980.
60. **Hossain, M. A., Nakano, Y., and Asada, K.,** Monodehydroascorbate reductase in spinach chloroplasts and its participation in regeneration of ascorbate for scavenging hydrogen peroxide, *Plant Cell Physiol.*, 25, 385, 1984.
61. **Hossain, M. A. and Asada, K.,** Monodehydroascorbate reductase from cucumber is a flavin adenine dinucleotide enzyme, *J. Biol. Chem.*, 260, 12920, 1985.
62. **Dalton, D. A., Langeberg, L., and Robbins, M.,** Purification and characterization of monodehydroascorbate reductase from soybean root nodules, *Arch. Biochem. Biophys.*, 292, 281, 1992.
63. **Bielski, B. H. J.,** Chemistry of ascorbic acid radicals, in *Ascorbic Acid: Chemistry, Metabolism and Uses*, Seib, P. A. and Tolbert, B. M., Eds., American Chemical Society, Washington, D.C., 1982, 81.
64. **Winterbourn, C. C.,** Superoxide as an intracellular radical sink, *Free Radical Biol. Med.*, 14, 85, 1993.

65. **Formi, L. G., Mönig, J., Mora-Arellano, V. O., and Willson, R. L.,** Thiyl free radicals: direct observations of electron transfer reactions with phenothiazines and ascorbate, *J. Chem. Soc., Perkin Trans. 2*, 961, 1983.
66. **Demmig-Adams, B. and Adams, W. W., III,** Photoprotection and other responses of plants to high light stress, *Annu. Rev. Plant Physiol. Plant Mol. Biol.*, 43, 599, 1992.
67. **Hossain, M. A. and Asada, K.,** Purification of dehydroascorbate reductase from spinach and its characterization as a thiol enzyme, *Plant Cell Physiol.*, 25, 85, 1984.
68. **Wells, W. W., Xu, D. P., Yang, Y., and Rocque, P. A.,** Mammalian thioltransferase (glutaredoxin) and protein disulfide isomerase have dehydroascorbate reductase activity, *J. Biol. Chem.*, 265, 15361, 1990.
69. **Ahn, B.-Y. and Moss, B.,** Glutaredoxin homologue encoded by vaccina virus is a virion-associated enzyme with thioltransferase and dehydroascorbate reductase activities, *Proc. Natl. Acad. Sci. U.S.A.*, 89, 7060, 1992.
70. **Hayakawa, T., Kanematsu, S., and Asada, K.,** Purification and characterization of thylakoid-bound Mn-superoxide dismutase in spinach chloroplasts, *Planta*, 166, 111, 1985.
71. **Okada, S., Kanematsu, S., and Asada, K.,** Intracellular distribution of manganic and ferric superoxide dismutase in blue-green algae, *FEBS Lett.*, 103, 106, 1979.
72. **Herbert, S. K., Samson, G., Fork, D. C., and Laudenbach, D. E.,** Characterization of damage to photosystem I and II in a cyanobacterium lacking detectable iron superoxide dismutase activity, *Proc. Natl. Acad. Sci. U.S.A.*, 89, 8716, 1992.
73. **Miyake, C., Michihata, F., and Asada, K.,** Scavenging of hydrogen peroxide in prokaryotic and eukaryotic algae: acquisition of ascorbate peroxidase during the evolution of cyanobacteria, *Plant Cell Physiol.*, 32, 33, 1991.
74. **Kanematsu, S. and Asada, K.,** Superoxide dismutase from an anaerobic photosynthetic bacterium, *Chromatium vinosum, Arch. Biochem. Biophys.*, 185, 473, 1978.
75. **Kanematsu, S. and Asada, K.,** Crystalline ferric superoxide dismutase from an anaerobic green sulfur bacterium, *Chlorobium thiosulfatophilum, FEBS Lett.*, 91, 94, 1978.
76. **Kanematsu, S. and Asada, K.,** CuZn-superoxide dismutases from the fern *Equisetum arvense* and the green alga *Spirogya* sp.: occurence of chloroplast and cytosol types of enzyme, *Plant Cell Physiol.*, 30, 717, 1989.
77. **Shigeoka, S., Nakano, Y., and Kitaoka, S.,** Metabolism of hydrogen peroxide in *Euglena gracilis* z. by L-ascorbic acid peroxidase, *Biochem. J.*, 186, 377, 1980.
78. **Yokota, A., Shigeoka, S., Onishi, T., and Kitaoka, S.,** Selenium as inducer of glutathione peroxidase in low-CO_2-grown *Chlamydomonas reinhardtii, Plant Physiol.*, 86, 649, 1988.
79. **Farber, J. M. and Levine, R. L.,** Sequence of a peptide susceptible to mixed-function oxidation: probable cation binding site in glutamine synthetase, *J. Biol. Chem.*, 261, 4547, 1986.
80. **Kaiser, W.,** The effect of hydrogen peroxide on CO_2 fixation of isolated intact chloroplasts, *Biochim. Biophys. Acta*, 440, 476, 1976.
81. **Tanaka, K., Otsubo, T., and Kondo, N.,** Participation of hydrogen peroxide in the inactivation of Calvin-cycle SH enzymes in SO_2-fumigated spinach leaves, *Plant Cell Physiol.*, 23, 1009, 1982.
82. **Charles, B. and Halliwell, B.,** Light activation of fructose bisphosphatase in isolated spinach chloroplasts and a deactivation by hydrogen peroxide — a physiological role for the thioredoxin system, *Planta*, 151, 242, 1981.
83. **Gardner, P. R. and Fridovich, I.,** Superoxide sensitivity of the *Escherichia coli* 6-phosphogluconate dehydratase, *J. Biol. Chem.*, 266, 1478, 1991.
84. **Asada, K. and Kanematsu, S.,** Reactivity of thiols with superoxide radicals, *Agric. Biol. Chem.*, 40, 1891, 1976.
85. **Kono, Y. and Fridovich, I.,** Superoxide radical inhibits catalase, *J. Biol. Chem.*, 257, 5751, 1982.

86. **Kono, Y. and Fridovich, I.,** Inhibition and reactivation of Mn-catalase: implication for valence changes at the active site manganese, *J. Biol. Chem.*, 258, 13646, 1983.
87. **Blum, J. and Fridovich, I.,** Inactivation of glutathione peroxidase by superoxide radical, *Arch. Biochem. Biophys.*, 240, 500, 1985.
88. **Asada, K. and Kiso, K.,** Initiation of aerobic oxidation of sulfite by illuminated spinach chloroplasts, *Eur. J. Biochem.*, 33, 253, 1973.
89. **Mottley, C. and Mason, R. P.,** Sulfate ion free radical formation by the peroxidation of (Bi) sulfite and its reaction with hydroxyl radical scavengers, *Arch. Biochem. Biophys.*, 257, 681, 1988.
90. **Feierabend, J. and Enger, S.,** Photoinactivation of catalase in vitro and in leaves, *Arch. Biochem. Biophys.*, 251, 567, 1986.
91. **Kyle, D. J.,** The biochemical basis for photoinhibition of photosystem II, in *Photoinhibition,* Kyle, D. J., Osmond, C. B., and Artnzen, C. J., Eds., Elsevier, Amsterdam, 1987, 197.
92. **Krause, G. H. and Cornic, G.,** Co_2 and O_2 interactions in photoinhibition, in *Photoinhibition,* Kyle, D. J., Osmond, C. B., and Arntzen, C. J., Eds., Elsevier, Amsterdam, 1987, 169.
93. **Andersson, B., Salter, A. H., Virgin, I., Vass, I., and Styring, S.,** Photodamage to photosystem II — primary and secondary events, *J. Photochem. Photobiol. B.*, 15, 15, 1992.
94. **Takahashi, M., Shiraishi, T., and Asada, K.,** COOH terminal residues of D1 and the 44-kDa CPa-2 at spinach photosystem II core complex, *FEBS Lett.*, 240, 6, 1988.
95. **Mi, H., Endo, T., Schreiber, U., Ogawa, T., and Asada, K.,** Electron donation from cyclic and respiratory flows to the photosynthetic intersystem chain is mediated by pyridine nucleotide dehydrogenase in the cyanobacterium *Synechocystis* PCC 6803, *Plant Cell Physiol.*, 33, 1233, 1992.
96. **Asada, K., Heber, U., and Schreiber, U.,** Electron flow to the intersystem chain from stromal components and cyclic electron flow in maize chloroplasts, as detected in intact leaves by monitoring redox change of P700 and chlorophyll fluorescence, *Plant Cell Physiol.*, 34, 39, 1993.
97. **Ruban, A. V., Rees, D., Pascal, A. A., and Horton, P.,** Mechanism of pH-dependent dissipation of absorbed excitation energy by photosynthetic membranes. II. The relationship between LHCII aggregation in vitro and qE in isolated thylakoids, *Biochim. Biophys. Acta,* 1102, 39, 1992.
98. **Krieger, A., Moya, I., and Weis, E.,** Energy-dependent quenching of chlorophyll a fluorescence: effect of pH on stationary fluorescence and picosecond-relaxation kinetics in thylakoid membranes and photosystem II preparations, *Biochim. Biophys. Acta,* 1102, 167, 1992.
99. **Heber, U. and Walker, D.,** Concerning a dual function of coupled cyclic electron transport in leaves, *Plant Physiol.*, 100, 1621, 1992.
100. **Neubauer, C. and Yamanoto, H. Y.,** Mehler-peroxidase reaction mediates zeaxanthine formation and zeaxanthine-related fluorescence quenching in intact chloroplasts, *Plant Physiol.*, 99, 1354, 1992.
101. **Miyake, C. and Asada, K.,** unpublished observations.

Chapter 4

LIGHT STRESS AND PHOTOPROTECTION RELATED TO THE XANTHOPHYLL CYCLE

Barbara Demmig-Adams and William W. Adams, III

TABLE OF CONTENTS

0-8493-5443-9/94/$0.00 + $.50

I. INTRODUCTION

Carotenoids have long been associated with a role in the photoprotection of the photosynthetic apparatus.[1-4] The precise function of each of the carotenoids present in plants has not, however, been clearly established. This chapter describes the involvement of a group of carotenoids, those of the xanthophyll cycle, in a novel photoprotective process, the dissipation of excess excitation energy in the light-harvesting chlorophyll pigment bed. The three xanthophylls of this xanthophyll cycle, violaxanthin, antheraxanthin, and zeaxanthin, are the only carotenoids of the photosynthetic membrane that undergo very rapid and reversible changes in their concentration in response to light stress.

It has recently been reported [5-7] that the major change in the carotenoid composition of leaves during an acclimation to higher light also concerns the levels of the three xanthophylls involved in the xanthophyll cycle (Figure 1). Therefore, sun leaves possess much larger xanthophyll cycle pools than do shade leaves. This is true of sun and shade leaves of the same species, or even the same plant, as is shown in Figure 1, and the difference is even greater between leaves of species which are typically found growing in shaded vs. sun-exposed habitats. Thus, there is a twofold response of the xanthophyll cycle to light stress: (1) a conversion of the existing pool, mainly to zeaxanthin, within minutes upon exposure of leaves to high light and (2) a slower increase in the pool size of the xanthophyll cycle upon an increase in the growth light regime. The only other carotenoid that also exhibits a considerable increase in its level during acclimation to high light is β-carotene, which is the immediate precursor of the three xanthophylls of the xanthophyll cycle in their biosynthesis.[8,9] In other words, there appears to be a preferential stimulation of β,β-carotenoid synthesis in leaves exposed to high light.[5,7] The other major carotenoids of the photosynthetic apparatus do not show such increases in high light, and α-carotene is usually present in lower levels in high-light-acclimated leaves (Figure 1).

In this chapter we focus on the photoprotective role of the xanthophyll cycle, i.e., those carotenoids that show the strongest accumulation in response to conditions under which photoprotection of the photosynthetic apparatus is required. Several aspects of this involvement of the xanthophyll cycle in photoprotection have also been discussed in other recent reviews.[10-13]

II. THE OPERATION OF THE XANTHOPHYLL CYCLE

In sun-exposed leaves, pronounced diurnal changes in the concentration of the three xanthophylls of the xanthophyll cycle occur in response to changes in incident sunlight (Figure 2).[14,15] Upon an increase in incident sunlight, de-epoxidation of the diepoxide, violaxanthin, to the monoepoxide,

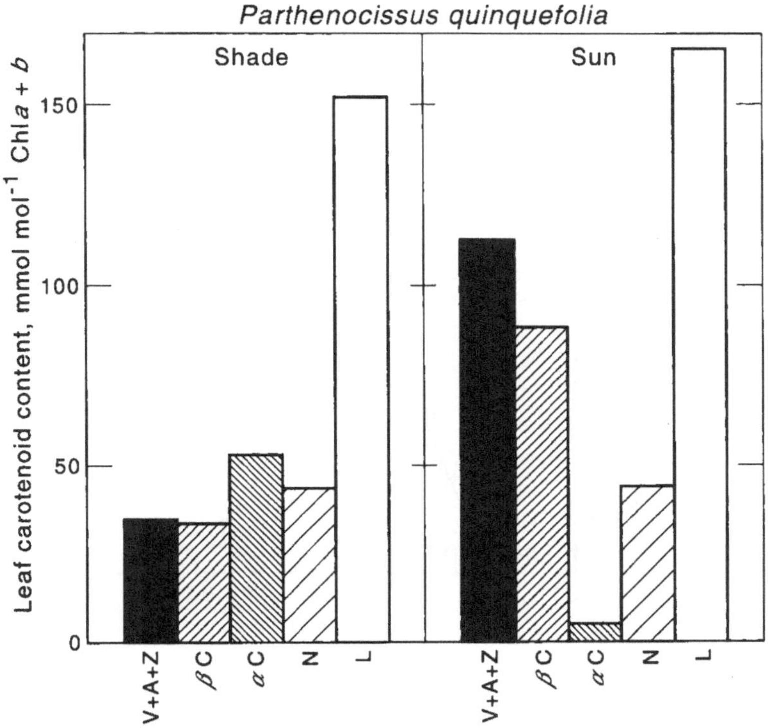

FIGURE 1. Carotenoid contents of shade and sun leaves of Virginia creeper (*Parthenocissus quinquefolia* [L.] Planch.). V + A + Z = violaxanthin, antheraxanthin, and zeaxanthin; βC, β-carotene; αC, α-carotene; N, neoxanthin; L, lutein. The chlorophyll content per leaf area was 433 and 632 μmol (Chl a + b) m^{-2} in sun and shade leaves, respectively. The incident PFD at the time of collection was 1950 for sun and 17 μmol m^{-2} s^{-1} for shade leaves.[5]

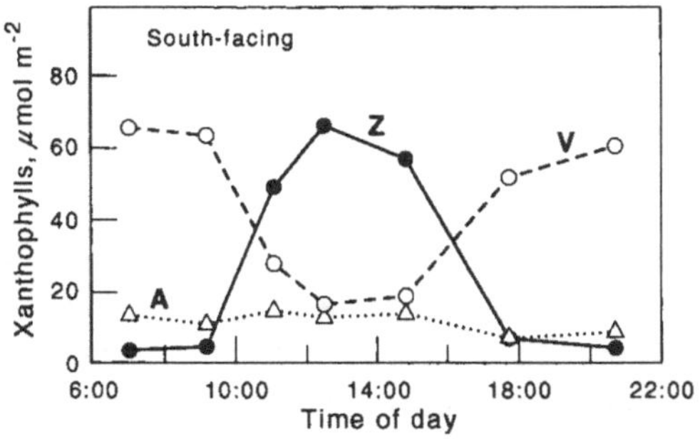

FIGURE 2. Diurnal changes in the composition of the xanthophyll cycle of a sun-exposed (south-facing) leaf of *Euonymus kiautschovicus* Loesener. V, violaxanthin; A, antheraxanthin; Z, zeaxanthin.[15]

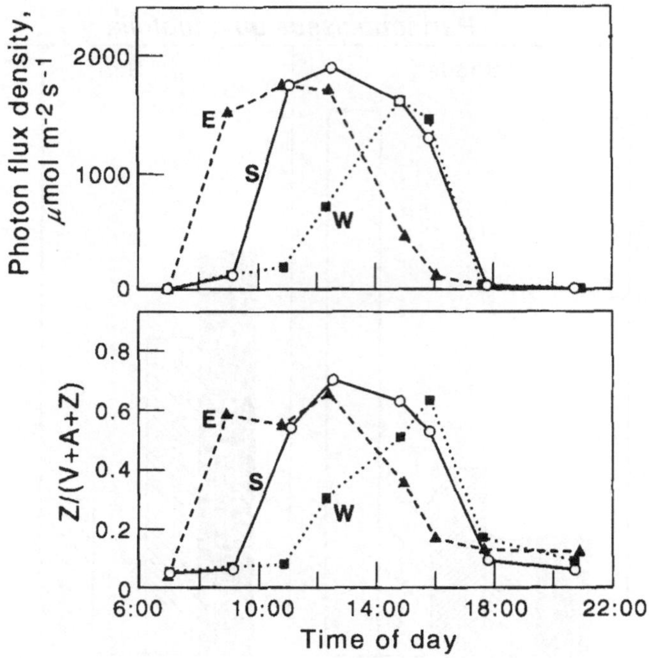

FIGURE 3. Diurnal changes in incident PFD and the fraction of the xanthophyll cycle pool present as zeaxanthin, Z/(V + A + Z), in three leaves of *Euonymus kiautschovicus* with different orientations. E, east-facing; S, south-facing; W, west-facing.[15]

antheraxanthin, and, further, to the epoxide-free form zeaxanthin occurs. The reverse sequence takes place when light levels decrease again. As the intermediate in both the forward and the backward reaction sequences, antheraxanthin does not exhibit the pronounced changes in concentration that one observes in violaxanthin and zeaxanthin (Figure 2). It had previously been shown that under laboratory conditions, leaves of apparently all higher plants undergo such a violaxanthin de-epoxidation when exposed to light levels in excess of those that can be utilized in photosynthesis.[16-19] Furthermore, all species in sun-exposed habitats examined to date undergo a reversible de-epoxidation to zeaxanthin in the course of the day.[14,15] The absolute levels of zeaxanthin formed in a leaf under high light are inversely related to the photosynthetic capacity of that leaf,[5,7] i.e., the capacity to dissipate excitation energy through photosynthetic electron transport. Thus, more zeaxanthin is formed in leaves with a lower rate of photosynthesis and less in leaves with a higher rate of photosynthesis when such leaves absorb the same amount of light.

Maximal levels of zeaxanthin in any given leaf are formed during that portion of the day when the leaf receives maximal exposure to sunlight. This is illustrated in Figure 3, which shows the photon flux density incident upon the upper surface of leaves with three different orientations and the level of zeaxanthin as a fraction of the total xanthophyll cycle pool for those three

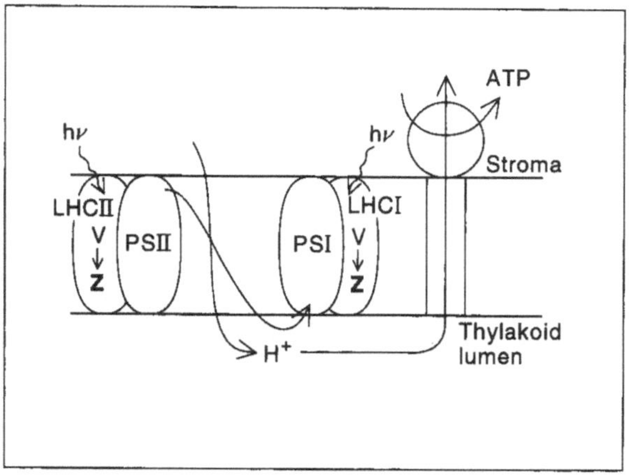

FIGURE 4. Schematic depiction of the localization of the xanthophyll cycle in the light-harvesting pigment complexes of photosystems I and II. In the light, an acidic thylakoid lumen pH is required for zeaxanthin formation (violaxanthin de-epoxidation). The thylakoid lumen pH is a measure of the balance between the generation and consumption of products of electron transport. LHCI, light-harvesting complex of PSI; LHCII, light-harvesting complex of PSII.

leaves. Each leaf exhibits an increase in zeaxanthin content followed by a decrease in zeaxanthin which parallels the diurnal increase and decrease in light absorption by each leaf. This demonstrates the very dynamic nature of the response of the xanthophyll cycle to changes in the level of excess light being absorbed by a leaf.

This response of the xanthophyll cycle to excess light is achieved through regulation of the enzymes responsible for the interconversions of the three xanthophylls. The xanthophyll cycle consists of a de-epoxidation and an epoxidation sequence that are catalyzed by two different enzymes. (For reviews on the biochemistry and regulation of the xanthophyll cycle, see References 20 to 22.) The two main factors involved in promoting the de-epoxidation sequence of the xanthophyll cycle that is catalyzed by violaxanthin de-epoxidase are an acidic pH on the lumen side of the thylakoid membrane (Figure 4) and reduced ascorbate. The reducing equivalents for the reductive de-epoxidation of violaxanthin are apparently supplied by reduced ascorbate,[23] which in turn receives reducing equivalents ultimately from NADPH (see Reference 24). A low pH is required for optimal violaxanthin de-epoxidase activity, since this enzyme has an acidic pH optimum.[25] The intrathylakoid pH is lowered when light induces electron transport-mediated proton transport into the lumen. This transthylakoid pH gradient decreases again when ATP is formed and consumed in carbon metabolism. Thus, the magnitude of the transthylakoid pH gradient, or the intrathylakoid pH, can serve as a quantitative measure of the balance between light absorption/primary photochemistry and carbon metabolism. The back reaction of the xanthophyll cycle, from zeaxanthin to violaxanthin, has a pH optimum at a nonacidic pH of 7.5.[20,22] This epoxidation reaction furthermore requires molecular oxygen and NADPH.[26]

A recent study[61] indicates that the xanthophylls of the xanthophyll cycle are bound to the light-harvesting chlorophyll protein complexes of both photosystem I and II, and that a similar de-epoxidation state is maintained in both photosystems under excess light.

III. INVOLVEMENT OF THE XANTHOPHYLL CYCLE IN THE REGULATION OF THE EFFICIENCY OF PHOTOCHEMISTRY

A. REGULATION OF THE EFFICIENCY OF PSII PHOTOCHEMISTRY AS ASSESSED FROM CHLOROPHYLL FLUORESCENCE

The interpretation of changes in chlorophyll fluorescence emission from leaves has been an essential tool in the identification of a novel regulatory process that allows an adjustment of electron transport rates to the demand from carbon metabolism in an environment of fluctuating light.[11,27,28] This process is the safe dissipation of excess excitation energy directly within the chlorophyll pigment bed.

The understanding of the usefulness of chlorophyll fluorescence measurements has increased immensely in recent years. It had long been recognized that the yield of chlorophyll fluorescence should be inversely related to the rate of photochemistry. However, it had been puzzling that a low yield of actual chlorophyll fluorescence (F) was observed in the light under conditions that allowed high rates of electron transport to proceed (Figure 5A and B) and also under conditions when only minimal rates of electron transport were possible (Figure 5C; see also Reference 29).

New technical developments[30] allow for a determination of the yield of maximal fluorescence (F_m') from a leaf that is absorbing sunlight or light from an actinic light source, i.e., when photosynthetic electron transport and carbon metabolism are operating. This approach demonstrated that the actual fluorescence yield (F) can be lowered by a high rate of photochemistry or, alternatively, by a previously unrecognized energy dissipation process. This process dissipates excess excitation energy directly within the photochemical apparatus under conditions when the products of electron transport are not consumed at the same rate at which they are generated. Maximal fluorescence yield at closed (reduced) PSII centers remains high as long as this energy dissipation activity in the photochemical system is low (Figure 5A and B), and is lowered as this dissipation activity increases (Figure 5C). The energy dissipation activity in the photochemical system increases when electron transport rates decrease, such as in the absence of either CO_2 fixation or photorespiration (or other O_2-dependent reactions), or when increases in absorbed

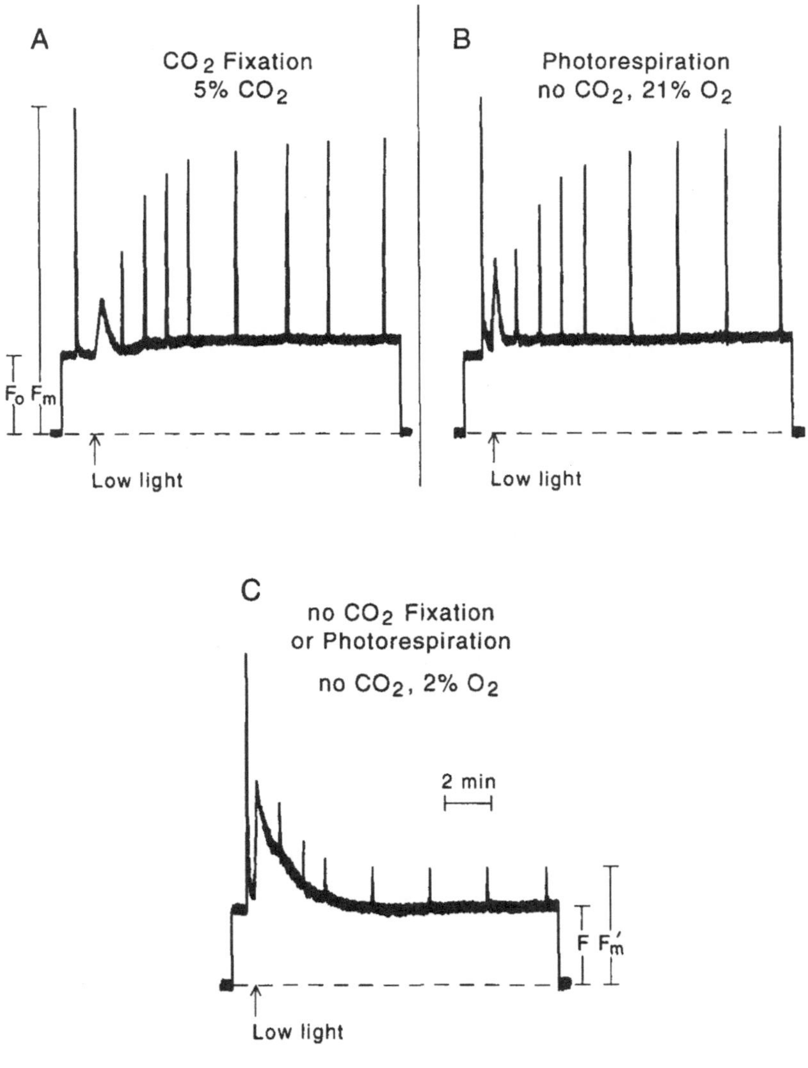

FIGURE 5. Original traces of chlorophyll fluorescence changes from spinach (*Spinacia oleracea* L.) leaves at a low PFD of 200 μmol m^{-2} s^{-1} under conditions allowing either only CO_2 fixation (A, 5% CO_2), only photorespiration and other O_2-dependent reactions (B, 21% O_2, no CO_2), or neither CO_2 fixation nor photorespiration (C, 2% O_2, no CO_2). These spinach leaves were grown at 1100 μmol (photons) m^{-2} s^{-1} under a 10 h/14 h day/night photoperiod in a growth chamber at 24°C/17.5°C day/night temperature. F_o, maximal possible yield of initial fluorescence at open PSII traps in darkness; F_m, maximal possible yield of maximal fluorescence at closed PSII traps in darkness; F, actual fluorescence yield during illumination with actinic light; F_m', actual (and quenched) yield of maximal fluorescence at closed PSII traps during ongoing illumination. The zeaxanthin content of the leaf in C was 53.3 ± 3.0 mmol (zeaxanthin) mol^{-1} Chl *a* + *b*, which corresponds to a ratio of Z/(V + A + Z) of 0.594 ± 0.033. (From Adams, W. W., III and Demmig-Adams, B., unpublished data.)

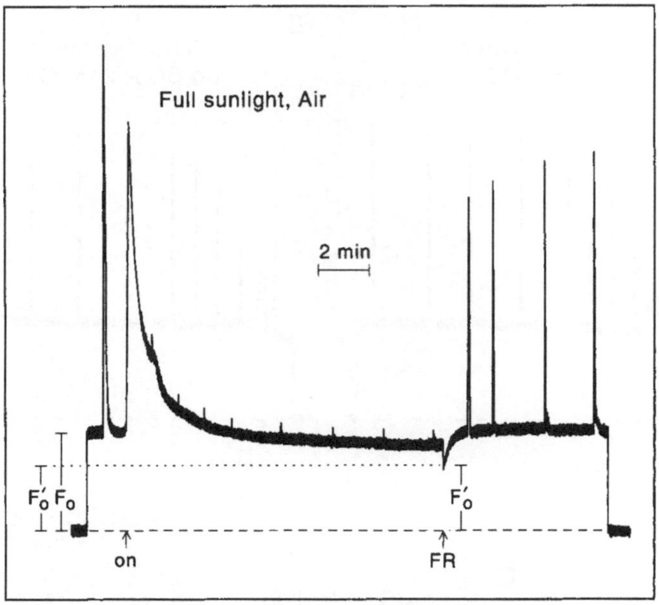

FIGURE 6. Original trace of chlorophyll fluorescence changes from a sun leaf of *Vinca minor* L. (periwinkle) upon exposure to full sunlight in ambient air and a subsequent transfer to very low intensity far-red (FR) light. Newly expanded leaf collected from outside in April 1992 in Boulder, CO. This treatment causes a very strong decrease in the fluorescence yield where the actual fluorescence yield (F) falls below the preexposure F_o. The yield of initial fluorescence during illumination, F_o', is decreased to 69% of the control (F_o) after 12 min of illumination, with F_m' being decreased to 18% of the control (F_m) after this period at a PFD equivalent to full sunlight. For corresponding zeaxanthin content, see Figure 9. (From Adams W. W., III and Demmig-Adams, B., unpublished data.)

energy cannot be matched by concomitant increases in electron transport rates. It has been suggested that this energy dissipation process occurs within the light-harvesting pigment bed of the photosynthetic membrane. Such has been concluded[10,12,18,31-35] from the fact that this process lowers the yield of fluorescence at both open (F_o') and closed (F_m') PSII centers (Figure 6; see Figure 5C for F_m').

The above fluorescence parameters allow a whole array of quantitative estimates of photosynthetic parameters. They permit a quantification of the actual PSII efficiency under illumination and, from that, estimation of the actual rate of PSII photochemistry. Figure 7 provides examples of these parameters in two species, one acclimated to high light (sunflower) and one acclimated to low light *(Oxalis)*. Concomitant measurements of the changes in the rate of photosynthetic O_2 evolution reveal that the latter changes are quantitatively similar to those in PSII activity, as estimated from chlorophyll fluorescence changes (Figure 7). Thus, chlorophyll fluorescence can serve as an indicator of the rate of electron transport under many conditions. The

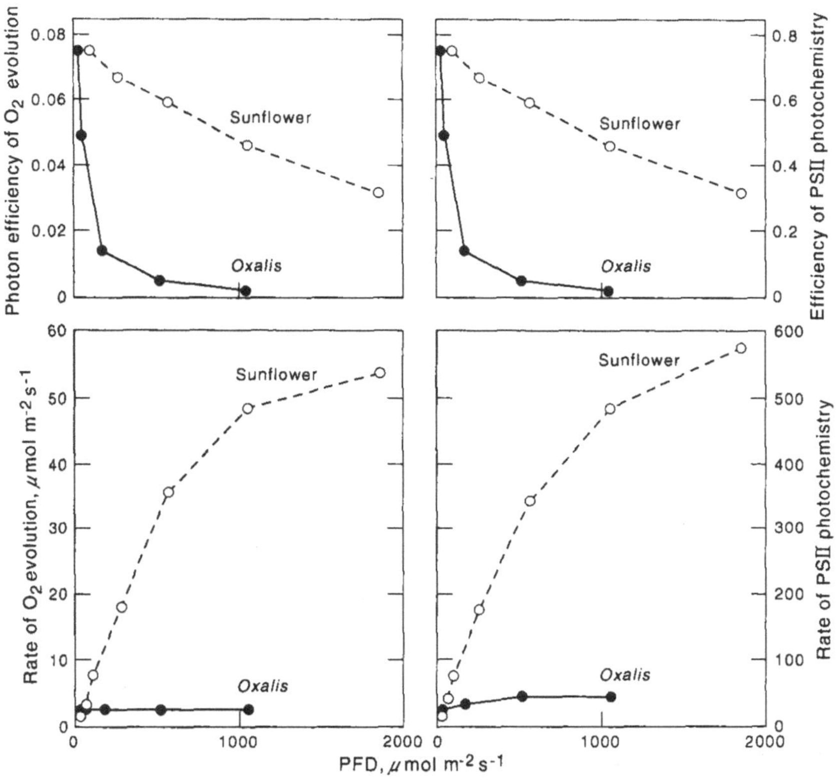

FIGURE 7. Actual photon efficiency of O_2 evolution and PSII photochemistry (upper panels) and rates of O_2 evolution and PSII photochemistry (lower panels) of leaves of sunflower (*Helianthus annuus* L.) and *Oxalis* cf. *trilliifolia* Hook. at various PFDs. The actual efficiency of PSII photochemistry is calculated from chlorophyll fluorescence after $F_m' - F/F_m'$. $F_m' - F/F_m'$ can also be expressed as the product of the intrinsic efficiency of PSII (F_v'/F_m') and the fraction of open PSII centers ($1 - Q_r/Q_t$), where the reduction state of PSII ($Q_r/Q_t = Q_{reduced}/Q_{total}$) is given by $(F - F_o')/(F_m' - F_o')$. The rate of PSII photochemistry is the actual efficiency of PSII photochemistry × PFD. Sunflower was grown at 1300 μmol m^{-2} s^{-1} in a growth chamber under a 10 h/14 h day/night photoperiod at a day/night temperature of 27.5°C/22.5°C. *Oxalis* leaves had developed at room temperature in a shaded location by a window, receiving low levels of natural sunlight of 10 μmol (photons) m^{-2} s^{-1} on average, and not exceeding 50 μmol (photons) m^{-2} s^{-1}. (From Adams, W. W., III and Demmig-Adams, B., unpublished data.)

actual efficiency of PSII photochemistry is quantified from the parameter $(F_m' - F)/F_m'$, as suggested by Genty et al.,[36] and equals the product of PSII photochemical efficiency at open centers in the light (F_v'/F_m') and the percentage of centers that are actually open.

Figure 8 shows the activity of energy dissipation in the pigment bed as quantified from nonphotochemical fluorescence quenching (NPQ), i.e., from the lowering of fluorescence emission from the level of F_m to a new (quenched) level F_m'. This particular quantification uses the Stern-Volmer equation

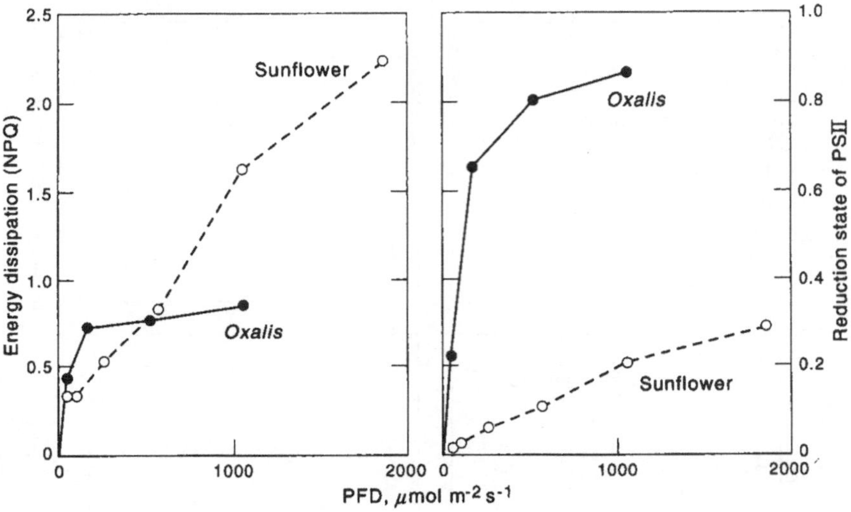

FIGURE 8. Energy dissipation activity in the pigment bed (expressed as nonphotochemical fluorescence quenching, NPQ = $F_m/F_m' - 1$) and reduction state of PSII from leaves of sunflower and *Oxalis* at various PFDs. For growth conditions, see legend of Figure 7. See Figure 9 for corresponding values of zeaxanthin content. (From Adams, W. W., III and Demmig-Adams, B., unpublished data.)

(see Reference 16), after which energy dissipation is given by $F_m/F_m' - 1$ ($F_{m\ unquenched}/F_{m\ quenched} - 1$). This expression is proportional to another parameter that has been used to quantify the dissipation activity, i.e. the rate constant for energy dissipation in the pigment bed as derived by Kitajima and Butler[34] (see also References 10, 12, 18, 31, 32, 37). As light becomes increasingly excessive, the energy dissipation activity in the pigment bed increases. This allows the removal of excess excitation energy and counteracts a potentially dangerous build-up of excitation energy in the photosynthetic system. As a side effect, this dissipation of excitation energy in the pigment bed contributes to the decrease in PSII photochemical efficiency that is observed with increasing PFD (Figure 7), since excitation energy is removed before it reaches the photosynthetic reaction centers. However, if energy dissipation in the pigment bed did not occur, the PSII centers would become more strongly reduced, i.e., more centers would "close" at a given PFD, and the actual PSII efficiency during illumination would decrease to the same extent (see Reference 16), but for a different reason.

A comparison of sun and shade leaves reveals that sun leaves possess a greater capacity not only to utilize light through photosynthesis[38] (Figure 7), but also to dissipate any excess excitation energy directly within the pigment bed (Figure 8). It is this combination of greater photosynthesis rates and greater rates of energy dissipation in the pigment bed that allows sun leaves to maintain a much lower reduction state of PSII (Figure 8) than shade leaves

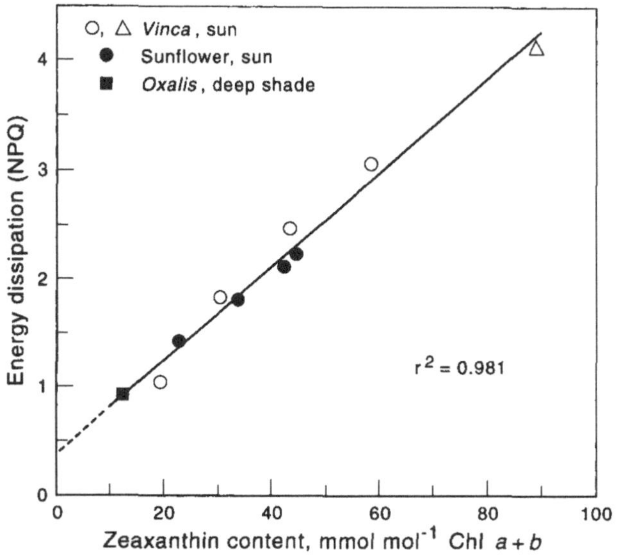

FIGURE 9. Relationship between the zeaxanthin content and the energy dissipation activity in the pigment bed (expressed as NPQ = $F_m/F_m' - 1$) in sun leaves of *Vinca minor* and sunflower as well as deep shade leaves of *Oxalis*. Both parameters were varied through variation of the incident PFD on the leaves. The highest values shown for *Vinca* and *Oxalis* represent the maximal possible, i.e., saturation levels of NPQ and zeaxanthin, which were obtained at 1050 μmol photons $m^{-2} s^{-1}$ in *Oxalis* at 2050 μmol (photons) $m^{-2} s^{-1}$ in *Vinca*. The highest levels of NPQ and zeaxanthin for sunflower were obtained at 2600 μmol (photons) $m^{-2} s^{-1}$ and probably do not represent the maximal possible levels. The control values of maximal fluorescence yield F_m were obtained after 10 min in darkness prior to the high light exposures. (○) From PFD response curve in *Vinca* sun leaves in air; (△) *Vinca* sun leaf from Figure 6 after 12 min at 2050 μmol (photons) $m^{-2} s^{-1}$; (■) *Oxalis* (from Figure 8): 1050 μmol (photons) $m^{-2} s^{-1}$ in 5% CO_2; (●) sunflower (from Figure 8): PFD response curve in 5% CO_2. (From Demmig-Adams, B. and Adams, W. W., III unpublished data.)

under exposure to high light. A high reduction state of PSII is thought to lead to damage to the photosynthetic apparatus.

B. THE XANTHOPHYLL CYCLE AND ENERGY DISSIPATION IN THE PIGMENT BED

Numerous reports[16-18,39-41] have indicated that energy dissipation in the pigment bed in leaves, as quantified from the lowering of the yield of maximum chlorophyll fluorescence, can be greater the higher the zeaxanthin level, or the more de-epoxidized the xanthophyll cycle (Figure 9). While there have been several approaches to quantifying the energy dissipation activity in the pigment bed (see above) that yielded proportional changes in this parameter, several possibilities have also been proposed to quantify the state of the xanthophyll cycle. In addition to using the level of the main product of violaxanthin de-epoxidation, zeaxanthin, on a chlorophyll basis (Figure 9), the epoxidation state of the xanthophyll cycle has also been used.[7,39] This epoxidation state is defined as the fraction of the xanthophyll cycle pool that is present in the form of de-epoxidized components, and thus (Z + 0.5A)/(V + A + Z). At this time it is unknown whether the monoepoxide, antheraxanthin, also has a role similar to that of zeaxanthin in energy dissipation in the pigment bed. However, a recent study indicates that it may.[64]

Correlations similar to the one shown in Figure 9 have been obtained for leaves of a wide range of higher plant species from different growth light environments[18] and for isolated chloroplast preparations.[33,42] The 2 parameters were varied by altering the PFD[16-18], or at a constant PFD, through changes in photosynthetic activity in response to changes in the availability of CO_2 and O_2[40,43] (see also Figure 5). Furthermore, a similar concomitant increase in energy dissipation activity and level of zeaxanthin was obtained when plants were subjected to long-term water stress in constant high light.[12,41]

Evidence for an involvement of the xanthophyll cycle in this energy dissipation process comes from studies using an inhibitor of violaxanthin de-epoxidation[16,17,33,44-46] as well as studies using photosynthetic systems that form zeaxanthin extremely slowly.[45,47,48] These studies indicated that prevention of violaxanthin de-epoxidation (and/or a very slow formation of zeaxanthin) was associated with a prevention of the rapid development of the nonphotochemical fluorescence quenching that is indicative of energy dissipation in the pigment bed.[12] The inhibitor of violaxanthin de-epoxidation used in these studies, dithiothreitol (DTT), inhibits the violaxanthin de-epoxidase[49] but does not prevent NPQ development in leaves that contain and retain zeaxanthin subsequent to a treatment with DTT.[62]

C. CONTROL OF ENERGY DISSIPATION IN THE PIGMENT BED THROUGH pH-RELATED CHANGES

Although the presence of zeaxanthin, or a de-epoxidized state of the xanthophyll cycle, is a necessary precondition for energy dissipation to occur in the pigment bed of leaves, a second factor is also required. The presence of both zeaxanthin and a low thylakoid lumen pH are necessary for the actual development of NPQ.[33,46] Thus, a low lumen pH is required for the biochemical conversion of violaxanthin to zeaxanthin in the xanthophyll cycle as well as for the actual development of the zeaxanthin-associated energy dissipation process (Figure 10). This pH requirement for the actual development of NPQ had been recognized from the studies of Krause et al.[50] and was referred to as "pH-dependent" or "high-energy state" nonphotochemical fluorescence quenching. A linear relationship between NPQ ($F_m/F_m' - 1$) and the intrathylakoid pH has been reported for isolated chloroplast preparations in which the majority of NPQ was associated with the xanthophyll cycle.[33] The observation that the zeaxanthin-associated NPQ requires activation through a low lumen pH can explain the fact that zeaxanthin-containing leaves often show a low fluorescence yield only under actual exposure to an

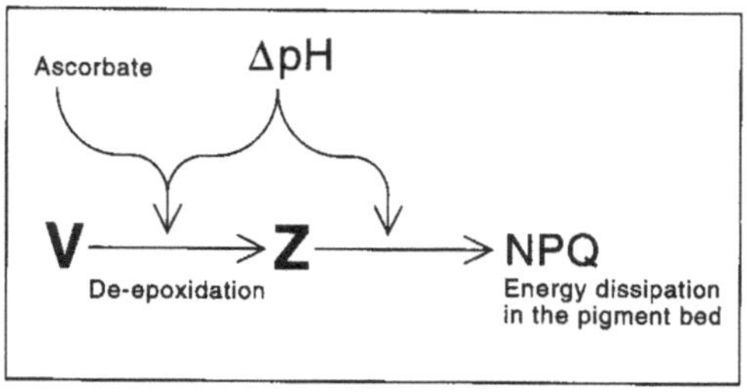

FIGURE 10. Schematic depiction of the regulation of zeaxanthin formation (via de-epoxidation of violaxanthin) and the actual development of zeaxanthin-associated NPQ.[46]

excess of light.[43,44] These zeaxanthin-containing leaves can show a rapid increase in the fluorescence yield upon darkening that precedes the removal of zeaxanthin.

NPQ has been observed to develop more rapidly in leaves[43] or chloroplasts[33] that contained considerable levels of zeaxanthin prior to the actual illumination period. In an elegant study by Gilmore and Yamamoto,[46] it was shown that in chloroplasts that had retained zeaxanthin, NPQ could be induced even in darkness, by treatments with ATP that resulted in the build-up of a transthylakoid pH gradient in darkness (via reverse proton pumping through the ATP synthase). Under some conditions, the reversal of high-light induced zeaxanthin-associated NPQ upon return to darkness or low light can be quite slow, on the order of hours instead of minutes. Gilmore and Yamamoto[46] have suggested that such a slow reversal of the increase in NPQ may be explained by an ATP-induced maintenance of a low thylakoid lumen pH even in darkness.

In photosynthesizing leaves, the major portion if not all of the rapidly reversible NPQ under excess light is associated with the xanthophyll cycle.[16,17] In the studies by Gilmore and Yamamoto,[33] the same was true for isolated chloroplasts as well. However, we do not wish to infer that all NPQ in isolated chloroplasts is associated with zeaxanthin under all conditions (see, e.g., Reference 17).

Furthermore, a low intrathylakoid pH does not appear to induce NPQ directly, since NPQ can be inhibited by treatments (with antimycin) that do not dissipate the pH gradient[51] or prevent zeaxanthin formation.[52] Thus, a low intrathylakoid pH induces an additional change that, in turn, promotes NPQ. The formation of a pH gradient across the thylakoid membrane results in a conformational change in the membrane[44,53,54] that may link zeaxanthin to NPQ.

The precise mechanism of the involvement of the xanthophyll cycle in energy dissipation in the pigment bed is still unknown. However, the association between zeaxanthin (or the de-epoxidized state of the xanthophyll cycle) and a lowering of the yield of maximal chlorophyll fluorescence suggests an involvement of the singlet excited state of chlorophyll that gives rise

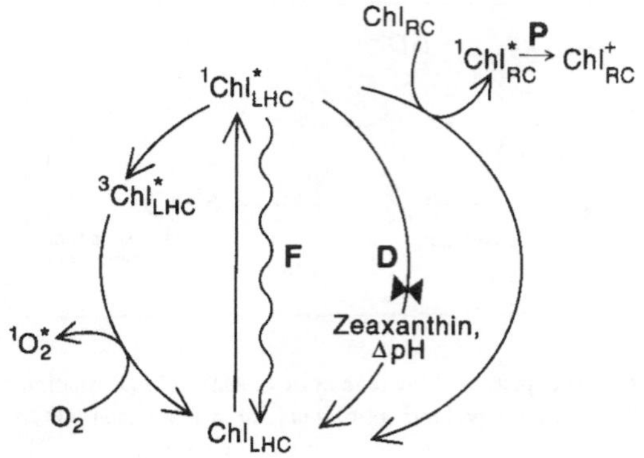

FIGURE 11. Schematic depiction of the possible fates of excitation energy in the form of the excited singlet state of chlorophyll in the light-harvesting complexes ($^1Chl^*_{LHC}$). F, fluorescence; D, dissipation of energy in the pigment bed, regulated by the presence of zeaxanthin and the thylakoid lumen pH; P, photochemistry; RC, reaction center.

to fluorescence (Figure 11). There are several hypothetical possibilities for either a direct or an indirect involvement of zeaxanthin in the deactivation of this singlet excited state of chlorophyll. A direct interaction between zeaxanthin and chlorophyll could occur, e.g., via a "nonclassical" mechanism such as rapidly reversible electron transfer reactions (see Reference 55 and 64). An indirect interaction could be via zeaxanthin-induced conformational changes of the chlorophyll-binding complexes, facilitating an otherwise unfavorable internal conversion of the singlet excited state of chlorophyll to the ground state (see Reference 27).

As postulated in Figure 11, the zeaxanthin-associated energy dissipation process in the pigment bed competes with photochemistry and fluorescence emission, presumably through an interaction with the singlet excited state of chlorophyll that is used predominantly for photochemistry when light is not excessive. Furthermore, and most importantly, the zeaxanthin-associated dissipation process is thus thought to prevent the formation of the dangerous reactive singlet oxygen that could be formed if the singlet excited state of chlorophyll were allowed to accumulate and convert into the triplet excited state.

IV. PHOTOPROTECTION AND THE XANTHOPHYLL CYCLE

A. THE REDUCTION STATE OF PSII

From measurements of chlorophyll fluorescence, it can be demonstrated that exposure of leaves to excess light leads to an increase in the reduction state of PSII (Figure 12). Ongoing energy dissipation in the pigment bed should remove some of the excess excitation energy before it reaches the PSII reaction centers. Thus, this dissipation process should counteract the accu-

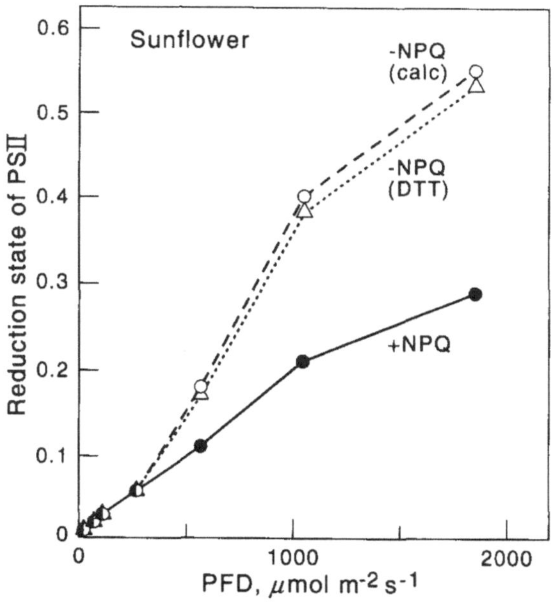

FIGURE 12. Reduction state of PSII at various PFDs in sunflower leaves in the presence and absence of energy dissipation in the pigment bed (NPQ). +NPQ, measured values of the reduction state of PSII in untreated control leaves; −NPQ(DTT), measured values of the reduction state of PSII in DTT-treated leaves in which NPQ was inhibited; −NPQ(calc), calculated values of the reduction state in control leaves under the assumption that NPQ was absent. The latter was calculated from $(F_m' - F)/F_m' = F_v'/F_m' \times (1 - Q_r/Q_t)$, rearranged to $Q_r/Q_t = 1 - ([F_m' - F]/F_m'$ at each PFD)/(F_v'/F_m' at 50 μmol (photons) $m^{-2}\ s^{-1}$). F_v'/F_m' at 50 μmol (photons) $m^{-2}\ s^{-1}$ is the intrinsic photon efficiency of PSII in the absence of energy dissipation in the pigment bed. (From Adams, W. W., III and Demmig-Adams, B., unpublished data.)

mulation of excess excitation energy and maintain a lower reduction state than in the absence of such dissipation. In Figure 12, this is demonstrated through two different approaches. The first technique involves the experimental use of dithiothreitol (DTT), which inhibits zeaxanthin formation and NPQ, and causes the reduction state of PSII to rise considerably at a given PFD compared to a control leaf which has not been treated with DTT. Second, the effect of NPQ on the reduction state of PSII can be assessed theoretically from the measured efficiency of PSII under the assumption that there were no contributions of energy dissipation to the decreases in efficiency, and that these decreases were caused exclusively by the closing of PSII traps, i.e., by increases in the reduction state of PSII (see legend of Figure 12). Figure 12 shows a similar and considerable increase in the reduction state of PSII as assessed by these two approaches. In the sunflower leaf grown at 1200 μmol (photons) $m^{-2}\ s^{-1}$, a 25% increase in the reduction state of PSII was observed in the absence of NPQ at PFDs approaching full sunlight.

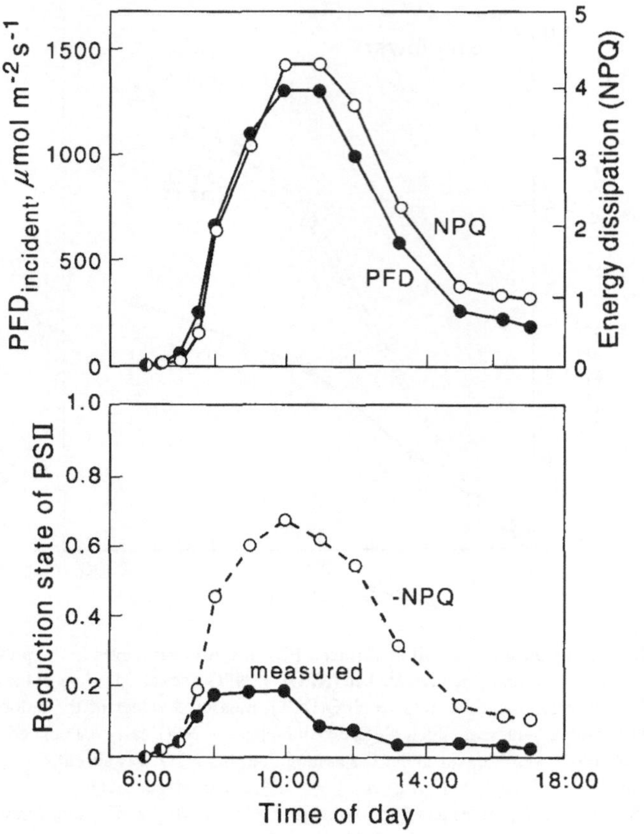

FIGURE 13. Diurnal changes in incident PFD (upper left panel), the actual efficiency of PSII photochemistry (upper right panel), the energy dissipation activity in the pigment bed (NPQ; lower left panel), and the reduction state of PSII (lower right panel) in the cactus *Nopalea cochenillifera* (L.) SD. in the field. (●) Measured actual levels of the reduction state; (○) hypothetical levels that would have been reached had there been no NPQ. For calculations, see the legend of Figure 12.[58]

Even greater differences in the reduction state of PSII in the absence and presence of NPQ have been observed in another species in the field (Figure 13). The cactus *Nopalea* maintained a very low reduction state of PSII while exposed to peak incident PFDs of about 1300 μmol (photons) m^{-2} s^{-1}, where the photon efficiency of PSII reached rather low values of 0.25 to 0.3. The maintenance of a low reduction state of only 20% was due to a very high level of energy dissipation in the pigment bed (NPQ = 4.25 at peak PFD). Calculations of the hypothetical reduction state of PSII in the absence of NPQ revealed that the reduction state would have increased to as high as 68% in the absence of energy dissipation in the pigment bed. This amounts to a difference in the reduction state of almost 50% due to NPQ.

B. THE PHOTON EFFICIENCY OF PHOTOSYNTHESIS

Adverse effects on the photosynthetic apparatus are thought to arise when PSII becomes highly reduced. The increase in the reduction state of PSII that results from inhibition of the energy dissipation process in the pigment bed leads to such adverse effects on the photosynthesis process. When no energy dissipation occurred in the pigment bed of leaves subjected to excess light, sustained depressions of the photon efficiency of PSII photochemistry as well as that of the whole photosynthesis process, were observed upon return to low light levels.[16,48,56,57] Such sustained effects were absent in control treatments in which energy dissipation in the pigment bed was allowed to develop. In these studies, the photon efficiency of photosynthesis was compared after a period of exposure to low light, following the high-light treatments, that was sufficient to allow the complete relaxation of NPQ (indicative of energy dissipation in the pigment bed) and thus permit a complete reversal of any effects of zeaxanthin-associated energy dissipation on the photon efficiency of photosynthesis.

It can be concluded that the zeaxanthin-associated energy dissipation process can prevent sustained depressions of the photosynthesis rate that might otherwise result from exposure to high light. This is evidence for a photoprotective role of the xanthophyll cycle.

C. ENERGY DISSIPATION IN THE PIGMENT BED VS. DAMAGE

As noted above, the regulatory process involving the dissipation of excess excitation energy in the pigment bed results in a transient lowering of the photon efficiency of photosynthesis, and is characterized by a lowering of the chlorophyll fluorescence yield at both the F_o and the F_m levels. Thus, decreases in the photon efficiency of PSII (F_v/F_m), or of the whole photosynthesis process, that are associated with a decrease in F_o have been attributed to this regulatory process.[16,32,37,48,56,58-60] Under other conditions, e.g., where energy dissipation in the pigment bed was inhibited or insufficient to dissipate all of the excess excitation energy, decreases in F_v/F_m were found to be associated with increases in F_o.[32,37,48] Depressions in PSII photochemical efficiency of the latter type have been attributed to some form of damage to PSII. However, this interpretation, based upon increases in F_o and a sustained depression in the efficiency of photosynthesis, requires further investigation. Two suggestions for the change in PSII that is responsible for this increase in F_o are (1) a decrease in the rate in PSII photochemistry and (2) a disconnection between the PSII reaction center and its antenna complexes, leading to a decrease in energy transfer between the two.[31,32,37]

D. CAPACITY FOR ENERGY DISSIPATION IN THE PIGMENT BED IN SUN AND SHADE LEAVES

As pointed out at the beginning of this chapter, leaves which have developed in full sunlight contain much larger xanthophyll cycle pools than do leaves which develop in the shade (Figure 1). Since the xanthophyll cycle pool is larger, one might expect that sun-acclimated leaves would have a larger capacity to form zeaxanthin and thus to employ the energy dissipation process involved in the photoprotection of the photosynthetic apparatus. This is, indeed, the case, as leaves which have developed in full sunlight have a greater maximal capacity for energy dissipation in the pigment bed than do shade leaves (Figures 8 and 9, and unpublished data).

Among sun leaves, however, there appears to be some variation in the degree to which the xanthophyll cycle pool is utilized that depends upon the growth form.[5] In sun leaves with high rates of photosynthesis (typically annual plants with mesophytic leaves), the maximal capacity for this energy dissipation process is typically not yet reached at full sunlight. Such leaves often convert only 30% of the xanthophyll cycle pool into zeaxanthin at full sunlight. On the other hand, plants which have evergreen leaves with lower rates of photosynthesis often convert 80 to 90% of their xanthophyll cycle pool into zeaxanthin at peak irradiance during the course of the day. Such leaves possess a xanthophyll cycle pool that appears to be just sufficient to mitigate the potentially harmful accumulation of excess excitation energy which they might otherwise experience under natural conditions.

Furthermore, it seems that energy dissipation in the pigment bed associated with the xanthophyll cycle is also involved in the acclimation of plants to an interaction between high light and other stress factors such as water stress, salinity, or chilling temperatures. Under such conditions, a sustained high activity of energy dissipation in the pigment bed and a retention of zeaxanthin in leaves[41] has been observed throughout the day and night cycle. This is in contrast to the pronounced diurnal changes in the de-epoxidation state of the xanthophyll cycle in sun leaves under favorable environmental conditions.

V. CONCLUSIONS

1. Zeaxanthin, or the de-epoxidized state of the xanthophyll cycle, is involved in the dissipation of excess excitation energy in the pigment bed of the photosynthetic membranes of all higher plants.
2. This zeaxanthin-associated energy dissipation process acts to protect the photosynthetic membranes from damage that might otherwise occur in the absence of this process.
3. The xanthophyll cycle responds rapidly to changes in the light environment, and plants which grow in sun-exposed habitats experience major diurnal changes in the de-epoxidation state of the xanthophyll cycle (i.e., large amounts of zeaxanthin are formed and removed on a daily basis) in response to the diurnal changes in PFD. The energy dissipation process associated with the de-epoxidation of violaxanthin is employed to a large extent during maximal exposure to sunlight.

4. Leaves which have developed in full sunlight possess larger xanthophyll cycle pools than do shade leaves, and this is accompanied by a greater maximal capacity for the zeaxanthin-associated energy dissipation process.

REFERENCES

1. **Krinsky, N. I.,** Carotenoid protection against oxidation, *Pure Appl. Chem.*, 51, 649, 1979.
2. **Ridley, S.,** Carotenoids and herbicide action, in *Carotenoid Chemistry and Biochemistry*, Britton, G. and Goodwin, T. W., Eds., Pergamon Press, Oxford, 1982, 353.
3. **Siefermann-Harms, D.,** The light-harvesting and protective functions of carotenoids in photosynthetic membranes, *Physiol. Plant.*, 69, 561, 1987.
4. **Young, A. J.,** The photoprotective role of carotenoids in higher plants, *Physiol. Plant.*, 83, 702, 1991.
5. **Demmig-Adams, B. and Adams, W. W., III,** Carotenoid composition in sun and shade leaves of plants with different life forms, *Plant Cell Environ.*, 15, 411, 1992.
6. **Demmig-Adams, B., Winter, K., Winkelmann, E., Krüger, A., and Czygan, F.-C.,** Photosynthetic characteristics and the ratios of chlorophyll, β-carotene, and the components of the xanthophyll cycle upon a sudden increase in growth light regime in several plant species, *Bot. Acta*, 102, 319, 1989.
7. **Thayer, S. S. and Björkman, O.,** Leaf xanthophyll content and composition in sun and shade determined by HPLC, *Photosynth. Res.*, 23, 331, 1990.
8. **Jones, B. L. and Porter, J. W.,** Biosynthesis of carotenes in higher plants, *CRC Crit. Rev. Plant Sci.*, 3, 295, 1986.
9. **Sandmann, G.,** Biosynthesis of cyclic carotenoids: biochemistry and molecular genetics of the reaction sequence, *Physiol. Plant.*, 83, 186, 1991.
10. **Demmig-Adams, B.,** Carotenoids and photoprotection in plants: a role for the xanthophyll zeaxanthin, *Biochim. Biophys. Acta*, 1020, 1, 1990.
11. **Demmig-Adams, B. and Adams, W. W., III,** Photoprotection and other responses of plants to high light stress, *Annu. Rev. Plant Physiol. Plant Mol. Biol.*, 43, 599, 1992.
12. **Demmig-Adams, B. and Adams, W. W., III,** The xanthophyll cycle, in *Carotenoids in Photosynthesis*, Young, A. and Britton, G., Eds., Chapman and Hall, London, 1993.
13. **Demmig-Adams, B. and Adams, W. W., III,** The xanthophyll cycle, in *Antioxidants in Higher Plants*, Alscher, R. G. and Hess, J. L., Eds., CRC Press, Boca Raton, FL, in press.
14. **Adams, W. W., III and Demmig-Adams, B.,** Operation of the xanthophyll cycle in higher plants in response to diurnal changes in incident sunlight, *Planta*, 186, 390, 1992.
15. **Adams, W. W., III, Volk, M., Hoehn, A., and Demmig-Adams, B.,** Leaf orientation and the response of the xanthophyll cycle to incident light, *Oecologia*, 90, 404, 1992.
16. **Bilger, W. and Björkman, O.,** Role of the xanthophyll cycle in photoprotection elucidated by measurements of light-induced absorbance changes, fluorescence and photosynthesis in leaves of *Hedera canariensis*, *Photosynth. Res.*, 25, 173, 1990.
17. **Demmig-Adams, B., Adams, W. W., III, Heber, U., Neimanis, S., Winter, K., Krüger, A., Czygan, F.-C., Bilger, W., and Björkman, O.,** Inhibition of zeaxanthin formation and of rapid changes in radiationless energy dissipation by dithiothreitol in spinach leaves and chloroplasts, *Plant Physiol.*, 92, 293, 1990.

18. **Demmig-Adams, B., Winter, K., Krüger, A., and Czygan, F.-C.,** Light response of CO_2 assimilation, dissipation of excess excitation energy, and zeaxanthin content of sun and shade leaves, *Plant Physiol.*, 90, 881, 1989.
19. **Siefermann, D.,** Kinetic studies on the xanthophyll cycle of *Lemna gibba* L. Influence of photosynthetic oxygen and supplied reductor, in *2nd Int. Congr. Photosynthesis,* Vol. 1, Forti, G., Avron, M., and Melandri, A., Eds., Dr. Junk, The Hague, 1972, 629.
20. **Hager, A.,** The reversible, light-induced conversions of xanthophylls in the chloroplast, in *Pigments in Plants,* 2nd. ed., Czygan, F.-C., Ed., Fischer, Stuttgart, 1980, 57.
21. **Siefermann-Harms, D.,** The xanthophyll cycle in higher plants, in *Lipids and Lipid Polymers in Higher Plants,* Tevini, M. and Lichtenthaler, H. K., Eds., Springer-Verlag, Berlin, 1977, 218.
22. **Yamamoto, H. Y.,** Biochemistry of the violaxanthin cycle in higher plants, *Pure Appl. Chem.*, 51, 639, 1979.
23. **Siefermann, D. and Yamamoto, H. Y.,** Light-induced de-epoxidation of violaxanthin in lettuce chloroplasts. V. Dehydroascorbate, a link between photosynthetic electron transport and de-epoxidation, in *Proc. 3rd Int. Congr. Photosynthesis,* Vol. 3, Avron, M., Ed., Elsevier, Amsterdam, 1974, 1991.
24. **Foyer, C. H., Lelandais, M., Edwards, E. A., and Mullineaux, P. M.,** The role of ascorbate in plants, interactions with photosynthesis and regulatory significance, in *Active Oxygen/Oxidative Stress and Plant Metabolism,* Pell, E. J. and Steffen, K. L., Eds., American Society of Plant Physiologists, Rockville, MD, 1991, 131.
25. **Hager, A.,** Lichtbedingte pH-Erniedrigung in einem Chloroplasten-Kompartiment als Ursache der enzymatischen Violaxanthin → Zeaxanthin-Umwandlung; Beziehungen zur Photophosphorylierung, *Planta,* 89, 224, 1969.
26. **Siefermann, D. and Yamamoto, H. Y.,** Properties of NADPH and oxygen-dependent zeaxanthin epoxidation in isolated chloroplasts. A transmembrane model for the violaxanthin cycle, *Arch. Biochem. Biophys.*, 171, 70, 1975.
27. **Horton, P., Ruban, A. V., Rees, D., Pascal, A. A., Noctor, G., and Young, A. J.,** Control of the light-harvesting function of chloroplast membranes by aggregation of the LHCII chlorophyll-protein complex, *FEBS Lett.*, 292, 1, 1991.
28. **Weis, E. and Berry, J. A.,** Quantum efficiency of photosystem II in relation to 'energy'-dependent quenching of chlorophyll fluorescence, *Biochim. Biophys. Acta,* 894, 198, 1987.
29. **Bilger, W. and Schreiber, U.,** Energy-dependent quenching of dark-level chlorophyll fluorescence in intact leaves, *Photosynth. Res.*, 10, 303, 1986.
30. **Schreiber, U., Schliwa, U., and Bilger, W.,** Continuous recording of photochemical and non-photochemical chlorophyll fluorescence quenching with a new type of modulation fluorometer, *Photosynth. Res.*, 10, 51, 1986.
31. **Björkman, O.,** Low-temperature chlorophyll fluorescence in leaves and its relationship to photon yield of photosynthesis in photoinhibition, in *Photoinhibition,* Kyle, D. J., Osmond, C. B., and Arntzen, C. J., Eds., Elsevier, Amsterdam, 1987, 123.
32. **Björkman, O.,** High-irradiance stress in plants and its interaction with other stress factors, in *Progress in Photosynthesis Research,* Biggins, J., Ed., Martinus Nijhoff, Dordrecht, 1987, 11.
33. **Gilmore, A. M. and Yamamoto, H. Y.,** Zeaxanthin formation and energy-dependent fluorescence quenching in pea chloroplasts under artificially-mediated linear and cyclic electron transport, *Plant Physiol.*, 96, 635, 1991.
34. **Kitajima, M. and Butler, W. L.,** Quenching of chlorophyll fluorescence and primary photochemistry in chloroplasts by dibromothymoquinone, *Biochim. Biophys. Acta,* 376, 105, 1975.
35. **Rees, D., Noctor, G. D., and Horton, P.,** The effect of high-energy-state excitation quenching on maximum and dark level chlorophyll fluorescence yield, *Fert. Res.*, 25, 199, 1990.

36. **Genty, B., Briantais, J.-M., and Baker, N. R.,** The relationship between the quantum yield of photosynthetic electron transport and quenching of chlorophyll fluorescence, *Biochim. Biophys. Acta,* 990, 87, 1989.
37. **Demmig, B. and Björkman, O.,** Comparison of the effect of excessive light on chlorophyll fluorescence (77K) and photon yield of O_2 evolution in leaves of higher plants, *Planta,* 171, 171, 1987.
38. **Björkman, O.,** Responses to different quantum flux densities, in *Encyclopedia of Plant Physiology, NS, Physiological Plant Ecology I,* Vol. 12A, Lange, O. L., Nobel, P. S., Osmond, C. B., and Ziegler, H., Eds., Springer-Verlag, Berlin, 1981, 57.
39. **Bilger, W. and Björkman, O.,** Temperature dependence of violaxanthin de-epoxidation and non-photochemical fluorescence quenching in intact leaves of *Gossypium hirsutum* L. and *Malva parviflora* L., *Planta,* 184, 226, 1991.
40. **Demmig, B., Winter, K., Krüger, A., and Czygan, F.-C.,** Photoinhibition and zeaxanthin formation in intact leaves. A possible role of the xanthophyll cycle in the dissipation of excess light energy, *Plant Physiol.,* 84, 218, 1987.
41. **Demmig, B., Winter, K., Krüger, A., and Czygan, F.-C.,** Zeaxanthin and the heat dissipation of excess light energy in *Nerium oleander* exposed to a combination of high light and water stress, *Plant Physiol.,* 87, 17, 1988.
42. **Neubauer, C. and Yamamoto, H. Y.,** Mehler-peroxidase reaction mediates zeaxanthin formation and zeaxanthin-related fluorescence quenching in intact chloroplasts, *Plant Physiol.,* in press.
43. **Demmig-Adams, B., Winter, K., Krüger, A., and Czygan, F.-C.,** Zeaxanthin and the induction and relaxation kinetics of the dissipation of excess excitation energy in leaves in 2% O_2, 0% CO_2, *Plant Physiol.,* 90, 887, 1989.
44. **Bilger, W., Björkman, O., and Thayer, S. S.,** Light-induced spectral absorbance changes in relation to photosynthesis and the epoxidation state of xanthophyll cycle components in cotton leaves, *Plant Physiol.,* 91, 542, 1989.
45. **Demmig-Adams, B., Adams, W. W., III, Czygan, F.-C., Schreiber, U., and Lange, O. L.,** Differences in the capacity for radiationless energy dissipation in green and blue-green algal lichens associated with differences in carotenoid composition, *Planta,* 180, 182, 1990.
46. **Gilmore, A. M. and Yamamoto, H. Y.,** Dark induction of zeaxanthin-dependent non-photochemical fluorescence quenching mediated by ATP, *Proc. Natl. Acad. Sci. U.S.A.,* 89, 1899, 1992.
47. **Demmig, B., Winter, K., Krüger, A., and Czygan, F.-C.,** Zeaxanthin synthesis, energy dissipation, and photoprotection of photosystem II at chilling temperatures, *Plant Physiol.,* 90, 894, 1989.
48. **Demmig-Adams, B., Adams, W. W., III, Green, T. G. A., Czygan, F.-C., and Lange, O. L.,** Differences in the susceptibility to light stress in two lichens forming a phycosymbiodeme, one partner possessing and one lacking the xanthophyll cycle, *Oecologia,* 84, 451, 1990.
49. **Yamamoto, H. Y. and Kamite, L.,** The effects of dithiothreitol on violaxanthin de-epoxidation and absorbance changes in the 500-nm region, *Biochim. Biophys. Acta,* 267, 538, 1972.
50. **Krause, G. H., Vernotte, C., and Briantais, J.-M.,** Photoinduced quenching of chlorophyll fluorescence in intact chloroplasts and algae. Resolution into two components, *Biochim. Biophys. Acta,* 679, 116, 1982.
51. **Oxborough, K. and Horton, P.,** Characterization of the effects of antimycin A upon high energy state quenching of chlorophyll fluorescence (q_E) in spinach and pea chloroplasts, *Photosynth. Res.,* 12, 119, 1987.
52. **Gilmore, A. M. and Yamamoto, H. Y.,** Zeaxanthin formation in q_E inhibited chloroplasts, in *Current Research in Photosynthesis,* Vol. 2, Baltscheffsky, M., Ed., Kluwer, Dordrecht, 1990, 495.

53. **Heber, U.,** Conformational changes of chloroplasts induced by illumination of leaves in vivo, *Biochim. Biophys. Acta,* 180, 302, 1969.
54. **Packer, L. and Crofts, A. R.,** The energized movement of ions and water by chloroplasts, in *Current Topics in Bioenergetics,* Vol. 2, Sanadi, D. R., Ed., Academic Press, New York, 1967, 23.
55. **Liddell, P. A., Nemeth, G. A., Lehman, W. R., Joy, A. M., Moore, A. L., Bensasson, R. V., Moore, T. A., and Gust, D.,** Mimicry of carotenoid function in photosynthesis: synthesis and photophysical properties of a carotenopyropheophorbide, *Photochem. Photobiol.,* 36, 641, 1982.
56. **Adams, W. W., III, Demmig-Adams, B., and Winter, K.,** Relative contributions of zeaxanthin-related and zeaxanthin-unrelated types of "high-energy-state" quenching of chlorophyll fluorescence in spinach leaves exposed to various environmental conditions, *Plant Physiol.,* 92, 302, 1990.
57. **Winter, K. and Königer, M.,** Dithiothreitol, an inhibitor of violaxanthin de-epoxidation, increases the susceptibility of leaves of *Nerium oleander* L. to photoinhibition of photosynthesis, *Planta,* 180, 24, 1989.
58. **Adams, W. W., III, Díaz, M., and Winter, K.,** Diurnal changes in photochemical efficiency, the reduction state of Q, radiationless energy dissipation, and non-photochemical fluorescence quenching in cacti exposed to natural sunlight in northern Venezuela, *Oecologia,* 80, 553, 1989.
59. **Björkman, O., Demmig, B., and Andrews, T. J.,** Mangrove photosynthesis: response to high-irradiance stress, *Aust. J. Plant Physiol.,* 15, 43, 1988.
60. **Demmig-Adams, B., Adams, W. W., III, Winter, K., Meyer, A., Schreiber, U., Pereira, J. S., Krüger, A., Czygan, F.-C., and Lange, O. L.,** Photochemical efficiency of photosystem II, photon yield of O_2 evolution, photosynthetic capacity, and carotenoid composition during the midday depression of net CO_2 uptake in *Arbutus unedo* growing in Portugal, *Planta,* 177, 377, 1989.
61. **Thayer, S. S. and Björkman, O.,** Carotenoid distribution and de-epoxidation in thylakoid pigment-protein complexes from cotton leaves and bundle-sheath cells of maize, *Photosynth. Res.,* 33, 213, 1992.
62. **Bilger, W. and Björkman, O.,** unpublished data.
63. **Gilmore, A. M. and Yamamoto, H. Y.,** Linear models relating xanthophylls and lumen acidity to non-photochemical fluorescence quenching. Evidence that antheraxanthin explains zeaxanthin-independent quenching, *Photosynth. Res.,* 35, 67, 1993.
64. **Owens, T. G., Shreve, A. P., and Albrecht, A. C.,** Dynamics and mechanism of singlet energy transfer between carotenoids and chlorophylls: light harvesting and non-photochemical fluorescence quenching, in *Research in Photosynthesis,* Vol. I., Marata, N., Ed., Kluwer, Dordrecht, 1992, 179.

Chapter 5

CHILLING STRESS AND PHOTOSYNTHESIS

Neil R. Baker

TABLE OF CONTENTS

0-8493-5443-9/94/$0.00+$.50

I. INTRODUCTION

Temperature is one of the major environmental determinants of plant distribution and productivity. Over 70% of the world's landmass experiences mean temperatures below 5°C, and in many habitats plants can be subjected to both wide daily and seasonal variations in temperature. In cold environments, plants can experience two distinct types of low-temperature stress. At temperatures below 0°C, freezing of both extra- and intracellular water can occur and result in cellular dehydration and disruption. Temperatures above 0°C, but which are suboptimal for growth, can disrupt metabolism in many plants, a phenomenon commonly referred to as chilling stress. It is the effects of chilling stress on the photosynthetic processes that are considered in this chapter.

Chilling temperatures are a major factor in depressing the yields of many crops.[1] This is particularly true for species of tropical and subtropical origin, such as *Zea mays,* when grown in temperate regions. Extensive efforts have been made to develop crop varieties with improved tolerance to chilling temperatures. However, success has been severely limited by a lack of information on the mechanistic bases of chill stress. This information is essential not only for the selection of appropriate characteristics to facilitate rapid screening of new varieties in conventional breeding programs, but is also required to identify key molecular sites to be manipulated in genetic engineering programs.

II. TEMPERATURE AND BIOLOGICAL SYSTEMS

Generally, metabolic reactions are markedly influenced by temperature, with the rate of most enzymatic reactions approximately doubling for each 10°C rise in temperature within the temperature range of physiological operation, giving a temperature coefficient, Q_{10}, of ≈ 2. The Q_{10} is defined by

$$Q_{10} = \sqrt{\frac{T + 10}{T}}\, e^{\frac{10\, U_B}{kT(T + 10)}} \qquad (1)$$

where T is temperature in °K, *k* is Boltzmann's constant, and U_B is the minimum kinetic energy required to produce a reaction.[2] Physical processes, such as light absorption, have no energy barrier to overcome, i.e., $U_B = 0$, and consequently have a Q_{10} close to 1.0. The majority of enzymatic reactions only occur when the reactants have considerable kinetic energy, and consequently the reactions are very temperature dependent. Although a Q_{10} of 2.0 or above is generally considered to be indicative of a metabolic reaction, it should be emphasized that passive processes having an appreciable energy barrier (U_B), e.g., diffusion of ions across biological membranes, can have a large Q_{10}.[2] The minimum kinetic energy (U_B) required for a reaction corresponds to the activation energy (E) of that reaction, i.e., the amount of energy needed to bring a mole of reactant at a given temperature to the transition state at the top of the activation barrier. Consequently, the relationship between the rate constant, *k*, of the reaction and temperature is defined in the Arrhenius equation:

$$k = Ae^{-E/RT} \tag{2}$$

where A is a constant and R is the gas constant. This equation can also be written as:

$$\ln k = \ln A - \frac{E}{RT} \tag{3}$$

which demonstrates that the logarithm of the rate constant is directly proportional to temperature. It should be noted that A and E are not entirely independent of temperature, and this is particularly true for reactions occurring in solution. However, since for many reactions both A and E remain essentially constant over a wide range of physiologically relevant temperatures, for the purposes of considering the effects of temperature on photosynthesis, they can be considered as being constant.[3] Many important reactions in cells have large values of E and are therefore extremely sensitive to temperature change. It is also of note that passive processes that do not involve enzymes can also have large E values. For example, E for the passive efflux of K^+ from many cells is ≈63 J mol^{-1} at 20°C, while E for the diffusion of K^+ in water is only ≈17 J mol^{-1}.[2] The high E for passive K^+ efflux from cells is due to the cell membrane constituting a considerable energy barrier to the diffusion of K^+. It should be emphasized that this is reflected in the passive efflux of K^+ from cells having a Q_{10} of ≈2.5, while the Q_{10} for K^+ diffusion in water is only ≈1.3.[2]

It is evident from the above discussion that changes in temperature will have a major effect on the rate of reaction of many biological processes. This is true not only for reactions catalyzed by enzymes, but also for many physicochemical processes which have high E values, such as passive diffusion of ions and solutes across membranes.

III. DIRECT EFFECTS OF TEMPERATURE ON PHOTOSYNTHESIS

Photosynthesis, defined as the light-induced assimilation of carbon, consists of a complex series of physicochemical and metabolic reactions, primarily occurring in the chloroplasts of leaves. Clearly, changes in temperature would be expected to have a major effect on photosynthesis. The typical hyperbolic

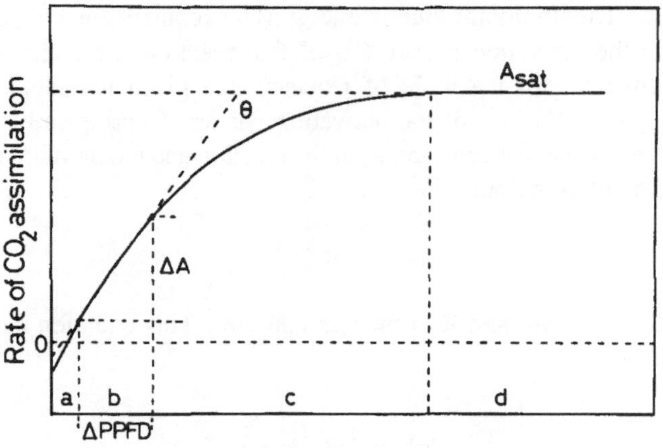

FIGURE 1. The response of the rate of CO_2 assimilation (A) in a leaf to absorbed photosynthetically active light (PPFD). Phases a, b, c, and days define specific regions of the response curve. Phases a and b designate the portion of the response curve in which CO_2 assimilation increases linearly with increasing PPFD. The transition from phase a to phase b is defined by a decrease in slope which is attributed to a light intensity-dependent inhibition of respiration.[6] In phase c, the hyperbolic dependence of CO_2 assimilation on PPFD can be seen, with the upper asymptote (phase d) defining the light-saturated rate of CO_2 assimilation, A_{sat}. The convexity coefficient, θ, quantitates the sharpness of the transition from light limitation to light saturation. The maximum quantum efficiency of CO_2 assimilation (ϕ_{max}) is determined from the slope of the response curve in phase b, i.e., ΔA/ΔPPFD. The quantum efficiency of CO_2 assimilation (ϕ_{CO2}) at any given point on the curve is given by A/absorbed PPFD.

dependence of the photosynthetic response of a leaf to increasing light intensity (Figure 1) can be defined by the quadratic equation:

$$\theta A^2 - A(A_{sat} + \phi_{max} I) + \phi_{max} I A_{sat} = 0 \quad (3)$$

where A is the rate of CO_2 assimilation, A_{sat} is the light-saturated rate of CO_2 assimilation, ϕ_{max} is the maximum quantum efficiency of CO_2 assimilation, θ is the convexity coefficient of the curve which quantitates the sharpness of the transition from light limitation to light saturation (θ can vary from 0 for a rectangular hyperbola to 1.0 for a "Blackman" response of two straight lines), and I is the light absorbed by the leaf.[4] This equation can be solved for A:

$$A = \frac{\phi_{max} I + A_{sat} - \sqrt{\{(\phi_{max} I + A_{sat})^2 - 4\theta(\phi_{max} I A_{sat})\}}}{2\theta} \quad (4)$$

Consequently, the photosynthetic performance of a leaf experiencing fluctuating light levels in the field will be determined by three physiological parameters: A_{sat}, ϕ_{max} and θ, and the light intensity (I). A_{sat} and ϕ_{max} reflect very different aspects of photosynthesis. A_{sat} is partially limited by diffusion of CO_2 from the atmosphere through the leaf's boundary layer and stomata. Within the mesophyll, A_{sat} is limited by the carboxylase activity of ribulose-1,5-bisphosphate carboxylase oxygenase (RuBisCo) at low internal CO_2 concentrations, and by the rate of regeneration of ribulose-1,5-bisphosphate (RuBP), the acceptor substrate for CO_2, at high internal CO_2 levels.[9] The rate of RuBP regeneration may be affected by a range of processes associated with the maximum capacity for production of ATP and NADPH and the reactions of the photosynthetic carbon reduction cycle. The maximum quantum yield of CO_2 assimilation (ϕ_{max}) will be determined by the factors influencing the utilization of absorbed light for ATP and NADPH production and the allocation of these products of the "light" reactions for carbon assimilation relative to other metabolic sinks.

It is becoming increasingly evident that the majority of photosynthesis in many crops is rarely light saturated.[6] Consequently, the parameters of the light response curve of leaves that define the photosynthetic performance at subsaturating light levels, i.e., ϕ_{max} and θ, will be particularly important factors to consider when evaluating the bases of depressions in photosynthesis at low temperatures.

The response of A_{sat}, ϕ_{max}, and θ to temperature varies markedly. A_{sat} is strongly dependent on temperature, typically exhibiting a bell-shaped response with a clear optimum and upper and lower thresholds,[7] whereas ϕ_{max} and θ are generally considered to be considerably less dependent on temperature.[4] In C_4 plants, ϕ_{max} is independent of immediate changes in temperature over a wide range,[8,9] although prolonged periods at low and high temperatures will result in stress-induced decreases.[7] In C_3 plants, ϕ_{max} increases with decreasing temperature due to the decreasing affinity of RuBisCo for O_2 relative to CO_2,[10] although this effect can be removed by minimizing the oxygenase activity by reducing the atmospheric O_2 to 1%.[8,9] Clearly, in C_3 plants at low temperatures in normal atmospheric O_2 levels, the decrease in the allocation of ATP and NADPH to photorespiration relative to photosynthetic carbon assimilation will result in an increase in ϕ_{max}.

Depressions in A_{sat} on chilling leaves do not appear to be attributable to increases in stomatal limitation to CO_2 diffusion, since, generally, chilling treatments have not been found to increase stomatal resistance.[11-13] In tomato, chill-induced depression in photosynthesis has been shown to be associated with a reduction in RuBP levels, indicating that RuBP regeneration, not the utilization of RuBP, was limiting CO_2 assimilation, even though RuBisCo activity decreased.[14] The *in situ* catalytic rates of the stromal phosphatases, fructose-1,6-bisphosphatase (FBPase) and sedoheptulose-1,7-bisphosphatase (SBPase), were reduced following the chill treatment, and the extractable enzyme activity of the stromal FBPase was found to be far below that needed to sustain the rate of photosynthesis observed in control leaves.[14] The chill-induced loss of FBPase activity was attributed to an inhibition of the thioredoxin-mediated activation of the FBPase, possibly due to increased levels of toxic oxygen species being generated under the high light and low temperature treatment.[14] The possibility of a light dependency of the inhibition of photosynthesis is considered more extensively in Section IV. Clearly, in this context there are two important, but separate, issues to consider: (1) the consequences of the thermodynamic effects of a temperature decrease on the constituent reactions of photosynthesis and (2) the possibility of a reduction in photosynthetic metabolism increasing the rate of production of potentially toxic active oxygen species.

In C_4 plants, pyruvate orthophosphate dikinase and RuBisCo have been identified as potential chilling-sensitive sites, and also the transport of metabolic intermediates between compartments and cells could be affected.[15] However, it seems unlikely that any single metabolic process in C_4 metabolism is particularly cold sensitive and the sole cause of the decline in A_{sat}.[16] It is more likely that many metabolic processes are affected, possibly with enzymes responding differently to incremental decreases in temperature. It should be remembered that all the reactions associated with photosynthetic metabolism are limited by the rate of the slowest component reaction, and it is likely that the slowest reaction will be different at different temperatures.

IV. INDIRECT EFFECTS OF TEMPERATURE ON PHOTOSYNTHESIS

Besides the direct thermodynamic effects of temperature change on photosynthetic reactions, perturbations of metabolic reactions and biological organization by low temperatures can also result in important secondary effects on photosynthetic capacity and efficiency.

A. PHOTOINHIBITION OF PHOTOSYNTHESIS

It is now well established that when plants are experiencing an environmental stress which restricts the ability to utilize absorbed light for carbon assimilation, or other metabolic processes such as photorespiration, a light-induced depression of CO_2 assimilation can occur. Such photoinhibition can reduce both ϕ_{max} and A_{sat}, and result in large decreases in the photosynthetic productivity of plants. There are two distinct processes that contribute to stress-induced photoinhibition of photosynthesis: (1) downregulation and (2) photodamage to the reaction center of PSII.

1. Downregulation of Photosynthesis

When a healthy, nonstressed leaf is exposed to increasing light levels, changes occur in the photosynthetic apparatus of the thylakoids which decrease the efficiency of utilizing absorbed photons for photochemistry. This response can be demonstrated clearly in an experiment in which ϕ_{max} was lowered by

TABLE 1
Relationship Between Maximum Quantum Yield of CO_2 Assimilation (ϕ_{max}), Quantum Yield of PSII Photochemistry (ϕ_{PSII}), and Chlorophyll-Quenching Parameters in Leaves Pretreated in the Dark and in High Light

Parameter	Dark	High light	Change (%)
ϕ_{max}	0.051	0.044	−14
ϕ_{PSII}	0.49	0.43	−12
F_v'/F_m'	0.71	0.53	−26
qP	0.69	0.82	+19

Note: Leaves of *Silene dioica* were kept in the dark or in a PPFD of 1500 μmol m^{-2} s^{-1} (high light) prior to exposure to a PPFD of 34 μmol m^{-2} s^{-1} to measure ϕ_{max}, ϕ_{PSII}, F_v'/F_m', and the photochemical-quenching coefficient, qP, at steady-state photosynthesis. The leaves were exposed to 400 ppm CO_2 and 1% O_2 in nitrogen to minimize photorespiration. The decreases in ϕ_{max}, ϕ_{PSII}, and F_v'/F_m' after the high-light pretreatment were fully reversible within 1 h of the leaves being placed in the dark.

Data from Genty, B., Briantais, J-M., and Baker, N. R., *Biochim. Biophys. Acta,* 990, 87, 1989. With permission.

about 15% by exposure of a dark-adapted leaf to 30 min of saturating light (Table 1). This decrease in the efficiency of light utilization with increasing light intensity is commonly referred to as a downregulation of photosynthesis, with the implication that the effect is regulatory and not caused by damage to the photosynthetic apparatus. It would appear that this regulatory mechanism operates to redirect an increasing proportion of absorbed photons away from the PSII reaction centers, dissipating them directly as heat, and in doing so, protecting the PSII reaction centers from photodamage. It may be more appropriate to consider the phenomenon of downregulation as a protective adjustment rather than a regulation, since the photosynthetic apparatus does not appear to actively regulate the quantum efficiency of photosynthesis in a way analogous to the regulation of metabolic pathways by allosteric enzymes. It would appear that this protective adjustment of the photosynthetic apparatus is an ubiquitous feature of oxygenic photosynthetic metabolism, operating in the thylakoid membranes of algae as well as lower and higher vascular plants. It is possible that downregulation is an inevitable consequence of processing photons at high rates and arises from intrinsic physical aspects of the photosynthetic apparatus itself.

Recently it has been demonstrated that simple chlorophyll fluorescence parameters can be used as a straightforward, rapid measure of the quantum efficiency of PSII photochemistry, ϕ_{PSII}.[17] This parameter depends upon (1) the efficiency with which an absorbed photon can reach a PSII reaction center to perform photochemistry and (2) the proportion of reaction centers which are capable of transferring an electron to a quinone acceptor at that point in time, i.e., photochemically "open" reaction centers.[17] Since the efficiency of excitation energy capture by open PSII reaction centers is defined by the ratio of variable to maximal fluorescence, F_v/F_m,[18] and the photochemical quenching coefficient, qP, is a measure of the proportion of open PSII reaction centers, Genty et al.[17] noted that the relative quantum efficiency of PSII photochemistry (ϕ_{PSII}) in a leaf under a given operating condition is defined by:

$$\phi_{PSII} = \frac{Fv'}{Fm'}\, qP \tag{5}$$

In practice ϕ_{PSII} for leaves under any given environmental condition can be made by monitoring the steady-state fluorescence level (F_s) under the conditions of interest, exposing the leaves to a saturating flash, and measuring the maximal fluorescence level (F_m') that is attained.[6,17] The parameter $(F_m' - F_s)/F_m'$ equates to $\{(F_v'/F_m') \cdot qP\}$ for the leaf under the given conditions, and thus is a measure of ϕ_{PSII}.[17] Decreases in F_v'/F_m' are indicative of increases in the rate of nonphotochemical quenching of excitation energy, and can be used to demonstrate the downregulation of photosynthesis (see below).

In maize leaves, which are C_4 and consequently have minimal photorespiratory activity, changes in the rate of CO_2 assimilation in response to changes in light intensity and atmospheric CO_2 concentration, or during the induction of photosynthesis in a dark-adapted leaf, were observed to be directly proportional to the changes observed in ϕ_{PSII}.[17] This demonstrated that the quantum efficiency of CO_2 assimilation (ϕ_{CO_2}) of these leaves under a wide range of operating conditions was directly related to ϕ_{PSII}. Linear relationships between ϕ_{CO_2} and ϕ_{PSII} have also been observed in barley and ivy leaves exposed to a range of light intensities under conditions in which photorespiration had been minimized by reducing the atmospheric O_2 concentration to 1%.[17,19] Similar relationships have also been found between ϕ_{CO_2},[20] the quantum yield of O_2 evolution,[21-23] and ϕ_{PSII} in the leaves of a range of species. These data demonstrate that under a wide range of physiological conditions, provided that the allocation of ATP and reductants produced by photosynthetic electron transport to processes other than CO_2 assimilation is minimal, a direct proportionality exists between ϕ_{CO_2} and ϕ_{PSII}, which can be defined by:

$$\phi_{CO_2} = \phi_{PSII}\, \frac{I_{PSII}}{I}\, \frac{1}{k} \tag{6}$$

where I_{PSII} and I are the photon flux densities absorbed by PSII antennae and the leaf, respectively, and k is the number of electron equivalents required to reduce one mole of CO_2.[16] As would be expected for leaves of C_3 plants under conditions favorable for photorespiratory activity where the products of electron transport are apportioned between carboxylation and oxygenation reactions, the strict linear relationship between ϕ_{CO_2} and ϕ_{PSII} breaks down.[19] Simultaneous measurements of ϕ_{PSII} and the quantum yield of PSI photochemistry (ϕ_{PSI}) made over a wide range of light intensities on leaves under both photorespiratory and nonphotorespiratory conditions have shown a good correlation between these parameters.[19,24,25] This observation, together with the linear relationship found between ϕ_{PSII} and ϕ_{CO_2}, indicates that ϕ_{PSII} is a good measure of the quantum efficiency of noncyclic electron transport.

It is evident from many experiments that quenching of excitation energy in PSII antennae by nonphotochemical processes is a major factor in producing large light-induced decreases in the efficiency of light utilization for photosynthesis in leaves,[26] and which constitute a downregulation of photosynthesis. Light-induced increases in nonphotochemical quenching of excitation energy in a leaf and the observed consequences for ϕ_{PSII} and ϕ_{CO_2} are shown in Table 1. The parameters F_v'/F_m', q_P, ϕ_{PSII}, and ϕ_{CO_2} were determined for dark- and light-adapted (30 min at a photosynthetically active photon flux density, PPFD, of 1500 μmol m^{-2} s^{-1}) leaves. Then the leaves were exposed to a PPFD of 34 μmol m^{-2} s^{-1}, which is sufficiently low to produce the maximum quantum yield of photosynthesis, and allowed to reach steady state. Pretreatment of the leaves in high light induced a 26% decrease in F_v'/F_m', indicative of a large increase in nonphotochemical quenching, with concomitant decreases of 12 and 14% in ϕ_{PSII} and ϕ_{CO_2}, respectively. The decrease in ϕ_{PSII} must be attributed to an increase in nonphotochemical quenching, since high-light pretreatment results in an increase in qP, thus demonstrating that in high-light-treated leaves, a larger proportion of PSII reaction centers are oxidized than in the dark-adapted leaves when the leaves are exposed to the low actinic light of 34 μmol m^{-2} s^{-1} and measurements made at steady-state photosynthesis. Such an increase in qP would increase ϕ_{PSII} (see Equation 5). Clearly, the greater decrease in F_v'/F_m', rather than the increase in qP, is the dominating factor in determining the change in ϕ_{PSII}, and consequently in ϕ_{CO_2}.

The molecular mechanism of this light-induced, nonphotochemical quenching has not yet been resolved. It is generally considered that the quenching occurs in the antennae of PSII[27,28] and is associated with the acidification of the thylakoid lumen associated with an increased *trans*thylakoid ΔpH.[26] The xanthophyll, violaxanthin, has been implicated in the mechanism of quenching, since the formation of zeaxanthin from violaxanthin in the thylakoids is favored by the formation of a ΔpH and a correlation between zeaxanthin content and the extent of nonphotochemical quenching has been

TABLE 2
Effect of Chilling on the Quantum Efficiencies of CO_2 Assimilation (ϕ_{CO_2}) and PSII Photochemistry (ϕ_{PSII}) and Efficiency of Excitation Energy Transfer to PSII Reaction Centers (F_v'/F_m') in Wheat Leaf

Temperature	ϕ_{CO_2}	ϕ_{PSII}	F_v'/F_m'
19°C	0.052	0.53	0.63
15°C	0.042	0.44	0.59
10°C	0.034	0.36	0.54

Note: A wheat leaf was placed in a gas exchange chamber and exposed to a PPFD of 300 μmol m^{-2} s^{-1} and temperature of 19°C until steady-state photosynthesis was achieved. The leaf was maintained in an atmosphere containing 1% O_2 in order to minimize photorespiration. The parameters were then determined for the leaf, before lowering the temperature to 15°C and repeating the measurement procedure after the leaf had reached the steady-state for photosynthesis. The temperature was then lowered further to 10°C and the measurement procedures repeated.

Data from Groom, Q. J. and Baker, N. R., *Plant Physiol.*, 100, 1217, 1992.

found frequently in leaves.[29] It has also been suggested that aggregation of the light-harvesting chlorophyll *a/b* protein complex (LHCII) associated with PSII occurs as a result of the acidification of the thylakoid lumen and may be associated with light-induced increases in the rate of nonphotochemical quenching in the PSII antennae.[30]

a. Chilling and Downregulation of Photosynthesis

As discussed in Section II, lowering the temperature of a leaf will decrease the rate of photosynthesis, provided the leaf is operating at a photon flux density in excess of that required to achieve the maximum quantum yield of photosynthesis. This is shown clearly from the large reduction in the operating quantum yield (ϕ_{CO_2}) of a wheat leaf when the temperature is lowered from 19 to 15°C and then to 10°C while being maintained at a PPFD of 300 μmol m^{-2} s^{-1} (Table 2). The leaf was also exposed to 1% O_2 in order to minimize photorespiration and prevent the effects of temperature on photosynthesis being masked by different effects on photorespiration. The maximum quantum yield of CO_2 assimilation of this leaf at 19°C was 0.070. On increasing the

PPFD from below 100 μmol m^{-2} s^{-1} to 300 μmol m^{-2} s^{-1}, the operating quantum yield decreased from the maximum to 0.052; a concurrent decrease in F_v'/F_m' from 0.70 to 0.63 indicated that a light-induced downregulation of photosynthesis was occurring.[31] On lowering the temperature of the leaf while keeping the PPFD constant, ϕ_{CO_2}, ϕ_{PSII}, and F_v'/F_m' decreased (Table 2). Clearly, the decrease in F_v'/F_m' indicates that downregulation is contributing to the chill-induced decreases in the efficiency of light utilization for PSII photochemistry and CO_2 assimilation. As temperature decreases, there will be a reduction in the rate of carbon metabolism, thus reducing the effective sink size for the reductants and ATP generated from photosynthetic electron transport. This will result in a reduction in noncyclic electron flow and an increase in the transthylakoid ΔpH, which will lead to an increase in nonphotochemical quenching and a reduction in the efficiency of light use for PSII photochemistry.

Light-induced, nonphotochemical quenching has been demonstrated to occur in the leaves of a number of species under natural conditions in the field.[31-34] Depressions of F_v/F_m, measured after a short period of dark adaptation, when leaves are exposed to high-light and moderate stress conditions are indicative of a light-induced decrease in the maximum quantum efficiency of PSII photochemistry. In a study of wheat leaves in the field during the winter, it was found that F_v/F_m values measured on leaves at 12.00 GMT on 40 d from November to April correlated significantly with both the daily integrated PPFD and the temperature between dawn and midday.[31] This would support the hypothesis that reductions in PSII photochemical efficiency are related to exposure to low temperatures in the presence of high light.

The close relationship between chilling temperatures, high light, and inefficiency of PSII photochemistry has been shown even more dramatically by following the changes in F_v/F_m throughout the day on 4 days when leaves of winter wheat were experiencing contrasting climatic conditions (Figure 2).[31] This study also demonstrated the rapid response of the photosynthetic apparatus to climate change. On January 29 and February 5, PPFDs were relatively low and similar (<600 μmol m^{-2} s^{-1}), yet the temperatures on these days were very different; subzero temperatures were recorded on January 29, but the temperature only once fell below 5°C on February 5 when F_v/F_m remained close to its maximum level throughout the day. Whereas on January 23 there was a significant drop in F_v/F_m during the morning followed by a recovery during the afternoon; at about 16.00 GMT, F_v/F_m had almost recovered to the level observed on February 5. Clearly, on the colder of these two days, January 29, PSII photochemical efficiency was lower than on February 5. On both February 23 and March 30, the PPFD was greater than on January 29 and February 5, exceeding 800 μmol m^{-2} s^{-1} at midday (Figure 2B). February 23 was a cold day, with the temperature varying between 0°C at dawn to 8°C at 14.00 GMT, while March 30 was a relatively warm day, reaching 17.5°C at 15.00 GMT (Figure 2B). The greatest drops

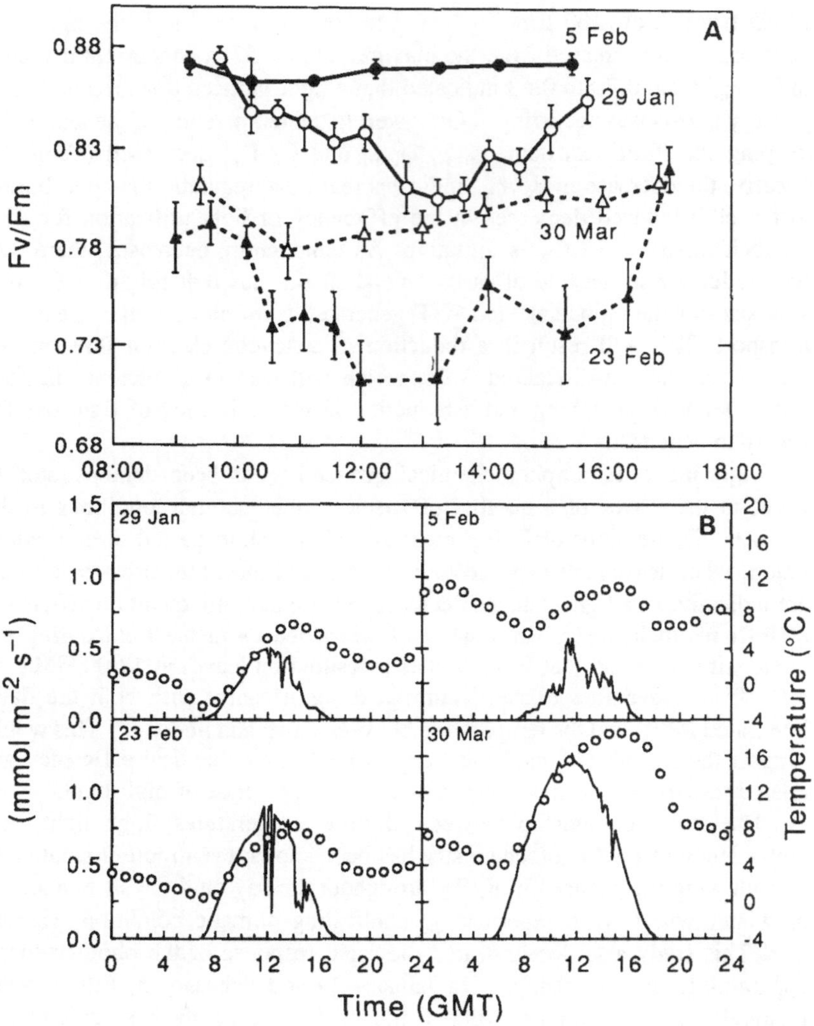

FIGURE 2. The effect of climate on the efficiency of PSII photochemistry in leaves of a winter wheat crop. A: F_v/F_m of leaves of a winter wheat crop on 4 days during the winter and early spring of 1989 with contrasting climatic conditions; January 29 (○), February 5 (●), February 23 (▲), and March 30 (△). Standard errors of the means are shown. F_v/F_m was measured after 10 min dark adaptation of the leaves to ensure that the primary quinone acceptor of PSII (Q_A) was maximally oxidized. B: The changes in PPFD (solid line) and air temperature (○) experienced by the crop during these days. (Data from Groom, Q. J. and Baker, N. R., *Plant Physiol.*, 100, 1217, 1992. With permission.)

in F_v/F_m were recorded on February 23. March 30 was the brightest and warmest day. Again, there were significant reductions in the F_v/F_m during the day, though not as large as on February 23. The reductions in F_v/F_m followed a rather different pattern from those observed on January 29 and February

23. F_v/F_m was already low when the first measurement was taken at 09.25 GMT; it then dropped further at 10.50 GMT, from where it gradually recovered during the rest of the day. It is evident that the large depressions in F_v/F_m that occurred during the morning on January 29 and February 23 are associated with low temperatures. However, it should be noted that these depressions in F_v/F_m are light dependent, with a recovery from the midday depression following the decline in the light level during the afternoon.

The possibility that light-induced decreases in F_v/F_m may be partially attributable to photodamage to PSII reaction centers (see Section IV.A.2) should not be ignored. However, in many cases the depressions in F_v/F_m in the field can be reversed within hours as light levels decrease or the stress is removed. Consequently, in these cases it is unlikely that photodamage to PSII is a major contributor to the decreases; repair of damaged PSII reaction centers would be expected to require considerably longer periods than the relaxation of light-induced downregulation.[26] Also, the absence of any change in the ability of thylakoids isolated from leaves of a winter wheat crop to bind 3-(3′,4′-dichlorophenyl)-1,1-dimethyl urea (DCMU) during the course of a day when changes in F_v/F_m occurred is indicative of a lack of photodamage to the PSII reaction centers (see Section IV.A.2). However, it must be emphasized that loss of photosynthetic efficiency due to photodamage to PSII may occur in the field in chilling-sensitive plants, such as maize, at moderate chilling temperatures and in chilling-tolerant plants, such as winter wheat, on exposure to severe low-temperature conditions, although this has not yet been demonstrated unequivocally.

2. Photodamage to PSII Reaction Centers

When the rate of excitation of the photosynthetic apparatus exceeds the rate at which energy can be dissipated by photochemical and nonphotochemical processes associated with the thylakoids, damage to PSII reaction centers can occur.[35,36] Excitation of PSII normally results in a reduction of primary and secondary quinone electron acceptors, Q_A and Q_B, respectively. Reduction of Q_B involves a two-electron transfer process requiring two sequential photochemical events; this mechanism is commonly termed the two-electron gate and is illustrated in Figure 3. Photoinhibitory damage involves impairment of quinone reduction followed by degradation of the 32-kDa D1 protein of the PSII reaction center.[38-40] Such damage is almost certainly taking place, albeit at a slow rate, in the majority of nonstressed leaves under normal growth conditions, since the D1 protein is continually and rapidly turning over in leaves, often with a half-time as short as 60 min.[40] However, no net accumulation of inactivated PSII complexes will occur in such cases, provided that the rate of repair of the damaged PSII population is equal to the rate of damage. Only if the capacity for repair is exceeded by the rate of photodamage will a loss in PSII activity become evident. This situation arises when leaves are exposed to light levels above those normally experienced or at considerably

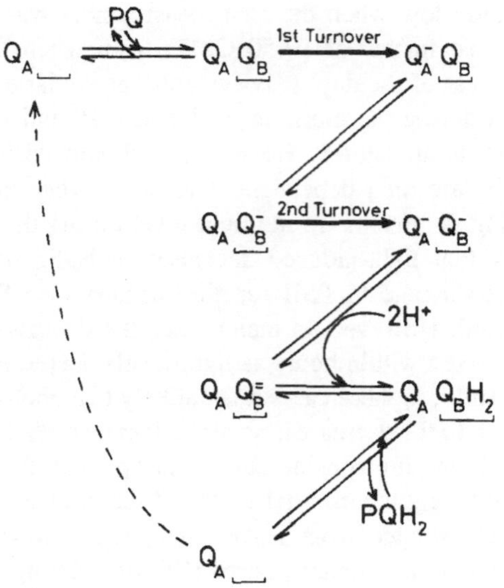

FIGURE 3. The reactions associated with the two-electron gate mechanism of quinone reduction by PSII. Excitation of PSII will generate the first turnover of the reaction center chlorophyll, P_{680}, and result in the transfer of an electron from P_{680} to the primary quinone acceptor, Q_A. The electron will then be transferred rapidly from Q_A^- to the bound plastoquinone, Q_B. The second turnover of the reaction center will produce a repeat of these events, with a second electron being transferred to the semiquinone, Q_B^-, to produce the doubly reduced quinone, Q_B^{2-}. Protonation of Q_B^{2-} produces plastoquinol, PQH_2, which readily dissociates from the binding site on the D1 protein of the reaction center. (Adapted from Ort, D. R., in *Encyclopedia of Plant Physiology, New Series,* Vol. 19, *Photosynthesis* III, Staehlin, L. A. and Arntzen, C. J., Eds., Springer-Verlag, Berlin, 1986, 143. With permission.)

lower light levels if photosynthetic carbon metabolism is inhibited. In both cases, the leaf does not have the capacity to dissipate the excess excitation energy by nonphotochemical processes associated with the light-induced downregulation discussed above (Section IV.A.1). A net loss of photochemically competent PSII complexes will result in a reduction in the maximal quantum efficiencies and possibly the light-saturated capacity for CO_2 assimilation in leaves.[7] This is seen clearly when maize leaves, grown at 25°C, are exposed to a chilling regime of 5°C for 6 h at a PPFD of 1500 μmol m^{-2} s^{-1}. Decreases of about 50% were induced in ϕ_{max}, F_v/F_m, and the ability of thylakoids isolated from the leaves to bind atrazine.[41] Since atrazine, like DCMU, binds to a site on the D1 protein close to the site for plastoquinone binding,[42] the capacity of thylakoids to bind this herbicide provides a good quantitative measure of the photoinhibitory damage to the plastoquinone-reductase sites on the D1 proteins. Recovery of the maize leaf from a 50% damage to its PSII population when returned to normal temperature and light regime (25°C and a PPFD of 250 μmol m^{-2} s^{-1}) took in excess of 48 h,[41] indicating that repair of the damaged D1 proteins does not occur as rapidly as recovery from light-induced downregulation of photosynthesis.

Although the inability of the PSII reaction center to photoreduce the bound quinone (Q_B) is symptomatic of the conditions leading to the onset of photodamage, it is now considered that the Q_B site on the D1 protein has an initiating rather than an active role in the events leading to the photodamage of the protein.[40] In the normal course of events, plastoquinone binds to the Q_B site and, on excitation of PSII, is reduced initially by a single electron transfer from Q_A to a plastosemiquinone; a second electron transfer event then effects a reduction of the semiquinone anion to produce plastoquinol, the fully reduced form of plastoquinone, PQH_2 (Figure 3).[37] The affinity of plastoquinol for the Q_B site is low in comparison with that of plastosemiquinone.[42] Consequently, when the free plastoquinone pool in the thylakoids is highly reduced, as will be the case under many stress conditions, the Q_B sites will have a low probability of binding plastoquinone. However, excitation of the PSII reaction center chlorophyll, P_{680}, will still occur and result in photochemical reduction of the bound primary quinone acceptor, Q_A, to not only the semiquinone anion, but also to a doubly reduced, protonated state, Q_AH_2.[40] Further excitation of the reaction center cannot result in any further quinone reduction; primary charge separation will occur and give rise to the formation of the radical pair P_{680^+}/pheophytin$^-$.[40] Recombination of charge will generate the triplet state of P_{680}; this can react rapidly with O_2, producing the highly reactive and toxic singlet O_2, which leads to the destruction of P_{680} followed by degradation of the D1 protein.[40] A number of studies have suggested the possibility that perturbations to the donor-side reactions of the PSII reaction center could be associated with photoinhibition.[6] Reduction in the rate of electron donation to P_{680^+} will increase the lifetime of this species and lead to the oxidation and destruction of neighboring chlorophyll and carotenoid species, followed by damage to the D1 protein.[38] Consequently, stress-induced decreases in water splitting, even under moderate light conditions, could result in damage to PSII reaction centers.

The loss of leaf photosynthetic activity due to photoinhibitory damage to PSII complexes will not only be dependent upon the amount of damage to D1 proteins, but will also be determined by the ability to degrade the damaged D1 proteins and synthesize and integrate new D1 proteins into the damaged complexes. Consequently, stress-induced modifications to the mobility of PSII complexes in the thylakoid membrane, to the protease activity required to degrade the damaged D1 protein, and to the ability to synthesize and integrate D1 protein into PSII complexes could be factors in determining the degree and persistence of loss of photosynthetic competence during and after the removal of a photoinhibitory stress. There is evidence that recovery from photoinhibition in leaves requires protein synthesis on chloroplast ribosomes,[43] consistent with a need to synthesize D1 protein, which is encoded by the chloroplast genome.

It should be emphasized that photodamage to PSII reaction centers will only occur under environmental conditions when downregulation is unable to dissipate excess excitation energy in the PSII antennae as heat. Recovery of a photodamaged PSII population in a leaf when an environmental stress condition is removed will be dependent on protein synthesis and processing, and consequently the leaf will require considerably longer (generally on the order of hours) to recover from such damage compared to the time required for the relaxation of downregulation (on the order of minutes). Proteases involved in the degradation of D1 protein have been found to be temperature sensitive, with the rate of breakdown of damaged D1 protein being considerably reduced at low temperatures.[44,45] This is consistent with the large effects of temperature on other chloroplast peptidases.[46]

Although PSI can be inhibited by high-light treatments of isolated chloroplasts and thylakoids, photoinhibition and damage to PSI does not appear to occur at any significant level in leaves under physiological conditions. This is not surprising since both P_{700} and its oxidized form, P_{700^+}, can act as efficient quenchers of excitation energy in the PSI antennae, dissipating the energy as heat.

3. Role of Oxygen in Photoinhibition

Oxygen can play an important role in determining the nature and extent of photoinhibition and photodamage to the photosynthetic apparatus in three ways: (1) as a substrate in photorespiration, (2) as an acceptor of electrons from components of the thylakoid membrane, and (3) in excitation energy transfer from excited chlorophyll molecules.

a. Photorespiration

In C_3 plants, the photosynthetic carbon oxidation cycle is an important sink for the reductants and ATP generated from photosynthetic electron transport.[47] Consequently photorespiratory metabolism has the potential to maintain photosynthetic electron transport at high rates when photosynthetic carbon metabolism is inhibited, and thus protect the photosynthetic apparatus from photoinhibition and photodamage.[48] Such protection will only occur if the inhibition of photosynthetic carbon metabolism is not accompanied by inhibition of photorespiratory metabolism. It is unlikely that this would be the case under chilling conditions, since there is no reason to suppose that the Q_{10} values for the enzymes of the photosynthetic carbon reduction (PCR) and oxidation (PCO) cycles are very different. Thus photorespiration is unlikely to play a major role in providing protection from photoinhibition and photodamage at low temperatures. It should also be noted that decreases in temperature will increase the solubility of CO_2 and the specificity of RuBisCo for CO_2 relative to O_2, and consequently increase the potential rate of carboxylation relative to the oxygenation of RuBP by RuBisCo.[49]

b. Photoreduction of Oxygen

It has been known for some 40 years that thylakoids can photoreduce O_2 with the subsequent production of H_2O_2, rather than water, in a reaction generally termed the Mehler reaction.[50-52] There is evidence that this reaction can occur *in vivo* in nonstressed leaves at rates approaching 30% of total noncyclic electron transport.[53,54] It would also appear that pseudocylic electron flow from water to O_2 is essential to increase the stoichiometry of ATP:NADPH production by the thylakoids for operation of the PCR cycle.[54] The ATP:NADPH ratio required for CO_2 assimilation is 1.5, yet noncyclic electron transport to NADP will result in an ATP:NADPH ratio of only 1.33, assuming a requirement of 3 H^+ per ATP synthesized, and be unable to sustain operation of the PCR cycle; the requirement for ATP relative to reductant will become even greater if photorespiration or starch metabolism is also occurring.[55] It is generally argued that the extra ATP required to sustain photosynthetic metabolism is provided by PSI-driven cyclic electron flux, although there is no convincing evidence that this is the case in leaves operating under normal physiological conditions. Simultaneous measurements of the quantum efficiencies of PSI and PSII in leaves under a wide range of operating conditions have indicated that if cyclic electron transport does occur in leaves, it does so at a rate that is always a constant proportion of the rate of noncyclic electron flux, and cannot be modulated to meet any changing ATP demand.[6] Consequently, pseudocyclic electron flux to O_2 becomes a very attractive explanation for the mechanism to increase the ATP:NADPH ratio generated by photosynthetic electron transport. Clearly, under photoinhibitory conditions there is the potential for pseudocyclic electron flux to O_2 to act as a protective sink for electrons and prevent overexcitation of PSII reaction centers, providing the toxic products of O_2 reduction, i.e., superoxide anion (O_2^-) and H_2O_2, can be effectively scavenged. However, it should be realized that for such a protection to occur, there must be available sinks for the ATP which will be produced as a result of the pseudocyclic electron flux from water to O_2, or the leaf must possess a mechanism for uncoupling photosynthetic electron flux from ATP synthesis, which is unlikely since no evidence for such uncoupling *in vivo* exists. If the ATP produced from pseudocyclic electron flux cannot be utilized, the resulting high transthylakoid ΔpH will restrict further electron flow and lead to overexcitation of the PSII reaction centers.

The most likely site in the thylakoids for the photoreduction of O_2 is the reducing side of PSI, although O_2 can also accept electrons via ferredoxin and from the reducing side of PSII.[54] Analysis of fluorescence transients has suggested that O_2 may be accepting electrons directly from Q_A in leaves.[56] It has been suggested that the inhibition of O_2 photoreduction by DCMU indicates that Q_A is not the site of photoreduction;[54] however, it is quite possible that when DCMU binds to the Q_B site on the D1 protein of the PSII reaction center, it prevents electron donation from Q_A to O_2. If PSII reduces O_2 directly at significant rates in leaves, this would be particularly important as a mechanism for protection from photodamage, since this reaction would involve proton accumulation in the thylakoid lumen from only the water oxidation site and not the plastoquinone shuttle. Consequently, the number of H^+ accumulated in the thylakoid lumen from PSII-mediated electron transfer from water to O_2 will be half, or less than half if a *Q* cycle operates through the cytochrome *b*/*f* complex,[37] of that accumulated for electron transfer from water to O_2 via both PSII and PSI in the classical Mehler reaction. The reduced rate of H^+ accumulation associated with direct reduction of O_2 by PSII, compared to PSII- and PSI-mediated O_2 reduction, will reduce the trans-thylakoid ΔpH and result in reduced restriction on the rate of electron transfer, which will effectively decrease the probability of photodamage to PSII.

The effectiveness of O_2 photoreduction as a mechanism to protect from photoinhibition will also be dependent on the ability to scavenge rapidly the resulting toxic products. The $O_2^{\cdot-}$ generated by the photoreduction of O_2 can reduce cellular components, such as cytochrome *f* and plastocyanin, or undergo a dismutation reaction:

$$O_2^{\cdot-} + O_2^{\cdot-} + 2H^+ \rightarrow H_2O_2 + O_2 \quad (7)$$

which will be accelerated by about four orders of magnitude in the presence of superoxide dismutase (SOD) enzymes.[57] Although the bulk of the H_2O_2 produced in leaves is considered to be from SOD-catalyzed reactions, small amounts can also be generated from the reduction of $O_2^{\cdot-}$ by chloroplast components such as reduced ferredoxin, glutathione, and ascorbate, and Mn^{2+}.[54] Chloroplasts can detoxify H_2O_2 using a metabolic cycle involving ascorbate and glutathione, and the enzymes ascorbate peroxidase (APX), monodehydroascorbate reductase (MDHAR), dehydroascorbate reductase (DHAR), and glutathione reductase (GR).[54] Clearly, the activity of these enzymes and SOD is crucial to the leaf being able to utilize photoreduction of O_2 as a mechanism for protection from photodamage. Recently, the effects of chilling temperature on these enzymes extracted from *Z. mays* and *Z. diploperennis* have been examined (Table 3).[59] Depression of the temperature from 19 to 5°C reduced the activity of SOD and APX by less than 40%, whereas MDHAR, DHAR, and GR were considerably more sensitive to the low temperature. The very large depressions in the activities of these enzymes would result in a decrease in the ability of the chloroplasts to scavenge H_2O_2, and consequently be potentially damaging for the chloroplasts. Such a situation would not be favorable for the protection of PSII from photodamage by photoreduction of O_2. However, it should be noted that these *Zea* species are considered to be sensitive to chilling, and the effects of chilling on the scavenging enzymes of chilling-resistant species may be quite different.

TABLE 3
Effect of Temperature Depression on the Specific Activity of Enzymes Involved in the Scavenging of Active O_2 Species in *Zea mays* and *Z. diploperennis*

Enzyme	Assay temperature (°C)	Specific activity (μmol mg^{-1} [protein] h^{-1}) *Z. mays*	*Z. diploperennis*
Superoxide dismutase[a]	19	17.2	39.5
	5	15.0	33.2
		(13%)	(16%)
Ascorbate peroxidase	19	52.5	107.2
	5	32.2	72.9
		(39%)	(32%)
Monodehydroascorbate reductase	19	12.2	10.3
	5	4.2	3.5
		(66%)	(67%)
Dehydroascorbate reductase	19	11.6	13.4
	5	0	0.5
		(100%)	(96%)
Glutathione reductase	19	5.5	4.7
	5	1.2	1.7
		(78%)	(63%)

Note: Plants were grown at 25°C during the 18-h photoperiod and at 20°C during the night. Figures in parentheses indicate the percentage decrease in enzyme-specific activity at 5°C relative to 19°C.

[a] Activity of SOD is expressed in mg^{-1} (protein) h^{-1}, as defined by Beyer and Fridovich.[58] Data from Janke, L. S., Hull, M. R., and Long, S. P., *Plant Cell Environ.*, 14, 97, 1991. With permission.

c. Singlet Oxygen

Excitation of a chlorophyll molecule in the ground state, ^{1}Chl, generates the excited singlet state, ^{1}Chl*, which can then be converted to the triplet state, ^{3}Chl*. Oxygen is capable of being excited to the highly reactive singlet state, $^1O_2$, on interaction with chlorophyll in a triplet state:

$$^1\text{Chl} \xrightarrow{h\nu} {}^1\text{Chl}^* \rightarrow {}^3\text{Chl}^*$$

$$^3\text{Chl}^* + O_2 \rightarrow {}^1\text{Chl} + {}^1O_2 \quad (8)$$

Although O_2 can accept excitation energy from chlorophyll in this way, $^1O_2$ is an extremely reactive species and will interact particularly with compounds containing conjugated double bonds.[54,57] For example, $^1O_2$ can stimulate the peroxidation of fatty acids and membrane lipids.[57] Carotenoids (Car) give protection to components of the thylakoid membrane from the potential oxidative damage by quenching the energy directly from the triplet-state chlorophylls:

$$^3Chl^* + {}^1Car \rightarrow {}^1Chl + {}^3Car^* \qquad (9)$$

Carotenoids can also be converted to the triplet state by reacting with $^1O_2$ to produce O_2:

$$^1Car + {}^1O_2 \rightarrow {}^3Car^* + O_2 \qquad (10)$$

Carotenoids in the triplet state will decay harmlessly to the ground state, with the energy being dissipated as heat:

$$^3Car^* \rightarrow {}^1Car + heat \qquad (11)$$

At low temperatures, carotenoids will continue to quench excitation energy from chlorophylls and inactivate $^1O_2$. However, it is unlikely in most plants that carotenoids will have the capacity to quench at a rate sufficient to prevent an increase in the excitation density of the pigment matrices and buildup of $^1O_2$ when photosynthetic metabolism is severely inhibited at low temperatures.

B. MEMBRANE PHASE TRANSITIONS

In the past, it has been frequently speculated in the literature that the inhibition of photosynthesis by chilling may be associated with a transition in the physical state of thylakoid lipids from a liquid crystal to a gel state. However, chilling-induced inhibition of photosynthesis will often persist for considerable periods after the return of the plants to optimal growth temperatures. Consequently, a change in the physical state of the membrane lipids, which would be rapidly reversible, cannot be the sole cause of the inhibition of photosynthesis, although it is possible that lipid phase transitions could induce secondary effects which inhibit photosynthesis and do not recover rapidly with a rise in temperature. It is now generally believed that the bulk membrane lipids of chilling-sensitive plants do not undergo gel-liquid crystal phase transitions; however, the possibility that minor lipid components may undergo such physical changes with temperature change and induce membrane dysfunction should not be excluded.[60] Subtle changes in the interactions between nonbilayer-forming lipids and protein complexes in the thylakoids could result in temperature-induced changes in the activity of membranes[60] and contribute to depressions in photosynthetic activities, although as yet there is no definitive evidence for this.

V. DEVELOPMENT AND CHILLING

The majority of studies on the effects of low temperatures on photosynthesis have involved chilling of mature leaves. However, for many plants in the natural environment perhaps the most important, and least studied, effects of low temperatures are on the development of a physiologically competent photosynthetic apparatus. Such effects can often be mediated by exposure of leaves to low temperatures in the dark, and consequently are not associated with photoinhibition of chloroplast systems.

A. CHILLING-SENSITIVE PLANTS

Many chilling-sensitive plants, such as maize, develop small, chlorotic leaves at low temperatures. This has important implications for the ability of the leaf to capture light and for its photosynthetic capacity and efficiency. Reduction in growth temperature results not only in decreases in the rate of expansion and size of leaf attained, but also in chlorophyll and thylakoid contents per unit leaf area, which in turn contribute to the observed decreases in the light-saturated capacity for photosynthesis (Table 4).

It is evident that in maize, depressions in growth temperature can have a dramatic effect on the amount of photosynthetic apparatus per unit area accumulated in the leaf, as indicated by the large decreases in chlorophyll and thylakoid protein contents per unit leaf area, as growth temperature is lowered (Table 4). Such reductions in the quantity of thylakoid components would obviously reduce the capacity of the leaf to generate reductants and ATP and inevitably lead to decreases in A_{sat}. Decreases in the activities per unit leaf area of some key enzymes involved in carbon metabolism also occur at suboptimal growth temperatures.[63,64] A reduction in growth temperature from 25 to 14°C produced decreases in the activity per unit leaf area of FBPase, pyruvate orthophosphate dikinase, phosphoenol pyruvate carboxylase, and NADP-dependent malate dehydrogenase.[63] These decreases were considerably less than, and could not account for, the accompanying 80% decrease in A_{sat}.[63] However, the possibility of RuBisCo content and activity limiting carbon assimilation when leaves are grown at low temperatures should not be discounted; decreases in excess of 70% have been observed in RuBisCo levels when developing maize leaves were chilled at 6°C for 1 or 2 d.[65]

In maize, low growth temperatures not only result in a decrease in the amount of photosynthetic apparatus per leaf area, but also in the preferential loss of specific thylakoid proteins. Although there was a decrease in the content of all thylakoid proteins per unit leaf area as the growth temperature was lowered, a selective inhibition on the accumulation of thylakoid proteins encoded by the chloroplast genome, relative to proteins encoded in the nuclear genome, was observed.[61] The reason for this selective effect on chloroplast-encoded proteins is not clear. Studies of protein synthesis in leaves grown at 14 and 25°C indicated that low temperature did not preferentially inhibit

TABLE 4
Effects of Growth Temperature on Expansion, Chlorophyll Content, and Photosynthetic Activities of Maize Leaves

Parameter	Growth temperature		
	14°C	17°C	25°C
Maximum leaf length (cm)	9.9	13.3	16.2
Time to reach maximum length (days)	28	19	6
Chlorophyll per leaf area ($\mu g\ cm^{-2}$)	12.9	43.0	46.2
Thylakoid protein per leaf area ($\mu g\ cm^{-2}$)	85	284	305
A_{sat}($\mu mol\ m^{-2}\ s^{-1}$)	5.4	16.3	30.7
ϕ_{max}	0.013	0.044	0.058
Maximum PSI activity per chlorophyll (nmol O_2 mg^{-1} [Chl] s^{-1})	19.0	30.7	33.1
Maximum PSII activity per chlorophyll (nmol O_2 mg^{-1} [Chl] s^{-1})	12.2	16.8	26.1
Maximum PSI activity per leaf area (% relative to 25°C sample)	16	87	100
Maximum PSII activity per leaf area (% relative to 25°C sample)	13	60	100

Note: Maize variety LG11 was grown in controlled environment cabinets with air temperatures of 14, 17, or 25°C during a 16-h photoperiod with a PPFD of 230 $\mu mol\ m^{-2}\ s^{-1}$. Night temperatures were 13, 14, and 22°C, respectively.

Baker, N. R. and Nie, G. Y., in *Biotechnology in Agriculture and Forestry: Maize*, Bajaj, Y. P. S., Ed., Springer-Verlag, Berlin, in press; Nie, G. Y. and Baker, N. R., *Plant Physiol.*, 95, 184, 1991; Nie, G. Y., Long, S. P., and Baker, N. R., *Physiol. Plant.*, 85, 554, 1992. With permission.

chloroplasts with respect to cytoplasmic, protein synthesis; chloroplast-encoded proteins, particularly the D1 protein, were synthesized in the 14°C-grown leaves at 14°C.[61] The possibility that chill-induced changes in thylakoid lipid composition may alter the stability of chloroplast-encoded proteins in the macromolecular protein complexes in the thylakoid membrane has been suggested,[61] although there is no clear evidence for this as yet. The deficiency in chloroplast-encoded, relative to the nuclear-encoded, proteins in the thylakoids of leaves grown at suboptimal growth temperatures has important implications for the quantum efficiency of photosynthesis. The decreases in the chloroplast-encoded reaction centers of PSI and PSII relative to the nuclear-encoded light-harvesting chlorophyll *a*/*b* proteins almost certainly contribute to the decreases in ϕ_{max} at low growth temperatures (Table 4). Inhibition of peptidases and proteases involved in protein processing and breakdown also appears to be a factor associated with perturbations of chloroplast development in maize leaves exposed to low temperatures.[45]

Severe impairment of the synthesis of chloroplast proteins has been observed at chilling temperatures in rice[66] and rape,[67] and on rewarming after chilling in tomato.[68] Recently, it has been shown in tomato that chilling-induced changes in the synthesis of LHCII polypeptides and RuBisCo activase are attributable to interference with the circadian rhythm controlling the expression of the genes coding for these proteins.[69] Clearly, many factors are potentially involved with the chilling sensitivity of chloroplast development and the attainment of photosynthetic competence at low temperatures. Parallel studies of chloroplast development and photosynthetic performance in maize, cucumber, and tomato have indicated that there are distinctive species-dependent differences in the response to low temperature and recovery on return to ambient growth temperature,[70] and consequently suggest that there may be no common molecular basis to chilling sensitivity between species.

B. CHILLING-TOLERANT PLANTS

Many plants can tolerate and rapidly recover from periods of chilling.[71] Tolerance to chilling has frequently been associated with acclimatory responses of the plants to suboptimal growth temperatures, which involve complex biochemical and physiological changes.[71,72] Many plants exhibit an increased resistance to low temperatures when they have been exposed to periods of sublethal chilling temperatures. This process of acclimation to chilling conditions is generally termed hardening. Although an extensive consideration of cold acclimation is beyond the scope of this chapter, a number of points relating specifically to chill-induced inhibitions of photosynthesis should be considered in the context of photoinhibition.

Many changes, albeit often rather small, have been found to occur in thylakoid membranes during cold acclimation of leaves.[71] These structural and functional changes induced in the photosynthetic apparatus do not modify ϕ_{max}[71] and consequently appear to be of no major physiological significance in terms of the efficiency of light utilization. However, cold acclimation involves biochemical changes associated with increased protection of the photosynthetic apparatus from photoinhibition.[73] Spinach that has been cold hardened has been shown to be more resistant to chill-induced photoinhibition than nonhardened control plants.[74,75] Cold-acclimated leaves were found to have increased activities of SOD, APX, MDHAR, and GR in the chloroplasts, and thylakoids contained increased levels of the xanthophylls, lutein, zeaxanthin, and violaxanthin, compared to nonhardened controls.[73,76] Also, the affinity of GR for its oxidized glutathione substrate was increased fourfold by cold hardening.[77] The carotenoid changes will increase the capacity of the thylakoids for downregulation, while increased capacity for scavenging active O_2 species and their products may decrease the potential for photodamage to PSII. Interestingly, in maize, a chilling-sensitive plant, exposure of leaves to 10°C produced immediate transient increases in the specific activities of APX and MDHAR; however, after 5 d at 10°C, decreases in the activities of APX and GR were found, and the activity of MDHAR was little different from that in control leaves.[59] When maize was grown at 14°C, large increases in the specific activities of SOD and GR were found compared to the activities in 25°C-grown leaves; however, MHDAR and DHAR activities were not significantly different and were considerably lower on a unit-leaf-area basis in leaves grown at 14°C compared to those grown at 25°C.[63] It would appear that maize leaves, unlike spinach leaves, are unable to increase their capacity to scavenge active O_2 species on chilling, and this may be an important factor contributing to the high susceptibility of maize to photoinhibition and photodamage at low temperatures.

VI. CONCLUDING REMARKS

It is evident from the discussions above that decreases in temperature can affect the photosynthetic performance of plants in many ways, and different species (and often different varieties) can exhibit quite different responses to similar decreases in temperature. Chilling inevitably has immediate and major consequences for the photosynthetic performance of leaves, via thermodynamic effects on the constituent physicochemical and metabolic reactions, photoinhibition, and possibly photodamage. Perturbations of the processes involved in the development and maintenance of the photosynthetic apparatus may not be so immediately apparent and require a considerable period, hours and often days, to be expressed. However, it would seem that such effects on transcription, translation, and posttranslational processes are potentially more important in chill-induced depressions in the photosynthetic productivity of plants in the longer term. Unfortunately, considerably less effort has been given to determining the effects of low temperatures on developmental and maintenance processes, and consequently we are almost certainly ignorant of many of the influences of temperature on the developmental biology of leaves. Hopefully, the rapid and impressive developments in plant molecular biology over the past decade will provide the basis for resolving the mechanistic bases of low-temperature-induced perturbations of leaf development and loss of photosynthetic productivity. Such information will be crucial to the future development of crop varieties with improved tolerance to chilling, since it will allow identification of appropriate characteristics for improved screening procedures and of the molecular sites to be manipulated in genetic engineering programs.

REFERENCES

1. **Boyer, J. S.,** Plant productivity and the environment, *Science,* 218, 443, 1982.
2. **Nobel, P. S.,** *Introduction to Biophysical Plant Physiology,* W. H. Freeman, San Francisco, 1974, chap. 3.

3. **Morris, J. G.,** *A Biologist's Physical Chemistry,* Edward Arnold, London, 1974, chap. 10.
4. **Thornley, H. M. and Johnson, I. R.,** *Plant and Crop Modelling: A Mathematical Approach to Plant and Crop Physiology,* Oxford Science Publications, Oxford, 1990, 228.
5. **Farquhar, G. D., von Cammaerer, S., and Berry, J. A.,** A biochemical model of photosynthetic CO_2 assimilation in leaves of C3 species, *Planta,* 149, 78, 1980.
6. **Baker, N. R. and Ort, D. R.,** Light and crop photosynthetic performance, in *Crop Photosynthesis: Spatial and Temporal Determinants,* Baker, N. R. and Thomas, H., Eds., Elsevier, Amsterdam, 1992, 289.
7. **Baker, N. R., Long, S. P., and Ort, D. R.,** Photosynthesis and temperature, with particular reference to effects on quantum yield, in *Plants and Temperature,* Long, S. P. and Woodward, F. I., Eds., Company of Biologists, Cambridge, 1988, 347.
8. **Ehleringer, J. and Björkman, O.,** Quantum yields for CO_2 uptake in C_3 and C_4 plants: dependence on temperature, CO_2 and O_2 concentrations, *Plant Physiol.,* 59, 86, 1977.
9. **Pearcy, R. W. and Ehleringer, J.,** Comparative ecophysiology of C_3 and C_4 plants, *Plant Cell Environ.,* 7, 1, 1984.
10. **Servaites, J. C. and Ogren, W. L.,** Oxygen inhibition of photosynthesis and stimulation of photorespiration in soybean leaf cells, *Plant Physiol.,* 61, 62, 1978.
11. **Martin, B. and Ort, D. R.,** The recovery of photosynthesis in tomato subsequent to chilling exposure, *Photosynth. Res.,* 6, 121, 1985.
12. **Strand, M. and Öquist, G.,** Inhibition of photosynthesis by freezing temperatures and high light levels in cold-acclimated seedlings of Scots pine *(Pinus sylvestris).* I. Effects on light-limited and light-saturated rates of CO_2 assimilation, *Physiol. Plant.,* 64, 425, 1985.
13. **Bongi, G. and Long, S. P.,** Light-dependent damage to photosynthesis in olive leaves during chilling and high temperature stress, *Plant Cell Environ.,* 11, 481, 1987.
14. **Sassenrath, G. F., Ort, D. R., and Portis, A. R.,** Impaired reductive activation of stromal bisphosphatases in tomato leaves following low-temperature exposure at high light, *Arch. Biochem. Biophys.,* 282, 302, 1990.
15. **Long, S. P.,** C_4 photosynthesis at low temperatures, *Plant Cell Environ.,* 6, 345, 1983.
16. **Labate, C. A., Adcock, M. D., and Leegood, R. C.,** Effects of temperature on the regulation of photosynthetic carbon assimilation in leaves of maize and barley, *Planta,* 181, 547, 1990.
17. **Genty, B., Briantais, J.-M., and Baker, N. R.,** The relationship between the quantum yield of photosynthetic electron transport and quenching of chlorophyll fluorescence, *Biochim. Biophys. Acta,* 990, 87, 1989.
18. **Butler, W. L. and Kitajima, M.,** Fluorescence quenching in photosystem II of chloroplasts, *Biochim. Biophys. Acta,* 376, 116, 1975.
19. **Genty, B., Harbinson, J., and Baker, N. R.,** Relative quantum efficiencies of the two photosystems of leaves in photorespiratory and non-photorespiratory conditions, *Plant Physiol. Biochem.,* 28, 1, 1990.
20. **Krall, J. P. and Edwards, G. E.,** Quantum yields of photosystem II electron transport and carbon dioxide fixation in C4 plants, *Aust. J. Plant Physiol.,* 17, 579, 1990.
21. **Keiler, D. R. and Walker, D. A.,** The use of chlorophyll fluorescence to predict CO_2 fixation during photosynthetic oscillations, *Proc. R. Soc. London Ser. B,* 241, 59, 1990.
22. **Seaton, G. G. R. and Walker, D. A.,** Chlorophyll fluorescence as a measure of photosynthetic carbon assimilation, *Proc. R. Soc. London Ser. B,* 242, 29, 1991.
23. **van Wijk, K. J. and van Hasselt, P. R.,** The quantum efficiency of photosystem II and its relationship to non-photochemical quenching of chlorophyll fluorescence; the effect of measuring and growth temperature, *Photosynth. Res.,* 25, 233, 1990.
24. **Harbinson, J., Genty, B., and Baker, N. R.,** Relationship between the quantum efficiencies of photosystems I and II in pea leaves, *Plant Physiol.,* 90, 1029, 1989.

25. **Harbinson, J., Genty, B., and Baker, N. R.,** The relationship between CO_2 assimilation and electron transport in leaves, *Photosynth. Res.*, 25, 213, 1990.
26. **Krause, G. H. and Weis, E.,** Chlorophyll fluorescence and photosynthesis: the basics, *Annu. Rev. Plant Physiol.*, 42, 313, 1991.
27. **Genty, B., Harbinson, J., Briantais, J.-M., and Baker, N. R.,** The relationship between non-photochemical quenching of chlorophyll fluorescence and the rate of photosystem 2 photochemistry in leaves, *Photosynth. Res.*, 25, 249, 1990.
28. **Rees, D., Noctor, G. D., and Horton, P.,** The effect of high-energy-state excitation quenching on maximum and dark level fluorescence, *Photosynth. Res.*, 25, 199, 1990.
29. **Demmig-Adams, B.,** Carotenoids and photoprotection in plants: a role for the xanthophyll zeaxanthin, *Biochim. Biophys. Acta*, 1020, 1, 1990.
30. **Horton, P., Ruban, A. V., Rees, D., Pascal, A. A., Noctor, G., and Young, A. J.,** Control of the light-harvesting function of chloroplast membranes by aggregation of the LHCII chlorophyll-protein complex, *FEBS Lett.*, 292, 1, 1991.
31. **Groom, Q. J. and Baker, N. R.,** Analysis of light-induced depressions of photosynthesis in leaves of a wheat crop during the winter, *Plant Physiol.*, 100, 1217, 1992.
32. **Adams, W. W., Terashima, I., Brugnoli, E., and Demmig, B.,** Comparisons of photosynthesis and photoinhibition in the CAM vine *Hoya australis* and several C3 vines on the coast of eastern Australia, *Plant Cell Environ.*, 11, 173, 1988.
33. **Ögren, E.,** Photoinhibition of photosynthesis in willow leaves under field conditions, *Planta*, 175, 229, 1988.
34. **Groom, Q. J., Baker, N. R., and Long, S. P.,** Photoinhibition of holly *(Ilex aquifolium)* in the field during the winter, *Physiol. Plant.*, 83, 585, 1991.
35. **Kyle, D. J.,** The biochemical basis for photoinhibition of photosystem II, in *Photoinhibition*, Kyle, D. J., Osmond, C. B., and Arntzen, C. J., Eds., Elsevier, Amsterdam, 1987, 197.
36. **Critchley, C.,** The molecular mechanism of photoinhibition — facts and fiction, *Aust. J. Plant Physiol.*, 15, 27, 1988.
37. **Ort, D. R.,** Energy transduction in oxygenic photosynthesis: an overview of structure and mechanism, in *Encylopedia of Plant Physiology, New Series*, Vol. 19, *Photosynthesis III*, Staehelin, L. A. and Arntzen, C. J., Eds., Springer-Verlag, Berlin, 1986, 143.
38. **Adir, N., Shochat, S., Inoue, Y., and Ohad, I.,** Mechanism of the light dependent turnover of the D1 protein, in *Current Research in Photosynthesis*, Vol. 2, Balscheffsky, M., Ed., Kluwer, Dordrecht, 1990, 409.
39. **Styring, S., Jägerschöld, C., Virgin, I., Ehrenberg, A., and Adersson, B.,** On the mechanisms for the photoinhibition of the electron transfer and the light-induced degradation of the D1 protein in photosystem II, in *Current Research in Photosynthesis*, Vol. 2, Balscheffsky, M., Ed., Kluwer, Dordrecht, 1990, 349.
40. **Barber, J. and Andersson, B.,** Too much of a good thing: light can be bad for photosynthesis. *Trends Biochem. Sci.*, 17, 61, 1992.
41. **Ortiz-Lopez, A., Nie, G. Y., Ort, D. R., and Baker, N. R.,** The involvement of the photoinhibition of photosystem II and impaired membrane energization in the reduced quantum yield of carbon assimilation in chilled maize, *Planta*, 181, 78, 1990.
42. **Bowyer, J. R., Camilleri, P., and Vermaas, W. F. J.,** Photosystem II and its interaction with herbicides, in *Herbicides*, Baker, N. R. and Percival, M. P., Eds., Elsevier, Amsterdam, 1991, 27.
43. **Greer, D. H., Berry, J. A., and Björkman, O.,** Photoinhibition of photosynthesis in intact bean leaves: role of light and temperature, and requirement for chloroplast protein synthesis during recovery, *Planta*, 168, 253, 1986.
44. **Aro, E.-M., Hundal, T., Carlberg, I., and Anderssson, B.,** *In vitro* studies on light-induced inhibition of photosystem II and D1-protein degradation at low temperatures, *Biochim. Biophys. Acta*, 1019, 269, 1990.

45. **Baker, N. R. and Nie, G. Y.,** Chilling sensitivity of photosynthesis in maize, in *Biotechnology in Agriculture and Forestry: Maize,* Bajaj, Y. P. S., Ed., Springer-Verlag, Berlin, in press.
46. **Covello, P. S., Hayden, D. B., and Baker, N. R.,** The roles of low temperature and light in accumulation of a 31 kDa polypeptide in the light-harvesting apparatus of maize leaves, *Plant Cell Environ.,* 11, 481, 1988.
47. **Lorimer, G. H. and Andrews, T. J.,** The C_2 chemo- and photorespiratory carbon oxidation cycle, in *The Biochemistry of Plants,* Vol. 8, Hatch, M. D. and Boardman, N. K., Eds., Academic Press, New York, 1981, 329.
48. **Osmond, C. B.,** Photorespiration and photoinhibition. Some implications for the energetics of photosynthesis, *Biochim. Biophys. Acta,* 639, 77, 1981.
49. **Long, S. P. and Drake, B. G.,** Photosynthetic CO_2 assimilation and rising atmospheric CO_2 concentrations, in *Crop Photosynthesis: Spatial and Temporal Determinants,* Baker, N. R. and Thomas, H., Eds., Elsevier, Amsterdam, 1992, 69.
50. **Mehler, A. H.,** Studies on reactions of illuminated chloroplast. I. Mechanism of the reduction of oxygen and other Hill reagents, *Arch. Biochem. Biophys.,* 33, 65, 1951.
51. **Mehler, A. H.,** Studies on reactions of illuminated chloroplast. II. Stimulation and inhibition of the reaction with molecular oxygen, *Arch. Biochem. Biophys.,* 34, 339, 1951.
52. **Good, N. and Hill, R.,** Photochemical reduction of oxygen in chloroplast preparations. II. Mechanisms of the reaction with oxygen, *Arch. Biochem. Biophys.,* 57, 355, 1955.
53. **Badger, M. R.,** Photosynthetic oxygen exchange, *Annu. Rev. Plant Physiol.,* 36, 27, 1985.
54. **Asada, K. and Takahashi, M.,** Production and scavenging of active oxygen in photosynthesis, in *Photoinhibition,* Kyle, D. J., Osmond, C. B., and Arntzen, C. J., Eds., Elsevier, Amsterdam, 1987, 227.
55. **Horton, P.,** Interactions between electron transfer and carbon assimilation, in *Photosynthetic Mechanisms and the Environment,* Barber, J. and Baker, N. R., Eds., Elsevier, Amsterdam, 1985, 135.
56. **Bradbury, M., Ireland, C. R., and Baker, N. R.,** An analysis of the chlorophyll-fluorescence transients from pea leaves generated by changes in atmospheric concentrations of CO_2 and O_2, *Biochim. Biophys. Acta,* 806, 357, 1985.
57. **Halliwell, B.,** Oxygen radicals: their formation in plant tissues and their role in herbicide damage, in *Herbicides,* Baker, N. R. and Percival, M. P., Eds., Elsevier, Amsterdam, 1991, 87.
58. **Beyer, W. F. and Fridovich, I.,** Assaying for superoxide dismutase: some large consequences for minor changes in conditions, *Anal. Biochem.,* 161, 559, 1987.
59. **Jahnke, L. S., Hull, M. R., and Long, S. P.,** Chilling stress and oxygen metabolising enzymes in *Zea mays* and *Zea diploperennis, Plant Cell Environ.,* 14, 97, 1991.
60. **Quinn, P. J. and Williams, W. P.,** Environmentally induced changes in chloroplast membranes and their effects on photosynthetic function, in *Photosynthetic Mechanisms and the Environment,* Barber, J. and Baker, N. R., Eds., Elsevier, Amsterdam, 1985, 1.
61. **Nie, G. Y. and Baker, N. R.,** Modifications to thylakoid composition during development of maize leaves at low growth temperatures, *Plant Physiol.,* 95, 184, 1991.
62. **Nie, G. Y., Long, S. P., and Baker, N. R.,** The effects of development at suboptimal-growth temperatures on photosynthetic capacity and susceptibility to chilling dependent photoinhibition, *Physiol. Plant.,* 85, 554, 1992.
63. **Hull, M. R.,** The Activity of Active Oxygen Scavenging and C_4 Cycle Enzymes in Relation to Photosynthesis of Two *Zea* Genotypes at Chilling Temperatures, Ph.D. thesis, University of Essex, Colchester, U.K., 1991.
64. **Stamp, P.,** Photosynthetic traits of maize genotypes at constant and fluctuating temperatures, *Plant Physiol. Biochem.,* 25, 729, 1987.

65. **Masih, W.**, Effects of Temperature on *In Vivo* Levels of Rubisco (Ribulose 1,5-Bisphosphate Carboxylase/Oxygenase) in Maize: An Immunoprecipitation Study, M.Sc. thesis, University of Essex, Colchester, U.K., 1986.
66. **Hahn, M. and Walbot, V.**, Effects of cold-treatment on protein synthesis and mRNA levels in rice leaves, *Plant Physiol.*, 91, 930, 1989.
67. **Meza-Basso, L., Alberdi, M., Raynal, M., Ferrero-Cardinanos, M.-L. and Delseny, M.**, Changes in protein synthesis in rapeseed *(Brassica napus)* seedlings during a low temperature treatment, *Plant Physiol.*, 82, 733, 1986.
68. **Cooper, P. and Ort, D. R.**, Changes in protein synthesis induced in tomato by chilling, *Plant Physiol.*, 88, 454, 1988.
69. **Martino-Catt, S. and Ort, D. R.**, Low temperature interrupts circadian regulation of transcriptional activity in chilling-sensitive plants, *Proc. Natl. Acad. Sci., U.S.A.*, 89, 3731, 1992.
70. **Brooks, S., Nie, G. Y., and Baker, N. R.**, unpublished data.
71. **Öquist, G. and Martin, B.**, Cold climates, in *Photosynthesis in Contrasting Environments*, Baker, N. R. and Long, S. P., Eds., Elsevier, Amsterdam, 1986, 237.
72. **Berry, J. and Björkman, O.**, Photosynthetic response and adaptation to temperature in higher plants, *Annu. Rev. Plant Physiol.*, 31, 491, 1980.
73. **Schöner, S. and Krause, G. H.**, Protective mechanisms against active oxygen species in spinach: response to cold acclimation in excess light, *Planta*, 180, 383, 1990.
74. **Somersalo, S. and Krause, G. H.**, Photoinhibition at chilling temperature. Fluorescence characteristics of unhardened and cold-acclimated spinach leaves, *Planta*, 177, 409, 1989.
75. **Somersalo, S. and Krause, G. H.**, Reversible photoinhibition of unhardened and cold-acclimated spinach leaves at chilling temperatures, *Planta*, 180, 181, 1990.
76. **de Kok, L. J. and Oosterhuis, F. A.**, Effects of frost-hardening and salinity on glutathione and sulfhydryl levels and on glutathione reductase activity in spinach leaves, *Physiol. Plant.*, 58, 47, 1983.
77. **Guy, C. L. and Carter, J. V.**, Characterisation of partially purified glutathione reductase from cold-hardened and non-hardened spinach leaf tissue, *Cryobiology*, 21, 454, 1984.

Chapter 6

MAN-INDUCED CAUSES OF FREE RADICAL DAMAGE TO PLANTS: O_3 AND OTHER GASEOUS POLLUTANTS

Horst Mehlhorn and Alan R. Wellburn

TABLE OF CONTENTS

0-8493-5443-9/94/$0.00 + $.50

I. INTRODUCTION

The potential for atmospheric pollutants to damage vegetation has been recognized for centuries.[1] However, injury to vegetation directly attributable to photochemical air pollution was only described for the first time this century. In the early 1940s, increases in population, road transport, and industrial activity were responsible for large emissions of nonmethane hydrocarbons (NMHC) as well as CO_2, SO_2, and nitrogen oxides, first around Los Angeles[2] and then elsewhere. Reactions between NMHC and nitrogen oxides in bright sunlight formed photochemical smog,[3,4] which caused nonpathogenic leaf injury to plants in the field, especially leafy vegetable crops.[5] Similarly, in the 1950s, Ponderosa pine trees within the San Bernardino National Forest in Southern California showed symptoms of decline.[6] Photochemical air pollution, and particularly O_3, was found to be responsible.[7-10]

Concentrations of O_3 in Europe were much lower over this period. However, as industrial activity, transport, etc. increased, so did concentrations of O_3[11] and O_3-induced injury to plants.[12] Later, during the late 1970s, widespread forest decline affecting many species was observed in many parts of Europe. While the exact causes of forest decline are still not clearly established, atmospheric pollution is thought to be responsible and a contributory role of O_3 is strongly suspected.[13,14]

II. ATMOSPHERIC CHEMISTRY AND THE ROLE OF FREE RADICALS

Concentrations of NO in the troposphere of the Northern Hemisphere have increased more than tenfold to 2 to 5 nl l^{-1}, while those in remote areas in the Southern Hemisphere are still less than 0.04 nl l^{-1}.[15] This has caused considerable change in the photochemistry of the lower troposphere, and levels of hydroperoxides as well as O_3 have risen.[11,16] It is estimated that concentrations of O_3 may have doubled over the last century,[17] and a similar rise in the concentrations of hydroperoxides is strongly suspected.[16,18,19]

In the troposphere, O_3 originates from three different sources: from the stratosphere, as a result of slow photochemistry upon CO and CH_4 in the free atmosphere, and from the photochemical reactions of NO_2.[15,20] Of these, the photochemical reactions of NO_2 contribute most to tropospheric concentrations of O_3:

$$NO_2 + h\nu \rightarrow NO + O(^3P) \quad (1)$$

$$O(^3P) + O_2 \rightarrow O_3 \quad (2)$$

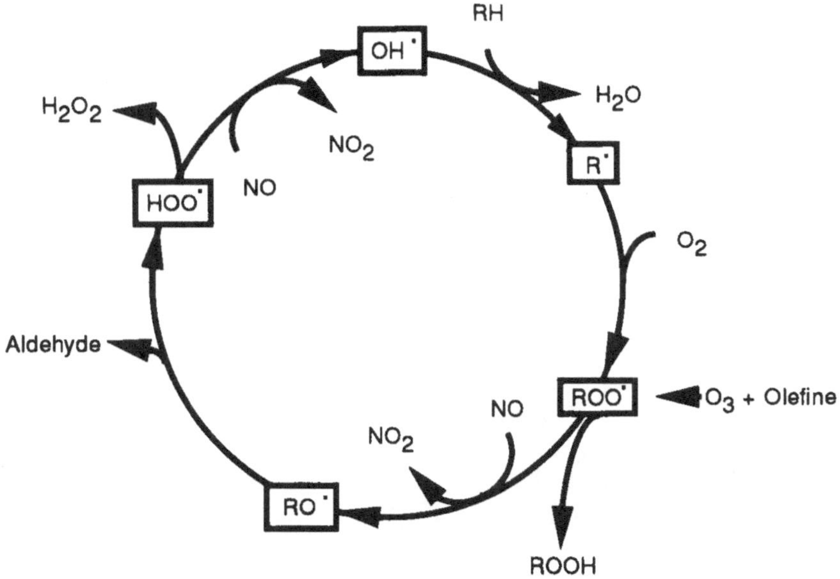

FIGURE 1. Oxidation of organic compounds by ozone and the ˙OH radical in the atmosphere in the presence/absence of NO.

However, both natural hydrocarbons and those released by the activities of man also participate in the reactions leading to tropospheric O_3 formation.[15] O_3 and the hydroxy (˙OH) radical are the dominant reactive species in the degradation of organic compounds in both the natural and polluted troposphere during the day (Figure 1). At night, their place is partially taken by the nitrate radical.[15,21,22]

Free radicals are atoms or molecules with an unpaired electron occupying an outer orbital. The unpaired electrons of free radicals have a strong tendency to interact with other electrons to form an electron pair and thereby a chemical bond.[23] This makes these chemical species highly reactive. Reaction constants for the ˙OH radical, for example, suggest that these reactions are essentially controlled by diffusion. Indeed, the ˙OH radical is so energetic that it reacts almost with the first molecule with which it collides.[22] Whenever a free radical has been formed in an initiation reaction, it may react with a nonradical and form a new radical, which then enters into a chain reaction that is only terminated when two radicals react with each other to form nonradical products. The generalized reaction sequence below illustrates these reactions (see also Figure 1):

$$\text{Initiation:} \qquad X + h\nu \rightarrow R\cdot \qquad (3)$$

Propagation:	$R\cdot + O_2$	$\rightarrow ROO\cdot$		(4)
	$ROO\cdot$	$\rightarrow ROOH + R\cdot$		(5)
	$2ROOH$	$\rightarrow ROO\cdot + RO\cdot + H_2O$		(6)
	$RO\cdot + ROOH$	$\rightarrow ROO\cdot + ROH$		(7)
Termination:	$2ROO\cdot$	$\rightarrow O_2$ + stable products		(8)
	$2R\cdot$	$\rightarrow$ stable products		(9)
	$ROO\cdot + R\cdot$	$\rightarrow ROOR$		(10)

III. EARLY FORMATION OF FREE RADICALS IN PLANTS DURING O_3 TOXICITY

When Haagen-Smit and co-workers[7] first described those chemical species responsible for leaf injury around Los Angeles, they already suspected that they were formed from reactions between O_3 and unsaturated hydrocarbons. In their experiments, they found that, while O_3 in combination with lower olefins (ethylene, propene, and normal butenes) produced visible leaf injury in fumigated plants, higher olefins, especially those with branched chains (e.g., *iso*-butene), increased toxicity. The visible injury caused also resembled more closely that associated with smog damage. Most smog-like damage occurred when the chain length of the reaction products was from three to six carbon atoms (e.g., with 1-hexene). They also concluded that "the active agent had to be found in the highly oxygenated organic compounds of peroxide nature". This was investigated further with experiments using 2-pentene,[24] and it was concluded that the phytotoxicant was an unstable species. Furthermore, it was suggested that this unstable species could be either an O_3-olefin addition complex or a zwitterion to which it decomposes.[24,25]

It is now accepted that O_3 toxicity may be mediated by the formation of free radicals.[26-30] This assertion is based on experiments showing increased formation of free radicals in plants exposed to O_3[31] and by the protection offered to plants exposed to O_3 by pretreatment with antioxidants.[32-43]

However, it still remains unclear which reactions inside plants may initiate free-radical formation. An increasing number of experiments suggest reactions between O_3 and biogenic unsaturated hydrocarbons, such as the isoprenes or ethylene.[44-49] Experiments with the inhibitor of ethylene biosynthesis, aminoethoxyvinylglycine (AVG), for example, show a reduction in O_3 toxicity, but only if ethylene biosynthesis is inhibited.[44,49,50] This rules out the possibility of an antioxidant function for AVG.[49-51] In addition, leaf injury caused by O_3 correlates strongly with ethylene formation by exposed

FIGURE 2. Visible leaf injury on 3-week-old pea plants (*Pisum sativum* cv. Feltham First), 2 d after a 7-h fumigation with ozone in the presence or absence of α-pinene. Treatment of plants with α-pinene alone caused no visible leaf injury and plants were identical to control plants. Fumigations with ozone and α-pinene were carried out as described earlier.[7,49]

plants,[44,48,49,52-56] while O_3 toxicity is enhanced in the presence of unsaturated hydrocarbons[7,8,24,57] (see also Figure 2). Furthermore, in plants pretreated with ethylenediurea (EDU), a chemical known to protect plants in photochemical smog conditions, stress ethylene formation is prevented and, consequently, O_3-mediated foliar injury is reduced.[50]

Reaction products of O_3 and unsaturated hydrocarbons include not only free radicals and peroxides,[31,46,58] but aldehydes and carbonic acids may also be formed, both in the outside atmosphere[7,15] and in plants exposed to O_3.[59,60] Meanwhile, damage to trees in areas affected by forest decline has been shown to be correlated with rates of ethylene formation, and exposure to O_3, alone or in combination with other air pollutants, increases ethylene formation in conifers.[44,53,56,61,62]

In recent years, evidence has shown that the nature of O_3 toxicity depends on the way O_3 and NMHCs are formed and how plants are exposed to them.[63-65] In experiments where plants were treated with O_3 added to charcoal-filtered air, for example, O_3 toxicity was found to be greater than in plants exposed to the same concentrations of O_3 in nonfiltered air.[65] Similarly, in experiments where O_3 was added to charcoal-filtered air or to charcoal/Purafil-filtered air, O_3 toxicity was greater when O_3 was added to air only filtered through charcoal.[64] In both cases, involvement of background concentrations of NO was suggested. While the mechanism behind these effects in biological systems is still unknown, it is possible that NO acts as a chain-propagating agent in the O_3-initiated oxidation of unsaturated hydrocarbons naturally emitted by plants.[64] The possible sequence of reactions involved is shown in Figure 1. This reaction sequence was first described by Leighton[66] and later in more detail by Finlayson-Pitts and Pitts.[15] In the absence of NO, chain reactions are terminated by the self-reaction between organic peroxy- and hydroperoxy radicals. Consequently, further reactions of free radicals thought to be responsible for O_3 toxicity are prevented as long as the resulting hydroperoxides are detoxified by plants.[67] The involvement of low concentrations of NO in the mechanisms of O_3 toxicity may also explain observations of reduced O_3 toxicity at higher elevations, where concentrations of NO are lower.[68,69]

A. FORMATION OF FREE RADICALS IN THE LIQUID PHASE DURING O_3 TOXICITY

Concentrations of O_3 in the intracellular air spaces of the leaf are close to zero.[70] This means that rapid decomposition of O_3 takes place in this region. While O_3 readily dissolves in water,[71] it only does so in the presence of organic solutes, and thereby initiates formation of superoxide anions ($O_2^{\cdot-}$) and $^{\cdot}OH$ radicals.[72,73] Thus, even at moderately high gas phase concentrations of O_3 (70 nl l^{-1}), concentrations of O_3 in atmospheric aerosols are only 0.02 n*M*.[71] Similarly, if O_3 is passed over a simple aqueous solution of potassium phosphate buffer (pH 7 to 8) containing the spin trap DMPO, no electron spin resonance (ESR) signals, indicating the formation of free radicals, are detected. Moreover, only a weak ESR signal is recorded when O_3 is bubbled directly into this solution. However, if the buffer solution also contains 0.1 m*M* caffeic acid, phenol, ferulic acid, or cinnamic acid, a strong ESR signal develops.[74] Further studies, investigating O_3 toxicity to carrot protoplasts, failed to cause damage to the plasma membrane even when the protoplasts were exposed to O_3 concentrations that led to the development of $^{\cdot}OH$ radical concentrations of 0.1 m*M*. Furthermore, experiments with the 5DS fatty acid spin label indicated that free radical degradation products were unable to penetrate the lipid bilayer. Consequently, Grimes and co-workers[74] concluded that, while O_3 or the $^{\cdot}OH$ radical are primary initiating agents in O_3 injury, the cell wall may be an additional site of extensive free radical chemistry during exposure to these agents.[74] Indeed, the possibility of reactions of O_3 with the multitude of double bonds provided by the phenolic acids within the cell wall was also suggested by ESR studies investigating O_3 toxicity in pea and bean plants.[31] Measurements of a decreased bulk modulus of elasticity,[75] increased formation of caffeic acid,[76] 4,7-dihydroxyflavone,[77-79] total phenolics,[80] stilbenes,[81] and an accumulation of tannins or changes in allocation within the phenolic groups[82,83] in plants exposed to O_3 also point in this direction.

It has also been suggested that O_3 may react with ascorbate dissolved in the apoplast.[84] Part of this assertion is based on experiments by Castillo and co-workers.[85] However, while high levels of ascorbate in the apoplast have been reported,[86] no experimental evidence was found for the presence of nonenzymatic systems in the apoplast that may be able to reduce ascorbate which has been oxidized by O_3.[86] The involvement of ascorbate in protecting leaves against toxic reactions induced by O_3 has already been investigated. Miller and colleagues[10] and Menser[87] found no correlation between endogenous levels of ascorbic acid and resistance to O_3, but others did.[88-90] Similarly, treatment of mitochondria or application of ascorbate to leaves or roots of plants increases their tolerance to O_3.[32,33,35] Further studies using K- and Ca-ascorbate have also shown that the effectiveness of ascorbate depended on its uptake into the leaf.[34] However, the protection offered by ascorbate was generally low,[89,91,92] and O_3 has no significant effect on the total quantities of ascorbate in leaves.[49,93-95] Therefore, further experiments to test this interesting suggestion[84] are still required, especially alternatives to high-performance liquid chromatography (HPLC) for ascorbate determination[85,86] because HPLC alone does not provide structural confirmation.

B. IONIC IMBALANCES

When O_3 dissolves in water in the presence of organic solutes, an ozonide anion is formed which decomposes to form $^{\bullet}OH$. This involves consumption of protons and may lead to a more alkaline pH of the apoplast, depending on its buffering capacity.[28,72] Under more acidic conditions, formation of $^{\bullet}OH$ radicals is greatly reduced,[74] and this may explain why no effect on O_3 toxicity has been observed in experiments where plants had been treated with both O_3 and acid rain[75,96-112] except by Chappelka and Chevone.[113]

During the initial stages of O_3 toxicity, K^+ efflux from cells has also been observed.[114] It has been suggested[28] that this may lead to displacement of Ca^{2+} in the cell wall and alter rigidity.[115,116] This was recently confirmed in long-term experiments with Norway spruce using more realistic concentrations of O_3.[75] Price,[117] using strips of epidermis taken from *Commelina communis,* also showed that the Ca^{2+} channel blocker, verapamil, or the ion chelator, EGTA, were able to prevent a rise in cytosolic calcium after treatment with H_2O_2 and, thereby, stop stomatal closure. The influence of O_3 on the transport of Ca^{2+} and its intracellular accumulation has also been shown by experiments by Fink[118] and Castillo and Heath.[119] While active efflux driven by an ATP-dependent system is still not impaired by O_3,[120] Ca^{2+} may be responsible for the release of extracellular peroxidases[121] and, thereby, detoxification of H_2O_2 during O_3 toxicity.[67,122,123]

C. EFFECTS ON CELL MEMBRANES AND LIPIDS

The development of visible injury symptoms by O_3 and associated physiological changes result from chemical reactions of cellular constituents with

O_3. Thus, early visible injury observed in leaves exposed to O_3 may include wilting, filling of intracellular spaces with water, and localized epidermal collapse.[124] This is preceded by changes in membrane permeability[114,118,125-133] and, later, accumulation of malondialdehyde or ethane — the products of lipid peroxidation.[44,49,128,134-136]

Because O_3 is a strong oxidizing agent and reacts readily with the double bonds of polyunsaturated fatty acids of cell membranes, destruction of membrane lipids could be one of the primary events of O_3 toxicity.[137,138] However, when the fatty acid composition of O_3-exposed plants was analyzed, changes in desaturation were minimal or absent.[139-143] They only occur at unrealistically high concentrations of O_3[144-146] or after prolonged periods of O_3 treatment.[147,148] Consequently, O_3 does not directly oxidize unsaturated fatty acids during the initial stage of exposure.[140,147,148]

However, rapid changes in the composition of leaf lipids have been found after O_3 fumigation. Galactolipid and phospholipid amounts are greatly decreased after treatment with O_3, and this is accompanied by marked increases in triacylglycerols.[140,145,146,148-150] Furthermore, O_3-treated plants also have decreased levels of free sterols,[146,151-155] increased sterol glycosides, and esterified sterol glycosides.[139,145,146,151] Levels of free fatty acids remain unchanged unless treatment with O_3 is severe.[140,150] Similar changes in lipid composition have also been suggested by recent studies investigating the levels of free hydrophilic head groups of phospholipids after O_3 treatment. In these experiments, increased quantities of free serine and also myoinositol were found to be present, particularly in the older leaves,[156,157] which are those most sensitive to O_3[44,49] (see also Figure 2). Similarly, O_3-treated leaves also have increased levels of raffinose and stachyose. Both sugars contain bound galactose, and it has been suggested that the biosynthesis of raffinose and stachyose may be a mechanism to detoxify enzymatically or free radical-removed galactose units from monogalactosyldiacylglycerols (MGDG) or digalactosyldiacylglycerols (DGDG) during O_3 toxicity.[156,157]

Apart from changing the lipid composition, O_3 may also severely affect the catabolism and biosynthesis of fatty acids and lipids. The ability to metabolize fatty acids, acetate, or acetyl-CoA for lipid biosynthesis is impaired by O_3 treatment, and more MGDGs are converted to triacylglycerols.[135,149,150] This has been attributed to the oxidation of essential SH groups, accumulation of hydrogen peroxide, enhancement of galactolipid:galactolipid galactosyltransferase by temporarily accumulated free fatty acids, or to increased concentrations of Mg^{2+}.[135,158] Recently, Wolfenden and Wellburn[143] reported reduced formation of ω5,9-18:2 and ω9,12-18:2 fatty acids in O_3-treated Norway spruce needles during winter hardening. This may account for the failure of summer O_3-treated spruces to harden effectively to very low temperatures, which has been repeated by other studies.[64,143,159-167] Changes in lipid composition may also explain enhanced rates of senescence,[27] reduced amount of thylakoid membranes and length of chloroplasts, and swelling of thylakoids and mitochondria.[168-171]

D. PROTECTION BY CELLULAR ANTIOXIDANTS

Reactions of O_3 with unsaturated hydrocarbons form hydroperoxides and free radicals.[31,44,46,49,74] As H_2O_2 is toxic, even at low concentrations, its accumulation in living tissues is undesirable. For this reason, plants have enzymic and nonenzymatic antioxidant systems to scavenge free radicals directly, and to prevent the formation of additional free radicals from hydroperoxide decomposition.[172,173] These antioxidants include polyamines, glutathione, ascorbate, and α-tocopherol, as well as the enzymes, superoxide dismutase, ascorbate peroxidase, and glutathione reductase.

Treatment of plants with polyamines,[42,174] ascorbate,[32,33,35] and glutathione[32] protects them from ozone damage. Similarly, the endogenous metabolism of polyamines,[48,175-178] ascorbate,[67,85,93,94,175,179-182] glutathione,[94-96,122,123,177,179,181-184] and α-tocopherol[93,94] are responsive in plants exposed to ozone. Furthermore, Rao and Dubey[185] reported correlations between the levels of these antioxidants and the resistance of plants to exposure to ambient levels of air pollution in the field. All these changes have been suggested to be cellular repair processes which alleviate the initial oxidative damage caused by ozone.[186,187]

Changes in the levels of superoxide dismutase after exposure to O_3 are more difficult to interpret. Increased,[85,184,188,189] decreased,[122] and unaltered levels of superoxide dismutase activity[190-193] have been reported in plants fumigated with O_3, although it was initially suggested that superoxide dismutase may be a protective enzyme against O_3 injury.[194] However, superoxide dismutase only converts one toxic derivative of oxygen (O_2^-) into another (H_2O_2). Consequently, superoxide dismutase may only provide cells with enhanced resistance against oxidative stress if the H_2O_2 produced is subsequently detoxified by ascorbate peroxidase.[67] This interpretation is supported by recent experiments with transgenic tobacco plants supplemented with genes encoding for glutathione reductase[195,196] and by experiments with SOD-transformed human HeLa cells,[197] which all show only limited protection against exposure to oxidative stress (for further discussion of this topic, see Chapters 12 and 13 of this volume).

E. GENETIC ASPECTS OF SENSITIVITY AND RESISTANCE

The sensitivity of plants to O_3 differs not only between species, but also between cultivars, provenances, and even individuals within a given species.[27] For example, different populations of *Populus tremuloides* Michx from five different areas in North America, regularly exposed to differing air quality, showed corresponding differences in their resistance to O_3, with plants from the most polluted site showing the highest resistance to O_3 in controlled environment fumigation experiments.[198] Similarly, Barnes and co-workers[199] have compared the sensitivity to O_3 of old and modern Greek cultivars of spring wheat. In their experiments, there was a highly significant ($p < 0.002$) negative relationship between plant sensitivity to O_3 and the year of introduction; the more modern the cultivar, the greater its tolerance of O_3. This compares well with increases in the concentrations of O_3 during the last century,[11,17] and shows that plants adapt to specific environmental stresses such as O_3.

For adaptation to occur, genetic variations in O_3 tolerance, differences in ambient O_3 levels among locations, and differential mortality of O_3-sensitive individuals are involved.[200-206] Furthermore, the genes responsible for O_3 resistance or sensitivity must be inherited from parent plants[207] and selection for O_3 sensitivity must be continuous. This is evident from experiments with O_3-sensitive cultivars of tobacco (Bel W3), which lose their sensitivity to O_3 unless it is continuously selected for.[49]

Interestingly, correlations between O_3 sensitivity and the suspected mode of action of O_3 have been shown in relation to both uptake by stomata[208] and rates of ethylene formation.[48,49] If such selection occurs, resistance to O_3 could be transferred to agronomically important cultivars within a short time[209] because even individual plants are able to adapt to exposure to O_3.[44,49,210]

F. INTERACTION OF O_3 WITH OTHER ENVIRONMENTAL STRESSES

In nature, plants are always exposed to variable environmental stresses such as O_3. Exposure to other atmospheric pollutants, pesticides or heavy metals, lack of nutrients, or attack by pathogens are normally reflected by enhanced rates of ethylene release.[211,212] Consequently, plants should be more sensitive to O_3 if they are also exposed to other environmental stresses. This is indeed the case. Infestation with greenhouse whitefly,[213] bacteria,[214] pretreatment with NO_2,[215] exposure to NO or NO_2,[44] SO_2,[56] aluminum,[216] cadmium or nickel,[217-219] zinc,[217] or pesticides[40,41,43] all enhance O_3 toxicity. Similarly, soil moisture stress[220] and infestation with fungal pathogens[221,222] are exacerbated by simultaneous exposure to O_3. All these results may be explained by the hypothesis that stress ethylene formation is responsible for O_3 toxicity,[44,49] but cannot be explained by the effects on stomatal opening.[51]

REFERENCES

1. **Wellburn, A. R.,** *Air Pollution and Acid Rain. The Biological Impact,* Longmans, London, 1988.
2. **Haagen-Smit, A. J.,** Chemistry and physiology of Los Angeles smog, *Ind. Eng. Chem.*, 44, 1342, 1952.

3. **Haagen-Smit, A. J. and Fox, M. M.,** Ozone formation in photochemical oxidation of organic substances, *Ind. Eng. Chem.*, 48, 1484, 1956.
4. **Stephens, E. R., Darley, E. F., Taylor, O. C., and Scott, W. E.,** Photochemical reaction products in air pollution, *Int. J. Air Water Pollut.*, 4, 79, 1961.
5. **Middleton, J. T., Kendrick, J. B., Jr., and Schwalm, H. W.,** Injury to herbaceous plants by smog or air pollution, *Plant Dis. Rep.*, 34, 245, 1950.
6. **Parmeter, J. R., Bega, R. V., and Neff, T.,** A chlorotic decline of Ponderosa pine in southern California, *Plant Dis. Rep.*, 46, 269, 1962.
7. **Haagen-Smit, A. J., Darley, E. F., Zaitlin, M., Hull, H., and Noble, W.,** Investigation on injury to plants from air pollution in the Los Angeles area, *Plant Physiol.*, 27, 18, 1952.
8. **Middleton, J. T.,** Photochemical air pollution damage to plants, *Annu. Rev. Plant Physiol.*, 12, 431, 1961.
9. **Richards, B. L., Taylor, O. C., and Edmunds, G. F.,** Ozone needle mottle of pine in southern California, *J. Air Pollut. Control Assoc.*, 18, 73, 1968.
10. **Miller, P. R., Parmeter, J. R., Flick, B. H., and Martinez, C. W.,** Ozone dosage response of Ponderosa pine seedlings, *J. Air Pollut. Control Assoc.*, 19, 435, 1969.
11. **Warmbt, W.,** Ergebnisse langjähriger Messungen des bodennahen Ozons in der DDR, *Z. Meteorol.*, 29, 24, 1979.
12. **Ashmore, M. R., Bell, J. N. B., Dalpra, C., and Runeckles, V. C.,** Visible injury to crop species by ozone in the United Kingdom, *Environ. Pollut.*, 21A, 209, 1980.
13. **McLaughlin, S. B.,** Effects of air pollution on forests. A critical review, *J. Air Pollut. Control Assoc.*, 35, 512, 1985.
14. **Schulze, E. D.,** Air pollution and forest decline in a spruce *(Picea abies)* forest, *Science*, 244, 776, 1989.
15. **Finlayson-Pitts, B. J. and Pitts, J. N.,** *Atmospheric Chemistry: Fundamentals and Experimental Techniques*, John Wiley & Sons, New York, 1986.
16. **Sigg, A. and Neftel, A.,** Evidence of a 50% increase in H_2O_2 over the past 200 years from a Greenland ice core, *Nature (London)* 351, 557, 1991.
17. **Volz, A. and Kley, D.,** Evaluation of the Montsouris series of ozone measurements made in the nineteenth century, *Nature (London)*, 332, 240, 1988.
18. **Sakugawa, H., Kaplan, I. R., Tsai, W., and Cohen, Y.,** Atmospheric hydrogen peroxide. Does it share a role with ozone in degrading air quality?, *Environ. Sci. Technol.*, 24, 1452, 1990.
19. **Thompson, A. M., Huntley, M. A., and Stewart, R. W.,** Perturbations to tropospheric oxidants, 1985–2035. II. Calculations of hydrogen peroxide in chemically coherent regions, *Atmos. Environ.*, 25A, 1837, 1991.
20. **Liu, S. C., Trainer, M., Fehsenfeld, F. C., Parrish, D. D., Williams, E. J., Fahey, D. W., Hübler, G., and Murphy, P. C.,** Ozone production in the rural troposphere and the implications for regional and global ozone distributions, *J. Geophys. Res.*, 92, 4191, 1987.
21. **Atkinson, R. and Carter, W. P. L.,** Kinetics and mechanisms of the gas-phase reactions of ozone with organic compounds under atmospheric conditions, *Chem. Rev.*, 84, 437, 1984.
22. **Atkinson, R.,** Kinetics and mechanisms of the gas-phase reactions of the hydroxyl radical with organic compounds under atmospheric conditions, *Chem. Rev.*, 86, 69, 1986.
23. **DelMaestro, R. F.,** An approach to free radicals in medicine and biology, *Acta Physiol. Scand. Suppl.*, 492, 153, 1980.
24. **Darley, E. F., Stephens, E. R., Middleton, J. T., and Hanst, P. L.,** Oxidant plant damage from ozone-olefin reactions, *Int. J. Air Pollut.*, 1, 155, 1959.
25. **Herron, J. T., Martinez, R. I., and Huie, R. E.,** Kinetics and energetics of the Criegee intermediate in the gas phase. I. The Criegee intermediate in ozone-alkene reactions, *Int. J. Chem. Kin.*, 14, 201, 1982.

26. **Heath, R. L.,** Initial events in injury to plants by air pollutants, *Annu. Rev. Plant Physiol.*, 31, 395, 1980.
27. **Guderian, R., Tingey, D. T., and Rabe, R.,** Effects of photochemical oxidants on plants, *Ecol. Stud.*, 52, 126, 1985.
28. **Heath, R. L.,** Biochemical mechanisms of pollutant stress, in *Assessment of Crop Loss from Air Pollutants,* Heck, W. W., Taylor, O. C., and Tingey, D. T., Eds., Elsevier, London, 1989, 259.
29. **Krupa, S. V. and Manning, W. J.,** Atmospheric ozone: formation and effects of vegetation, *Environ. Pollut.*, 50, 101, 1988.
30. **Mehlhorn, H. and Wellburn, A. R.,** Ozone toxicity mechanisms in plants and animals, in *Free Radicals: Chemistry, Pathology and Medicine,* Rice-Evans, C. and Dormandy, T., Eds., Richilieu Press, London, 253, 1988.
31. **Mehlhorn, H., Tabner, B. J., and Wellburn, A. R.,** Electron spin resonance evidence for the formation of free radicals in plants exposed to ozone, *Physiol. Plant.*, 79, 377, 1990.
32. **Freebairn, H. T.,** Reversal of inhibitory effects of ozone on oxygen uptake of mitochondria, *Science,* 126, 303, 1957.
33. **Freebairn, H. T.,** The prevention of air pollution damage to plants by use of vitamin C sprays, *J. Air Pollut. Control Assoc.*, 10, 314, 1960.
34. **Freebairn, H. T.,** Uptake and movement of 1-^{14}C ascorbic acid in bean plants, *Physiol. Plant.*, 16, 517, 1963.
35. **Freebairn, H. T. and Taylor, O. C.,** Prevention of plant damage from air-borne oxidizing agents, *Proc. Am. Soc. Hortic. Sci.*, 76, 693, 1960.
36. **Lee, T. T.,** Chemical regulation of ozone susceptibility in *Nicotiana tabacum, Can. J. Bot.*, 44, 487, 1966.
37. **Elfving, D. C., Gilbert, M. D., Edgerton, L. F., Wilde, M. H., and Lisk, D. J.,** Antioxidant and antitranspirant protection of apple foliage against ozone injury, *Bull. Environ. Contam. Toxicol.*, 15, 336, 1976.
38. **Gilbert, M. D., Elfving, D. C., and Lisk, D. J.,** Protection of plants against ozone injury using the antioxidant N-(1,3-dimethylbutyl)-N′-phenylene-diamine, *Bull. Environ. Contam. Toxicol.*, 18, 783, 1977.
39. **Rubin, B., Leavitt, J. R. C., Penner, D., and Saettler, A. W.,** Interaction of antioxidants with ozone and herbicide stress, *Bull. Environ. Contam. Toxicol.*, 25, 623, 1980.
40. **Hatzios, K. K.,** Interactions of the herbicides EPTC and EPTC plus R-25788 with ozone and antioxidants in corn, *J. Agric. Food Chem.*, 31, 1187, 1983.
41. **Hatzios, K. K.,** Effects of CGA-43089 on responses of sorghum *(Sorghum bicolor)* to metolachlor combined with ozone or antioxidants, *Weed Sci.*, 31, 280, 1983.
42. **Bors, W., Langebartels, C., Michel, C., and Sandermann, H., Jr.,** Polyamines as radical scavengers and protectants against ozone damage, *Phytochemistry,* 28, 1589, 1989.
43. **Mersie, W., Mebrahtu, T., and Rangappa, M.,** Response of corn to combinations of atrazine, propyl gallate and ozone, *Environ. Exp. Bot.*, 30, 443, 1990.
44. **Mehlhorn, H. and Wellburn, A. R.,** Stress ethylene formation determines plant sensitivity to ozone, *Nature (London),* 327, 417, 1987.
45. **Möller, D.,** Production of free radicals by an ozone-olefin reaction — a possible factor in the new-type forest decline?, *Atmos. Environ.*, 22, 2607, 1989.
46. **Hewitt, C. N., Kok, G. L., and Fall, R.,** Hydroperoxides in plants exposed to ozone mediate air pollution damage to alkene emitters, *Nature (London),* 344, 56, 1990.
47. **Heller, W., Rosemann, D., Osswald, W. F., Benz, B., Schönwitz, R., Lohwasser, K., Kloos, M., and Sandermann, H.,** Biochemical response of Norway spruce (*Picea abies* (L.) Karst.) towards 14-month exposure to ozone and acid mist. I. Effects on polyphenol and monoterpene metabolism, *Environ. Pollut.*, 64, 353, 1990.

48. **Langebartels, C., Kerner, K., Leonardi, S., Schraudner, M., Trost, M., Heller, W., and Sandermann, H.,** Biochemical plant responses to ozone. I. Differential induction of polyamine and ethylene biosynthesis in tobacco, *Plant Physiol.*, 95, 882, 1991.
49. **Mehlhorn, H., O'Shea, J. M., and Wellburn, A. R.,** Atmospheric ozone interacts with stress ethylene formation by plants to cause visible leaf injury, *J. Exp. Bot.*, 42, 17, 1991.
50. **Zilinskas, B. A., Greenhalgh-Weidman, B., and Brennan, E.,** The relationship between EDU pre-treatment and C_2H_4 evolution in ozonated pea plants, *Environ. Pollut.*, 65, 241, 1990.
51. **Unsworth, M. H.,** Adding ethylene to injury, *Nature (London)*, 327, 364, 1987.
52. **Craker, L. E.,** Ethylene production from ozone injured plants, *Environ. Pollut.*, 1, 299, 1971.
53. **Tingey, D. T. Standley, C., and Field, R. W.,** Stress ethylene evolution: a measure of ozone effects on plants, *Atmos. Environ.*, 10, 969, 1976.
54. **Stan, H. J., Schicker, S., and Kassner, H.,** Stress ethylene evolution of bean plants — a parameter indicating ozone pollution, *Atmos. Environ.*, 15, 391, 1981.
55. **Stan, H. J. and Schicker, S.,** Effect of repetitive ozone treatment on bean plants, stress ethylene production and necrosis, *Atmos. Environ.*, 16, 2267, 1982.
56. **Mehlhorn, H., Francis, B. J., and Wellburn, A. R.,** Prediction of forest decline damage to Norway spruce using three simple site-independent diagnostic parameters, *New Phytol.*, 110, 525, 1988.
57. **Adepipe, N. O. and Ormrod, D. P.,** Hormonal regulation of ozone phytotoxicity in *Raphanus sativus*, *Z. Pflanzenphysiol.*, 68, 254, 1972.
58. **Becker, K. H., Brockmann, K. J., and Bechara, J.,** Production of hydrogen peroxide in forest air by reaction of ozone with terpenes, *Nature (London)*, 346, 256, 1990.
59. **Mader, P. P., Cann, G., and Palmer, L.,** Effect of polluted atmospheres on organic acid composition in plant tissues, *Plant Physiol.*, 30, 318, 1955.
60. **Kimmerer, T. W. and Kozlowski, T. T.,** Ethylene, ethane, acetaldehyde and ethanol production by plants under stress, *Plant Physiol.*, 69, 840, 1982.
61. **Fuhrer, J.,** Ethylene production and premature senescence of needles from fir trees *(Abies alba)*, *Eur. J. For. Pathol.*, 15, 227, 1985.
62. **Hohendorff, U. and Vogels, K.,** Einfluss von Emissionen auf die Ethenproduktion und den Primärstoffwechsel von Fichten (*Picea abies* (L.) Karst.), *Eur. J. For. Pathol.*, 20, 219, 1990.
63. **Brown, K. A. and Roberts, T. M.,** Effects of ozone on foliar leaching in Norway spruce (*Picea abies* L. Karst.): confounding factors due to NOx production during ozone generation, *Environ. Pollut.*, 55, 55, 1988.
64. **Neighbour, E. A., Pearson, M., and Mehlhorn, H.,** Purafil-filtration prevents the development of ozone-induced frost injury: a potential role for nitric oxide, *Atmos. Environ.*, 24A, 711, 1990.
65. **Olszyk, D. M., Dawson, P. J., Morrison, C. L., and Takemoto, B. K.,** Plant response to nonfiltered air vs. added ozone generated from dry air or oxygen, *J. Air Pollut. Control Assoc.*, 40, 77, 1990.
66. **Leighton, P. A.,** *Photochemistry of Air Pollution*, Academic Press, New York, 1961.
67. **Mehlhorn, H.,** Ethylene-promoted ascorbate peroxidase activity protects plants against hydrogen peroxide, ozone and paraquat, *Plant Cell Environ.*, 13, 971, 1990.
68. **Fuhrer, J., Egger, A., Lehnherr, B., Grandjean, A., and Tschannen, W.,** Effects of ozone on the yield of spring wheat (*Triticum aestivum* L., cv. Albis) grown in open-top field chambers, *Environ. Pollut.*, 60, 273, 1989.
69. **Grandjean, A. and Fuhrer, J.,** Growth and leaf senescence in spring wheat *(Triticum aestivum)* grown at different ozone concentrations in open-top field chambers, *Physiol. Plant.*, 77, 389, 1989.

70. **Laisk, A., Kull, O., and Moldau, H.,** Ozone concentration in leaf intercellular air spaces is close to zero, *Plant Physiol.*, 90, 1163, 1989.
71. **Graedel, T. E. and Weschler, C. J.,** Chemistry within aqueous atmospheric aerosols and raindrops, *Rev. Geophys. Space Phys.*, 19, 505, 1981.
72. **Staehelin, J. and Hoigne, J.,** Decomposition of ozone in water in the presence of organic solutes as promoters and inhibitors of radical chain reactions, *Environ. Sci. Technol.*, 19, 1206, 1985.
73. **Lelieveld, J. and Crutzen, P. J.,** Influences of cloud photochemical processes on tropospheric ozone, *Nature (London)*, 343, 227, 1990.
74. **Grimes, H. D., Perkins, K. K., and Boss, W. F.,** Ozone degrades into hydroxyl radical under physiological conditions. A spin trapping study, *Plant Physiol.*, 72, 1016, 1983.
75. **Barnes, J. D., Eamus, D., and Brown, K. A.,** The influence of ozone, acid mist and soil nutrient status on Norway spruce (*Picea abies* (L.) Karst). I. Plant-water relations, *New Phytol.*, 114, 713, 1990.
76. **Howell, R. K.,** Influence of air pollution on quantities of caffeic acid isolated from leaves of *Phaseolus vulgaris*, *Phytopathology*, 60, 1626, 1970.
77. **Keen, N. T. and Taylor, O. C.,** Ozone injury in soybean: isoflavonoid accumulation is related to necrosis, *Plant Physiol.*, 55, 731, 1975.
78. **Hurwitz, B., Pell, E. J., and Sherwood, R. T.,** Status of coumestrol and 4′,7′-dihydroxyflavone in alfalfa foliage exposed to ozone, *Phytopathology*, 69, 810, 1979.
79. **Skärby, L. and Pell, E. J.,** Concentrations of coumestrol and 4′,7′-dihydroxyflavone in four alfalfa cultivars after exposure to ozone, *J. Environ. Qual.*, 8, 285, 1979.
80. **Tingey, D. T., Wilhow, R. G., and Standley, C.,** The effect of chronic ozone exposures on the metabolite content of Ponderosa pine seedlings, *For. Sci.*, 22, 234, 1976.
81. **Rosemann, D., Heller, W., and Sandermann, H., Jr.,** Biochemical plant responses to ozone. II. Induction of stilbene biosynthesis in Scots pine (*Pinus sylvestris* L.) seedlings, *Plant Physiol.*, 97, 1280, 1991.
82. **Howell, R. K.,** Phenols, ozone and their involvement in pigmentation and physiology of plant injury, in *Air Pollution Effects on Plant Growth*, Dugger, M., Ed., American Chemical Society, Washington, D.C., 1974, 94.
83. **Jordan, D. N., Green, T. H., Chappelka, A. H., Lockaby, B. G., Meldahl, R. S., and Gjerstad, D. H.,** Response of total tannins and phenolics in loblolly pine foliage exposed to ozone and acid rain, *J. Chem. Ecol.*, 17, 505, 1991.
84. **Chameides, W. L.,** The chemistry of ozone deposition to plant leaves: role of ascorbic acid, *Environ. Sci. Technol.*, 23, 595, 1989.
85. **Castillo, F. J., Miller, P. R., and Greppin, H.,** Extracellular markers of photochemical oxidant air pollution damage to Norway spruce, *Experientia*, 43, 111, 1987.
86. **Polle, A., Chakrabarti, K., Schürmann, W., and Rennenberg, H.,** Composition and properties of hydrogen peroxide decomposing system in extracellular and total extracts from needles of Norway spruce (*Picea abies* L., Karst.), *Plant Physiol.*, 94, 312, 1990.
87. **Menser, H. A.,** Response of plants to air pollutants. III. A relation between ascorbic acid levels and ozone susceptibility of light preconditioned tobacco plants, *Plant Physiol.*, 39, 564, 1964.
88. **Hanson, G. P., Thorne, L., and Jativa, C. D.,** Vitamin C — a natural smog resistance mechanism in plants?, *Lasca Leaves*, 20, 6, 1970.
89. **Barnes, R. L.,** Effects of chronic exposure to ozone on soluble sugar and ascorbic acid contents of pine seedlings, *Can. J. Bot.*, 50, 215, 1972.
90. **Lee, E. H., Jersey, J. A., Gifford, C., and Bennett, J.,** Differential ozone tolerance in soybean and snapbeans: analysis of ascorbic acid in O_3-susceptible and O_3-resistant cultivars by high-performance liquid chromatography, *Environ. Exp. Bot.*, 24, 331, 1984.
91. **Siegel, S. M.,** Protection of plants against airborne oxidants: cucumber seedlings at extreme ozone levels, *Plant Physiol.*, 37, 261, 1962.
92. **Dass, H. C. and Weaver, G. M.,** Modification of ozone damage to *Phaseolus vulgaris* by antioxidants, thiols and sulfhydryl reagents, *Can. J. Plant Sci.*, 48, 569, 1968.

93. **Mehlhorn, H., Seufert, G., Schmidt, A., and Kunert, K. J.,** Effect of SO_2 and O_3 on production of antioxidants in conifers, *Plant Physiol.*, 82, 336, 1986.
94. **Kunert, K. J. and Hofer, G.,** Geben Veränderungen des antioxidativen Systems von Pflanzen Hinweise auf die Wirkung von Luftschadstoffen?, *Allg. Forstz.*, 27/28/29, 697, 1987.
95. **Bermadinger, E., Guttenberger, H., and Grill, D.,** Physiology of young Norway spruce, *Environ. Pollut.*, 68, 319, 1990.
96. **Norby, R. J. and Luxmoore, R. J.,** Growth analysis of soybean exposed to simulated acid rain and gaseous air pollutants, *New Phytol.*, 95, 277, 1983.
97. **Troiano, J., Colavito, L., Heller, L., McCune, D. C., and Jacobson, J. S.,** Effects of acidity of simulated rain and its joint action with ambient ozone on measures of biomass and yield in soybean, *Environ. Exp. Bot.*, 23, 113, 1983.
98. **Rebbeck, J. and Brennan, E.,** The effect of simulated acid rain and ozone on the yield and quality of glasshouse-grown alfalfa, *Environ. Pollut.*, 36A, 7, 1984.
99. **Norby, R. J., Ricter, D. D., and Luxmoore, R. J.,** Physiological processes in soybean inhibited by gaseous pollutants but not by acid rain, *New Phytol.*, 100, 79, 1985.
100. **Crang, R. E. and McQuattie, C. J.,** Qualitative and quantitative effects of acid misting and two air pollutants on foliar structures of *Liriodendron tulipifera, Can. J. Bot.*, 64, 1237, 1986.
101. **Johnston, J. W. and Shriner, D. S.,** Yield response of Davis soybeans to simulated acid rain and gaseous pollutants in the field, *New Phytol.*, 103, 695, 1986.
102. **Johnston, J. W., Shriner, D. S., and Kinnerley, C. K.,** The combined effects of simulated acid rain and ozone on injury, chlorophyll, and growth of radish, *Environ. Exp. Bot.*, 26, 107, 1986.
103. **Reich, P. B., Schoettle, A. W., and Amundson, R. G.,** Effects of O_3 and acidic rain on photosynthesis and growth in sugar maple and northern red oak seedlings, *Environ. Pollut.*, 40, 1, 1986.
104. **Taylor, G. E., Jr., Norby, R. J., McLaughlin, S. B., Johnson, A. H., and Turner, R. S.,** Carbon dioxide assimilation and growth of red spruce (*Picea rubens* Sarg.) seedlings in response to ozone, precipitation chemistry, and soil type, *Oecologia*, 70, 163, 1986.
105. **Reich, P. B., Schoettle, A. W., Stroo, H. F., Troiano, J., and Amundson, R. G.,** Effects of ozone and acid rain on white pine *(Pinus strobus)* seedlings grown in five soils. I. Net photosynthesis and growth, *Can. J. Bot.*, 65, 977, 1987.
106. **Reich, P. B., Schoettle, A. W., Stroo, H. F., and Amundson, R. G.,** Effects of ozone and acid rain on white pine *(Pinus strobus)* seedlings grown in five soils. III. Nutrient relation, *Can. J. Bot.*, 66, 1517, 1988.
107. **Temple, P. J.,** Injury and growth of Jeffrey pine and giant sequoia in response to ozone and acidic mist, *Environ. Exp. Bot.*, 28, 323, 1988.
108. **Laurence, J. A., Kohut, R. J., and Amundson, R. G.,** Response of red spruce seedlings exposed to ozone and simulated acidic precipitation in the field, *Arch. Environ. Contam. Toxicol.*, 18, 285, 1989.
109. **Simmons, G. L. and Kelly, J. M.,** Influence of O_3, rainfall acidity, and soil Mg status on growth and ectomycorrhizal colonization of loblolly pine roots, *Water Air Soil Pollut.*, 44, 159, 1989.
110. **Barnes, J. D., Eamus, D., and Brown, K. A.,** The influence of ozone, acid mist and soil nutrient status on Norway spruce (*Picea abies* (L.) Karst). II. Photosynthesis, dark respiration and soluble carbohydrates of trees during late autumn, *New Phytol.*, 115, 149, 1990.
111. **Edwards, G. S., Edwards, N. T., Kelly, J. M., and Mays, P. A.,** Ozone, acidic precipitation, and soil Mg effects on growth and nutrition of loblolly pine seedlings, *Environ. Exp. Bot.*, 31, 67, 1991.
112. **Patton, R. L., Jensen, K. F., and Schier, G. A.,** Responses of red spruce seedlings to ozone and acid deposition, *Can. J. For. Res.*, 21, 1354, 1990.

113. **Chappelka, A. H. and Chevone, B. I.,** White ash seedling growth response to ozone and simulated acid rain, *Can. J. For. Res.*, 16, 786, 1986.
114. **Chimiklis, P. E. and Heath, R. L.,** Ozone-induced loss of intracellular potassium ion from *Chlorella sorokiniana*, *Plant Physiol.*, 56, 723, 1975.
115. **Morvan, B. P., Demarty, M., and Thellier, M.,** Titration of isolated cell walls of *Lemna minor* L., *Plant Physiol.*, 63, 1117, 1979.
116. **Cosgrove, D. J.,** Wall relaxation and the driving forces for cell expansive growth, *Plant Physiol.*, 84, 561, 1987.
117. **Price, A. H.,** A possible role for calcium in oxidative stress, *Free Rad. Res. Commun.*, 10, 345, 1990.
118. **Fink, S.,** Unusual patterns in the distribution of calcium oxalate in spruce needles and their possible relationships to the impact of pollutants, *New Phytol.*, 119, 41, 1991.
119. **Castillo, F. J. and Heath, R. L.,** Ca^{2+} transport in membrane vesicles from Pinto bean leaves and its alteration after ozone exposure, *Plant Physiol.*, 94, 788, 1990.
120. **Dominy, P. J. and Heath, R. L.,** Inhibition of the K^+-stimulated ATPase of the plasmalemma of Pinto bean leaves by ozone, *Plant Physiol.*, 77, 43, 1985.
121. **Castillo, F. J., Penel, C. L., and Greppin, H.,** Peroxidase release induced by ozone in *Sedum album* leaves. Involvement of Ca^{2+}, *Plant Physiol.*, 74, 846, 1984.
122. **Tanaka, K., Suda, Y., Kondo, N., and Sugahara, K.,** O_3 tolerance and the ascorbate-dependent H_2O_2 decomposing system in chloroplasts, *Plant Cell Physiol.*, 26, 1425, 1985.
123. **Mehlhorn, H., Cottam, D. A., Lucas, P. W., and Wellburn, A. R.,** Induction of ascorbate peroxidase and glutathione reductase by interactions of air pollutants, *Free Rad. Res. Commun.*, 3, 193, 1987.
124. **Bobrov, R. A.,** The effect of smog on the anatomy of oat leaves, *Phytopathology*, 42, 558, 1952.
125. **Evans, L. S. and Ting, I. P.,** Ozone sensitivity of leaves: relationship to leaf water content, gas transfer, resistance, and anatomical characteristics, *Am. J. Bot.*, 61, 592, 1974.
126. **Evans, L. S. and Ting, I. P.,** Ozone-induced membrane permeability changes, *Am. J. Bot.*, 60, 155, 1973.
127. **Perchorowicz, J. T. and Ting, I. P.,** Ozone effects on plant cell permeability, *Am. J. Bot.*, 61, 787, 1974.
128. **Frederick, P. E. and Heath, R. L.,** Ozone-induced fatty acid viability changes in *Chlorella*, *Plant Physiol.*, 55, 15, 1975.
129. **Heath, R. L. and Frederick, P. E.,** Ozone-induced alterations of K^+ influx and efflux from *Chlorella sorokiniana*, *Plant Physiol.*, 59 (Suppl.), 124, 1977.
130. **Heath, R. L. and Frederick, P. E.,** Ozone alteration of membrane permeability in *Chlorella*. I. Permeability of potassium ion as measured by ^{86}Rubidium tracer, *Plant Physiol.*, 64, 455, 1979.
131. **Elkiey, T. and Ormrod, D. P.,** Ozone and/or sulfur dioxide effects on tissue permeability of petunia leaves, *Atmos. Environ.*, 13, 1165, 1979.
132. **Beckerson, D. W. and Hofstra, G.,** Effects of sulfur dioxide and ozone, singly or in combination, on membrane permeability, *Can. J. Bot.*, 58, 451, 1980.
133. **Keitel, A. and Arndt, U.,** Ozoninduzierte Turgeszenzverluste bei Tabak (*Nicotiana tabacum* var Bel W3) — ein Hinweis auf schnelle Permeabilitätsveränderungen der Zellmembranen, *Angew. Bot.*, 57, 193, 1983.
134. **Tomlinson, H. and Rich, S.,** Disulfides in bean leaves exposed to ozone, *Phytopathology*, 60, 1842, 1970.
135. **Mudd, J. B., McManus, T. T., Ongun, A., and McCullogh, T. E.,** Inhibition of glycolipid biosynthesis in chloroplasts by ozone and sulfhydryl reagents, *Plant Physiol.*, 48, 335, 1971.
136. **Nakamura, H. and Saka, H.,** Photochemical oxidant injury in rice plants. III. Effects of ozone on physiological activities in rice plants, *Jpn. J. Crop Sci.*, 47, 707, 1978.

137. **Scott, D. B. M. and Lesher, E. C.**, Effects of ozone and survival and permeability of *Escherichia coli*, *J. Bacteriol.*, 85, 567, 1963.
138. **Mudd, J. B.**, Effects of oxidants on metabolic function, in *Effects of Gaseous Air Pollution in Agriculture and Horticulture*, Unsworth, M. H. and Ormrod, D. P., Eds., Butterworth, London, 189, 1982.
139. **Swanson, E. S., Thomson, W. W., and Mudd, J. B.**, The effect of ozone on leaf cell membranes, *Can. J. Bot.*, 51, 1213, 1973.
140. **Nouchi, I. and Toyama, S.**, Effects of ozone and peroxyacetyl nitrate on polar lipids and fatty acids in leaves of morning glory and kidney bean, *Plant Physiol.*, 87, 638, 1988.
141. **Fangmeier, A., Kress, L. W., Lepper, P., and Heck, W. W.**, Ozone effects on the fatty acid composition of loblolly pine needles (*Pinus taeda* L.), *New Phytol.*, 115, 639, 1990.
142. **Kyburz, S., Eichenberger, W., Keller, T., Rennenberg, H., and Schröder, P.**, Fatty acids of ozone-exposed Scots pine and poplar trees and of damaged field-grown Norway spruce trees, *Eur. J. For. Pathol.*, 21, 49, 1991.
143. **Wolfenden, J. and Wellburn, A. R.**, Effects of summer ozone on membrane lipid composition during subsequent frost hardening in Norway spruce (*Picea abies* (L.) Karst), *New Phytol.*, 118, 323, 1991.
144. **Tomlinson, H. and Rich, S.**, Relating lipid content and fatty acid synthesis to ozone injury of tobacco leaves, *Phytopathology*, 59, 1284, 1969.
145. **Pauls, K. P. and Thompson, J. E.**, Effects of in vitro treatment with ozone on the physical and chemical properties of membranes, *Physiol. Plant.*, 53, 255, 1981.
146. **Whitaker, B. D., Lee, E. H., and Rowland, R. A.**, EDU and ozone protection: foliar glycerolipids and steryl lipids in snapbean exposed to ozone, *Physiol. Plant.*, 80, 286, 1990.
147. **Sakaki, T., Kondo, N., and Sugahara, K.**, Breakdown of photosynthetic pigments and lipids in spinach leaves with ozone fumigation: role of active oxygens, *Physiol. Plant.*, 59, 28, 1983.
148. **Sakaki, T., Ohnishi, J., Kondo, N., and Yamada, M.**, Polar and neutral lipid changes in spinach leaves with ozone fumigation: triacylglycerol synthesis from polar lipids, *Plant Cell Physiol.*, 26, 253, 1985.
149. **Sakaki, T., Saito, K., Kawaguchi, A., Kondo, N., and Yamada, M.**, Conversion of monogalactosyldiacylglycerols to triacylglycerols in ozone-fumigated spinach leaves, *Plant Physiol.*, 94, 766, 1990.
150. **Sakaki, T., Kondo, N., and Yamada, M.**, Pathway for the synthesis of triacylglycerols from monogalactosyldiacyglycerols in ozone-fumigated spinach leaves, *Plant Physiol.*, 94, 773, 1990.
151. **Tomlinson, H. and Rich, S.**, Effects of ozone on the steroid content of bean leaves, *Phytopathology*, 61, 132, 1971.
152. **Tomlinson, H. and Rich, S.**, Anti-senescent compounds reduce injury and steroid changes in ozonated leaves and their chloroplasts, *Phytopathology*, 63, 903, 1973.
153. **Spotts, R. A., Lukezic, F. L., and LaCasse, N. L.**, The effect of benzimidazole, cholesterol, and a steroid inhibitor on leaf sterols and ozone resistance of bean, *Phytopathology*, 65, 45, 1975.
154. **Menser, H. A., Chaplin, J. F., Cheng, A. L. S., and Sorokin, T.**, Polyphenols, polysterols, and reducing sugars in air-cured tobacco leaves injured by ozone air pollution, *Tobacco Sci.*, 21, 35, 1977.
155. **Trevathan, L. E., Moore, L. D., and Orcutt, D. M.**, Symptom expression and free sterol and fatty acid composition of flue-cured tobacco plants exposed to ozone, *Phytopathology*, 69, 582, 1979.
156. **Bücker, J.**, *Immissionsbedingte Störungen im Kohlenhydrathaushalt junger Pappeln und Fichten*, Westarp Wissenschaften, Essen, 1991.

157. **Bücker, J., Strobel, P., and Guderian, R. L.,** Biochemical investigations on accelerated senescence of plants induced by air pollutants, in Proceedings of the 9th World Clean Air Congress (Aug. 30–Sept. 4, 1992) Montreal, Canada.
158. **Sakaki, T., Kondo, N., and Yamada, M.,** Free fatty acids regulate two galactosyltransferases in chloroplast envelope membranes isolated from spinach leaves, *Plant Physiol.*, 94, 781, 1990.
159. **Brown, K. A., Roberts, T. M., and Blank, L. W.,** Interaction between ozone and cold sensitivity in Norway spruce: a factor contributing to the forest decline in Central Europe, *New Phytol.*, 105, 149, 1987.
160. **Barnes, J. D. and Davison, A. W.,** The influence of ozone on the winter hardiness of Norway spruce (*Picea abies* (L.) Karst.), *New Phytol.*, 108, 159, 1988.
161. **Barnes, J. D., Reiling, K., Davison, A. W., and Renner, C. J.,** Interaction between ozone and winter stress, *Environ. Pollut.*, 53, 235, 1988.
162. **Lucas, P. W., Cottam, D. A., Sheppard, L. J., and Francis, B. J.,** Growth responses and delayed winter hardening in Sitka spruce following summer exposure to ozone, *New Phytol.*, 108, 495, 1988.
163. **Fincher, J., Cumming, J. R., Alscher, R. G., Rubin, G., and Weinstein, L.,** Long-term ozone exposure affects winter hardiness of red spruce seedlings, *New Phytol.*, 113, 85, 1989.
164. **Chappelka, A. H., Kush, J. S., Meldahl, R. S., and Lockaby, B. G.,** An ozone-low temperature interaction in loblolly pine (*Pinus taeda* L.), *New Phytol.*, 114, 721, 1990.
165. **Edwards, G. S., Pier, P. A., and Kelly, J. M.,** Influence of ozone and soil magnesium status on the cold hardiness of loblolly pine (*Pinus taeda* L.) seedlings, *New Phytol.*, 115, 157, 1990.
166. **Eissenstat, D. M., Syvertsen, J. P., Dean, D. J., Yelenosky, G., and Johnson, J. D.,** Sensitivity of frost resistance and growth in citrus and avocado to chronic ozone exposure, *New Phytol.*, 118, 139, 1991.
167. **Eissenstat, D. M., Syvertsen, J. P., Dean, D. J., Johnson, J. D., and Yelenosky, G.,** Interaction of simulated acid rain with ozone on freeze resistance, growth, and mineral nutrition in citrus and avocado, *J. Am. Soc. Hortic. Sci.*, 116, 838, 1991.
168. **Huttunen, S. and Soikkeli, S.,** Effects of various gaseous pollutants on plant cell ultrastructure, in *Gaseous Air Pollutants and Plant Metabolism*, Koziol, M. J. and Whatley, F. R., Eds., Butterworths, London, 1984, 117.
169. **Vogels, K. and Masuch, G.,** Metabolic and substructural changes of bean leaves caused by separate and simultaneous ozone and sulphur dioxide exposures, in Proc. Int. Workshop Evaluation and Assessment of Effects of Photochemical Oxidants on Human Health, Agricultural Crops, Forestry, Materials and Visibility, February 29 to March 2, Göteborg, Sweden, 1984.
170. **Sutinen, S.,** Ultrastructure of mesophyll cells of spruce needles exposed to O_3 alone and together with SO_2, *Eur. J. For. Pathol.*, 17, 362, 1987.
171. **Sutinen, S., Skärby, L., Wallin, G., and Sellden, G.,** Long-term exposure of Norway spruce, *Picea abies* (L.) Karst., to ozone in open-top chambers, *New Phytol.*, 115, 345, 1990.
172. **Salin, M. L.,** Toxic oxygen species and protective systems of the chloroplast, *Physiol. Plant.*, 72, 681, 1987.
173. **Alscher, R. G.,** Biosynthesis and antioxidant function of glutathione in plants, *Physiol. Plant.*, 77, 457, 1989.
174. **Ormrod, D. P. and Beckerson, D. W.,** Polyamines as antiozonants for tomato, *HortScience*, 21, 1070, 1986.
175. **Peters, J. L., Castillo, J., and Heath, R. L.,** Alteration of extracellular enzymes in Pinto bean leaves upon exposure to air pollutants, ozone and sulfur dioxide, *Plant Physiol.*, 89, 159, 1989.

176. **Rowland-Bamford, A. J., Borland, A. M., Lea, P. J., and Mansfield, T. A.,** The role of arginine decarboxylase in modulating the sensitivity of barley to ozone, *Environ. Pollut.*, 61, 95, 1989.
177. **Dohmen, G. P., Koppers, A., and Langebartels, C.,** Biochemical response of Norway spruce (*Picea abies* (L.) Karst.) towards 14-month exposure to ozone and acid mist: effects on amino acid, glutathione and polyamine titers, *Environ. Pollut.*, 64, 375, 1990.
178. **Kramer, G. F., Lee, E. H., and Rowland, R. A.,** Effects of elevated CO_2 concentration on the polyamine levels of field-grown soybean at three O_3 regimes, *Environ. Pollut.*, 73, 137, 1991.
179. **Guri, A.,** Variation in glutathione and ascorbic acid content among selected cultivars of *Phaseolus vulgaris* prior to and after exposure to ozone, *Can. J. Plant Sci.*, 63, 733, 1983.
180. **Castillo, F. J. and Greppin, H.,** Balance between anionic and cationic extracellular peroxidase activities in *Sedum album* leaves after ozone exposure. Analysis by high-performance liquid chromatography, *Physiol. Plant.*, 68, 201, 1986.
181. **Castillo, F. J. and Greppin, H.,** Extracellular ascorbic acid and enzyme activities related to ascorbic acid metabolism in *Sedum album* L. leaves after ozone exposure, *Environ. Exp. Bot.*, 28, 231, 1988.
182. **Tanaka, K., Saji, H., and Kondo, N.,** Immunological properties of spinach glutathione reductase and inductive biosynthesis of the enzyme with ozone, *Plant Cell Physiol.*, 29, 637, 1988.
183. **Price, A., Lucas, P. W., and Lea, P. J.,** Age dependent damage and glutathione metabolism in ozone fumigated barley: a leaf section approach, *J. Exp. Bot.*, 41, 1309, 1990.
184. **Sen Gupta, A. S., Alscher, R. G., and McCune, D.,** Response of photosynthesis and cellular antioxidants to ozone in *Populus* leaves, *Plant Physiol.*, 96, 650, 1991.
185. **Rao, M. V. and Dubey, P. S.,** Biochemical aspects (antioxidants) for development of tolerance in plants growing at different low levels of ambient air pollutants, *Environ. Pollut.*, 64, 55, 1990.
186. **Chevrier, N., Sarhan, F., and Chung, Y. S.,** Oxidative damages and repair in *Euglena gracilis* exposed to ozone. I. SH groups and lipids, *Plant Cell Physiol.*, 29, 321, 1988.
187. **Chevrier, N., Chung, Y. S., and Sarhan, F.,** Oxidative damages and repair in *Euglena gracilis* exposed to ozone. II. Membrane permeability and uptake of metabolites, *Plant Cell Physiol.*, 31, 987, 1990.
188. **Decleire, M., DeCat, W., DeTemmermann, L., and Baeten, H.,** Changes of peroxidase, catalase, and superoxide dismutase activities in ozone-fumigated spinach leaves, *J. Plant Physiol.*, 116, 147, 1984.
189. **Klumpp, G., Guderian, R., and Küppers, K.,** Peroxidase- und Superoxiddismutase-Aktivität sowie Prolingehalte von Fichtennadeln nach Belastung mit O_3, SO_2 und NO_2, *Eur. J. For. Pathol.*, 19, 84, 1989.
190. **Matters, G. and Scandalios, J.,** Synthesis of isozymes of superoxide dismutase in maize leaves in response to O_3, SO_2 and elevated O_2., *J. Exp. Bot.*, 38, 842, 1987.
191. **Becker, K., Saurer, M., Egger, A., and Fuhrer, J.,** Sensitivity of white clover to ambient ozone in Switzerland, *New Phytol.*, 112, 235, 1989.
192. **Hausladen, A., Madamanchi, N., Fellows, S., Alscher, R., and Amundson, R.,** Seasonal changes in antioxidants in red spruce as affected by ozone, *New Phytol.*, 115, 447, 1990.
193. **Polle, A. and Rennenberg, H.,** Superoxide dismutase activity in needles of Scots pine and Norway spruce under field and chamber conditions: lack of ozone effects, *New Phytol.*, 117, 335, 1991.
194. **Lee, E. H. and Bennett, J. H.,** Superoxide dismutase. A possible protective enzyme against ozone injury in snap beans (*Phaseolus vulgaris* L.), *Plant Physiol.*, 69, 1444, 1982.

195. **Aono, M., Kubo, A., Saji, H., Natori, T., Tanaka, K., and Kondo, N.,** Resistance to active oxygen toxicity of transgenic *Nicotiana tabacum* that expresses the gene glutathione reductase from *Escherichia coli, Plant Cell Physiol.*, 32, 691, 1991.
196. **Foyer, C., Lelandais, M., Galap, C., and Kunert, K. J.,** Effects of elevated cytosolic glutathione reductase activity on the cellular glutathione pool and photosynthesis in leaves under normal and stress conditions, *Plant Physiol.*, 97, 863, 1991.
197. **Elroy-Stein, O., Bernstein, Y., and Groner, Y.,** Overproduction of human Cu/Zn-superoxide dismutase in transfected cells: extenuation of paraquat-mediated cytotoxicity and enhancement of lipid peroxidation, *EMBO J.*, 5, 615, 1986.
198. **Berrang, P., Karnosky, D. F., Mickler, R. A., and Bennett, J. P.,** Natural selection for ozone tolerance in *Populus tremuloides, Can. J. For. Res.*, 16, 1214, 1986.
199. **Barnes, J. D., Velissariou, D., Davison, A. W., and Holevas, C. D.,** Comparative ozone sensitivity of old and modern Greek cultivars of spring wheat, *New Phytol.*, 116, 707, 1990.
200. **Miller, P. R.,** Oxidant-induced community change in a mixed conifer forest, *Adv. Chem. Ser.*, 122, 101, 1973.
201. **Karnosky, D. F.,** Evidence for genetic control of response to sulfur dioxide and ozone in *Populus tremuloides, Can. J. For. Res.*, 7, 437, 1977.
202. **Karnosky, D. F. and Steiner, K. C.,** Provenance and family variation in response of *Fraxinus americana* and *F. pennsylvanica* to ozone and sulfur dioxide, *Phytopathology*, 71, 804, 1981.
203. **Wolff, G. T. and Lioy, P. J.,** Development of an ozone river associated with synoptic scale episodes in eastern United States, *Environ. Sci. Technol.*, 14, 1257, 1980.
204. **Wolff, G. T., Kelly, N. A., and Ferman, M. A.,** Source regions of summertime ozone and haze episodes in the eastern United States, *Water Air Soil Pollut.*, 18, 65, 1982.
205. **Vukovich, F. M., Fishman, J., and Browell, E. V.,** The reservoir of ozone in the boundary layer of the eastern United States and its potential impact on the global tropospheric ozone budget, *J. Geophys. Res.*, 90, 5687, 1985.
206. **Heagle, A. S., McLaughlin, M. R., Miller, J. E., Joyner, R. L., and Spruill, S. E.,** Adaptation of a white clover population to ozone stress, *New Phytol.*, 119, 61, 1991.
207. **DeVos, N. E., Hill, R. R., Pell, E. J., and Cole, R. H.,** Quantitative inheritance of ozone resistance in potato, *Crop Sci.*, 22, 992, 1982.
208. **Knudson-Butler, L. and Tibbits, T. W.,** Stomatal mechanisms determining genetic resistance to ozone in *Phaseolus vulgaris* L., *J. Am. Soc. Hortic. Sci.*, 104, 213, 1979.
209. **Mebrahtu, T., Mersie, W., and Rangappa, M.,** Inheritance of ambient ozone insensitivity in common bean (*Phaseolus vulgaris* L.), *Environ. Pollut.*, 67, 79, 1990.
210. **Held, A. A., Mooney, H. A., and Gorham, J. N.,** Acclimation to ozone stress in radish: leaf demography and photosynthesis, *New Phytol.*, 118, 417, 1991.
211. **Abeles, F. B.,** *Ethylene in Plant Biology,* Academic Press, New York, 1973.
212. **Wang, S. Y., Wang, C. J., and Wellburn, A. R.,** Role of ethylene under stress conditions, in *Stress Responses in Plants: Adaptation and Acclimation Mechanisms,* Plant Biology Vol. 12, Alscher, R. G. and Cumming, J. R., Eds., Wiley-Liss, New York, 1990, 147.
213. **Rosen, P. M. and Runeckles, V. C.,** Interaction of ozone and greenhouse whitefly in plant injury, *Environ. Conserv.*, 3, 70, 1976.
214. **Pell, E. J., Lukezic, F. L., Levine, R. G., and Weissberger, W. C.,** Response of soybean foliage to reciprocal challenges by ozone and a hypersensitive-response-inducing pseudomonad, *Phytopathology*, 67, 1342, 1977.
215. **Runeckles, V. C. and Palmer, K.,** Pretreatment with nitrogen dioxide modifies plant response to ozone, *Atmos. Environ.*, 21, 717, 1987.
216. **Schier, G. A., McQuattie, C. J., and Jensen, K. F.,** Effect of ozone and aluminium on pitch pine *(Pinus rigida)* seedlings: growth and nutrient relations, *Can. J. For. Res.*, 20, 1714, 1990.

217. **Czuba, M. and Ormrod, D. P.,** Effects of cadmium and zinc on ozone-induced phytotoxicity in cress and lettuce, *Can. J. Bot.*, 52, 645, 1974.
218. **Czuba, M. and Ormrod, D. P.,** Cadmium concentrations in cress shoots in relation to cadmium-enhanced ozone phytotoxicity, *Environ. Pollut.*, 25A, 67, 1981.
219. **Ormrod, D. P.,** Cadmium and nickel effects on growth and ozone sensitivity of pea, *Water Air Soil Pollut.*, 8, 263, 1977.
220. **Heggestad, H. E., Gish, T. J., Lee, E. H., Bennett, J. H., and Douglass, L. W.,** Interaction of soil moisture stress and ambient ozone on growth and yields of soybeans, *Phytopathology,* 75, 472, 1985.
221. **von Tiedemann, A., Ostländer, P., Firsching, K. H., and Fehrmann, H.,** Ozone episodes in Southern Lower Saxony (FRG) and their impact on the susceptibility of cereals to fungal pathogens, *Environ. Pollut.*, 67, 43, 1990.
222. **von Tiedemann, A., Weigel, H. J., and Jäger, H. J.,** Effects of open-top chamber fumigations with ozone on three fungal leaf diseases of wheat and the mycoflora of the phyllosphere, *Environ. Pollut.*, 72, 205, 1991.

Chapter 7

INVOLVEMENT OF SUPEROXIDE IN SIGNAL TRANSDUCTION: RESPONSES TO ATTACK BY PATHOGENS, PHYSICAL AND CHEMICAL SHOCKS, AND UV IRRADIATION

Noriyuki Doke, Yoshio Miura, Leandro M. Sanchez, and Kazuhito Kawakita

TABLE OF CONTENTS

0-8493-5443-9/94/$0.00 + $.50

I. INTRODUCTION

The discovery of superoxide dismutase (SOD), which dismutates O_2^- to H_2O_2 in aerobic cells, was an outstanding event in understanding new aspects of the role of oxygen in cell metabolism.[1] The nature, occurrence, and role of SOD in animal and plant cells, as well as the biochemical implication of oxygen radical production and reactivity, have been well studied.[2-4] The respiration burst in mammalian phagocytic cells following recognition of alien microorganisms, or certain chemicals, has been shown to depend on an O_2^--generating NADPH oxidase in the plasma membrane, which produces a flux of oxygen radicals outside the membrane that is lethal for pathogenic microorganisms as well as the activated phagocytic cells themselves.[5,6] The contribution of O_2^- generation in animal phagocytic cells to defense is now beyond doubt.

In contrast to these differentiated cells that specialize in defense in animals, cells in plant tissues do not seem to be differentiated specifically for defense against invaders. However, the epidermal cells, and even cells in the inner tissues, display some role in the recognition and localization of invaders, generally showing rapid cell death and necrosis, the so-called, hypersensitive reaction (HR).[7] Such a function of plant cells has been well documented and contributes to the localization of invading microorganisms by the induction of chemical and physical barriers such as the accumulation of antimicrobial substances (phytoalexins), chitinase, glucanase, lignin, etc.[8,9] These are also induced by ultraviolet (UV) light irradiation, biotic and abiotic elicitor molecules, or sometimes physical shocks.[10,11] Research on the molecular mechanism of *de novo* induction of such new metabolic changes is now in fashion, especially in relation to the expression and regulation of defense genes.[9,12] However, details of the signal and transduction mechanisms that trigger HR or stress metabolism remain obscure.

New insight for understanding the mechanism of triggering HR was provided in the field of activated oxygen and its associated metabolism in plants after Doke[13] found that SOD-sensitive cytochrome *c* or NBT (nitroblue tetrazolium)-reducing activities were a very early response in the hypersensitively reacting cells in potato tissues against infection with incompatible races of *Phytophthora infestans*. This finding seems to have triggered a new wave of research on the role of active oxygen in plant defense or adaptation to attack by microorganisms and various stresses. UV-light exposure, which produces activated oxygen species in living cells, has been known to induce metabolic changes similar to pathogen attack, wounding, or temperature shock.[10,11]

In this chapter, we review and discuss recent research on the involvement and implication of superoxide and associated active oxygen species in the initial induction events of the plant defense reaction against microbial organisms and other stresses, with special reference to the signaling process in plant responses to pathogen attack or stresses.

II. O_2^- GENERATION AND THE HYPERSENSITIVE RESPONSE OF PLANT TISSUES TO PATHOGEN ATTACK

A. FUNGAL INFECTION

In the case of potato cultivars with the major *R*-gene against *P. infestans* and the races of the fungus corresponding to each *R*-gene, very clear differences in the host cell responses to infection have been demonstrated between incompatible and compatible interactions.[7] Attempted infection with zoospores on aged (18 h) or fresh potato tuber slices allowed the fungus to initiate its penetration into the surface cells 1 to 3 h after inoculation. Dramatic changes in the physiology of aged, but not fresh, tissue slices occurred just after penetration by incompatible, but not compatible, races (hypersensitive cell reactivity was established *de novo* during aging after wounding).[7,15] The initial events in HR are characterized by cytoplasmic aggregation,[15] depolarization of the membrane diffusion potential,[16] leakage of electrolytes, loss of selective membrane permeability, and loss of the plasmolyzing ability of the cells penetrated by the fungus,[7,18] followed by the induction of lipid peroxidation, phytoalexin synthesis,[14,19] and/or polyphenol oxidation in tissues at later stages of infection. In the case of infection with a compatible race, however, such events scarcely occur from the very early to the late stages of infection. Thus, the key reaction responsible for this race-cultivar-specific interaction has been suggested to exist at the time the invading hyphae made contact with the host plasma membrane in the case of the *P. infestans* potato interaction.[15,20]

Why and how such dramatic events occur at this very early stage of infection has long been under consideration. Doke[13] found a novel biochemical reaction, O_2^- generation, which was enhanced only around the contact site in plant cells with invading hyphae of incompatible, but not compatible, races where cytoplasmic aggregation concomitantly appeared. Figure 1 illustrates

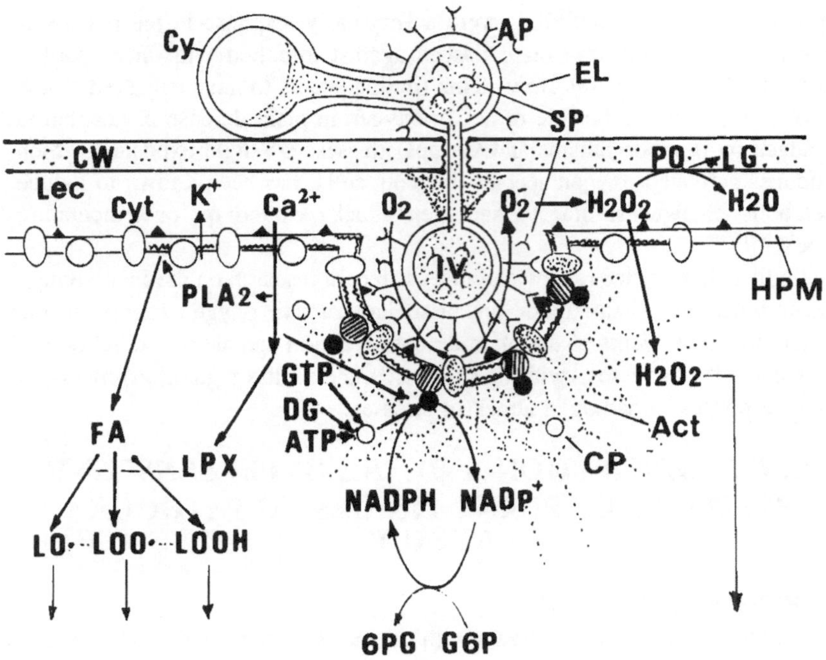

FIGURE 1. Diagram of the concept of the activation of the NADPH-dependent superoxide anion-generating system in potato tissues just after penetration by *P. infestans*. The fungal cystospore (Cy) germinates, forms an appressorium (Ap) on the host cell surface, invades and makes an infection thread in the host cell wall (CW), and then forms the primary infection vesicle (IV), on the surface of which an elicitor moiety (EL) may exist. The fungus produces a suppressor (SP). The putative receptor for the EL and potato lectin (Lec) for binding to the fungal surface exist on the host plasma membrane (HPM). The NADPH-dependent O_2^--generating system on the host plasma membrane is shown to be activated, depending on Ca^{2+}, GTP, diacylglycerol (DG), ATP, and soluble protein components, immediately after contact (CP) with the hyphal elicitor. Ca^{2+} influx, aggregation of actin molecules (Act), and activation of phospholipase A_2 (PLA_2) occur during this process. Lipid peroxidation in the cell begins with the activation of lipoxygenase (LPX). G6P, glucose-6-phosphate; 6PG, 6-phosphogluconate.

the superoxide anion-generating reaction in potato cells just after penetration by an incompatible race.

The O_2^- generation was visualized by rapid accumulation of blue formazan around the infection peg of incompatible, but not compatible, races when the inoculated tissues were supplied with NBT solution. In addition, enhancement of the SOD-sensitive reducing activity of extracellular cytochrome *c* in the tissues was detected after the initiation of penetration by the incompatible race.[13] These were detected to a lesser extent in the case of aged tuber slices infected with compatible races or fresh ones infected with incompatible races.[13] These observations led to the idea of a possible triggering effect of O_2^- in the initiation of a cascade reaction in HR. As described below, a transient O_2^--generating reaction also appeared just after slicing the potato tubers. This reaction could be associated with the wound induction of new metabolic pathways associated with the acquisition of the hypersensitive potential in plant cells.[21]

Some trials on the effect of superoxide scavengers, in host tissues vacuum-infiltrated with SOD[13] or by the surface application of tiron to inoculated tissues,[22] significantly reduced or delayed the occurrence of HR, as evidenced by rapid cell death (early cell response) or phytoalexin production (late tissue response). Direct application of O_2^--generating systems such as the xanthine-xanthine oxidase system or H_2O_2 to potato tuber slices resulted in an induction of phytoalexin production similar to that found with infection with incompatible races.[23] These results also support the hypothesis that O_2^- or a derivative of superoxide may play an important signaling role in the induction of HR.

In intact potato leaves, *P. infestans* initiated its invasion of epidermal cells 4 to 8 h after inoculation with zoospore suspensions and caused rapid cell death within about 1 h after fungal penetration (loss of plasmolyzing ability) in the case of infection with incompatible, but not compatible, races. In the former case, the infected leaf cells also showed enhanced O_2^- generation with an increase in fungal penetration for at least 8 h after inoculation.[23] The epidermal cells penetrated by the incompatible fungi responded to localized invasion by browning. In this case, the principles contributing to the localization of invading fungi are not well explained since the production of enough antifungal substances to inhibit fungal growth was not detected at the same time. However, O_2^- generation may be closely associated with the occurrence of HR, as well as localized invasion.

Enhanced expression of phenylalanine ammonia lyase (PAL) and 4-coumarate: CoA ligase (4-CL) genes has been shown to occur in leaves before penetration by the fungi in a race-nonspecific manner.[24] Intact leaf systems respond to infection by transient O_2^- generation in a race-nonspecific manner at infection stages before fungal penetration. During this period of infection, epidermal cells have been shown to acquire a hypersensitive potential.[20] Such transient activation of O_2^- generation and induction of a hypersensitive potential were reproduced by the application of spore germination fluid.[23] Thus, O_2^- generation may be considered to be closely associated with the induction of hypersensitive potential[23] or gene expression of PAL and 4-CL, which are activated in intact leaves during the prepenetration stages of infection.[24] Iwanova et al.[25] reported O_2^- generation in the leaves of potato cultivars with field resistance which were inoculated with a compatible race of *P. infestans*. This occurred in addition to the enhancement of O_2^- generation upon infection with incompatible races.[25] However, the transient generation of O_2^- in leaves before the period of fungal penetration following inoculation was not detected.

There are many problems with the quantitative measurement of O_2^- generation in plant tissues. The detection of O_2^- generation in these systems has often been performed by assaying extracellular cytochrome *c* reduction by

O_2^- released from the plant cell walls. Therefore, only part of the total O_2^- generated in the cells can be detected because of its capacity for nonenzymatic or enzymatic dismutation. Since it is not thought to permeate the plasma membrane, O_2^- must be produced primarily between the cell wall and the plasma membrane. Interestingly, not all potato tissue cells respond hypersensitively to invasion, even by incompatible races.[19] The degree of resistance in vertically resistant cultivars depends primarily on the ratio of hypersensitively reactive cells present. Thus, detection of O_2^- generated by the tissues of low-resistance cultivars can sometimes be missed. It should also be noted that O_2^- generated inside tissues may be hard to detect effectively by measuring the reduction of extracellular cytochrome *c*.

In other plant systems challenged by the fungi, experimental data showing the involvement of O_2^- in HR are now accumulating. In preliminary experiments, Doke et al.[21,23] showed that leaves of soybean, French bean, broad bean, and cowpea also produced O_2^- at a detectable level following inoculation with *P. infestans* as a nonpathogenic fungus or after treatment with the hyphal wall component (HWC) elicitor.[21,23] Sekizawa et al.[26,27] demonstrated the NADPH-dependent generation of O_2^- in leaf tissues of rice plants inoculated with an incompatible race of *Pyricularia oryzae,* a pathogen of rice blast. Press-injured inoculation of rice leaf tissues with conidiospores of incompatible races caused a much more rapid and greater production of O_2^- than that of compatible races. The production of O_2^- oscillated during the course of infection. In rice plants, Aver'yanov et al.[28] also reported generation of the superoxide radical in relation to blast resistance.[28]

B. BACTERIAL INFECTION

In plant tissues infected with bacterial pathogens, Keppler and Baker[29] and Keppler et al.[30] have shown an involvement of O_2^- in the hypersensitive response to attempted inoculation with incompatible bacterial pathovars. A depolarization of the host plasma-membrane diffusion potential has been observed in the case of attempted infection of an incompatible host (cotton) by *P. syringae* pv. *tabaci.*[31] The depolarization was accompanied by membrane lipid peroxidation and electrolyte leakage from the cell in the cucumber (*P. syringae* pv. *pisi*) system.[32] The implication that O_2^- is important in these phenomena was suggested by experimental results showing that SOD infiltrated into plant tissues before bacterial infection resulted in a significant reduction in both membrane lipid oxidation and subsequent electrolyte leakage.[31,32] These findings provided additional information about an oxygen radical-mediated disruption of the host cell membrane during expression of the hypersensitive response to infection with an incompatible bacterial pathogen.

C. VIRAL INFECTION

Tobacco cultivars with the *N*-gene for resistance to tobacco mosaic virus (TMV) are well known for showing a hypersensitive response, with the formation of necrotic lesions at temperatures lower than 28°C, followed by localized virus multiplication in the lesions.[33] At temperatures higher than 30°C, this response disappeared, resulting in systemic infection without the formation of necrotic lesions.[34] Why and how the necrotic lesions are formed remains obscure. Leaves of the tobacco cultivars with the *N*-gene, which had been inoculated with TMV and incubated at 30°C for 36 h, failed to show any symptoms, regardless of normal virus multiplication. When transferred to temperatures lower than 28°C, however, necrotic lesions appeared within several hours.[34] The earliest change which appeared in infected cells after transferring infected leaf tissues from 30 to 25°C may be a kind of trigger reaction for the hypersensitive cell response. Doke and Ohashi[35] found that tobacco leaf tissues which had been inoculated and incubated at 30°C revealed SOD-sensitive reduction of extracellular cytochrome *c* immediately after transfer to 25°C.[35] This oscillated gradually and decreased in activity during a further incubation until necrotic lesions appeared. The initial level of O_2^- generation was almost proportional to the number of lesions appearing later.

Application of nitroblue tetrazolium to the inoculated tissues, just after transfer from 30 to 25°C, caused locally accumulated lesions showing blue formazan formation, which supported the suggestion that O_2^- was generated in close association with necrotic lesion formation.[35] When inoculated leaf tissues were returned to 30°C during the formation of necrotic lesions at 25°C (after the previous transfer from 30°C), the formation of necrotic areas stopped on the way to formation of complete lesions. Vacuum infiltration of the TMV-inoculated leaf tissues with SOD or $NADP^+$ delayed the appearance of necrotic lesions at 25°C, after transfer from 30°C. These experimental results showed that the generation of O_2^- depended on NADPH. The trans:superoxide produced may also play an important role in triggering the induction of necrotic lesion formation associated with the expression of defense genes in the case of virus-infected plant tissues.

III. O_2^- GENERATION IN PROTOPLASTS OR SUSPENSION CELL CULTURES FOLLOWING STIMULATION BY ELICITORS

As a model system for analyzing signals and transduction mechanisms in plant-microbe interactions, protoplasts or suspension cell cultures have been employed under some limited conditions. The HWC of *Phytophthora infestans* has been characterized in terms of the race-nonspecific elicitor of hypersensitive cell death and phytoalexin production in potato tuber slices.[36] Doke and Tomiyama[37] developed a system using protoplasts for analysis of the hypersensitive cell response to the HWC elicitor. The HWC also caused a rapid aggregation of protoplasts from potato tuber tissues similar to that formed in tissue cells following infection with incompatible races. Protoplasts isolated from various potato cultivars, which showed different degrees of true resistance, responded immediately to the HWC elicitor at different ratios of reactive-to-nonreactive cells, with a high correlation with the degree of resistance.[38] The reactive protoplasts showed rapid cytoplasmic aggregation.[39] This ratio was also highly correlated with that of hypersensitively dead/not dead cells in tissues infected with incompatible races. These observations emphasized that the interaction of protoplasts with HWC may include very early events which seem to be initiated in cells during penetration by incompatible races *in situ*.

Doke[39] reported that the protoplasts isolated from potato tuber tissues showed a rapid increase in the activity of SOD-sensitive extracellular cytochrome *c* reduction following application of the HWC elicitor. This was accompanied by the occurrence of cytoplasmic aggregation.[39] However, the activity was much increased in the presence of externally added NADPH. The SOD-sensitive reduction of extracellular cytochrome *c* by protoplasts demonstrates a means of producing O_2^- outside of the protoplasts. Interestingly, this activation of the O_2^--generating system was inhibited by pretreatment of protoplasts with suppressor glucans isolated from compatible, but not incompatible, races of *P. infestans*, which had been shown to be race-specific inhibitors of the induction of HR and a determinant for the establishment of compatible interaction.[39]

Doke et al.[40] also confirmed that in protoplasts isolated from soybean, French bean, cowpea, broadbean, tomato, and tobacco tissues (which also show a hypersensitive reaction to HWCs similar to that of potato tissues), enhanced generation of NADPH-dependent O_2^- occurred immediately after application of HWCs.

Linder et al.[41] reported that soybean suspension cultures exhibited an enhanced luminol-mediated chemiluminescence 20 to 30 min after exposure to an elicitor from *P. megasparma* f. sp. *glycinea* (soybean pathogen). The enhanced chemiluminescence was inhibited by scavengers of superoxide and H_2O_2, such as SOD and catalase, and also by the peroxidase inhibitors, salycylhydroxamic acid (SHAM) and cyanide. These authors[41] suggested that soybean cells may produce O_2^- and H_2O_2 as an early response to an elicitor and that the luminol-mediated chemiluminescence was generated via a peroxidase, and not directly by O_2^-. Sanchez et al.[42] detected a rapid enhancement of luminol-mediated chemiluminescence in suspension-cultured cells of tomato, sweet pepper, and tobacco within several minutes of exposure to elicitors of HWCs isolated from *P. infestans* (virulent to tomato), *P. capsici* (virulent to tomato and sweet pepper), and *P. nicotianae* var *nicotianae* (virulent to tomato and tobacco). These HWCs also caused rapid death of the cells in a species-nonspecific manner. In the above case, water-soluble glucans from *P. infestans, P. capsici,* and *P. nicotianae* var. *nicotianae,* which suppressed the occurrence of HWC-elicited cell death of susceptible host plants (similar to those of various races of *P. infestans*), also suppressed the occurrence of hypersensitive cell death in susceptible cultivars.[43] These glucans markedly inhibited the enhancement of HWC-elicited chemiluminescence in the suspension cells of the respective susceptible host plants only.[44] These experimental results indicate that the *Phytophthora* pathogens themselves may produce the suppressor glucans which inhibit the activation of superoxide generation that is responsible for induction of the resistance reaction in the corresponding susceptible cells of solanaceau plants, thus establishing successful susceptible interactions.

The rapidity of activation of the $O_2^{\cdot-}$-generating reaction following elicitation with an elicitor (within 1 or 2 min) in protoplasts or suspension cell cultures emphasizes the nature of this mechanism as an intracellular trigger responsible for the induction of a cascade of metabolic changes in induced resistance.

IV. NADPH DEPENDENT $O_2^{\cdot-}$-GENERATING SYSTEMS IN ISOLATED MEMBRANE FRACTIONS

The cellular localization of the NADPH-dependent $O_2^{\cdot-}$-generating system stimulated by incompatible microbes or elicitors, has been examined. Doke[45] reported that enhanced NADPH-dependent $O_2^{\cdot-}$-generating activity was present in the microsomal fraction isolated from potato tuber slices inoculated with incompatible, but not compatible, races of *P. infestans*. This activity, assayed *in vitro,* increased with infection time as the hypersensitive cell reaction proceeded. The assay was carried out using acetylated cytochrome *c,* which is not a substrate for contaminating enzymes such as NADPH, cytochrome *c* reductase, and cytochrome *c* oxidase. The reducing activity of the microsomal fraction was highly dependent on NADPH and inhibited by SOD and $NADP^+$, implying that the plant cell may possess a novel $O_2^{\cdot-}$-generating NADPH oxidase on its membranes that is activated by the incompatible interaction with pathogens.[45] The activity was later localized on the plasma membrane-rich fraction. After further isolation, it was found to be insensitive to EGTA.[46]

Doke and Ohashi[35] showed that the plasma membrane-rich fraction isolated from the *N*-gene tobacco leaves infected at 30°C with TMV, as stated above, also exhibited a NADPH-dependent $O_2^{\cdot-}$-generating activity. The $O_2^{\cdot-}$-generating activity from both potato and tobacco plant tissues was not inhibited by respiratory inhibitors, but was sensitive to thiol (SH) reagents.[35] Interestingly, the activity was reversibly inactivated at temperatures higher than the critical temperatures of 22 and 28°C, where the respective potato and tobacco tissues failed to express the hypersensitive response.[47] The temperature sensitivity of the superoxide-generating system in plants also suggests that this reaction is the key reaction responsible for the induction of resistance.

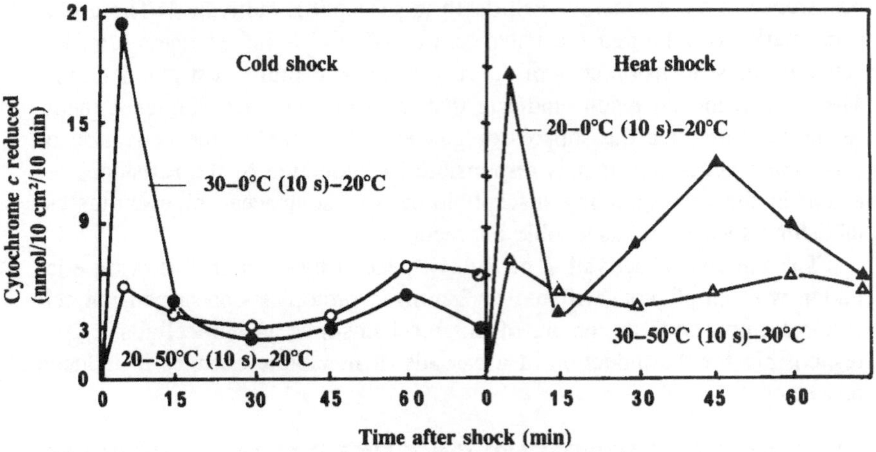

FIGURE 2. Aspects of superoxide anion generation from potato leaf tissues following exposure to heat or cold shocks. Ten leaf discs (10 mm in diameter) were incubated at 30 or 20°C before the shocks. In the case of cold shock, the discs were immersed in an ice-water bath for 10 s, blotted with absorbant paper, and immersed in 10 m*M* K-phosphate buffer (pH 7.8) containing 0.1 m*M* EGTA and 80 μ*M* cytochrome *c* at 20°C for 15 min after the shock, for 5 min. In the case of heat shock, the tissues were immersed in a hot-water bath at 50°C for 10 s and then placed in the assay medium, as stated above, at 20 or 30°C for 15 min after the shock, for 5 min.

V. O_2^- GENERATION IN PLANT TISSUES UPON COLD OR HEAT SHOCK

Both cold and heat shocks to plant tissues have been implicated in the induction of metabolic changes called shock responses. Doke and colleagues[21,48] detected a rapid, transient enhancement of SOD-sensitive cytochrome *c* reducing activity in tobacco and potato leaf tissues at 25 and 20°C within a few minutes of exposure to 50°C for 10 s (Figure 2). Both heat-shocked tissues, however, showed less O_2^--generating activity when assayed at 30°C. On the other hand, when tobacco and potato leaf tissues incubated at 30 and 20°C, respectively, were exposed to 0°C for 10 s, they also showed cytochrome *c* reducing activity at 20°C, but not at 30°C.

These preliminary experiments suggested that when plant tissues are exposed to heat or cold shock, within certain critical temperatures for each plant, they may activate a transient O_2^--generating reaction at optimal temperatures. This kind of enhancement of the O_2^--generating reaction by shock may be associated with changes in membrane fluidity. The transient O_2^- generation of plant tissues should be considered as a possible trigger for new metabolism induced by heat or cold shock.

VI. O_2^- GENERATION IN PLANT TISSUES FOLLOWING WOUNDING

The wounding of plant tissues has been shown to induce metabolic changes, together with novel gene expression. Doke[21,48] detected an immediate (within 2 to 3 min), transient (30-min duration) activation of the O_2^--generating reaction in potato tuber tissues after slicing. This was determined by the enhanced SOD-sensitive reduction of extracellular nitroblue tetrazolium, together with the corresponding appearance of enhanced O_2 consumption and chemiluminescence as measured with luminol. It is difficult to make an appropriate control for wounding in terms of quantitatively assaying the enhanced O_2^--generating reaction. Although further evidence is necessary, the fact that O_2^- generation occurs immediately after wounding suggests that it is associated with the induction of some metabolic changes in the response to wounding.

VII. MECHANISMS OF O_2^--GENERATING REACTIONS IN PLANTS UNDER STRESS

A. INVOLVEMENT OF Ca^{2+} IN ACTIVATION OF THE O_2^--GENERATING REACTION

Preinfection treatment of potato tissues with the Ca^{2+} chelator, EGTA, significantly inhibited activation of the O_2^--generating reaction, as well as hypersensitive cell death, which were caused by infection with incompatible races.[22] Activation of the NADPH-dependent O_2^--generating reaction in potato tuber protoplasts (see above) was also inhibited by pretreatment of protoplasts with EGTA before elicitation with HWC treatments.[39] However, the previously activated NADPH-dependent O_2^--generating reaction in the plasma membrane-rich fraction from infected potato tuber slices was insensitive to EGTA.[45] These lines of evidence suggest that Ca^{2+} may be involved in the activation process of the NADPH-dependent O_2^--generating system associated with the plasma membrane.

Doke[22,49] detected a rapid movement of Ca^{2+} through the plasma membrane of cells of potato tuber slices following application of HWCs. This was tested using chlortetracycline (CTC), which fluoresces when Ca^{2+} is chelated in the hydrophobic membrane (possible membrane-bound Ca^{2+}).[22,49] The fluorescence in cells, loaded with CTC, decreased immediately after HWC application, followed by activation of O_2^--generating systems. Thus, involvement of the release of plasma membrane-bound Ca^{2+} in the activation of the NADPH-dependent O_2^--generating reaction has been suggested.

B. *IN VITRO* ACTIVATION OF THE SUPEROXIDE-GENERATING SYSTEM

Doke and Nishimura[50] presented preliminary evidence showing the successful activation of the NADPH-dependent O_2^--generating reaction *in vitro* by using isolated plasma membrane fractions stimulated by HWC elicitor.[50] The reaction was marked principally in the presence of the soluble proteinaceous fraction from the potato tuber, Ca^{2+}, ATP, and $Mg.^{2+}$ Activation occurred several minutes after stimulation by the addition of HWC to the reaction mixture. Activation of O_2^- generation was monitored at the same time by spectrophotometrically determining the SOD-sensitive reduction of acetylated cytochrome *c* and polargraphically determining O_2 consumption. Furthermore, Miura et al.[51] reported that *in vitro* activation of the O_2^--generating reaction using microsomal fractions isolated from potato tuber tissues was more enhanced in the presence of GTP, diacylglycerol (DG), ATP, calmodulin (CaM), Ca^{2+}, and the soluble proteinaceous fraction of potato tuber slices.[51] Also, inhibitors of CaM (W-7) and protein kinases (staurospolin) were also shown to prevent activation of a complete reaction system.

These experimental results suggested that the O_2^--generating NADPH-oxidase of potato tissues may be regulated by a GTP-binding protein, Ca^{2+}, and a CaM-dependent protein kinase or C-kinase. Further investigation of the activation mechanism is in progress.

These phenomena suggest that the NADPH-dependent O_2^--generating system in plants, in response to various stresses, may closely resemble the O_2^--generating NADPH-oxidase system in leukocytes and macrophages, which is known to be activated during phagocytosis following recognition of alien substances such as bacteria.[52] The amount of O_2^- generated per amount of plant protoplasts was far less than that produced by leukocytes.

VIII. ULTRAVIOLET LIGHT IRRADIATION

UV light exposure has been known to induce the formation of phytoalexins and the enzymes associated with their synthesis in plants. Hadwiger and Schwochau[53] reported that the irradiation of excised pea pods with short wave UV (max = 254 nm) light, but not with long wave UV (max = 366 nm) light, promoted the subsequent increase in phenylpropanoid phytoalexin, pisatin, and PAL, the activity of which is usually enhanced in close association with phytoalexin synthesis. The pisatin-producing response was effectively triggered by relatively short exposures of UV light, and pisatin was detectable within 6 to 8 h after irradiation. The action spectrum of light quality enhancing PAL and pisatin formation is limited in the short-wave range. Long-wave UV or blue light (440 to 620 nm) exposure (5 to 10 min) immediately after short-wave UV exposure did not appear to negate the short-wave UV-induced formation of PAL in peas.[53] Bridge and Klarman[54] reported that the phenylpropanoid phytoalexin, hydroxyphaseollin, was detected in soybean hypocotyls 12 h after short-wave UV (max = 253 nm) irradiation when plants were maintained in darkness, but not in light immediately after irradiation. In this case, the photoreactivation obtained seems to be similar to that of UV light-induced biological damage in higher plants, which appears to be reduced or negated by immediate posttreatment with visible light.[55,56]

Hahlbrock et al.[57], showed that cell suspension cultures of parsley produced and accumulated flavonoids when treated with UV light under fluorescent lamps, and furanocoumarins when treated with fungal elicitor. These antifungal substances are not related to disease resistance, but, rather, to UV light resistance. Kuhn et al.[59] further showed that the mRNA encoding two enzymes of phenylpropanoid metabolism, PAL and 4-CL, were induced in cultured parsley cells either by irradiation with UV light or by treatment with elicitor, although there was a difference in their response times. This suggests that UV irradiation or addition of fungal elicitor to parsley cells enhances the transcription rates of genes encoding the respective enzymes, resulting in greater production of the compounds.

The sesquiterpenoid phytoalexins, rishitin and lubimin, which are induced in potato tissues along with the enhancement of 3-hydroxy-3-methylglutaryl CoA reductase (HMGR) in response to incompatible pathogens or their elicitors, are also promoted by exposure to short-wave UV light at optimum dose.[60] The initial activation of the enzymes by infection was shown to be independent of the enhanced transcription level of the corresponding mRNA.

UV light seems to give both direct and indirect signals for enhancement of the synthetic systems for phytoalexins or stress compounds. In addition, the assumed mechanism suggests that UV light, which can dimerize thymine bases and cause lesions in DNA, also influences the cellular response through direct action on DNA.[61] It will be very interesting to investigate if and how UV light generates intracellular signals for the expression of silent genes. It is well understood in the field of animal science that a proportion of the biological effects caused by UV light result from indirect mechanisms involving endogeneous sensitizers and reactive oxygen species.[62] The proportion seems to be greater at the longer wavelengths. Photosensitized breakage of DNA by sensitizers and long-wave UV light has been shown.[62] Induction of phytoalexin production in plants, however, preferentially appeared as a result of exposure to short-wave, rather than long-wave, UV light. If so, photosensitized breakage of DNA might not be dominantly associated with the induction of phytoalexin synthesis in plants.

Exposure of potato tuber slices to short wave UV light (254 nm) for a short period caused an immediate tissue leakage of electrolytes as a very early cell response, far before the induction of phytoalexin production,[63] similar to an initial cell response to infection with incompatible races or to elicitor treatment as described above. A Ca^{2+} chelator, EGTA, and an inhibitor of lipid peroxidases, SHAM, reduced not only the immediate leakage of electrolytes, but also the later induction of phytoalexin production in potato tuber slices exposed to UV light. From these preliminary results, exposure to UV light seems to activate the lipid peroxidation system, similar to the induction of the hypersensitive reaction caused by infection or elicitor, resulting in stimulation of the phytoalexin production system.

In relation to this, it has been reported that animal cells in culture release arachidonate metabolites into media immediately after irradiation with UV light in a dose-dependent manner and that Ca^{2+} chelators and phospholipase A_2 inhibitors prevented this release.[64] Hence, this early response to UV light does not seem to be dependent on DNA damage.

IX. IMPLICATIONS OF O_2^- GENERATION IN PLANTS UNDER VARIOUS STRESSES

Potato tuber slices which were fed, preinfection, with scavengers of oxygen radicals (tiron or SOD) showed a suppressed or delayed occurrence of hypersensitive cell death, as well as suppressed or delayed phytoalexin production, caused by infection with an incompatible pathogen. Such an inhibitory effect (more than 70%) of tiron on phytoalexin production was detected only by treatment during the early stages of infection (2 h after inoculation), before the O_2^--generating system had activated to the maximum extent.[22]

The rapidity of the response, from elicitation with HWC to enhancement of the O_2^--generating reaction in protoplasts or suspension cultures, indicates that superoxide may function as an initial signal leading to the induction of sequential metabolic changes in the hypersensitive response of plant tissues.[22] Exogeneous application of O_2^- or H_2O_2 to the surface of potato tuber slices induced some production of phytoalexins. Similarly, the enhancement of O_2^- generation in plant tissues upon cold, heat, or wounding shock is worth consideration as a factor involved in the intracellular signal of each response.[65]

In potato tuber slices inoculated with incompatible races of *P. infestans*, enhancement of the O_2^--generating reaction 1 h after inoculation was followed by a transient activation of phospholipase A_2 (PLA_2). Membrane-bound lipid peroxidase (LPX) was also activated about 1 h after activation of PLA_2. Along with the activation of LPX, lipid peroxidation in tissues was activated. The activation of HMGR, an essential enzyme for the synthesis of sesquiterpenoid phytoalexin in potato, occurred 1 h later than that of LPX.[66] Phytoalexin production was detectable about 7 h after inoculation.[9] Scavenging superoxide at an early stage in the tissue response resulted in suppression of these metabolic changes.[22] Inhibitors of LPX also suppressed the production of phytoalexin when they were applied before activation of LPX.[22]

Since little O_2^- generation occurred in the compatible interaction, there was no induction of the above sequential activation of enzymes.[13] Production of O_2^-, depending on NADPH and its derivatives, may be a direct or indirect factor in causing the sequential enzymatic changes necessary for phytoalexin production. Chai and Doke[67] demonstrated the systemic activation of the

$O_2^{\cdot-}$-generating reaction in potato plants by treating the lower leaves with HWC elicitor, which activated the reactions in the treated tissues, as described above. Interestingly, the primary treatment also enhanced the $O_2^{\cdot-}$-generating reaction in the upper leaves, far distant from the treated site.[68] Activation of SOD and peroxidase occurred 2 and 3 d later, respectively. These leaves systemically acquired resistance even to compatible races of *P. infestans*.[68] In the induction of systemic resistance, the $O_2^{\cdot-}$-generating reaction seems to be associated with both a primary and a secondary trigger.

The generation of $O_2^{\cdot-}$ does not necessarily mean that this molecule alone is responsible for the induction of stress responses. Some derivatives of the oxygen radical molecules, such as H_2O_2, $^{\cdot}OH$, and lipid peroxides, initially derived from $O_2^{\cdot-}$, might play a key role in the induced responses. Several papers have reported the involvement of H_2O_2, $^{\cdot}OH$, $^1O_2$, or lipid peroxides in HR.[69-71] In the hypersensitive response of the potato cell, activation of LPX and production of lipid peroxides followed activation of $O_2^{\cdot-}$ generation, and these were responsible for the induction of phytoalexin production. Deterioration of membrane lipids and lipid peroxidation in bacterially induced HR were also shown to be initiated by $O_2^{\cdot-}$ generation.[29] Various physiological changes during HR were effectively inhibited by application of SOD at the initial stages of induction.

A role for H_2O_2 has also been suggested. Apostol et al.[69] noted the rapid and SOD-insensitive quenching of fluorescence probes in suspension cell cultures from soybean after stimulation by an elicitor of phytoalexin induction or by exogeneously added H_2O_2. When catalase was added to the cultures prior to addition of the elicitor, quenching and phytoalexin induction were significantly inhibited. These authors[69] reasonably concluded that H_2O_2 may contribute as a second messenger in eliciting the resistance response. Doke et al.[40] found that soybean protoplasts initiated the generation of $O_2^{\cdot-}$, dependent on NADPH oxidation, outside the protoplast membrane. Thus, the $O_2^{\cdot-}$-generating system may be an enzyme system of primary importance, and H_2O_2 derived from the dismutation of $O_2^{\cdot-}$ may be the active molecule in soybean cells. In potato tissues, H_2O_2 has been shown to be a good elicitor of phytoalexin production even though SOD treatment of the $O_2^{\cdot-}$-generating tissues inoculated with incompatible races inhibited or delayed the HR, including phytoalexin production.[13] The question of what types of active oxygen molecule are involved in the above responses is critical, but the answer remains complicated.

The generation of $O_2^{\cdot-}$ in tissues under stress seems to be associated with the enhancement of lipid peroxidation, and has also been suggested to be associated with the induction of phytoalexin production in potato tissues. Free radical scavengers have been shown to inhibit the production of phytoalexins in plant tissues.[70,71] LPX is activated in such tissues. Lipid peroxides are potential molecules for elicitor activity with regard to phytoalexins. Whether superoxide alone or superoxide plus lipoxygenase are involved in the production of lipid peroxides as internal elicitors for phytoalexin production remains unsolved.

Exposure of plant tissues to UV light is also assumed to generate superoxide immediately after exposure. Potato tuber slices treated with nitroblue tetrazolium solutions showed a rapid formation of blue formazan in cells at the tissue surface directly exposed to the UV light beam.[72] Enhanced chemiluminescence was also detected in potato cell suspension cultures exposed to UV light in the presence of luminol.[72] This enhanced chemiluminescence was inhibited by the oxygen radical scavenger, tiron.

The aforementioned leakage of electrolytes from potato tissues during exposure to UV light may be associated with the generation of active oxygen, similar to that during infection with incompatible pathogens. It seems, therefore, that the generation of superoxide or oxygen radicals following treatment with UV light may trigger a cascade reaction similar to that of HR caused by infection or elicitor treatment. It is obscure, however, whether UV light photochemically generates active oxygen or, alternatively, activates the O_2^--generating system in plant cells.

X. CONCLUDING REMARKS

A novel NADPH-dependent O_2^--generating system is suggested to be activated on the plasma membrane immediately after stimulation by incompatible agents such as pathogens or elicitors in the Solanaceae and Leguminosae, and in rice plants. The rapidity of the activation in comparison with other inducible changes in plant cells may indicate the nature of the trigger of the following responses. The suppression or delay of occurrence of the sequential reaction sequence in HR by the inhibition of O_2^- generation or scavenging of the O_2^- generated also supports this argument.

The novel finding of transient O_2^- generation upon wounding, and heat and cold shock suggests that the generation of O_2^- may indicate a close association with the induction of cell responses against such stress reactions.

The generation of O_2^- in response to the aforementioned stresses may depend on a NADPH-dependent O_2^--generating system similar to that found in phagocytes. The success of *in vitro* activation of the O_2^--generating system by elicitor stimulation provides a further basis for detailed analysis. This O_2^--generating system is suggested to be activated by various molecules associated with signal transduction. In this sense, O_2^- or its derivative active oxygen molecules may be intracellular signals which mediate the exchange of extracellular stimuli for intracellular ones.

It was shown that plant tissues generate superoxide as a rapid response to infection, heat and cold shocks, or wounding, which generally cause some inducible metabolic changes. Exposure to UV light and exogeneous applications of H_2O_2, O_2^--generating systems, or lipid peroxides also cause similar inducible cell responses. Thus far, superoxide generation seems to be inti-

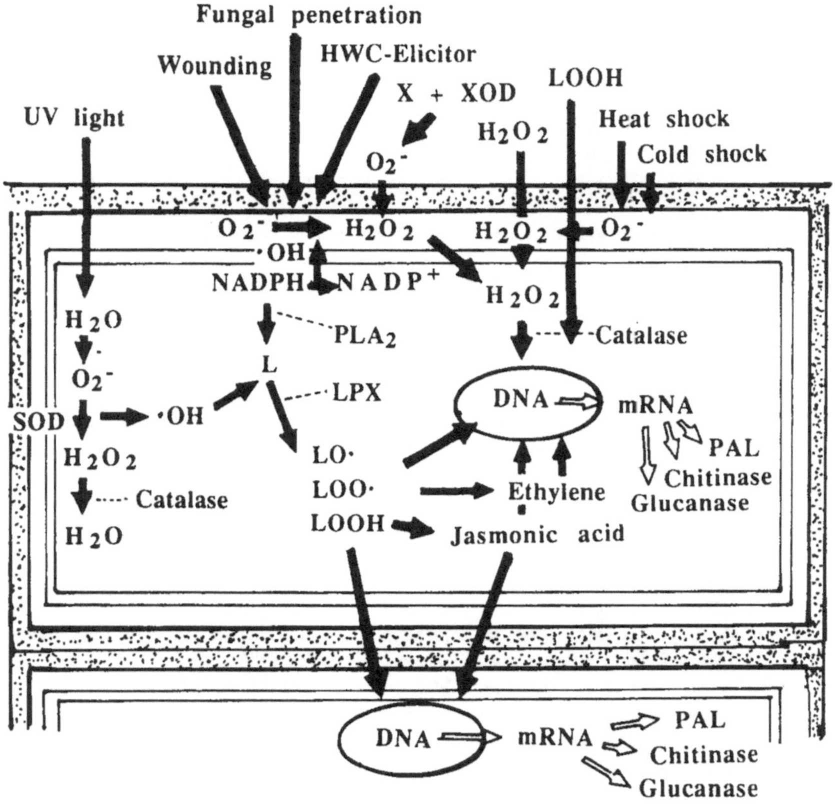

FIGURE 3. Diagram of the relationship between various stresses causing superoxide generation in plant cells and the signal transduction system for the expression of stress metabolism. Lipid peroxides, their derivatives (ethylene, jasmonic acid, etc.), and superoxide radicals may directly or indirectly regulate the expression of genes for defense against microbial invasion or physical and chemical stresses. Superoxide triggers the linoleic-linolenic cascade reaction.

mately associated with an activation of metabolism associated with lipid peroxidation, the so-called linoleic-linolenic cascade in plants, since activation of PLA_2 and LPX is usually associated with these stresses. A tentative concept related to the generation of superoxide during stresses, showing inducible metabolic changes, is illustrated in Figure 3. Further investigation of these and their relationships with the induction of metabolic changes will be necessary to evaluate the role of O_2^- generated during stress as a signal molecule for cellular responses.

ACKNOWLEDGMENTS

Our research was partly supported by Grant-in-Aid for Scientific Research (02454051) and for Developmental Scientific Research (03556007) from the Ministry of Education, Science, and Culture, Japan.

REFERENCES

1. **McCord, J. M. and Fridovich, I.,** Superoxide dismutase, *J. Biol. Chem.*, 244, 6049, 1969.
2. **Fridovich, I.,** Superoxide radical: an endogeneous toxicant, *Annu. Rev. Pharmacol. Toxicol.*, 23, 239, 1983.
3. **Fridovich, I.,** Biological effects of the superoxide radical, *Arch. Biochem. Biophys.*, 247, 1, 1986.
4. **Bygkowsky, J. Z. and Gessner, T.,** Biological role of superoxide ion-radical, *Int. J. Biochem.*, 20, 569, 1988.
5. **Bavior, B. M.,** The enzymatic basis for O_2^- production by human neutrophils, *Can. J. Physiol. Pharmacol.* 60, 1353, 1982.
6. **Bavior, B. M.,** The respiratory burst of phagocytes, *J. Clin. Invest.*, 73, 599, 1684.
7. **Tomiyama, K., Doke, N., Nozue, M., and Ishiguri, Y.,** The hypersensitive response of resistant plants, in *Recognition and Specificity in Plant Host-Parasite Interactions,* Daly, J. M. and Uritani, I., Eds., Japan Science Society Press, Tokyo, and University Park Press, Baltimore, 1979, 69.
8. **Doke, N., Tomiyama, K., and Furuichi, N.,** Elicitation and suppression of the hypersensitive response in host-parasite specificity, in *Plant Infection; The Physiological and Biochemical Basis,* Asada, Y., Bushnell, W. R., Ouchi, S., and Vance, C. P., Eds., Japan Science Society Press, Tokyo and Springer-Verlag, New York, 1982, 79.
9. **Dixon, R. A. and Lamb, C. J.,** Molecular communication in interactions between plants and microbial pathogens, *Annu. Rev. Physiol. Plant Mol. Biol.*, 41, 339, 1990.
10. **Chappell, J. and Hahlbrock, K.,** Transcription of plant defence genes in response to UV light or fungal elicitor, *Nature (London),* 3116, 76, 1984.
11. **Hadwiger, L. A. and Schwochau, M.,** Ultraviolet light-induced formation of pisatin and phenylalanine ammonia lyase, *Plant Physiol.*, 47, 588, 1971.
12. **Lamb, C. J., Lawton, M. A., Dron, M., and Dixon, R. A.,** Signals and transduction mechanisms for activation of plant defences against microbe attack, *Cell,* 56, 215, 1989.
13. **Doke, N.,** Involvement of superoxide anion generation in the hypersensitive response of potato tuber tissues to infection with an incompatible race of *Phytophthora infestans* and to the hyphal wall components, *Physiol. Plant Pathol.*, 23, 345, 1983.
14. **Sato, N., Kitazawa, K., and Tomiyama, K.,** The role of rishitin in localizing the invading hyphae of *Phytophthora infestans* in infection sites at the cut surface of potato tubers, *Physiol. Plant Pathol.*, 1, 288, 1971.
15. **Kitazawa, K., Inagaki, H., and Tomiyama, K.,** Cinephotomicrographic observation on the dynamic responses of protoplasm of a potato plant cell to infection by *Phytophthora infestans, Phytopathol. Z.*, 76, 80, 1973.
16. **Tomiyama, K., Okamoto, H., and Katou, K.,** Effect of infection by *Phytophthora infestans* on the membrane potential of potato cells, *Physiol. Plant Pathol.*, 22, 233, 1983.
17. **Matsumoto, N., Tomiyama, K., and Doke, N.,** Alteration of membrane permeability of potato tuber tissue infected by incompatible and compatible races of *Phytophthora infestans* during the initial phase of infection, *Ann. Phytopathol. Soc. Jpn.*, 42, 279, 1976.
18. **Tomiyama, K.,** Cytological and biochemical studies of the hypersensitive potato cells to *Phytophthora infestans,* in *Morphological and Biochemical Events in Plant-Parasite Interaction,* Akai, S. and Ouchi, S., Eds., Phytopathology Society of Japan, Tokyo, 1971, 387.
19. **Doke, N.,** A further study on the role of hypersensitivity in resistance in potato cultivars to infection by an incompatible race of *Phytophthora infestans, Physiol. Plant Pathol.* 21, 85, 1982.

20. **Doke, N., Tomiyama, K., and Furuichi, N.,** Elicitation and suppression of the hypersensitive response in host-parasite specificity, in *Plant infection: The Physiological and Biochemical Basis,* Asada, Y., Bushenell, W. R., Ouchi, S., and Vance, C. P., Eds., Japan Science Society Press, Tokyo and Springer-Verlag, Berlin, 1982, 79.
21. **Doke, N., Miura, Y., Chai, H. B., and Kawakita, K.,** Involvement of active oxygen in induction of plant defence response against infection and injury, in *Active Oxygen/Oxidative Stress and Plant Metabolism,* Pell, E. and Steffen, K., Eds., American Society of Plant Physiologists, 1991, 84.
22. **Doke, N.,** Hypersensitive reaction and its significance in signal transduction for plant active defence, in *Recent Advances in Physiological Plant Pathology,* Oku, H., Khomoto, K., Doke, N., Otani, H., and Kodama, M., Eds., publishing committee of Recent Advances in Physiology and Plant Pathology, Nagoya, 1991, 81 (in Japanese).
23. **Chai, H. B. and Doke, N.,** Superoxide anion generation: a response of potato leaves to infection with *Phytophthora infestans, Phytopathology,* 77, 645, 1987.
24. **Fritcemeter, K. H., Cretin, C. M., Komblink, E., Rohwer, F., Taylor, J., Scheel, D., and Hahlbrock, K.,** Transient induction of phenylalanine ammonia-lyase and 4-coumarate:CoA ligase mRNA in potato leaves infected with virulent or avirulent races of *Phytophthora infestans, Plant Physiol.,* 85, 34, 1987.
25. **Ivanova, D. G., Gughova, N. V., Merzlyak, M. N., and Rassadina, G. V.,** Effect of *Phytophtora infestans* infection on superoxide dismutase dependent cytochrome *c*-reducing activities of leaves as related to resistance of potato plants to late blight, *Plant Sci.,* 78, 151, 1991.
26. **Sekizawa, Y., Haga, M., Hirabayashi, E., Takeuchi, N., and Takino, Y.,** Dynamic behavior of superoxide generation in rice leaf tissues infected with blast fungus and its regulation by some substances, *Agric. Biol. Chem.,* 51, 763, 1987.
27. **Sekizawa, Y., Haga, M., and Kanoh, H.,** A superoxide anion forming enzyme, NADPH oxidase of rice blade tissue stimulated with a blast fungus elicitor, *Ann. Phytopathol. Soc. Jpn.,* 56, 565, 1990.
28. **Aver'yanov, A. A., Lapikova, B. P., Umonov, A. M., and Dzhavakhiya, V. G.,** Generation of superoxide radical by rice leaves in relation to blast resistance, Translated from *Fiziol. Rast.,* 34, 373, 1987.
29. **Keppler, L. D. and Baker, C. J.,** O_2^--initiated lipid peroxidation in a bacteria-induced hypersensitive reaction in tobacco cell suspension, *Phytopathology,* 79, 555, 1989.
30. **Keppler, L. D., Baker, C. J., and Atkinson, M. M.,** Active oxygen production during bacteria induced hypersensitive reaction in tobacco suspension cells, *Phytopathology,* 79, 974, 1989.
31. **Pavlovkin, J., Novacky, A., and Ullrich-Eberius, C. I.,** Membrane potential changes during bacterial-induced hypersensitive reaction, *Physiol. Mol. Plant Pathol.,* 28, 125, 1986.
32. **Keppler, L. D. and Novacky, A.,** The initiation of membrane lipid peroxidation during bacteria-induced hypersensitive reaction, *Physiol. Mol. Plant Pathol.,* 30, 233, 1987.
33. **Kassannis, B.,** Some effects of high temperature on the susceptibility of plants to infection with viruses, *Ann. Appl. Biol.,* 39, 358, 1952.
34. **Westeijin, E. A.,** Lesion growth and virus localization in leaves of *Nicotiana tabacum* cv. Xanthi nc. after inoculation with tobacco mosaic virus and inoculation alternatively at 22°C and 32°C., *Physiol. Plant Pathol.,* 18, 356, 1981.
35. **Doke, N. and Ohashi, Y.,** Involvement of an O_2^- generating system in the induction of necrotic lesion on tobacco leaves infected with tobacco mosaic virus, *Physiol. Mol. Plant Pathol.,* 32, 163, 1988.
36. **Doke, N., Garas, N. A., and Kuc, J.,** Partial characterization and aspects of mode of action of a hypersensitivity-inhibiting factor (HIF) isolated from *Phytophthora infestans, Physiol. Plant Pathol.,* 15, 127, 1979.

37. **Doke, N. and Tomiyama, K.,** Effect of hyphal wall components from *Phytophthora infestans* on protoplasts of potato tuber tissues, *Physiol. Plant Pathol.*, 16, 169, 1980.
38. **Doke, N. and Fruichi, N.,** Response of protoplasts to hyphal wall components in relation to resistance of potato to *Phytophthora infestans, Physiol. Plant Pathol.*, 21, 23, 1982.
39. **Doke, N.,** Generation of superoxide anion by potato tuber protoplasts during the hypersensitive response to hyphal wall components of *Phytophthora infestans* and specific inhibition of the reaction by suppressors of hypersensitivity, *Physiol. Plant Pathol.*, 23, 359, 1983.
40. **Doke, N., Hirano, R., and Miura, Y.,** An activation of NADPH dependent O_2^- generating reaction in protoplasts isolated from various *Leguminosae* and *Solanaceae* plants following treatment with elicitors of hyphal wall components from *Phytophthora* spp., *Ann. Phytopathol. Soc. Jpn.*, submitted.
41. **Linder, W. A., Hoffmann, C., and Grisebach, H.,** Rapid elicitor-induced chemiluminescence in soybean cell suspension cultures, *Phytochemistry,* 27, 2501, 1988.
42. **Sanchez, L. M., Miura, Y., Kawakita, K., and Doke, N.,** Elicitor-induced chemiluminescence in some solanaceous cultured cells and its inhibition by suppressors from *Phytophthora* spp., in Abstr. Pap. Annu. Meet. Phytopathology Society of Japan, 1992.
43. **Sanchez, L. M., Ohno, Y., Miura, Y., Kawakita, K., and Doke, N.,** Host selective suppression by water-soluble glucans from *Phytophthora* spp. on hypersensitive cell death of suspension cultured cells from Solanaceous plants caused by hyphal wall elicitors of the fungus, *Ann. Phytopathol. Soc. Jpn.*, 58, 1, 1992.
44. **Sanchez, L. M., Miura, Y., and Doke, N.,** unpublished data.
45. **Doke, N.,** NADPH-dependent O_2^- generation in membrane-fraction isolated from wounded potato tubers inoculated with *Phytophthora infestans, Physiol. Plant Pathol.*, 27, 311, 1985.
46. **Doke, N.,** unpublished data.
47. **Doke, N. and Miura, Y.,** unpublished data.
48. **Doke, N.,** Activation of a superoxide generating system in plant tissues following fungal infections and various stresses, *Ann. Phytopathol. Soc. Jpn.*, 54, 99, 1988 (in Japanese).
49. **Doke, N.,** Release of plasma membrane bounded Ca^{++} ion at an initiation stage of hypersensitive reaction of potato cells, *Ann. Phytopathol. Soc. Jpn.*, 53, 391, 1987 (in Japanese).
50. **Doke, N. and Nishimura, S.,** Signal transduction mechanisms in hypersensitively reacting cells of potato:superoxide generating NADPH oxidase and phospholipase A_2, Ann. *Phytopathol. Soc. Jpn.*, 55, 94, 1989.
51. **Miura, Y., Kawakita, K., Yoshioka, H., and Doke, N.,** In vitro activation of NADPH-O_2^- generating reaction by elicitor using microsomal fraction from potato tuber tissues, (Abstr. in Japanese), *Ann. Phytopathol. Soc. Jpn.*, 58, 564, 1992.
52. **Rossi, F.,** The O_2^- forming NADPH oxidase of the phagocytes: nature, mechanisms of activation and function, *Biochem. Biophys. Acta,* 186, 148, 1986.
53. **Hadwiger, L. A. and Schwochau, M. E.,** Ultraviolet light-induced formation of pisatin and phenylalanine ammonia lyase, *Plant Physiol.*, 47, 588, 1971.
54. **Bridge, M. A. and Klarman, W. A.,** Soybean phytoalexin, hydroxyphaseolin, induced by ultraviolet irradiation, *Phytopathology,* 63, 606, 1973.
55. **Ikenaga, M. and Mabuchi, T.,** Photoreactivation of endosperm mutations induced by ultraviolet light in maize, *Radiat. Bot.*, 6, 165, 1966.
56. **Trosuko, J. E. and Mansour V. H.,** Photoreactivation of ultraviolet light-induced pyrimidine dimers in Ginko cells grown in vitro, *Mutat. Res.*, 7, 120, 1969.
57. **Hahlbrock, K., Knobloch, K.-H., Kreuzaler, F., Potts, J. R. M., and Wellmann, E.,** Coordinated induction and subsequent activity changes of two groups of metabolically interrelated enzymes. Light-induced synthesis of flavonoid glycosides in cell suspension cultures of *Petroselinum hortense, Eur. J. Biochem.*, 61, 199, 1976.

58. **Tietjen, K. G., Hunkler, D., and Matern, U.,** Differential response of cultured parsley cells to elicitors from two non-pathogenic strains of fungi. I. Identification of induced products as coumarin derivatives, *Eur. J. Biochem.*, 131, 401, 1983.
59. **Kuhn, D. N., Chappell, J. C., Boudet, A., and Hahlbrock, K.,** Induction of phenylalanine ammonia-lyase and 4-coumarate:CoA ligase mRNA in cultured plant cells by UV light or fungal elicitor, *Proc. Natl. Acad. Sci. U.S.A.*, 81, 1120, 1984.
60. **Cheema, A. S. and Haard, N. F.,** Induction of rishitin and lubimine synthesis in potato tuber slices by non-specific elicitors, role of gene expression, *J. Food Prot.*, 42, 512, 1979.
61. **Smith, K. C. and Meun, D. H. C.,** Repair of radiation-induced damage in *Escherichia coli.* I. Effect of rec mutations on post-replication repair of damage due to ultraviolet radiation, *J. Mol. Biol.*, 51, 450, 1970.
62. **Peak, J. G., Peak, M. J., and Tuveson, R. W.,** Ultraviolet action spectra for aerobic and anaerobic inactivation of *Escherichia coli* strains specifically sensitive and resistant to near ultraviolet radiation, *Photochem. Photobiol.*, 38, 541, 1978.
63. **Doke, N.,** unpublished data.
64. **Vincent, A. D. L., Hanson, D., Weinstein, I. B., and Harber, L. C.,** Ultraviolet radiation stimulates the release of arachidonic acid from mammalian cells in culture, *Photochem. Photobiol.*, 41, 51, 1985.
65. **Doke, N. and Nishimura, S.,** Activation of O_2^- generating reaction in potato plants infected with *Phytophthora infestans:* its mechanism and significance in induced resistance, in Abstr. 5th Int. Congr. Plant Pathology, 1988, 229.
66. **Oba, K., Kondo, K., Doke, N., and Uritani, I.,** Induction of 3-hydroxy-3-methylglutaryl CoA reductase in potato tubers after slicing, fungal infection or chemical treatment, and some properties of the enzyme, *Plant Cell Physiol.*, 26, 873, 1985.
67. **Chai, H. B. and Doke, N.,** Systemic activation of O_2^- generating reaction, superoxide dismutase, and peroxidase in potato plants in relation to induction of systemic resistance to *Phytophthora infestans, Ann. Phytopathol. Soc.*, 53, 585, 1987.
68. **Doke, N., Ramirez, A. V., and Tomiyama, K.,** Systemic induction of resistance in potato against *Phytophthora infestans* by local treatment with hyphal wall components of the fungus, *J. Phytopathol.*, 119, 232, 1987.
69. **Apostol, I., Heistein, P. F., and Kow, P. S.,** Rapid stimulation of oxidative burst during elicitation of cultured plant cells, *Plant Physiol.*, 90, 109, 1989.
70. **Hoffman, R. M. and Heale, J. B.,** Effect of free radical scavengers on 6-methoxymellein accumulation and resistance to *Botrytis cinerea* in carrot root, *Mycol. Res.*, 92, 25, 1989.
71. **Epperlein, M. M., Noronha-Dutra, A. A., and Strange, R. N.,** Involvement of the hydroxy radical in the abiotic elicitation of phytoalexins in legumes, *Physiol. Plant Pathol.*, 28, 67, 1986.
72. **Doke, N.,** unpublished data.

Chapter 8

PHOTOOXIDATIVE STRESS IN TREES

Andrea Polle and Heinz Rennenberg

TABLE OF CONTENTS

0-8493-5443-9/94/$0.00 + $.50

I. INTRODUCTION

Tree species in the temperate zone are C_3 plants. This implies that their rate of light-saturated photosynthesis is limited by the uptake and reduction of CO_2. In coniferous species, maximum rates of photosynthesis were observed at photon flux densities in the range of 1000 to 1500 μmol m^{-2} s^{-1} (Scots pine,[1] Norway spruce,[2] and Black spruce[3]). Although the light requirement of photosynthesis is normally well-adapted to incident light intensity,[4] trees must occasionally cope with oversaturating light conditions up to 2000 μmol m^{-2} s^{-1}. For instance, in poplar leaves, only 135 μmol (quanta) m^{-2} s^{-1} would be needed for maximum O_2 production (15 μmol [O_2] m^{-2} s^{-1}), i.e., under high irradiation, up to 90% of the incoming light must be dissipated.[5] This shows that powerful protective systems are required in trees to prevent photooxidation under high-light conditions.

However, reactive oxygen species can also be produced under light-limiting conditions. Approximately 5 to 10% of the electrons transferred from water to their final electron acceptor $NADP^+$ "leak" to molecular oxygen.[6,7] Hodgson and Raison[8] used a specific electron spin probe for O_2^- with thylakoids from *Nerium oleander* and observed a light dependence for O_2^- production similar to that for photosynthetic oxygen production, although the concentration of the physiological electron acceptor, $NADP^+$, was not limiting. An important observation was that the rate of O_2^- production did not increase significantly when the light saturation of photosynthesis was exceeded.[8] At moderate light intensity (350 μmol m^{-2} s^{-1}), O_2^- was produced at a rate of 12.5 μmol (O_2^-) mg^{-1} (chlorophyll) h^{-1}.[8] Kröniger et al.[9] calculated that this production rate would yield a steady-state concentration of 15 μM O_2^- if O_2^- scavengers were absent. O_2^- causes H_2O_2 production by spontaneous dismutation or by catalysis by superoxide dismutase (SOD). Tanaka et al.[10] found that isolated spinach thylakoids generated 5 μmol H_2O_2 mg^{-1} chlorophyll h^{-1} in the absence of physiological electron acceptors and scavengers for H_2O_2. Under these conditions, a concentration of 10 μM H_2O_2, which inhibits chloroplastic enzymes by 50%,[11] would be exceeded after 250 ms. Apart from their own injurious effects, O_2^- and H_2O_2 may give rise to highly toxic $^{\cdot}OH$ radicals, either by the Haber-Weis reaction or by modifying Cu,Zn superoxide dismutase to produce free $^{\cdot}OH$ in the presence of excess hydrogen peroxide.[12] Apparently, the presence of protective systems against oxidative damage is indispensible in chloroplasts, even in the absence of distinct stress factors.

Tree species may encounter specific problems as compared to annual species, since they must cope with large changes in environmental conditions in their long life-span. In this chapter, we review protection from oxidative stress present in the leaves of trees and evaluate their role in preventing photooxidative damage.

TABLE 1
Reaction Constants for O_2^- and $^1O_2$ with Principal Lipophilic and Hydrophilic Antioxidants

Antioxidant	Oxidant	$K(M^{-1}\ s^{-1})$	Remarks	Ref.
Carotenoids	$^1O_2$	0.1 to 2×10^{10}	Dependent on no. conjugated double bonds (n = 8 to 11)	116
α-Tocopherol	$^1O_2$	2.5×10^8	In pyridine	117
Ascorbate	$^1O_2$	0.84 to 1.06×10^7	In pyridine	118
Glutathione	$^1O_2$	2.4×10^6	In D_2O, pH 7.4	119
SOD	O_2^-	2×10^9		6
Ascorbate	O_2^-	2×10^7		6
Glutathione	O_2^-	7×10^7		6
O_2^-	O_2^-	4×10^5	pH 7.0, pH dependent	6

II. ANTIOXIDANTS IN LEAVES OF DECIDUOUS AND EVERGREEN TREE SPECIES

In order to remove reactive oxygen species such as $^1O_2$, O_2^-, and H_2O_2, plant cells are equipped with lipophilic (carotenoids, tocopherols) and water-soluble antioxidants (glutathione, ascorbate). The contribution of any of these compounds to the reduction of harmful oxygen species is basically determined by two factors: their rate constant with the oxidant and their concentration in the compartment where the oxidant is generated. The bimolecular rate constants of important cellular antioxidants with $^1O_2$ and O_2^- are summarized in Table 1. Although antioxidant concentrations have been frequently determined in leaves of tree species, data for chloroplasts, the principal site of $^1O_2$ and O_2^- formation, are scarce. Estimates of subcellular antioxidant concentrations are difficult because the protective systems show large developmental and seasonal fluctuations in leaves of tree species. In the following sections, we discuss the complex changes which may be anticipated in the antioxidant status of chloroplasts in order to provide a better understanding of the role of different antioxidants for the removal of activated oxygen.

A. CAROTENOIDS

Primary carotenoids (this term does not refer to carotenoids in red fruits and flowers) are exclusively localized in chloroplasts.[13] They are organized in pigment-protein complexes in thylakoid membranes where they have light-harvesting and protective functions.[14] Carotenoids protect the photosynthetic apparatus from excess light by energy dissipation and by their ability to deactivate $^1O_2$ at extremely high rates (Table 1). Carotenoids are also involved in radiationless energy dissipation, i.e., are part of the mechanisms to prevent photoinhibition. Discussion of photoinhibition is beyond the scope of this chapter, but is presented in Chapter 2 of this volume. The role of carotenoids in photoinhibition is also the subject of several recent reviews.[15-18]

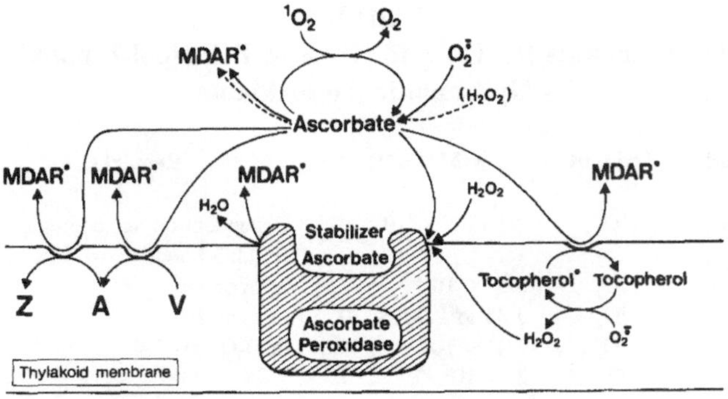

FIGURE 1. The central role of ascorbate for the removal of activated oxygen in chloroplasts. V, violaxanthin; A, antheraxanthin; Z, zeaxanthin; MDAR ·, monodehydroascorbate radical.

Among different tree species, the pigment composition of Norway spruce needles was probably most frequently investigated. Spruce needles contain about 200 μg of carotenoid per gram of fresh mass and about 1000 μg of chlorophyll per gram of fresh mass.[19,21] The molar ratio of carotenoids to chlorophyll is about 1:3.[14] As in other plant species,[15] β-carotene and lutein are dominant in spruce needles, comprising about 60% of the total carotenoids.[14] Neoxanthin amounts to about 15%, whereas antheraxanthin, violaxanthin, and zeaxanthin contribute together about 20% to the total carotenoid content in spruce needles.[14] The latter three pigments constitute the xanthophyll cycle, which is thought to be predominantly responsible for radiationless deactivation of light energy.[16] When leaves are irradiated, the ratio of the three xanthophyll cycle pigments is shifted by a two-step de-epoxidation of violaxanthin toward zeaxanthin.[16] In this reaction, ascorbate is used as cosubstrate (see Chapter 4 of this volume) (Figure 1).[22]

The above figures for pigment contents in conifer needles are rough estimates because the composition of carotenoids and chlorophyll is affected by various environmental factors. Sun- and shade-adapted needles show typical chlorophyll-to-carotenoid ratios of 4 to 5 (w/w) and 6 to 8 (w/w), respectively, i.e., in needles grown under high irradiance, the carotenoid content is relatively enhanced.[21] The total pigment content can vary in one season by a factor of 2.[19] The highest pigment content is generally found in the fall. The pigment content decreases in winter and increases again in the following season, normally showing a net increase with increasing needle age.[23] The pigment decrease observed in winter is larger for chlorophyll than for carotenoids, resulting in a relative increase in carotenoids of about 25%.[19,23] The enhancement is primarily caused by a relative increase in the lutein and xanthophyll content.[24] As absorbed light quanta are used for linear electron transport in winter-hardened thylakoid membranes to a limited extent,[25] the relative increase in xanthophyll may be necessary to satisfy an increased need for energy dissipation.

B. TOCOPHEROL

Among the four tocopherols (α-, β-, γ-, and δ-tocopherol) found in plants, α-tocopherol is the most abundant species in green tissues.[26] In one-year-old needles of 28 different spruce species investigated by Franzen et al.[27] and in red spruce,[28] α-tocopherol was found at concentrations of about 200 nmol g^{-1} fresh weight. The other tocopherol species were not detected in the needles, but were present in trace amounts in cotyledons (β-tocopherol) and seeds (γ-tocopherol) of Norway spruce.[27]

Typically, most of the leaf tocopherol is localized in chloroplasts, where it is freely mobile in membranes.[29,30] α-Tocopherol is an important antioxidant because it can deactivate $^1O_2$, reduce $O_2^{\bar{}}$, and terminate lipid peroxidation by reducing fatty acyl peroxy radicals. Which of these functions is prevalent *in vivo* is not known. However, $^1O_2$ is probably predominantly quenched by carotenoids, which are present at high local concentrations in light-harvesting systems and which react with activated oxygen at diffusion-controlled rates (Table 1). Circumstantial evidence suggests that a production site for $O_2^{\bar{}}$ may exist in the aprotic environment of thylakoid membranes,[8,31] thereby attributing a role to α-tocopherol in the protection from $O_2^{\bar{}}$. Another important function of α-tocopherol is its chain-breaking activity, preventing lipid peroxidation.[32,33] Reconstitution experiments with liposome membranes showed, however, that the pool of α-tocopherol was rapidly exhausted when an external supply with reductant such as ascorbate was lacking,[34] whereas ascorbate alone was too hydrophilic to interact directly with fatty acyl peroxy radicals.[35] The chemical reaction between α-tocopherol radicals and ascorbate occurs at a rate constant of $1.5 \times 10^6\ M^{-1}\ s^{-1}$ and yields monodehydroascorbate radicals (Figure 1).[36]

Spinach chloroplasts contain a molar tocopherol-to-chlorophyll ratio of 1:10.[37] In chloroplasts of tree species, this ratio is highly variable. The α-tocopherol content varied with the season[38] and increased with increasing needle age.[39,40] For example, 3-year-old spruce needles contained approximately 20-fold more α-tocopherol than young needles which had just emerged from the buds.[40] In the same period of time, chlorophyll increased only threefold.[23] A high α-tocopherol content in leaves of tree species was associated with leaf senescence in china trees and beech trees.[39,41] Rise et al.[41] suggested that degradation of chlorophyll might provide the precursors for an increased synthesis of tocopherol. However, when leaves of the herbaceous species *Cucumis sativa* were exposed to photooxidative stress, a depletion of the tocopherol pool preceded chlorophyll degradation.[42] This observation suggested that tocopherol protected chlorophyll from photooxidation. This idea is also supported by the observation that sun-exposed beech leaves contained a threefold higher α-tocopherol content than shaded leaves.[43]

C. ASCORBATE

In conifer needles, ascorbate has been investigated for half a century. Bukatsch[44] found higher ascorbate content in sun-exposed needles than in shaded needles. This finding was confirmed by Esterbauer et al.[45] The observed difference in ascorbate was relatively low (20%). The ascorbate content was also affected by needle age. The total ascorbate content in 6- to 7-year-old needles was 35% lower than in 5-year-old needles; the relative portion of oxidized ascorbate increased threefold.[45]

Seasonal changes of the ascorbate concentration are high and can amount to a factor of 4. In spruce needles, the highest ascorbate concentrations were found in spring before bud break, amounting to 11 to 14 μmol g^{-1} fresh weight in Norway spruce[45-47] and 14 to 15 μmol g^{-1} fresh weight in red spruce.[28,48] Minimum contents were present by the end of September and in October, amounting to 4 to 8 μmol g^{-1} fresh weight in Norway spruce[45-47] and 10 μmol g^{-1} fresh weight in red spruce.[28,48] In the cold season, needles often contained a concentration intermediate between the spring maximum and the fall minimum. In Scots pine *(Pinus sylvestris)* needles, the ascorbate content fluctuated between a minimum in October of 8 μmol g^{-1} fresh weight and a maximum of 15 μmol g^{-1} fresh weight at the beginning of April, with a similar pattern as in spruce needles.[46] In contrast, in Eastern White pine *(P. strobus)* needles, ascorbate increased continuously from bud break in June until January and decreased continuously from March until June, covering a concentration range of 6 to 23 μmol g^{-1} fresh weight.[49] A major difference between both pine species is that Eastern White pine needles are shed after about 16 months, whereas Scots pine needles are kept for about 3 years.

In spruce needles, dehydroascorbate contributed 10 to 15% to the total ascorbate[45,46,50] and 6% or less in pine.[46,49] By use of EPR spectroscopy, monodehydroascorbate radicals were qualitatively detected, but only in spruce needles subjected to oxidizing conditions.[51] Quantitative figures for the monodehydroascorbate radical have not been reported.

Large seasonal variations of ascorbate were also found in leaves of deciduous woody plants. Young beech leaves contained about 10 μmol of ascorbate per gram of dry weight,[39] which corresponds to approximately 2 μmol of ascorbate per gram of fresh weight. Mature beech leaves, in summer, contained 12 to 17 μmol of ascorbate per gram of dry weight, which corresponds to about 7.5 to 10.5 μmol of ascorbate per gram of fresh weight.[39,52] Leaves of six other woody plants (*Sambus nigra, Aesculus parviflora, Parrotia persica, Parthenociccus quinqefolia, Liquidambar styraciflua,* and *Ginko biloba*) showed species-dependent differences in ascorbate ranging from 10 *(Ginko biloba)* to 20 μmol g^{-1} fresh weight *(Sambucus nigra)* in mature leaves in summer.[52] In general, oxidized ascorbate contributed about 10 to 20% of the total ascorbate. Within a period of about 2 weeks before leaf fall, ascorbate decreased by about 30 to 80% in most deciduous tree species. In beech leaves, the decrease in total ascorbate was accompanied by an increase in the relative amount of dehydroascorbate of up to 80%.[52]

On the basis of a relative water content of 40 to 45%, average cellular ascorbate concentrations are in the range of 13 to 26 m*M* in mature conifer needles and beech leaves. However, figures for ascorbate concentrations in chloroplasts from tree species are lacking. Ascorbate fractions associated with pea and spinach chloroplasts amounted to about 30 to 40% of the total leaf ascorbate and corresponded to concentrations of 12 to 25 m*M*.[53-56] Based on these figures, a chloroplastic ascorbate concentration in conifers of 30 to 50 m*M* may be expected in summer and fall. In winter and spring, a rise in chloroplastic ascorbate may be anticipated, since ascorbate increased whereas chlorophyll decreased. In fact, the chloroplastic ascorbate concentration increased by about 30% after cold-hardening of spinach leaves, but this increase was smaller than the increase in total leaf tissue (twofold).[56]

In conclusion, ascorbate has a central role in the chloroplastic antioxidant system for the following reasons (Figure 1).

1. Because of its high concentration and high rate constant with $^1O_2$ and O_2^- (Table 1), it is an effective chemical scavenger for reactive oxygen species in the hydrophilic phase. In contrast, its chemical reaction with H_2O_2 is slow at alkaline pH values which are relevant in chloroplasts.
2. It serves as a mediator between the hydrophilic and lipophilic phase in order to maintain the antioxidant properties of membrane-localized protective systems (α-tocopherol and xanthophyll).
3. It is important for the removal of H_2O_2, since it is both substrate and stabilizer for an ascorbate-specific peroxidase.

D. GLUTATHIONE

In leaves of higher plants, glutathione (or its homologs such as homoglutathione) is the most abundant low molecular thiol.[57] Its concentration is generally tenfold higher than the concentrations of its precursors cysteine and γ-glutamylcysteine.[58] In order to evaluate its antioxidant functions in the leaves of tree species, it is important to keep in mind that glutathione also functions in sulfur metabolism as a storage and transport form of reduced sulfur,[59,60] and is also important in the detoxification of xenobiotics and heavy metals.[57,61]

The thiol content was found to be higher in sun-exposed than in shaded needles.[62] It undergoes light-dependent changes in spruce with an amplitude of about 200 nmol g^{-1} fresh weight.[63] Schupp et al.[64] investigated glutathione transport by feeding needles with radioactive glutathione through the cut needle tip and found that glutathione export from the needles was fivefold higher in the dark than in the light. This finding suggested that the light-dependent accumulation was caused (at least partially) by low rates of glutathione export.

Seasonal changes in the glutathione content of conifer needles have been observed frequently. Low glutathione contents were found from June until October: 100 to 200 nmol g^{-1} fresh weight in Norway spruce,[47,58,62,65] 200 to 250 nmol g^{-1} fresh weight in red spruce,[28,48] and 120 to 280 nmol g^{-1} fresh weight in Eastern White pine.[49] The glutathione content increased in the cold season. Maximum glutathione contents were generally present from January until bud break and amounted to 400 to 600 nmol g^{-1} fresh weight in Norway spruce[58,65] and 300 to 700 nmol g^{-1} fresh weight in Eastern White pine.[49] Buds contained up to 1000 nmol of glutathione per gram of fresh weight.[58] Schupp et al.[64] showed that glutathione was exported from the older needles to the young developing needles and suggested that elevated glutathione levels function to supply sinks with reduced sulfur.[64] This assumption was also supported by the observation that the activities of enzymes of sulfate reduction in young flushing needles with a high demand for reduced sulfur were relatively low.[58]

The year-to-year variability of glutathione in the needles was much higher than that observed for ascorbate.[47,49,58] This suggests that glutathione responds more sensitively to changes in environmental conditions than ascorbate. Changes in glutathione appeared to correspond to changes in temperature. Anderson et al.[49] observed a smaller amplitude of the seasonal glutathione fluctuation in a mild than in a severe winter. Esterbauer and Grill[65] assumed that glutathione was important for frost tolerance. However, statistical analysis of glutathione concentrations with average air temperatures revealed only a low correlation coefficient of 0.55.[58] Smith et al.[66] pointed out that we lack clear experimental evidence to determine whether or not elevated concentrations of glutathione in winter are a necessary prerequisite for frost tolerance.

In spruce needles, average tissue concentrations of glutathione ranged between 0.2 to 0.7 m*M*.[58] Pea and spinach chloroplasts contained about 10% of the total leaf glutathione.[54,67] Assuming that the chloroplastic volume amounts to 35 μl mg^{-1} chlorophyll,[6] chloroplastic glutathione concentrations of 0.2 to 2 m*M* can be calculated for spruce needles. This estimate shows that glutathione is apparently ten- to 50-fold less concentrated in conifer chloroplasts than ascorbate. As the rate constants of ascorbate and glutathione for $^1O_2$ and O_2^- are similar (Table 1), chemical removal of active oxygen species in the stroma is more likely to use ascorbate than glutathione.

The glutathione content in leaves of deciduous tree species has only rarely been investigated. In young, expanding beech leaves, the total glutathione content amounted to 500 nmol g^{-1} fresh weight and ranged from 300 to 500 nmol g^{-1} fresh weight in mature leaves in summer.[68] Peach leaves sampled in August contained 73 nmol of reduced and 58 nmol of oxidized glutathione per gram of dry weight.[69] This finding is surprising, since the ratio of oxidized to reduced glutathione is generally small and, e.g., amounted to 0.01 to 0.1 in red spruce needles.[28] This ratio was only substantially increased under oxidizing conditions.[28,70]

E. ENZYMATIC DEFENSE SYSTEMS

O_2^- and H_2O_2 are detoxified by a sequence of enzyme-catalyzed reactions. In this pathway, O_2^- radicals are scavenged by superoxide dismutases at almost diffusion-controlled rates (Table 1). The reaction generates molecular oxygen and hydrogen peroxide. Hydrogen peroxide is removed by a specific ascorbate peroxidase[71,72] which is probably localized in thylakoid membranes as well as in the stroma.[73,74] The products of the ascorbate peroxidase-catalyzed reaction are monodehydroascorbate radicals which can either be removed enzymatically by monodehydroascorbate radical reductase at the expense of NAD(P)H[75] or by spontaneous dismutation, yielding ascorbate and dehydroascorbate. The reduction of dehydroascorbate requires glutathione. The reaction has a substantial nonenzymatic rate, especially above pH 7.5,[67] but can also occur by an enzymatic reaction catalyzed by dehydroascorbate reductase.[76] The product, glutathione disulfide, is reduced by glutathione reductase, with the consumption of NADPH.[67] The benefits of this pathway with respect to the removal of O_2^-, regeneration of NADP as an electron acceptor for photosynthesis, and maintenance of the redox state of ascorbate and glutathione are obvious. Another consequence of the operation of this pathway is the alkalization of the stroma at a rate similar to that of, and the same stoichiometry as, $NADP^+$ reduction.[77] It has been suggested that a physiological function of O_2^- production and detoxification during normal photosynthesis (see Chapter 1 of this volume) is the generation of a proton gradient permitting synthesis of "extra" ATP uncoupled from $NADP^+$ reduction.[17]

The complete set of antioxidants and enzymes necessary to drive the superoxide dismutase-ascorbate-glutathione pathway has been demonstrated in spruce needles.[50,78] Seasonal changes of the activities of superoxide dismutase,[28,47,79,80,81] ascorbate peroxidase,[47,49] dehydroascorbate reductase,[45] and glutathione reductase[47,49,65] were analyzed in pine and spruce needles. Data for seasonal changes of monodehydroascorbate radical reductase activity in leaves of tree species are not available.

Dehydroascorbate reductase activity in spruce needles was high in the vegetation period, declined in winter, and remained low in the spring until bud break.[45] Minimum activities were found in April.[45]

The activities of the other protective enzymes were low in young needles (on the basis of fresh weight) which had just emerged and increased as the needles grew older. Superoxide dismutase and ascorbate peroxidase activities showed a maximum in the fall.[47,49,80] The maximum activities reported for ascorbate peroxidase were 30 and 500 nkat g^{-1} fresh weight in pine and spruce, respectively.[47,49] As superoxide dismutase activity was measured by different indirect methods which did not have the same sensitivity, it is not possible to compare the absolute activities found in different studies. Only a few measurements of superoxide dismutase and ascorbate peroxidase activity were performed in winter, but they suggest that the activities of both enzymes decreased or remained at the same level as in the fall.[47-49,81] Superoxide dismutase activity and dehydroascorbate reductase showed a net decrease with increasing needle age.[45,82,83]

Glutathione reductase activity increased more slowly than ascorbate peroxidase or superoxide dismutase activity and was generally present at maximum activity from January until bud break (45 to 75 nkat g^{-1} fresh weight in pine and spruce needles[47,49,65]). In beech, glutathione reductase activity showed two maxima, one in leaf buds in winter and the other in mature leaves in summer.[68]

In spruce needles,[50,78] beech leaves,[120] and a range of annual plant species (soybean root nodules,[84] cytoplasm of castor bean endosperm,[85] barley leaves,[86] maize,[87] and pea chloroplasts[54]), ascorbate peroxidase activity was 15- to 20-fold higher than glutathione reductase activity. Occasionally, a lower ratio was also reported (barley leaves,[88] pine needles[49]), but in those studies ascorbate peroxidase was extracted in the absence of ascorbate. This might have caused loss of activity, since ascorbate peroxidase from several species (spinach, tea, and spruce) was found to be inactivated in the absence of its substrate.[72,78,89] From the observation that glutathione reductase has generally lower activity than monodehydroascorbate radical reductase and ascorbate peroxidase *in vitro*,[54,78,84-87] it cannot be concluded that glutathione reductase is also rate limiting *in vivo*. The enzymes of the ascorbate-glutathione pathway appear to be regulated coordinately, since they were jointly increased in pea and bean leaves when the plants were grown under enhanced light intensities (400 instead of 100 μmol [quanta] m^{-2} s^{-} and 580 instead of 80 μmol [quanta] m^{-2} s^{-1}, respectively[90,91]). Chloroplastic superoxide dismutase activity also increased significantly in tomato plants transferred from weak (5 μmol m^{-2} s^{-1}) to moderate light conditions (67 μmol m^{-2} s^{-1}).[92] However, in barley leaves, protective enzymes (monodehydroascorbate reductase, dehydroascorbate reductase, and ascorbate peroxidase) did not respond to a transition from moderate to exceedingly high light conditions (280 to 1400 μmol [quanta] m^{-2} s^{-1}) for 3 h.[88] A rapid increase in glutathione reductase was observed only transiently (15 min). The effect of light conditions on the antioxidant enzymes was not investigated in leaves of tree species, except for dehydroascorbate reductase. The activity of this enzyme was 20% lower in shaded than in sun-exposed needles.[45]

Little information is available with respect to the characteristics and subcellular localization of protective enzymes in leaves of tree species. Apparently, protective enzymes have not yet been purified from leaves of deciduous trees. Pine and spruce needles were used for the purification and characterization of superoxide dismutase and glutathione reductase.[9,93-95] In both spruce and pine, the dominant isozyme of superoxide dismutase, which constituted about 2/3 of the total activity, was localized in chloroplasts.[9,95] A similar fraction (50 to 80%) of leaf superoxide dismutase was also found in chloroplasts of different herbaceous plants.[96] Kröniger et al.[9] estimated that the chloroplastic concentration of superoxide dismutase in spruce needles was in the order of 10 μM.

In Norway spruce as well as Scots pine, only one isozyme of glutathione reductase was found.[94,97] Eastern white pine needles contained two isozymes of glutathione reductase, which increased proportionally in winter.[49,93] The subcellular distribution of glutathione reductase activity was not investigated in tree species, but on the basis of results obtained with herbaceous plants,[54,67,98] 70 to 80% may be expected in the chloroplasts. A similar fraction of ascorbate peroxidase was localized in the chloroplasts of pea leaves.[54]

The high concentrations of antioxidant enzymes and substrates in chloroplasts and their modulation by light conditions emphasizes their importance for coping with photosynthetically generated oxidants, $O_2^{\cdot-}$ and H_2O_2, in the plant's everyday life. Conifer needles in winter must cope with increased stress since they are exposed to light at low temperatures. However, clear experimental evidence is lacking as to whether or not photostress in winter is also accompanied by enhanced $O_2^{\cdot-}$ or H_2O_2 production in the chloroplasts of conifers. High activities of protective enzymes were reported in the cold season, but the measurements have mostly been performed at a standard temperature (20 or 25°C) not relevant for winter conditions. The partial pressure of oxygen is higher in winter due to an enhanced water solubility of oxygen at low temperature, but linear electron transport, the predominant source of photosynthetic $O_2^{\cdot-}$ production, is reduced by 80% in thylakoids isolated from winter-hardened spruce needles.[99] Moreover, thylakoids from a cold-tolerant species, spinach, produced less $O_2^{\cdot-}$ when assayed at 5 rather than 20°C.[8] Experimental approaches are needed to distinguish the roles of $^1O_2$ and $O_2^{\cdot-}$ in photooxidative stress in winter.

F. OTHER SCAVENGERS FOR REACTIVE OXYGEN SPECIES

Among different activated oxygen species produced in plant tissue, $^{\cdot}OH$ has the highest reactivity, making specific protection measures impossible. So-called compatible solutes, i.e., compounds which are tolerated in plant tissues at high concentrations, such as proline, mannitol, myoinositol, etc., have been implicated in the defense against $^{\cdot}OH$ radicals. But even if they would react at diffusion-controlled rate constants ($K_{\cdot OH} = 10^{10}\ M^{-1}\ s^{-1}$), concentrations of about 1 *M* would be required in order to suppress $^{\cdot}OH$ reactivity by 90%.[100] In an *in vitro* system, myoinositol (33 m*M*) protected the activity of malate dehydrogenase from $^{\cdot}OH$ damage by 50%.[101] Myoinositol was used as a commercially available analog of pinitol. Since pinitol is present in spruce needles at a concentration of about 0.1 *M*,[102] it may play a significant role in the protection against $^{\cdot}OH$. Other compatible solutes, such as proline (33 m*M*), were less effective than myoinositol and protected malate dehydrogenase only by 20%.[101] In plant tissue, proline concentrations were lower than in the *in vitro* system, e.g., 0.3 to 15 m*M* in pine needles (calculated from Reference 103) and 0.8 to 15 m*M* in spruce needles (determined from Reference 104), and thus were unlikely to protect leaves from $^{\cdot}OH$

damage. The prevention of $^{\cdot}OH$ formation by removal of its precursors, $O_2^{\cdot -}$ and H_2O_2, in the superoxide dismutase-ascorbate-glutathione pathway appears to be the best protective measure.

In addition to the protective systems discussed above, plants contain numerous compounds with antioxidant properties, such as phenolics, polyamines, flavonoids, etc. (reviewed in Reference 105). Especially phenolics, which occur in spruce needles at concentrations >100 m*M*[106], may play an important role for scavenging activated oxygen. However, information with respect to their subcellular distribution and significance for photoprotection is lacking.

III. PHOTOOXIDATIVE DAMAGE

Stress factors such as drought, unseasonally low temperatures, nutrient limitations, air pollution, etc. can enhance the production of reactive oxygen species and injure cellular structures. However, in tree species, the role of light in chlorophyll degradation and the corresponding responses of the antioxidant system have only been investigated for a few stress factors, i.e., environmental stress at high altitude, nutrient limitations, and air pollutants. Oxidative stress by air pollutants, especially SO_2 and ozone, has been reviewed previously[107,108] (also see Chapters 6 and 14 of this volume) and therefore is not discussed here.

A. HIGH ALTITUDE

Tree species at high altitude are exposed to higher irradiation, lower average temperatures, and higher mean ozone concentrations than trees grown at valley level.[109,110] Needles from spruce trees grown at high altitude contained less chlorophyll than needles from trees grown at valley level.[109] In contrast, carotenoids were less reduced and are therefore likely to provide increased protection of the photosynthetic apparatus.[50] When needles from red spruce grown at increasing altitude (100 to 1000 m above sea level) were investigated, increased glutathione content was found, but ascorbate content and superoxide dismutase activity were unaffected.[48] Needles from Norway spruce trees grown at high altitude (1700 m above sea level) contained elevated ascorbate and glutathione content and increased glutathione reductase activity, but superoxide dismutase and ascorbate peroxidase activities similar to that of needles from valley level (700 m above sea level).[47,111] However, the mean enhancement in antioxidant capacity between valley and high altitude was modest on the basis of fresh weight (+25 to 35%). When antioxidants were compared in mature needles from valley and high altitude on the basis of chlorophyll, the relative increase in protective capacity was considerable: superoxide dismutase, +66%; ascorbate, +154%; glutathione, +192%; and glutathione reductase, +166% (calculated for June with data taken from Reference 47 and 110). Still, these observations give no clear-cut evidence on how chloroplasts are protected from photooxidative damage at high altitude, as the subcellular distribution of antioxidants in spruce needles is not known.

B. NUTRIENT LIMITATIONS

Recently, Marschner and Cakmak[91,112] showed that the development of chlorosis in bean leaves grown with deficiencies in Mg, K, or Zn was promoted by high light intensity and accompanied by an increase in the antioxidant substrates and enzyme activities of the superoxide dismutase-ascorbate-glutathione pathway. Several studies suggest that nutrient deficiencies have effects in spruce trees similar to those in herbaceous plants. Polle et al.[50] reported that Mn deficiency observed in chlorotic needles was correlated with increases in ascorbate, monodehydroascorbate radical reductase, and superoxide dismutase. Although the effect of light was not explicitly investigated, the observation that needles on northerly branches of the same trees were less chlorotic suggested a link between nutrient deficiency and photooxidation.

Mg-deficient spruce needles were found to be more chlorotic on the upper, sun-exposed side than on the lower side and contained enhanced glutathione and ascorbate content, especially in chlorotic segments.[113,114] The effect of light on the antioxidant contents of the needles was not investigated, but by reduction of incident radiation by 90% using red filters, chlorophyll destruction was prevented.[24]

In spruce as in beans, the increase in ascorbate under nutrient limitation was more pronounced than the increase in glutathione.[50,91,113,114] In chlorotic needles, enhanced activity of superoxide dismutase (but only in older needle age classes[83,115]) and enhanced activities of dehydroascorbate reductase and glutathione reductase were reported,[97] whereas Osswald et al.[97] found no effect on superoxide dismutase activity.

The increased contents of antioxidants and activities of protective enzymes present in chlorotic needles suggest that nutrient-deficient plants have to cope with increased production rates of O_2^{-}. Some lines of evidence indicate that the export of photosynthate might have been prevented in chlorotic tissues, thereby limiting the availability of $NADP^+$ and favoring oxygen reduction.[91,110,112] However, O_2^{-} formation in spruce needles has not yet been determined. It is also possible that the levels of activated oxygen in chlorotic tissue must be kept at lower-than-normal concentrations because light-harvesting proteins from bleached needles are more susceptible to photooxidation than those from healthy tissue.[20]

IV. SUMMARY

In leaves of evergreen trees, the antioxidative system shows large seasonal and developmental fluctuations. In contrast to general belief, high antioxidant concentrations do not always occur in winter, but were also found, e.g., for ascorbate in spring and summer. It is not known to which extent changes in the total antioxidant capacity also reflect changes in the antioxidant capacity of chloroplasts. High ascorbate concentrations can protect leaves of conifers from oxidizing compounds and may constitute an adaptation measure for survival of these species in harsh and unfavorable environments. Glutathione has, beside its antioxidant properties, an important role in the sulfur nutrition of coniferous trees. In contrast to conifers, only little information is available for the antioxidative systems in the leaves of deciduous trees. To date, most studies conducted with trees present only circumstantial evidence for an association between levels of protective systems, environmental changes, and damage symptoms. Therefore, investigations are needed to determine whether stress tolerance in susceptible tree species can be enhanced by increasing their antioxidative capacity.

ACKNOWLEDGMENTS

The research conducted in the authors' laboratory was financially supported by the Deutsche Forschungsgemeinschaft (Schwerpunkt Physiologie der Bäume) and by the Bayerisches Staatsministerium für Landesentwicklung und Umweltfragen.

REFERENCES

1. **Hällgren, J. E., Lundmark, T., and Strand, M.,** Photosynthesis of Scots pine in the field after night frost during summer, *Plant Physiol. Biochem.*, 28, 437, 1990.
2. **Benner, P., Sabel, P., and Wild, A.,** Photosynthesis and transpiration of healthy and diseased spruce trees in the course of three vegetation periods, *Trees,* 2, 223, 1988.
3. **Vowinckel, T., Oechel, W., and Boll, W.,** The effect of climate on photosynthesis in *Picea mariana* at the subartic tree line. I. Field measurements, *Can. J. Bot.*, 53, 604, 1975.
4. **Rühle, W. and Wild, A.,** Die Anpassung des Photosyntheseapparates höherer Pflanzen an die Lichtbedingungen, *Naturwissenschaften,* 72, 10, 1985.
5. **Farquhar, G., Wong, S., Evans, J., and Hubick, K.,** Photosynthesis and gas exchange, in *Plants Under Stress,* Jones, H., Flowers, T., and Jones, M., Eds., Cambridge University Press, Cambridge, 1989.
6. **Asada, K. and Takahashi, M.,** Production and scavenging of active oxygen species in photosynthesis, in *Photoinhibition,* Kyle, D., Osmond, C., and Arntzen, C., Eds., Elsevier, 1987.
7. **Robinson, J. M.,** Does O_2 photoreduction occur within chloroplasts in vivo?, *Physiol. Plant.*, 72, 666, 1988.
8. **Hodgson, R. and Raison, J.,** Superoxide production by thylakoids during chilling and its implication in the susceptibility of plants to chilling induced photoinhibition, *Planta,* 183, 222, 1991.

9. **Kröniger, W., Rennenberg, H., and Polle, A.,** Purification of two superoxide dismutase isozymes and their subcellular localization in needles and roots of Norway spruce (*Picea abies*, L.), *Plant Physiol.*, 100, 334, 1992.
10. **Tanaka, K., Kondo, N., and Sugahara, K.,** Accumulation of hydrogen peroxide in chloroplasts of SO_2-fumigated spinach leaves, *Plant Cell Physiol.*, 23, 999, 1982.
11. **Kaiser, W.,** Reversible inhibition of Calvin cycle and activation of oxidative pentose phosphate cycle in isolated intact chloroplasts by hydrogen peroxide, *Planta*, 145, 377, 1979.
12. **Yim, M., Chock, P., and Stadman, E.,** Copper,zinc superoxide dismutase catalyzes hydroxyl radical production from hydrogen peroxide, *Proc. Natl. Acad. Sci. U.S.A.*, 87, 5006, 1990.
13. **Lichtenthaler, H.,** Chlorophylls and carotenoids: pigments of photosynthetic biomembranes, *Methods Enzymol.*, 148, 350, 1987.
14. **Siefermann-Harms, D.,** Physiologische und biochemische Untersuchungen zur montanen Vergilbung von Fichten., *KfK PEF Ber.*, 50, 294, 1989.
15. **Demmig-Adams, B.,** Carotenoids and photoprotection in plants: a role for the xanthophyll zeaxanthin, *Biochim. Biophys. Acta*, 1020, 1, 1990.
16. **Demmig-Adams, B. and Adams, W., III,** Photoprotection and other responses of plants to high light stress, *Annu. Rev. Plant Physiol. Plant Mol. Biol.*, 43, 599, 1992.
17. **Krause, G. H.,** Photoinhibition of photosynthesis. An evaluation of damaging and protective mechanisms, *Physiol. Plant.*, 74, 566, 1988.
18. **Siefermann-Harms, D.,** The light-harvesting and protective functions of carotenoids in photosynthetic membranes, *Physiol. Plant.*, 69, 561, 1987.
19. **Linder, S.,** Seasonal variations of pigments in needles: a study of Scots pine and Norway spruce seedlings grown under different nursery conditions, *Stud. For. Suec.*, 100, 5, 1972.
20. **Siefermann-Harms, D.,** Physiologische und biochemische Untersuchungen zur montanen Vergilbung von Fichten, *KfK PEF Ber.*, 35, 173, 1988.
21. **Lichtenthaler, H., Schmuck, G., and Döll, M.,** Photosyntheseaktivität bei Nadeln gesunder und geschädigter Koniferen, *Lis Ber.*, 57, 87, 1985.
22. **Siefermann-Harms, D.,** The xanthophyll cycle in higher plants, in *Lipids and Lipid Polymers in Higher Plants*, Tevini, M. and Lichtenthaler, H., eds., Springer-Verlag, Berlin, 1977.
23. **Zimmer-Rinderle, U. and Lichtenthaler, H.,** Jahresverlauf der Photosyntheseaktivität geschädigter Fichten im Nordschwarzwald., *KfK PEF Ber.*, 94, 81, 1992.
24. **Siefermann-Harms, D.,** Physiologische und biochemische Untersuchungen zur montanen Vergilbung von Fichten. III. Licht und Vergilbung, *KfK PEF Ber.*, 61, 285, 1990.
25. **Öquist, G.,** Effect of low temperature on photosynthesis, *Plant Cell Environ.*, 6, 281, 1983.
26. **Fryer, M. J.,** The antioxidant effects of thylakoid vitamin E (α-tocopherol), *Plant Cell Environ.*, 15, 381, 1992.
27. **Franzen, J., Bausch, J., Glatzle, D., and Wagner, E.,** Distribution of vitamin E in spruce seedlings and mature tree organs, and within the genus, *Phytochemistry*, 30, 147, 1991.
28. **Hausladen, A., Madamachani, N., Fellows, S., Alscher, R., and Amundson, R.,** Seasonal changes in antioxidants in red spruce as affected by ozone, *New Phytol.*, 115, 447, 1990.
29. **Grumbach, K.,** Distribution of chlorophylls, carotenoids and quinones in chloroplasts of higher plants, *Z. Naturforsch.*, 38, 996, 1983.
30. **Yerin, A., Kormanovskyii, A., and Ivanov, I.,** Localization of α-tocopherol in chloroplasts, *Biophysics*, 29, 363, 1984.
31. **Takahashi, M. and Asada, K.,** Superoxide production in aprotic interior of chloroplast thylakoids, *Arch. Biochem. Biophys.*, 267, 714, 1988.

32. **Leung, H., Vang, M., and Mavis, R.,** The cooperative interaction between vitamin E and vitamin C in suppression of peroxidation of membrane phospholipids, *Biochim. Biophys. Acta,* 664, 266, 1981.
33. **Takenaka, Y., Miki, M., Yasuda, H., and Mino, M.,** The effect of α-tocopherol as an antioxidant on the oxidation of membrane protein thiols induced by free radicals generated in different sites, *Arch. Biochem. Biophys.,* 285, 344, 1991.
34. **Liebler, D., Kling, D., and Reed, D.,** Antioxidant protection of phospholipid bilayers by α-tocopherol, *J. Biol. Chem.,* 261, 12114, 1986.
35. **Nagao, A. and Terao, J.,** Antioxidant activity of 6-phosphatidyl-L-ascorbic acid, *Biochem. Biophys. Res. Commun.,* 172, 385, 1990.
36. **Packer, J., Slater, T., and Willson, R.,** Direct observation of a free radical interaction between vitamin E and vitamin C, Nature *(London),* 278, 737, 1979.
37. **Halliwell, B.,** Toxic effects of oxygen on plant tissues, in *Chloroplast Metabolism: The Structure and Function of Chloroplasts in Green Leaf Cells,* Claredon Press, Oxford, 1981, chap. 8.
38. **Martin, B., Martensson, O., and Öquist, G.,** Seasonal effects on photosynthetic electron transport and fluorescence properties in isolated chloroplasts of *Pinus sylvestris, Physiol. Plant.,* 44, 109, 1978.
39. **Kunert, K. and Ederer, M.,** Leaf ageing and lipid peroxidation: the role of the antioxidants vitamin C and E, *Plant Physiol.,* 65, 85, 1985.
40. **Franzen, J., Glatzle, D., Bausch, J., and Wagner, E.,** Age dependent accumulation of vitamin E in spruce needles, *Phytochemistry,* 30, 2147, 1991.
41. **Rise, M., Cojocaru, M., Gottlieb, H., and Goldschmidt, E.,** Accumulation of α-tocopherol in senescing organs as related to chlorophyll degradation, *Plant Physiol.,* 89, 1028, 1989.
42. **Wise, R. and Naylor, A.,** Chilling-enhanced photooxidation. Evidence for the role of singlet oxygen and superoxide in the breakdown of pigments and endogenous antioxidants, *Plant Physiol.,* 83, 278, 1987.
43. **Lichtenthaler, H.,** Occurrence and function of prenyl lipids in the photosynthetic membrane, in *Advances in the Biochemistry and Physiology of Plant Lipids,* Appelqvist, L. and Liljenberg, C., Eds., Elsevier, Amsterdam, 1979.
44. **Bukatsch, F.,** Über den Askorbinsäuregehalt der Coniferennadeln, *Vitam. Horm. (Leipzig),* 4, 192, 1943.
45. **Esterbauer, H., Grill, D., and Welt, R.,** Der jahreszeitliche Rhythmus des Ascorbinsäuresystems in Nadeln von *Picea abies, Z. Pflanzenphysiol.,* 98, 393, 1980.
46. **Franke, W.,** Der jahreszeitlich unterschiedliche Vitamin-C Gehalt in Coniferennadeln als Folge von Ascorbinsäurebiosynthese und -abbau, *Z. Pflanzenphysiol.,* 60, 30, 1968.
47. **Polle, A. and Rennenberg, H.,** Field studies on Norway spruce trees at high altitudes. II. Defence systems against oxidative stress in needles, *New Phytol.,* 121, 635, 1992.
48. **Madamachani, N., Hausladen, A., Alscher, R., Amundson, R., and Fellows, S.,** Seasonal changes in antioxidants in red spruce (*Picea rubens,* Sarg.) from three field sites in the northeastern United States, *New Phytol.,* 118, 331, 1991.
49. **Anderson, J., Chevone, B., and Hess, J.,** Seasonal variation in the antioxidant system of eastern white pine needles, *Plant Physiol.,* 98, 501, 1992.
50. **Polle, A., Chakrabarti, K., Chakrabarti, S., Seifert, F., Schramel, P., and Rennenberg, H.,** Antioxidants and manganese deficiency in needles of Norway spruce (*Picea abies,* L.) trees, *Plant Physiol.,* 99, 1084, 1992.
51. **Stegmann, H., Schuler, P., and Ruff, H.,** Ascorbic acid radicals as indicator of damage to forest — EPR investigations of spruce needles in vivo, *Z. Naturforsch.,* 45, 9, 1989.
52. **Franke, W.,** Über den Vitamin-C-Gehalt der Herbstblätter. I. Der Vitamin C Gehalt bis zum Laubabwurf, *Z. Pflanzenphysiol.,* 53, 289, 1965.
53. **Foyer, C., Rowell, J., and Walker, D.,** Measurement of the ascorbate content of spinach leaf protoplasts and chloroplasts during illumination, *Planta,* 157, 239, 1983.

54. **Gillham, D. and Dodge, A.,** Hydrogen peroxide scavenging systems within pea chloroplasts: a quantitative study, *Planta,* 167, 246, 1986.
55. **Law, M., Charles, S., and Halliwell, B.,** Glutathione and ascorbic acid in spinach *(Spinacea oleracea)* chloroplasts, *Biochem. J.,* 210, 899, 1983.
56. **Schöner, S. and Krause, G. H.,** Protective systems against active oxygen species in spinach: response to cold-acclimation in excess light, *Planta,* 180, 383, 1990.
57. **Rennenberg, H. and Lamoureux, G.,** Physiological processes that modulate the concentration of glutathione in plant cells, in *Sulfur Nutrition and Sulfur Assimilation in Higher Plants,* Rennenberg, H., Brunold, C., DeKok, L., and Stulen, I., Eds., SPB Academic Publishing, The Hague, 1990.
58. **Schupp, R. and Rennenberg, H.,** Changes in sulfur metabolism during needle development of Norway spruce, *Bot. Acta,* 105, 180, 1992.
59. **Rennenberg, H.,** Glutathione metabolism and possible biological roles in higher plants, *Phytochemistry,* 21, 2771, 1982.
60. **Rennenberg, H.,** The fate of excess sulfur in higher plants, *Annu. Rev. Plant Physiol.,* 35, 121, 1984.
61. **Steffens, J.,** Heavy metal stress and the phytochelatin response, in *Stress Responses in Plants: Adaptation and Acclimation Mechanisms,* Alscher, R. and Cumming, J., Wiley-Liss, New York, 1990.
62. **Grill, D., Pfeifhofer, H., and Esterbauer, H.,** Further investigations on the thiol content of Norway spruce needles, *Phyton,* 27, 311, 1987.
63. **Schupp, R. and Rennenberg, H.,** Diurnal changes in the glutathione content of spruce needles (*Picea abies,* L.), *Plant Sci.,* 57, 113, 1988.
64. **Schupp, R., Schatten, T., Willenbrink, J., and Rennenberg, H.,** Long-distance transport of reduced sulfur in spruce (*Picea abies,* L.), *J. Exp. Bot.,* 43, 1243, 1992.
65. **Esterbauer, H. and Grill, D.,** Seasonal variation of glutathione and glutathione reductase in needles of *Picea abies, Plant Physiol.,* 61, 119, 1978.
66. **Smith, I. K., Polle, A., and Rennenberg, H.,** Glutathione, in *Stress Responses in Plants: Adaptation and Acclimation Mechanisms,* Alscher, R. and Cumming, J., Wiley-Liss, New York, 1990.
67. **Foyer, C. and Halliwell, B.,** The presence of glutathione and glutathione reductase in chloroplasts: a proposed role in ascorbic acid metabolism, *Planta,* 133, 21, 1976.
68. **Polle, A., Morawe, B., and Rennenberg, H.,** Seasonal variations of glutathione reductase and glutathione in leaves and leaf buds of beech trees (*Fagus sylvatica,* L.), *Phyton,* 32, 99, 1992.
69. **Siller-Cepeda, J., Chen, T., and Fuchigami, L.,** High performance liquid chromatography analysis of reduced and oxidized glutathione in woody plant tissues, *Plant Cell Physiol.,* 32, 1179, 1991.
70. **Gupta, A., Alscher, R., and McCune, D.,** Response of photosynthesis and cellular antioxidants to ozone in *Populus* leaves, *Plant Physiol.,* 96, 650, 1991.
71. **Nakano, Y. and Asada, K.,** Hydrogen peroxide is scavenged by ascorbate-specific peroxidase in spinach chloroplasts, *Plant Cell Physiol.,* 22, 867, 1981.
72. **Chen, G. and Asada, K.,** Ascorbate peroxidase in tea leaves: occurrence of two isozymes and the difference in their enzymatic and molecular properties, *Plant Cell Physiol.,* 30, 987, 1989.
73. **Groden, D. and Beck, E.,** H_2O_2 destruction by ascorbate-dependent systems from chloroplasts, *Biochim. Biophys. Acta,* 546, 426, 1979.
74. **Miyake, C. and Asada, K.,** Thylakoid-bound ascorbate peroxidase in spinach chloroplasts and photoreduction of its primary oxidation product monodehydroascorbate radicals in thylakoids, *Plant Cell Physiol.,* 33, 541, 1992.
75. **Hossain, M. A., Nakano, Y., and Asada, K.,** Monodehydroascorbate reductase in spinach chloroplasts and its participation in regeneration of ascorbate for scavenging hydrogen peroxide, *Plant Cell Physiol.,* 25, 385, 1984.

76. **Foyer, C. and Halliwell, B.**, Purification and properties of dehydroascorbate reductase from spinach leaves, *Phytochemistry*, 16, 1347, 1977.
77. **Polle, A. and Junge, W.**, The slow rate of proton consumption at the reducing site of photosystem I is limited by the rate of redox reactions of extrinsic electron acceptors, but not by a diffusion barrier for protons, *Biochim. Biophys. Acta*, 848, 274, 1986.
78. **Polle, A., Chakrabarti, K., Schürmann, W., and Rennenberg, H.**, Composition and properties of hydrogen peroxide decomposing systems in extracellular and total extracts from needles of Norway spruce (*Picea abies* L., Karst.), *Plant Physiol.*, 94, 312, 1990.
79. **Huttunen, S. and Heiska, E.**, Superoxide dismutase activity in Scots pine (*Pinus sylvestris* L.) and in Norway spruce (*Picea abies* L. Karst.) needles in northern Finland, *Eur. J. For. Pathol.*, 18, 343, 1988.
80. **Tandy, N., DiGiulio, R., and Richardson, C.**, Assay and electrophoresis of superoxide dismutase from red spruce (*Picea rubens* Sarg.), Loblolly pine (*Pinus taeda* L.) and Scotch pine (*Pinus sylvestris* L.), *Plant Physiol.*, 90, 742, 1989.
81. **Schulz, H.**, Biochemische und faktoranalytische Untersuchungen zur Interpretation von SO_2-Indikationen in Nadeln von *Pinus sylvestris*, *Biochem. Physiol. Pflanz.*, 181, 241, 1986.
82. **Polle, A., Krings, B., and Rennenberg, H.**, Superoxide dismutase activity in needles of Norwegian spruce trees (*Picea abies*, L.), *Plant Physiol.*, 90, 1310, 1989.
83. **Polle, A. and Rennenberg, H.**, Superoxide dismutase activity in needles of Scots pine and Norway spruce under field and chamber conditions: lack of ozone effects, *New Phytol.*, 117, 335, 1991.
84. **Dalton, D., Russel, S., Hanus, F., Pascoe, G., and Evans, H.**, Enzymatic reactions of ascorbate and glutathione that prevent peroxide damage in soyabean root nodules. *Proc. Natl. Acad. Sci. U.S.A.*, 83, 3811, 1986.
85. **Klapheck, S., Zimmer, I., and Cosse, H.**, Scavenging of hydrogen peroxide in the endosperm of *Ricinus communis* by ascorbate peroxide, *Plant Cell Physiol.*, 31, 1005, 1990.
86. **Smirnoff, N. and Colombe, S.**, Drought influences the activity of enzymes of the chloroplast hydrogen scavenging system, *J. Exp. Bot.*, 39, 1097, 1988.
87. **Jahnke, L., Hull, M., and Long, S.**, Chilling stress and oxygen metabolizing enzymes in *Zea mays* and *Zea diploperennis*, *Plant Cell Environ.*, 14, 97, 1991.
88. **Foyer, C., Dujardyn, M., and Lemoine, Y.**, Responses of photosynthesis and the xanthophyll and ascorbate-glutathione cycles to changes in irradiance, photoinhibition and recovery, *Plant Physiol. Biochem.*, 27, 751, 1989.
89. **Nakano, Y. and Asada, K.**, Purification of ascorbate peroxidase in spinach chloroplasts, its inactivation in ascorbate depleted medium and reactivation by monodehydroascorbate radical, *Plant Cell Physiol.*, 28, 131, 1987.
90. **Gillham, D. and Dodge, A.**, Chloroplast superoxide and hydrogen peroxide scavenging systems from pea leaves: seasonal variations, *Plant Sci.*, 50, 105, 1987.
91. **Cakmak, I. and Marschner, H.**, Magnesium deficiency and high light intensity enhance activities of superoxide dismutase, ascorbate peroxidase, and glutathione reductase in bean leaves, *Plant Physiol.*, 98, 1222, 1992.
92. **Perl-Treves, R. and Galun, E.**, The tomato Cu,Zn superoxide dismutase genes are developmentally regulated and respond to light and stress, *Plant Mol. Biol.*, 17, 745, 1991.
93. **Anderson, J., Hess, J., and Chevone, B.**, Purification, characterization and immunological properties for two isoforms of glutathione reductase from Eastern White pine needles, *Plant Physiol.*, 94, 1402, 1990.
94. **Wingsle, G.**, Purification and characterization of glutathione reductase from Scots pine needles, *Physiol. Plant.*, 76, 24, 1989.
95. **Wingsle, G., Gardeström, P., Hällgren, J.-E., and Karpinski, S.**, Isolation, purification and subcellular localization of isozymes of superoxide dismutase from Scots pine (*Pinus sylvestris* L.) needles, *Plant Physiol.*, 95, 21, 1991.

96. **Foster, J. G. and Edwards, G. E.,** Localisation of superoxide dismutase in leaves of C_3 and C_4 plants, *Plant Cell Physiol.*, 21, 895, 1980.
97. **Osswald, W., Kraus, R., Hippeli, S., Benz, B., Volpert, R., and Elstner, E.,** Comparison of the enzymatic activities of dehydroascorbic acid reductase, glutathione reductase, catalase, peroxidase and superoxide dismutase of healthy and damaged spruce needles (*Picea abies* (L.) Karst.), *J. Plant Physiol.*, 139, 742, 1992.
98. **Edwards, E. A., Rawthorne, S., and Mullineaux, P.,** Subcellular distribution of multiple forms of glutathione reductase in leaves of pea (*Pisum sativum*, L.), *Planta*, 180, 278, 1990.
99. **Senser, M. and Beck, E.,** On the mechanisms of frost injury and frost hardening of spruce chloroplasts, *Planta*, 137, 195, 1977.
100. **Saran, M., Michel, C., and Bors, W.,** Reactivities of free radicals, in *Air Pollution and Plant Metabolism*, Schulte-Hostede, S., Darrell, N., Blank, L., and Wellburn, A., Eds., Elsevier, London, 1987.
101. **Smirnoff, N. and Cumbes, Q.,** Hydroxyl radical scavenging activity of compatible solutes, *Phytochemistry*, 28, 1057, 1989.
102. **Krasenbrink, A.,** Osmotisch wirksame Inhaltsstoffe in Fichtennadeln in Abhängigkeit von Standort und der Jahreszeit, Ph.D. thesis, Universität Münster, 1991.
103. **Vance, N. and Zaerr, J.,** Analysis by high-performance liquid chromatography of free amino acids extracted from needles of drought-stressed and shaded *Pinus ponderosa* seedlings, *Physiol. Plant.*, 79, 23, 1990.
104. **Bode, J., Kühn, H., and Wild, A.,** Die Akkumulation von Prolin in Nadeln geschädigter Fichten (*Picea abies* (L.) Karst.) Forstwiss, *Cbl.*, 6, 353, 1985.
105. **Larson, R.,** The antioxidants of higher plants, *Phytochemistry*, 27, 969, 1988.
106. **Esterbauer, H., Grill, D., and Beck, G.,** Untersuchungen über Phenole in Nadeln von *Picea abies*, *Phyton*, 17, 87, 1975.
107. **Rennenberg, H. and Polle, A.,** Metabolic consequences of atmospheric sulfur influx into plants, in *Air Pollution and Plant Metabolism*, Alscher, R. and Wellburn, A., Eds., in press.
108. **Polle, A. and Rennenberg, H.,** The significance of antioxidants in plant adaptation to environmental stress, in *Plant Adaptation to Environmental Stress*, Mansfield, T., Fowden, L., and Stoddard, F., Eds., James and James, London, in press.
109. **Tranquillini, W.,** The physiology of plants at high altitudes, *Annu. Rev. Plant Physiol.*, 15, 345, 1964.
110. **Polle, A., Mössnang, M., von Schönborn, A., Sladkovic, R., and Rennenberg, H.,** Field studies on Norway spruce trees at high altitudes. I. Mineral, pigment and soluble protein contents of the needles as affected by climate and pollution, *New Phytol.*, 121, 89, 1992.
111. **Grill, D., Pfeifhofer, H., Tschulik, A., Hellig, K., and Holzer, K.,** Thiol content of spruce needles at forest limits, *Oecologia*, 76, 294, 1988.
112. **Marschner, H. and Cakmak, I.,** High light intensity enhances chlorosis and necrosis in leaves of zinc, potassium and magnesium deficient bean (*Phaseolus vulgaris*) plants, *J. Plant Physiol.*, 134, 308, 1989.
113. **Osswald, W. and Elstner, E.,** Fichtenerkrankungen in den Hochlagen der Bayerischen Mittelgebirge, *Ber. Dtsch. Bot. Ges.*, 99, 313, 1986.
114. **Osswald, W., Senger, H., and Elstner, E.,** Ascorbic acid and glutathione contents of spruce needles from different locations in Bavaria, *Z. Naturforsch.*, 42, 879, 1987.
115. **Elstner, E., Osswald, W., and Youngman, R.,** Basic mechanisms of pigment oxygen bleaching and loss of structural resistance in spruce (*Picea abies*) needles: advances in phytomedical diagnostics, *Experientia*, 41, 591, 1985.
116. **Krinsky, N.,** Carotenoid protection against oxidation, *Pure Appl. Chem.*, 51, 649, 1979.

117. **Foote, C. S., Ching, T., and Geller, G.,** Chemistry of singlet oxygen. XVIII. Rates of reaction and quenching of α-tocopherol and singlet oxygen, *Photochem. Photobiol.*, 20, 511, 1974.
118. **Rooney, M.,** Ascorbic acid as photooxidation inhibitor, *Photochem. Photobiol.*, 38, 619, 1983.
119. **Devasagayam, T., Sundquist, A., Dimascio, P., Kaiser, S., and Sies, H.,** Activity of thiols as singlet molecular oxygen quenches, *Photochem. Photobiol.*, 9, 105, 1991.
120. **Polle, A. and Morawe, B.,** unpublished results.

Chapter 9

HERBICIDE ACTION AND EFFECTS ON DETOXIFICATION PROCESSES

Alan D. Dodge

TABLE OF CONTENTS

0-8493-5443-9/94/$0.00 + $.50

I. INTRODUCTION

Agriculture today relies heavily on the use of pesticides to protect crop plants from weeds, insect predators, and fungal pathogens. The selection of crop plants over thousands of years for productivity rather than survival has required cultivation, good husbandry, and, in this century, pesticides as crop protection agents.

In addition to these problems, all plants are faced with changes in the physical environment, such as temperature, light intensity, and water supply. Such changes can lead to internal toxic biochemical events that are damaging by free radical- or singlet oxygen-induced action. Under relatively normal circumstances, plants are able to withstand a wide range of environmental changes because they possess a number of protective or scavenging systems that function continuously or are latent until required.

Herbicides in use today initiate an event or a series of events that will lead to the death of weeds. In many instances, there are subtle reasons for selectivity between crop and weed due to differential uptake or metabolism, or sensitivity of a target site. However, in some cases, weeds are killed because the normally competent protective systems are unable to cope with the excessive demands induced or promoted by herbicides. In other words, the detoxification processes are overtaxed and unable to prevent cellular damage and death.

II. HERBICIDE-INDUCED RADICAL GENERATION AND CONTROL

Although the bipyridinium herbicides paraquat and diquat were discovered over 30 years ago, there has since that time been no major new developments of compounds that function in a similar manner. One key feature for the action of these herbicides is a negative redox potential, appropriate for interaction with the reducing side of photosystem I, and an ability to be autooxidized by molecular oxygen. In practice, paraquat is reduced from iron sulfur center B (FeS_B) primarily,[1] and binding to the chloroplast thylakoid is facilitated by the dicationic nature of the herbicide.[2] Early physiological studies with the bipyridinium herbicides indicated that their action was promoted by both light and oxygen,[3] and it was subsequently shown in pulse radiolysis studies that reduced paraquat will interact with oxygen with a reaction rate constant of 7.7×10^{-8} M^{-1} S^{-1}, which will give rise to a micromolar concentration of superoxide.[4] During "normal" photosynthesis, some electron flow will pass to oxygen, in what is generally termed as "endogenous Mehler reaction", and the superoxide thus generated is adequately scavenged by superoxide dismutase and the sequence of reactions incorporating ascorbate and glutathione, known as the ascorbate-glutathione cycle (see Chapter 1 of this volume). With the enhanced production of superoxide to a level ten- to

20-fold higher than normal,[5] the protective systems are inadequate, and a series of deleterious events will follow. Whereas both superoxide and hydrogen peroxide would normally be removed, they can interact via a metal catalyst, such as Fe^{2+}, to give hydroxyl radicals.

$$O_2^{\overline{\cdot}} + H_2O_2 \rightarrow O_2 + OH^- + OH\cdot \quad (1)$$

As more damage proceeds, iron is released from ferredoxin, cytochromes, and phtyoferritin,[6] thus enhancing this Fenton-type reaction.

The diversion of electron flow caused by the bipyridiniums will also prevent $NADP^+$ reduction. This is apparent as carbon dioxide fixation quickly ceases,[7] although this fixation is also affected directly by the inhibiting effect of hydrogen peroxide on some enzymes of the Calvin cycle, such as fructose bisphosphatase and sedoheptulose bisphosphatase.[8,9] NADPH is also required for the continuation of the ascorbate-glutathione cycle via glutathione reductase, as well as for the NADPH-dependent monodehydroascorbate reductase.[10] The net result of these events is that unscavenged hydroxyl radicals[11] will initiate a cascade of deleterious events by interacting with unsaturated fatty acids of membrane lipids to yield lipid peroxides, and with certain amino acids of proteins, such as methionine and histidine, and thymine of DNA. Thus, cellular disorganization and death will proceed quite rapidly, leading to the typical visible symptoms of chlorophyll bleaching.

The pivotal role of ascorbate in the scavenging processes in chloroplasts is clearly evident as it participates in the ascorbate-glutathione cycle, as well as having the potential to scavenge hydrogen peroxide, hydroxyl radicals, and singlet oxygen directly.[12] Glutathione, in addition to its role in the ascorbate-glutathione cycle is also an effective antioxidant.[13] Experiments with whole isolated chloroplasts showed that on the addition of paraquat in the light, there was a rapid oxidation of reduced glutathione[14] and ascorbate[15] (Figure 1). The magnitude of the effect was such that at least 95% oxidation of chloroplast stromal ascorbate was accomplished within 1 min of illumination.[15] Experiments with whole leaves showed that herbicidal concentrations of paraquat lead to a loss of activity of the scavenging enzymes superoxide dismutase, ascorbate peroxidase, and glutathione reductase as well as of the antioxidant ascorbate[16] (Figure 2). These changes were clearly due to an accumulation of a range of events including direct enzyme inactivation by radicals, to a loss of cellular compartmentation and functional organization.

It is evident that when excessive levels of superoxide are generated, cellular disruption and plant death will follow. On the other hand, a number of investigations have indicated that if the oxidative stress is less severe and cellular integrity is not destroyed, there will be an increase in scavenging enzymes. This was demonstrated in experiments with pea leaf discs for both ascorbate peroxidase and glutathione reductase[16] (Figure 2). A number of workers have shown enhanced levels of superoxide dismutase, for example,

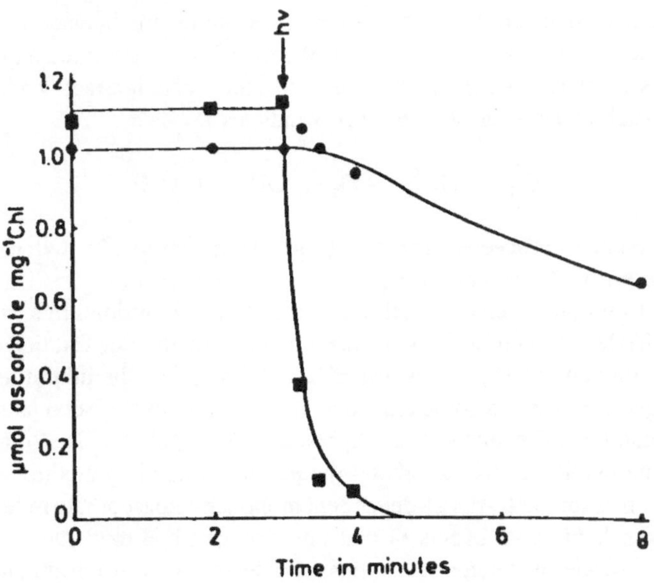

FIGURE 1. The effect of paraquat on the ascorbate content of intact spinach chloroplasts during illumination. The ascorbate content of intact chloroplasts was measured in the absence (●) and presence (■) of 0.2 m*M* paraquat. (From Foyer, C., Rowell, J., and Walker, D. A., *Planta*, 157, 239, 1983. With permission.)

in bean leaves,[17] *Chlorella sorokiniana,*[18] and maize leaves.[19] In the latter instance, a 40% increase in superoxide dismutase was induced with a lesser increase in catalase. The enzyme increase was related to enhanced expression of *Sod* genes, and this response was not specific to paraquat, but, rather, to the level of superoxide generated. In recent experiments with *Nicotiana plumbaginifolia,* Tsang et al.[20] showed that stress imposed upon leaves by paraquat and light led to an enhancement of superoxide dismutase mRNA in both chloroplastic and extrachloroplastic cellular compartments. This indicates that the cellular stress imposed by this herbicide was widespread, and appropriate signaling mechanisms led to enhanced protective systems. It is not surprising, therefore, that tolerance to one form of oxidative stress is apparent with other chemically unrelated stresses, such as paraquat, SO_2, and ozone. A paraquat-tolerant strain of *Conyza bonariensis* was also more tolerant to SO_2,[21] and this appeared to correlate with higher levels of superoxide dismutase, ascorbate peroxidase, and glutathione reductase. *Lolium* strains that were tolerant to paraquat and SO_2 possessed 70% more superoxide dismutase and 20% more glutathione reductase than the sensitive strain.[21] Cross tolerance between ozone and paraquat was demonstrated in a *N. tabacum* cultivar,[21] and in mung bean leaves, where a preincubation to ethylene enhanced the level of ascorbate peroxidase, which provided a greater protection to both toxic agents.[23] In the

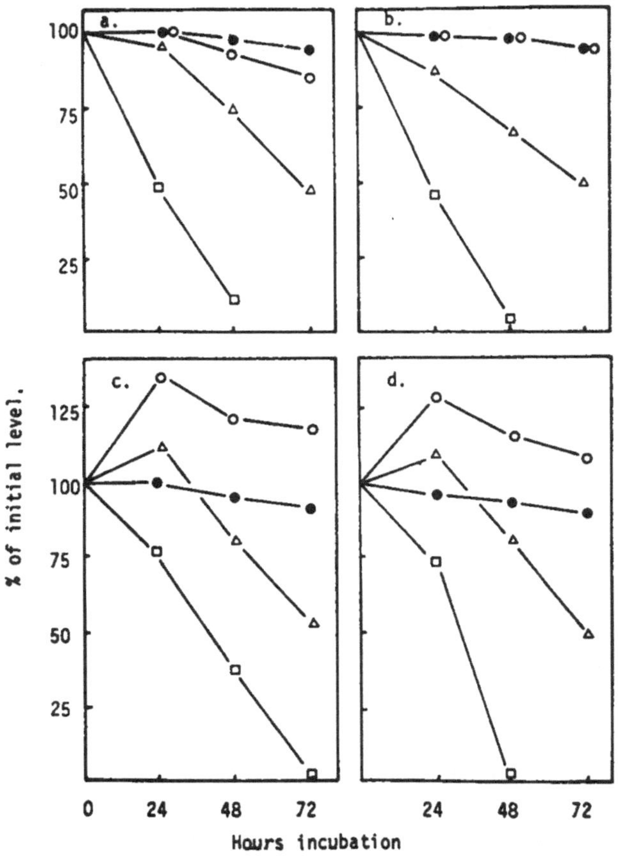

FIGURE 2. The effect of paraquat at 0.1 μM (○), 1 μM (△), or 10 μM (□) compared with a water control (●) on ascorbate (a), superoxide dismutase (b), ascorbate peroxidase (c), and glutathione reductase (d) in illuminated pea leaf discs. (From Gillham, D. J. Aspects of Chloroplast Protection Against Photo-Oxidative Damage, Ph.D. thesis, University of Bath, Bath, England, 1986.)

alga *Chlorella sorokiniana,* sulfite treatment enhanced superoxide dismutase, and this resulted in increased tolerance to paraquat.[24]

Attempts to restrict the action of paraquat by the addition of exogenous scavengers have met with limited success. The pseudo superoxide dismutase scavenger, copper penicillamine, retarded the action of paraquat as measured by pigment breakdown[25] or ethane generation,[26] a product of lipid peroxidation in pea leaves, while in *Ipomoea purpurea* cotyledons, the addition of ascorbate and the ascorbate precursor galactonolactone restrained the action of paraquat.[27] On the other hand, addition to pea leaves of the superoxide dismutase inhibitor diethyl dithiocarbamate (DDTC) (Figure 3) enhanced the action of the herbicide.

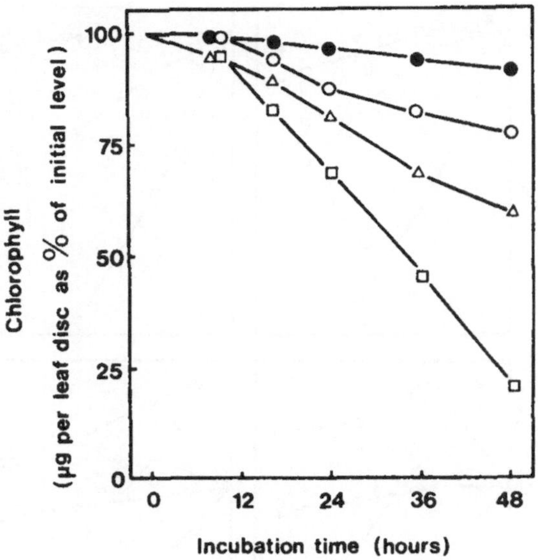

FIGURE 3. Chlorophyll bleaching in pea leaf discs incubated on water (●), 1 μ*M* DDTC (○), 1 μ*M* paraquat (△), or DDTC plus paraquat (□). (From Dodge, A. D., in *Herbicides and Plant Metabolism*, Dodge, A. D., Ed., Cambridge University Press, Cambridge, 1989, 37. With permission.)

Variations in the activity of bipyridinium herbicides might in some instances be related to major variations in the endogenous level of scavenging enzymes that could be linked to leaf age,[29] environmental treatment, or seasonal changes.[30] High light incubation of pea leaves led to an enhanced level of ascorbate peroxidase, glutathione reductase, dehydroascorbate reductase, and ascorbate, whereas there was little change in both superoxide dismutase and glutathione.[30] Thus, leaves with enhanced scavenging were more tolerant to paraquat (Figure 4). Although some paraquat-tolerant weed biotypes such as *Conyza*[22] and *Erigeron*[31] may possess enhanced levels of some scavenging systems, in many instances tolerance mechanisms are related to a failure of the herbicide to reach the target site.

The techniques of molecular biology have now shown the possibility of transferring desirable traits into crop plants. A major ambition in the field of crop protection is to enhance crop tolerance to herbicides. Considerable work has shown that the endogenous level of glutathione and glutathione-*S*-transferase can be enhanced by the addition of a so-called herbicide safener, such as *N,N*-diallyl-2,2-dichloroacetamide. This may be used in maize in conjunction with the lipid biosynthesis inhibitor thiocarbamate herbicide, EPTC. At present, no commercial safener has been devised to use with a free radical- or singlet oxygen-promoting herbicide. The alternative strategy is to transfer genes to crop plants that will enhance scavenging enzymes or scavengers.

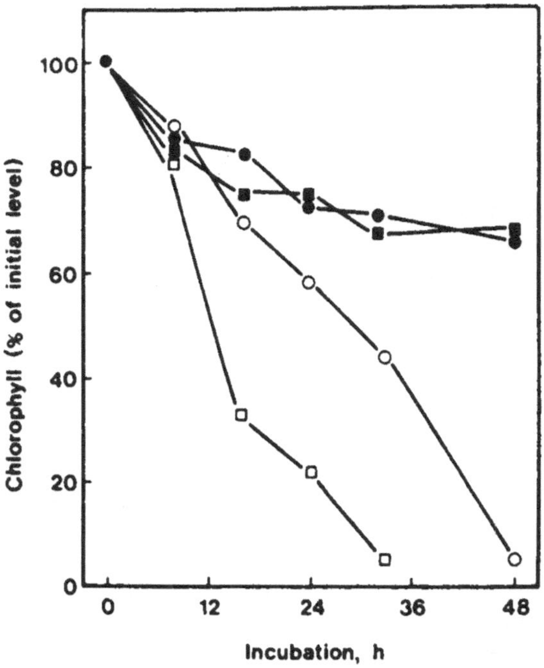

FIGURE 4. Paraquat-induced chlorophyll bleaching in pea leaf discs cut from plants grown in a constant environment chamber at 100 (■ □) or 400 (● ○) μmol m^{-2} s^{-1} photon flux density. Discs were then incubated in petri dishes containing water (■ ●) or 10 μM paraquat (□ ○). (From Gillham, D. J. and Dodge, A. D., *Plant Sci.*, 50, 105, 1987. With permission.)

Tepperman and Dunsmuir[33] were able to genetically engineer petunia plants with a 50 times higher level of superoxide dismutase; however, these plants were not more resistant to paraquat. *N. tabacum* plants were transformed using a bacterial gene that coded for glutathione reductase, and a two- to tenfold higher level of this enzyme was achieved. In the work of Aono et al.,[35] the transgenic plants were more tolerant to paraquat than control plants, but not more tolerant to ozone. In general, however, it would seem unlikely that genetically engineered plants would be more tolerant to oxidative-type herbicides unless a range of scavenging enzymes and scavengers are transferred;[34,36] otherwise, as in the case of abundantly enhanced superoxide dismutase,[33] the increased production of unscavenged H_2O_2 could, without adequately enhanced scavenging, promote the Fenton-type production of the more damaging hydroxyl radical.

III. HERBICIDE-INDUCED SINGLET OXYGEN GENERATION

Whereas superoxide is formed by the donation of electrons to oxygen via a redox intermediate, whether an endogenous chloroplast electron transport component or an artificial mediator such as a bipyridyl herbicide, singlet oxygen ($^1O_2$) is generated by the excitation of triplet (ground-state) oxygen by a triplet sensitizer. As the active, light-absorbing, photosynthesizing chloroplast possesses both sensitizer pigments (chlorophylls) and oxygen in abundance, the potential for $^1O_2$ formation is considerable. The fact that chloroplasts function under "normal" environmental conditions, without apparent damage, is a tribute to the efficacy of their protective systems. When environmental conditions change so that, e.g., high light is accompanied by drought, not only will photoinhibition occur, but progressively photodynamic damage will become evident as chloroplast bleaching. In such conditions, chlorophyll is excited to the singlet state (^{1}Chl), but if the excitation is not utilized by resonance transfer or released as fluorescence, intersystem crossing will occur, leading to the longer-lived triplet state (^{3}Chl). Interaction of triplet oxygen ($^3O_2$) with ^{3}Chl will, as previously noted, generate $^1O_2$. Damage to chloroplasts may be induced by type I reactions in which the triplet sensitizer reacts directly with biomolecules. In type II reactions, $^1O_2$ reacts with membrane unsaturated fatty acids, yielding hydroperoxides, and consequently membrane damage and cellular destruction will follow. As with free radical attack, certain amino acids such as histidine and methionine are particularly sensitive to oxidation, leading to protein damage and enzyme inactivation. Although it has not been possible to experimentally demonstrate the generation of $^1O_2$ *in vivo,* experiments with isolated chloroplast thylakoids,[37] or with whole leaves and scavenging systems, suggest that such a deleterious series of events will occur under appropriate stress conditions.

$^1O_2$ generation by herbicides is by (1) photosynthetic electron transport inhibitors, (2) herbicides that promote the formation of an excessive level of photosensitizers by either enzyme inhibition or proherbicide conversion, and (3) the inhibition of $^1O_2$ protective systems (see Section IV below).

A wide range of chemically diverse herbicides, including the phenylureas, triazines, and hydroxybenzonitriles, interact with the D1 polypeptide of the reaction center of photosystem II. This leads to the displacement of the redox quinone Q_B and thus the cessation of electron flow between Q_A and the plastoquinone pool. Plants in the field will continue to receive light energy, but, because electron transport is prevented, will be unable to safely dissipate the excitation energy. An increase in fluorescence emission is indicative of energy excess. The inevitable consequence is that ^{3}Chl and hence $^1O_2$ are generated, and cellular damage and death will ensue. This has been indicated experimentally by the fact that both increased illumination and the presence of oxygen enhance herbicidal action, whereas action is limited to an extent by the use of exogenous scavengers.[38]

In experiments using whole leaves, the sequence of pigment deterioration commenced with the carotenoids and was followed by the chlorophylls. Carotenoid pigments (see Section IV) are of considerable importance *in vivo* in quenching both ^{3}Chl and $^1O_2$ (Figure 5).[38] It is not surprising, therefore, that

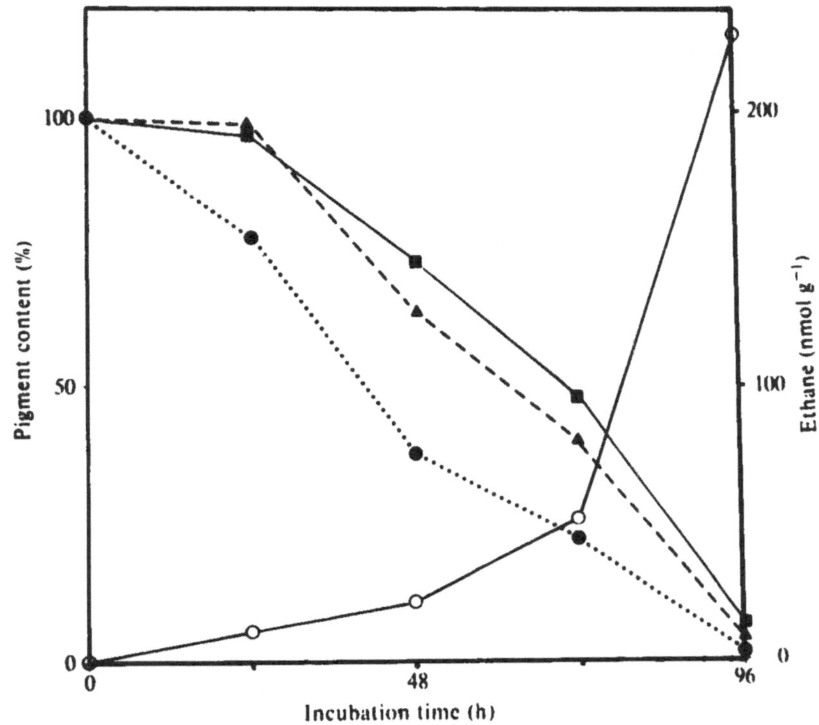

FIGURE 5. The effect of 1×10^{-3} monuron on pigment content and ethane generation of flax cotyledon leaves incubated at 30 Wm^{-2}. Pigment levels as a percentage of the levels at zero h: (●) α- and β-carotene, (▲) lutein, zeaxanthin, violaxanthin, and neoxanthin, (■) total chlorophyll, (○) ethane generation, a measure of lipid peroxidation. (From Pallett, K. E. and Dodge, A. D., *J. Exp. Bot.*, 31, 1051, 1980. With permission.)

a recently utilized herbicidal mixture, "panther", contains both a photosynthetic inhibitor, isoproturon, and a carotenoid biosynthesis inhibitor, diflufenican.

Thus, with photosynthetic inhibitor herbicides, cellular disruption and death appear to result from misdirected excitation energy rather than starvation. They rely upon light absorption by preexisting thylakoid pigments. With some of the dinitrophenol ether herbicides (DPE), particularly the *p*-nitro compounds acifluorfen and oxyfluorfen, the accumulation of a photodynamically active chlorophyll precursor is the toxic agent.

It had been known for many years that the action of DPE herbicides was promoted by light, and several ideas were formulated for action. Recently, Matringe and Scalla[39] showed that DPE action was due to the accumulation of protoporphyrin IX by the inhibition of the enzyme protoporphyrinogen oxidase. Although protoporphyrin IX is the product of this reaction, accumulation is suggested to proceed via a nonenzymic oxidation of protoporphyrinogen IX to protoporphyrin IX that promotes the generation of $^1O_2$.[40]

Although protoporphyrin IX is a normal intermediate of chlorophyll biosynthesis, with DPE treatment there is an excessive production of this compound away from the cellular site of subsequent conversion by magnesium chelatase to mg-protoporphyrin; hence, it is not organized into chlorophyll within the chloroplast thylakoid.

A number of studies have shown that there is a direct relationship between protoporphyrin IX accumulation and a herbicidal effect, as measured by the release of electrolytes as an indicator of cellular damage.[41] There is a concomitant consumption of oxygen and generation of $^1O_2$.[42] With *Scenedesmus* cells treated with the DPE herbicide oxyfluorfen, Kunert and Böger[43] were able to limit herbicidal damage, as measured by pigment breakdown or ethane loss, by application of the antioxidants ethoxyquin and α-tocopherol. Furthermore, the ability of α-tocopherol to limit oxyfluorfen action in the leaves of a number of higher plants was related to an optimum ascorbate-to-α-tocopherol ratio of 10 to 15:1 (wt/wt).[44] There was also a 40% higher level of ascorbate in tobacco biotypes that were tolerant to the DPE herbicides, acifluorfen and bifenox.[45]

In experiments with cucumber cotyledons, Kenyon and Duke[46] showed that acifluorfen caused a rapid decrease in ascorbate and glutathione content after light incubation, with a 50% decrease within 1 h (Figure 6). These experiments together indicate the effect of $^1O_2$ on cellular antioxidants and the importance of ascorbate not only as a scavenger of $^1O_2$ and free radicals in its own right, but also in the regeneration of the membrane-located scavenger α-tocopherol.

With stress imposed within plant cells by oxygen radicals, there was evidence of induction of some scavenging enzymes (see Section II above). With the DPE herbicide acifluorfen, Gullner and Dodge[47] showed that the level of pea leaf glutathione was enhanced if illuminated in the presence of subtoxic concentrations of acifluorfen (Figure 7).

Work with DPE herbicides has indicated the toxic potential of chlorophyll precursors to catalyze photodynamic reactions. Another approach that shows considerable potential is to use as a proherbicide, a molecule that will be converted by the metabolism of the plant to an active photosensitizer. Rebeiz and co-workers[48,49] introduced this concept by treating plants with δ-aminolevulinic acid (ALA). If application was followed by a period of darkness, ALA was converted to protochlorophyllide. On subsequent illumination, protochlorophyllide acted as a photosensitizer, generating $^1O_2$ and leading to rapid bleaching and plant death. (Figure 8). Further work showed that there were considerable species differences in the susceptibility of plants to ALA treatment, and that the effect could be modulated by the addition of chelators such as 2.2^1-dipyridyl.[49] Interestingly, if ALA is used in conjunction with a DPE herbicide such as acifluorfen, there is a reduction in the conversion of this precursor to protochlorphyllide.[50] As DPE herbicide treatment was shown to enhance leaf glutathione levels, experiments with similar leaves and ALA

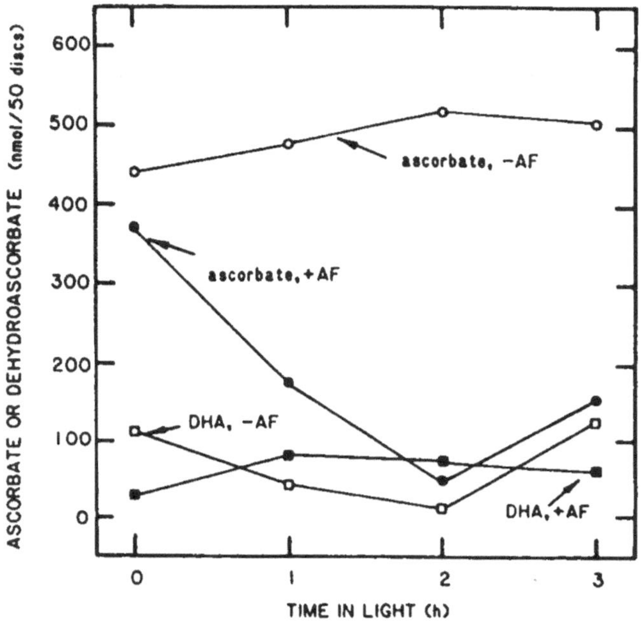

FIGURE 6. The effect of acifluorfen (AF) on the ascorbate or dehydroascorbate content (DHA) of green cucumber cotyledons (From Kenyon, W. H. and Duke, S. O., *Plant Physiol.*, 79, 862, 1985. With permission.)

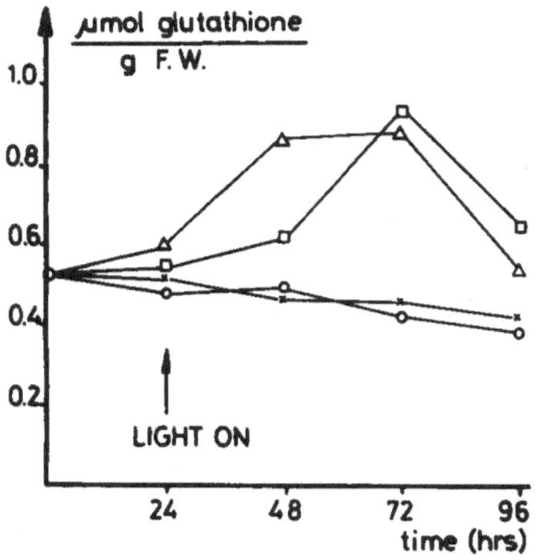

FIGURE 7. The effect of acifluorfen on the total glutathione level of pea leaf discs. Discs were incubated in darkness for 24 h, then exposed to light of 400 μM m^{-2} s^{-1}. Water control (○), 1 × 10^{-4} *M* acifluorfen (△), 1 × 10^{-5} *M* acifluorfen (□), 1 × 10^{-6} *M* acifluorfen (x). (From Gullner, G. and Dodge, A. D., in press.)

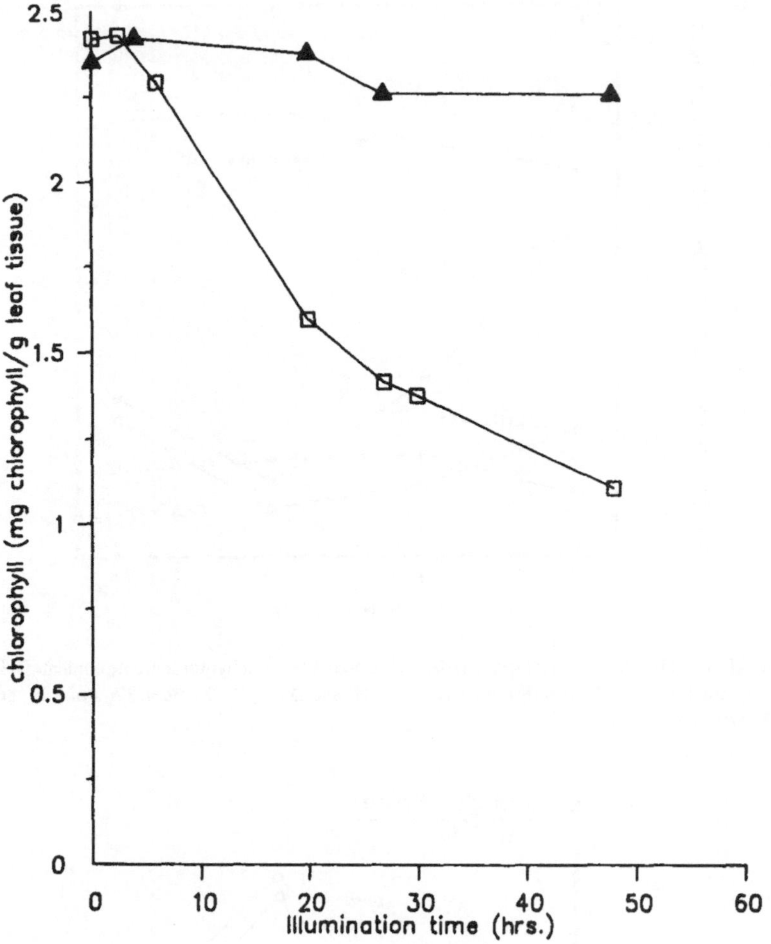

FIGURE 8. The effect of δ-amino levulinic acid (ALA) on the chlorophyll content of pea leaf discs. Discs were incubated in darkness for 12 h, then exposed to illumination of 400 μM m^{-2} s^{-1}. Leaf discs were incubated throughout on either water (▲) or 5 m*M* ALA (□). (From Herbert, D. and Dodge, A. D., unpublished data.)

feeding showed that, on illumination, there was also a considerable increase in leaf glutathione[47] (Figure 9).

As photosensitizers are effective phytotoxic agents, it is surprising that little work has been performed in the development of exogenously applied photosensitizers as herbicides.[51] This strategy is used by some plants as an effective defense system.[52] Some studies have shown that various xanthene dyes, such as rose bengal and eosin,[54] have a herbicidal effect that is promoted by both light and oxygen. Photodynamic damage was shown as pigment bleaching, and this correlated with lipid peroxidation.[55] Subtoxic

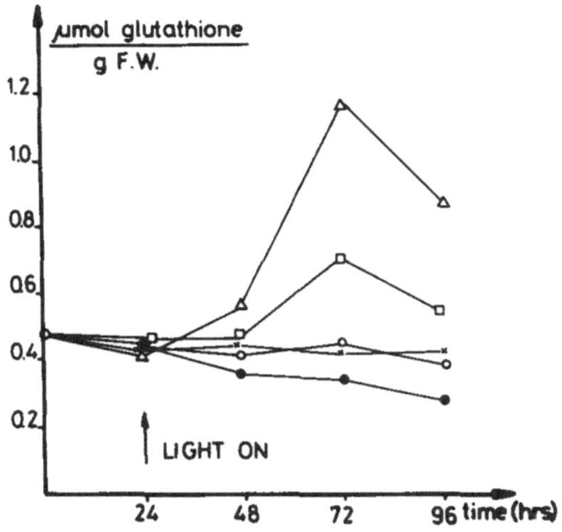

FIGURE 9. The effect of δ-amino levulinic acid (ALA) on the total glutathione level of pea leaf discs under illumination of 400 $\mu M\ m^{-2}\ s^{-1}$ after a 24-h dark preincubation. Water control (○); water control in continuous darkness (●); with the addition of ALA at 0.1 mM (x), 1 mM (□), or 3 mM (△) (From Gullner, G. and Dodge, A. D., in press.)

concentrations of some exogenously applied photosensitizers also enhance the level of leaf glutathione.[47]

IV. HERBICIDE-INDUCED INHIBITION OF ANTIOXIDANT ACTION AND FORMATION

During normal photosynthesis in C_3 plants, an appreciable amount of carbon flow proceeds through the photorespiratory pathway. The generation of glycolate within the chloroplast is followed by export to the peroxisomes, where glycolate oxidase catalyzes the conversion to glyoxylate. In addition to glyoxylate, the reaction also produces hydrogen peroxide. This is removed by the peroxisome-based enzyme, catalase. The herbicide aminotriazole has been used as a total-kill weedkiller for over 30 years, and after application leads to distinctive bleaching of leaf tissue. Although this herbicide may possess multiple sites of operation, its primary action is as a peroxisome catalase inhibitor.[56] It is of considerable interest that barley mutants that contained only 10% of the normal level of catalase were bleached rapidly in normal air (340 $\mu l\ l^{-1}\ CO_2$), but little affected in CO_2-enriched air where photorespiration was nonfunctional (5000 $\mu l\ l^{-1}\ CO_2$).[57] It is assumed that the accumulation of hydrogen peroxide will lead not only to direct enzyme inhibition, but also to the formation of hydroxyl radicals. In barley catalase less mutants and aminotriazole-treated plants, there was evidence of an enhanced level of leaf glutathione.[58,59] It was suggested that hydrogen peroxide oxidizes reduced glutathione and that this was a signal for enhanced glutathione synthesis.[60]

The role of the carotenoid pigments as scavengers of both $^1O_2$ and triplet chlorophyll has already been mentioned. It has been known for many years that carotenoidless mutants are extremely susceptible to photodynamic damage in the presence of oxygen.[61] Inhibitors of carotenoid biosynthesis are therefore effective nonselective herbicides. A number of pyridazinone compounds such as metflurazon and norflurazon, and others such as fluridone, difunone, and dichlormate, specifically inhibit phytoene desaturase reactions.[62] The intermediates that accumulate, phytoene and, to a lesser extent, phytofluene, are not sufficiently unsaturated to be effective as photoprotectants.[63] When such herbicides are used, there is relatively little bleaching of preexisting leaves, but major damage to developing leaves. These become completely chlorotic, as the lack of carotene biosynthesis during chloroplast development in such leaves renders them vulnerable to photodynamic damage. In the early stages of chloroplast biogenesis, a high carotenoid-to-chlorophyll ratio is of vital importance.[64]

In addition to a role in photoprotection, carotenoid pigments also function as accessory pigments in the light-harvesting apparatus. Exactly how the two functional roles are divided between the range of carotenoid pigments has not been clearly established. β-Carotene appears to be predominantly involved in light harvesting, and a number of xanthophyll pigments in photoprotection. The role of zeaxanthin in a linked cycle with violaxanthin appears to be of major importance (see Chapter 4 of this volume).[65] This so-called xanthophyll cycle is dependent upon the de-epoxidation of violaxanthin to a zeaxanthin via de-epoxidation reactions. The reactions are essential to maintain the level of the photoprotecting zeaxanthin and appear to utilize ascorbate as a reductant.[12] Therefore, the maintenance of adequate photoprotection relies not only on the presence of carotenoid pigments, but also on the level of individual components and on photosynthetic electron transport to maintain the reduction of $NADP^+$ for monodehydroascorbate reductase. It is not surprising, therefore, that photodynamic damage to leaves, promoted by photosynthetic electron transport inhibitor herbicides, is seen preferentially as carotenoid loss (Figure 5).

V. CONCLUSIONS

This review has shown that plant survival is dependent upon the adequate control of free radicals and singlet oxygen. It is essential that the chloroplast is well endowed with protective systems because of the close juxtaposition of excited pigments and oxygen. The use of herbicides pushes such protective systems to the limit, and they ultimately fail.

It has been shown that plants can respond to various chemical and environmental stresses by the induction or modulation of various scavengers.

Attempts to genetically engineer or chemically induce changes in protective systems of crop plants must recognize that our present understanding of such systems is incomplete. This is clearly demonstrated in the case of petunia plants with a 50-fold higher level of superoxide dismutase not being more resistant to paraquat.[33] Furthermore, modulation of the level of an apparently ubiquitous protectant, such as ascorbate, should take into consideration the fact that excess reducing capacity within the cell can maintain Fe^{2+} and, in so doing, enhance damage via the Fenton-type generation of hydroxyl radicals. An increased understanding of plant scavenging and photoprotective systems is not only a challenge to the academic scientist, but could also provide the basis for the site-directed synthesis of new herbicidal molecules.

REFERENCES

1. **Fujii, T., Yokoyama, E., Inoue, K., and Sakurai, H.,** The sites of electron donation of photosystem I to methyl viologen, *Biochim. Biophys. Acta,* 1015, 41, 1990.
2. **Cuendet P. and Gratzel, M.,** Interaction of hydrophilic and amphiphilic viologens with the thylakoid surface, *Photobiochem. Photobiophys.,* 5, 335, 1983.
3. **Mees, G. C.,** Experiments on the herbidical action of 1,1′-ethylene-2,2′-dipyridylium dibromide, *Ann. Appl. Biol.,* 48, 601, 1960.
4. **Farrington, J. A., Ebert, M., Land, E. J., and Fletcher, K.,** Bipyridylium quaternary salts and related compounds. V. Pulse radiolysis studies of the reaction of paraquat radical with oxygen. Implications for the mode of action of bipyridyl herbicides, *Biochim. Biophys. Acta,* 314, 372, 1973.
5. **Asada, K., Takahashi, M.-A., Tanaka, K., and Nakano, Y.,** Formation of active oxygen and its fate in chloroplasts, in *Biochemical and Medical Aspects of Active Oxygen.* Hayaishi, O. and Asada, K., Eds., University Park Press, Baltimore, 1977, 45.
6. **Suito, M., Thomas, C. E., and Aust, S. D.,** Paraquat and ferritin-dependent lipid peroxidation, *J. Free Rad. Biol. Med.,* 1, 179, 1985.
7. **Harris, N. and Dodge, A. D.,** The effect of paraquat on flax cotyledon leaves: physiological and biochemical changes, *Planta,* 104, 210, 1972.
8. **Kaiser, W.,** Reversible inhibition of the Calvin cycle and the activation of the pentose phosphate pathway in isolated chloroplasts by hydrogen peroxide, *Planta,* 145, 377, 1979.
9. **Heldt, H. W., Chon, C. J., Lilley, R. M., and Portis, A.,** The role of fructose and sedoheptulose bisphosphatase in the control of CO_2 fixation. Evidence for the effects of Mg^{2+} concentration, pH and H_2O_2, in *Photosynthesis 1977,* Hall, D. O., Coombs, J., and Goodwin, T. W., Eds., Biochemical Society, London, 1978, 469.
10. **Asada K. and Takahashi, M.,** Production and scavenging of active oxygen in photosynthesis, in *Photoinhibition,* Osmond, C. B. and Arntzen, C. J., Eds., Elsevier, Amsterdam, 1987, 227.
11. **Babbs, C. F., Pham, J. A., and Coolbaugh, R. C.,** Lethal hydroxyl radical production in paraquat-treated plants, *Plant Physiol.,* 90, 1267, 1989.
12. **Foyer, C. H., Lelandais, M., Edwards, E. A., and Mullineaux, P. M.,** The role of ascorbate in plants, interactions with photosynthesis, and regulatory significance, in *Active Oxygen/Oxidative Stress and Plant Metabolism,* Pell, E. and Steffen, K., Eds., American Society of Plant Physiologists, Rockville, MD, 1991, 131.

13. **Alscher, R. G.,** Biosynthesis and antioxidant function of glutathione in plants, *Physiol. Plant.*, 77, 457, 1989.
14. **Law, M. Y., Charles, S. A., and Halliwell, B.,** Glutathione and ascorbic acid in spinach *(Spinacia oleracea)* chloroplasts. The effect of hydrogen peroxide and paraquat, *Biochem. J.*, 210, 899, 1983.
15. **Foyer, C., Rowell, J., and Walker, D. A.,** Measurement of the ascorbate content of spinach leaf protoplasts and chloroplasts during illumination, *Planta*, 157, 239, 1983.
16. **Gillham, D. J.,** Aspects of Chloroplast Protection Against Photo-Oxidative Damage, Ph.D. thesis, University of Bath, Bath, England, 1986.
17. **Chia, L. S., McRae, D. G., and Thompson, J. E.,** Light-dependence of paraquat-initiated membrane deterioration in bean plants. Evidence for the involvement of superoxide, *Physiol. Plant.*, 56, 492, 1982.
18. **Robinowitch, H. D., Clare, D. A., Crapo, J. D., and Fridovich I.,** Positive correlation between superoxide dismutase and resistance to paraquat toxicity in the green algae *Chlorella sorokiniana, Arch. Biochem. Biophys.*, 225, 640, 1983.
19. **Matters, G. L. and Scandalios, J. G.,** Effect of the free radical-generating herbicide paraquat on the expression of the superoxide dismutase *(Sod)* genes in maize, *Biochim. Biophys. Acta*, 882, 29, 1986.
20. **Tsang, E. W. T., Bowler, C., Hérouart, D., Van Camp, W., Villarroel, R., Genetello, C., Van Montagu, M., and Inzé, D.,** Differential regulation of superoxide dismutases in plants exposed to environmental stress, *Plant Cell*, 3, 783, 1991.
21. **Shaaltiel, Y., Glazer A., Bocion, P. F., and Gressel, J.,** Cross tolerance to herbicidal and environmental oxidants of plant biotypes tolerant to paraquat, sulfur dioxide and ozone, *Pest. Biochem. Physiol.*, 31, 13, 1988.
22. **Shaaltiel, Y. and Gressel, J.,** Multienzyme oxygen radical detoxification system correlated with paraquat resistance in *Conyza bonariensis, Pest. Biochem. Physiol.*, 26, 22, 1986.
23. **Mehlhorn, H.,** Ethylene-promoted ascorbate peroxidase activity protects plants against hydrogen peroxide, ozone and paraquat, *Plant Cell Environ.*, 13, 971, 1990.
24. **Rabinowitch, H. D. and Fridovich, I.,** Sulfur dioxide and paraquat: dual protection by superoxide dismutase in *Chlorella sorokinana*, in *Oxy Radicals and their Scavenger Systems*, Vol. 1, Cohen, G. and Greenwald, R. A., Eds., Elsevier, New York, 1983, 222.
25. **Youngman, R. J. and Dodge, A. D.,** Mechanism of paraquat action: inhibition of the herbicidal effect by a copper chelate with superoxide dismutating activity, *Z. Naturforsch. Teil C*, 34, 1032, 1979.
26. **Youngman, R. J., Dodge, A. D., Lengfelder, E., and Elstner, E. F.,** Inhibition of paraquat phytoxicity by a novel copper chelate with superoxide dismutasing activity, *Experientia*, 35, 1295, 1979.
27. **Schmidt, A. and Kunert, K. J.,** Antioxidative systems: defense against oxidative damage in plants, in *Molecular Strategies for Crop Protection*, Vol. 48, UCLA Symp. Mol. Cell Biol., Arntzen, C. and Ryan, C., Eds., Alan R. Liss, New York, 1987, 401.
28. **Dodge, A. D.,** Herbicides interacting with photosystem I, in *Herbicides and Plant Metabolism*, Dodge, A. D., Ed., Cambridge University Press, Cambridge, 1989, 37.
29. **Lee, E. H. and Bennett, J. H.,** Superoxide dismutase. A possible protective enzyme against ozone injury in snap beans (*Phaseolus vulgaris* L.), *Plant Physiol.*, 69, 1444, 1982.
30. **Gillham, D. J. and Dodge, A. D.,** Chloroplast superoxide and hydrogen peroxide scavenging systems from pea leaves: seasonal variations, *Plant Sci.*, 50, 105, 1987.
31. **Matsunaka, S. and Ito, K.,** Paraquat resistance in Japan, in *Herbicide Resistance in Weeds and Crops*, Caseley, J. C., Cussans, G. W., and Atkin, R. K., Eds., Butterworth-Heinemann, Oxford, 1991, 77.

32. **Dodge, A. D.,** Mechanisms of paraquat tolerance, in *Herbicide Resistance in Weeds and Crops,* Caseley, J. C., Cussans, G. W., and Atkin, R. K., Eds., Butterworth-Heinemann, Oxford, 1991, 165.
33. **Tepperman, J. M. and Dunsmuir, P.,** Transformed plants with elevated levels of chloroplast superoxide dismutase are not more resistant to superoxide toxicity, *Plant Mol. Biol.,* 14, 501, 1990.
34. **Foyer, C., Lelandais, M., Galap, C., and Kunert, K. J.,** Effects of elevated cytosolic glutathione reductase activity on the cellular glutathione pool and photosynthesis in leaves under normal and stress conditions, *Plant Physiol.,* 97, 863, 1991.
35. **Aono, M., Kubo, A., Saji, H., Natori, T., Tanaka, K., and Kondo, N.,** Resistance to active oxygen toxicity of transgenic *Nicotiana tabacum* that expresses the gene for glutathione reductase from *Escherichia coli, Plant Cell Physiol.,* 32, 691, 1991.
36. **Slooten, L., Bowler, C., Sybesma, C., Van Montagu, M., and Inzé, D.,** Improvement of plant resistance to oxygen radical stress: the effect of over expression of manganese superoxide dismutase, in *Active Oxygen/Oxidative Stress and Plant Metabolism,* Pell, E. and Steffen, K., Eds., American Society of Plant Physiologists, Rockville, MD, 1991, 256.
37. **Percival, M. P. and Dodge, A. D.,** Photodynamic damage to isolated chloroplasts: a possible model for *in vivo* effects of photosynthetic inhibitor herbicides, *Z. Naturforsch. Teil C,* 39, 482, 1984.
38. **Pallett, K. E. and Dodge, A. D.,** Studies into the action of some photosynthetic inhibitor herbicides, *J. Exp. Bot.,* 31, 1051, 1980.
39. **Matringe, M. and Scalla, R.,** Studies on the mode of action of acifluorfen-methyl in nonchlorophyllous soybean cells. Accumulation of tetrapyroles, *Plant Physiol.,* 86, 619, 1988.
40. **Becerril, J. M. and Duke, S. O.,** Protoporphyrin IX content correlates with activity of photobleaching herbicides, *Plant Physiol.,* 90, 1175, 1989.
41. **Duke, S. O., Becerril, J. M., Sherman, T. D., Lydon, J., and Matsumoto, H.,** The role of photoporphyrin IX in the mechanism of action of diphenyl ether herbicides, *Pest. Sci.,* 30, 367, 1990.
42. **Haworth, P. and Hess, H.,** The generation of singlet oxygen by the nitrodyphenyl ether herbicide oxyfluorfen is independent of photosynthesis, *Plant Physiol.,* 86, 672, 1988.
43. **Kunert, K. J. and Böger, P.,** The diphenyl-ether herbicide oxyfluorfen: action of antioxidants, *J. Agric. Food Chem.,* 32, 725, 1984.
44. **Finckh, B. F. and Kunert, K. J.,** Vitamins C and E: an antioxidative system against herbicide-induced lipid peroxidation in higher plants, *J. Agric. Food Chem.,* 33, 574, 1985.
45. **Gullner, G., Kömíves, T., and Király, L.,** Enhanced inducibility of antioxidant systems in a *Nicotiana tabacum* L. biotype results in acifluorfen resistance, *Z. Naturforsch. Teil C,* 46, 875, 1991.
46. **Kenyon, W. H. and Duke, S. O.,** Effects of acifluorfen on endogenous antioxidants and protective enzymes in cucumber (*Cucumis sativus* L.) cotyledons, *Plant Physiol.,* 79, 862, 1985.
47. **Gullner, G. and Dodge, A. D.,** Effect of singlet oxygen generating substances on the ascorbate and glutathione content in pea leaves, in press.
48. **Rebeiz, C. A., Montazer-zouhoor, A., Hopen, H. J., and Wu, S. M.,** Photodynamic herbicides. I. Concept and phenomenology, *Enzyme Microbiol. Technol.,* 6, 390, 1984.
49. **Rebeiz, C. A., Montazer-zouhoor, A., Mayasich, J. M., Tripathy, B. C., Wu, S. M., and Rebeiz, C. C.,** Photodynamic herbicides and chlorophyll biosynthesis modulators, in *Light Activated Pesticides,* Heitz, R. J. and Downum, K. R., Eds., American Chemical Society, Washington, D. C., 1987, 295.
50. **Herbert, D. and Dodge, A. D.,** unpublished data, 1991.

51. **Zweig, A. and Nachtigall, G. W.,** Photosensitized herbicidal action, *Photochem. Photobiol.*, 22, 257, 1975.
52. **Knox, J. P. and Dodge, A. D.,** Isolation and activity of the photodynamic pigment hypericin, *Plant Cell Environ.*, 8, 19, 1985.
53. **Percival, M. P. and Dodge, A. D.,** Photodynamic effects of rose bengal on senescent flax cotyledons, *J. Exp. Bot.*, 34, 47, 1983.
54. **Knox, J. P. and Dodge, A. D.,** The photodynamic action of eosin, a singlet-oxygen generator. Some effects on leaf tissue of *Pisum sativum*, *Planta*, 164, 22, 1985.
55. **Dodge, A. D. and Gillham, D. J.,** Methods for assessing the action of light activated herbicides. I. The assessment of lipid peroxidation, *Aspects Appl. Biol.*, 11, 97, 1986.
56. **Margoliash, E. and Novogradsky, A.,** Irreversible reaction of 3-amino-1,2,4-triazole and related inhibitors with the protein of catalase, *Biochem J.*, 74, 339, 1960.
57. **Parker, M. L. and Lea, P. J.,** Ultrastructure of the mesophyll cells of leaves of a catalase-deficient mutant of barley (*Hordeum vulgare* L). *Planta*, 159, 512, 1983.
58. **Smith, I. K., Kendall, A. C., Keys, A. J., Turner, J. C., and Lea, P. J.,** Increased levels of glutathione in a catalase-deficient mutant of barley (*Hordeum vulgare* L), *Plant Sci. Lett.*, 37, 29, 1984.
59. **Smith, I. K.,** Stimulation of glutathione synthesis in photorespiring plants by catalase inhibitors, *Plant Physiol.*, 79, 1044, 1985.
60. **Smith, I. K., Kendall, A. C., Keys, A. J., Turner, J. C., and Lea, P. J.,** The regulation of the biosynthesis of glutathione in leaves of barley (*Hordeum vulgare* L.), *Plant Sci.*, 41, 11, 1985.
61. **Anderson, I. C. and Robertson, D. S.,** The role of carotenoids in protecting chlorophyll from photodestruction, *Plant Physiol.*, 35, 531, 1960.
62. **Britton, G., Barry, P., and Young, A. J.,** Carotenoids and chlorophylls: herbicidal inhibition of pigment biosynthesis, in *Herbicides and Plant Metabolism*, Dodge, A. D., Ed., Cambridge University Press, Cambridge, 1989, 51.
63. **Krinsky, N. I.,** Function, in *Carotenoids*, Isler, O., Ed., Birkhäuser, Basel, 1971, 669.
64. **Gillham, D. J. and Dodge, A. D.,** Chloroplast protection in greening leaves, *Physiol. Plant.*, 65, 393, 1985.
65. **Demmig-Adams, B.,** Carotenoids and photoprotection in plants. A role for the xanthophyll zeaxanthin, *Biochim. Biophys. Acta*, 1020, 1, 1990.

Chapter 10

GENETIC CONTROLS OF PHOTOOXIDANT TOLERANCE

Jonathan Gressel and Esra Galun

TABLE OF CONTENTS

0-8493-5443-9/94/$0.00 + $.50

I. THE POWERS OF CLASSICAL AND MOLECULAR GENETICS

A plethora of information has been presented in previous chapters on systems that detoxify or quench photochemically produced active oxygen species, and the sensitivity and tolerance of plants to these toxic species. These are mere correlations, a fact that must be remembered at all times. We must continuously guard ourselves against bias, and just because all the bits and pieces seem to fall into place, and the correlative data seem to make sense, we should still be wary. Genetics offer another needed dimension of proof.

Classical genetics allows us to follow whether or not the correlations between putative protective mechanisms and stress tolerance cosegregate through subsequent generations when crosses are performed. If there is cosegregation, the correlation between tolerance and the biochemical mechanism is strengthened. Still, classical genetics should not always be considered proof, unless the genes themselves are fully known. The typical reaction of an organism to a single stress is to turn on many defense and other pathways, not just the ones needed. For example, the attack of a fungus induced the synthesis of tens of phenolic compounds, only one of which is a phytoalexin.[1] There are cases where one regulatory gene controls oxidant tolerance as well as other stress tolerances and other seemingly unrelated pathways. These are pleiotropic control mechanisms, and we initially do not know precisely what such genes turn on. In such cases, mistaken correlations can be made due to the indirect nature in which a control gene works; too often an effect may be ascribed as being due to a change in a structural gene when actually a regulatory gene was involved.

Molecular genetics adds a further dimension to the picture. The novel techniques are not yet ripe to identify the pleiotropic gene(s) controlling photooxidative tolerance, but they do allow the insertion and deletion of single genes, as well as controlling the time, place, and levels of expression. Thus, if it is expected that a certain enzyme is responsible or is part of a stress tolerance mechanism, its structural gene can be isolated, engineered back into plants with control mechanisms that can target its activity to different parts of the cell, and have it expressed either constitutively or inducibly at different levels and at different stages of development or in different tissue types. Conversely, the gene can be engineered into plants in the antisense form, with the same types of expression-controlling elements, such that it can turn off the expression of the endogenous plant enzymes at specific times and places. Thus, molecular genetics provides us with an analytic tool to introduce and analyze each component separately (see other chapters in this volume).

There are caveats to the molecular approach. The most important one is to remember that here, too, correlations may be achieved and not direct proofs. If gene A is introduced and a tolerant phenotype is obtained, it could be because the product of expression of gene A induces the expression of unknown gene B, which in turn provides the direct photooxidant resistance. Thus, both classical and molecular genetics can strongly strengthen correlations, but total proof is, as always, elusive in biology.

These introductory remarks on the value of genetics, along with the caveats, are meant to help the reader put the data in the following sections in a realistic perspective. While the elements described above are elementary, they are so elementary that they are often forgotten while analyzing sophisticated experiments.

We hope that the few places where classical and molecular genetics have contributed to our understanding of photooxidative stress tolerance will stimulate researchers to further use these techniques. The genetic analysis of crosses between resistant and susceptible biotypes/strains/crosses, correlated with measurements of a variety of stress protective mechanisms, together with the use of gene probes and engineering, could teach much more than can be learned without genetics. This will be easy to see in reviewing the small body of data available. The reader is likely to say of many results throughout this volume, "If only they'd also performed a genetic analysis, we would better understand stress tolerance".

A. GENETIC CONTROLS — CONSTITUTIVENESS AND INDUCIBILITY

The types of photooxidant stresses a plant may encounter can arbitrarily be divided into acute and chronic. Acute oxidative stresses are encountered when agricultural levels of herbicides such as paraquat or the diphenyl ether types are used, or when an environmental catastrophe suddenly brings high levels of photooxidative pollutants. These can kill a plant within a few hours.

Paraquat interacts with the photosynthetic electron flow, producing photooxidants, so that a plant does not have the time nor the NADPH needed to produce stress-relieving enzyme systems. The plant must have high constitutive levels of such enzyme systems before encountering high levels of oxidants for it to survive. Conversely, photosystem II-inhibiting herbicides seem to produce chronic levels of oxidants, and much more time is required for a plant to succumb. A maize plant can spend a whole day detoxifying a herbicide such as atrazine, and not show photooxidative damage despite a total blockage of photosystem II. Many environmental toxicants are at high chronic levels, sometimes with additional acute peaks. Induced systems that cope with high chronic levels need not necessarily be useful against the higher acute peaks.

Two types of inducible or constitutive pathways that can interact, or act separately, are known to protect plants: (1) pathways that degrade the xenobiotics and (2) pathways that protect the plants from photooxidant products causing stress. Plants must have constitutive photooxidant protections to protect against acute toxicants; with chronic toxicants, there may be time for protective pathways to be induced.

The constitutiveness and inducibility of enzymes are both inherited, and populations can evolve stress tolerance by changing the levels and types of gene products.[2] The genes controlling stress tolerance seem to control whole pathways in a pleiotropic manner and not single enzymes. For example, a single gene controls the constitutive level of at least three enzymes of the oxidant detoxifying pathways.[3] Similarly, single stress inducers such as magnesium deficiency can vastly increase the same enzymes, as well as their substrates, especially in high light.[4]

Constitutiveness and inducibility are not as distinct properties as one would like, unless one lives for the moment. Constitutiveness is really an experimental term and relates only to baseline levels before an experimental manipulation, and does not consider the previous history of the system. The constitutive levels of a system are **not** fixed; they are just the levels of enzymes present **before** a stress is applied. The constitutive levels can change during development, even in a plant grown under constant controlled conditions.

One "economic" model shows that having high constitutive levels of photooxidative stress tolerance mechanisms has a high cost on plant productivity.[5] Thus, unless it is expected that a plant will encounter acute photooxidative stress, the evolution of mechanisms for high constitutive enzyme levels does not have practical applicability.

There can be different constitutive levels of a system at different times of development and in different organs, tissues, and organelles of a plant. The levels of variation can be huge, e.g., more than 100-fold seasonal variations of glutathione reductase and ascorbate peroxidase have been measured in pine needles[6] (Figure 1). The results shown in Figure 1 are of a much higher magnitude than the previous variations in these two enzymes found in

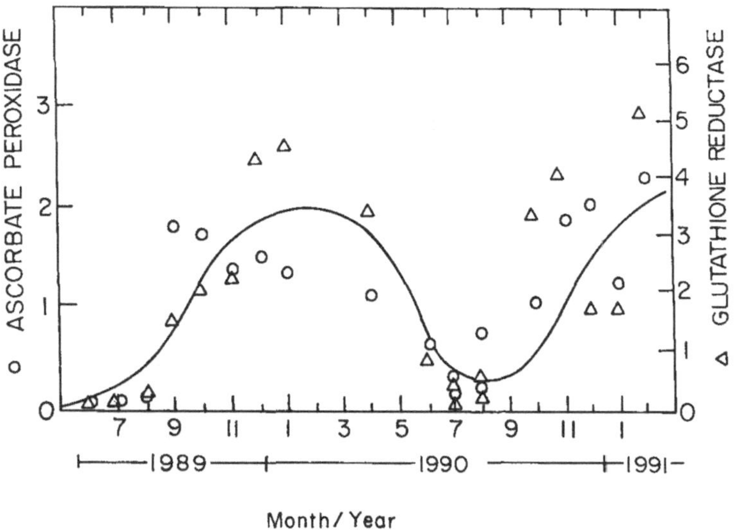

FIGURE 1. Seasonal variations in the constitutive levels of two photooxidant protective enzymes. The data shown are from pine (*Pinus strobus* L.). (Redrawn and condensed from Anderson, J. V., Chevone, B. I., and Hess, J. L., *Plant Physiol.*, 98, 501, 1992.)

peas.[7] Superoxide dismutase was also checked in peas, and no seasonal variations were found.[7] The constitutive levels can vary as a function of the environmental history of a plant. This means that constitutive levels can be affected by prior conditions. Thus, light flux (and possibly light quality), temperature, mineral nutrition, water relations, and low chronic levels of photooxidative stress can all affect the constitutive levels of a protective mechanism. When the levels of enzymes conferring tolerance to an acute stress are measured, different results can be found because of the cultural history of the plant. Thus, the variation in constitutive levels may be a function of evolution for preparedness for seasonal environmental changes, or due to a previous induction causing a change in level.

These problems can confound genetic analyses of tolerance mechanisms, just as they confound physiological and biochemical measurements. They even make it hard to compare experiments within the same laboratory, unless environments are strictly controlled. The variance between laboratories can often preclude predicting the utility of different genes and genotypes in the real world. A stress-tolerant genotype may not be expressed as a stress-tolerant phenotype in the "wrong" environment. For example, a mineral deficiency for a cofactor required for activity of photooxidant protective enzymes can easily preclude resistance.

Another issue confounding data interpretation is the problem of how much of a given system is needed to cope with a new photooxidative stress. The huge variations in the enzymes seen in Figure 1 show that plants vary the constitutive levels, probably as a selected evolutionary response to seasonal differences in housekeeping needs. How much added capacity is available in each case to cope with a new, sudden stress? Whether or not plants might be able to withstand a strong whiff of SO_2 in winter, when enzymes are far higher, cannot be estimated just by looking at the data in Figure 1.

B. PHYSIOLOGICAL CONSTRAINTS TO FULFILLING GENETIC POTENTIAL

In research on the genetics of photooxidant resistance, just as in all genetic research, the phenotype of a given genotype need not always be expressed. The lack of ability to express the genotype can lead to erroneous conclusions. Thus, the environment of plants cultivated for experimental purposes must be carefully controlled to allow phenotypic expression of genotypes. An example of a physiological constraint is zinc deficiency; many others can be envisaged. Zinc-deficient plants had less superoxide dismutase activity[8] than might be expected from their genotype, because superoxide dismutase requires zinc as a cofactor. Zinc-deficient plants contained more superoxide, which strongly inhibited their growth.[9] Similarly, copper deficiency caused a simultaneous loss of the Cu,Zn superoxide dismutase as well as ascorbate peroxidase,[10] both copper-containing enzymes. Physiological compensatory controls caused an elevation of the Mn superoxide dismutase, further complicating genetic understanding.

1. Indirect Correlations: Copper Poisoning, Oxidative Stress, and Glutathione

The symptoms of photooxidative damage to plants are clear, but this does not mean that every correlation between a stress and such damage, even when supported by genetics, is direct. An example is given below with copper toxicity, where plants are photooxidatively damaged. High levels of copper in soil solution cause lipid peroxidation in normal *Silene cucubalus* (=*S. vulgaris*). A resistant strain was found showing no signs of photooxidative stress[11] under control of a single dominant gene.[12] This was traced to an indirect effect of the copper. Copper stress induced the biosynthesis of a copper-chelating phytochelatin. Glutathione is a precursor of phytochelatin, and glutathione levels drop when plants are grown with high levels of copper. Glutathione is then not available for "housekeeping" quenching of oxidative stresses. There are no major changes in phytochelatin or glutathione levels when the resistant plants are challenged with the same levels of copper; resistance is due to an exclusionary mechanism that keeps the copper out.[11] Therefore, in this case, photooxidative stress tolerance is (indirectly) correlated with copper exclusion.

II. INHERITANCE OF OXIDANT TOLERANCE

Information on the inheritance of tolerance to various photooxidative stresses is sparce and is discussed before the inheritance of precise stress-protective mechanisms.

A. TOLERANCE TO GASEOUS POLLUTANTS

There are a number of studies on the genetics of pollution tolerance. Most of these studies are without biochemical analyses (Table 1). The number of types of inheritance can suggest that a considerable variety of biochemical systems, separately or together, are responsible for conferring resistance.

1. Ozone Tolerance

Ozone tolerance is clearly under genetic control, as there are differences among crop cultivars (Figure 2A). The genetics of this phenomenon have not been widely studied except in tobacco and beans, and there are only inklings of the mode of action. Atmospheric ozone pollution can totally reduce the value of cigar wrapper tobacco due to an extreme "weather flecking" of the leaves. This is "photooxidative", as the damage at any ozone concentration is a function of light intensity.[26] Sulfur dioxide causes different symptom patterns, yet is synergistic with ozone in causing damage. Genetic studies led to the use of a series of varieties for environmental monitoring: super ozone-sensitive Bel W3, sensitive Bel C, and tolerant Bel B.[26] Resistance to ozone in tobacco is inherited as an additive effect of more than two genes, with heredity estimates of 60 to 80%.[63-65] One ozone-tolerant variety (Florida) had higher superoxide dismutase levels, especially of one isozyme, than sensitive Bel C or tolerant Bel B, Pen-Bel69, and KY154.[20] This same variety, but not others, had a modicum of cross tolerance to photooxidant-generating paraquat.[20]

There is also a great deal of variability in ozone tolerance in common beans[29] and in soybeans.[24] A large number (410) of common bean accessions from various sources around the world were tested with an acute ozone stress (0.6 ppm for 2 h): 23 accessions were defoliated, while 17 were highly tolerant.[29] No geographic correlations with more highly toxic environments could be made. Conversely, poplar clones from ozone-polluted areas had much higher tolerance to an acute ozone treatment than those from pristine areas (Figure 2B), suggesting that long-term natural selection can work in this fast-growing tree species.[27] The inheritance of ozone tolerance must also be complex: one cross between a tolerant and susceptible variety yielded an intermediate phenotype, while another cross was more sensitive than either

TABLE 1
Intraspecific Genetic Variations in Tolerance to Photooxidative Stresses

Primary stress selection	Species[a]	Cross-resistances	Physiology/biochemistry[b]	Genetics of resistance	Ref.
		Environmental[c]			
Photoinhibitory light	*Dunaliella bardawil* (M)	None stated	High carotene	Unknown	13, 14
Hyperbaric O_2	*Plectonema boryanum* (M)	None stated	High H_2O_2-insensitive MnSOD	Unknown	15
Cold	*Chlorella ellipsoidea* (B)	None stated	High H_2O_2-insensitive MnSOD	Unknown	16
Transient drought	*Zea mays* (C)	Paraquat, acifluorfen, SO_2	High SOD with high GR	Unknown	17
Sulfur dioxide	*Pisum sativum* (C)	None stated	Inducible SOD, GR, GSH	Unknown	18
	Lolium perenne (B)	None stated	No information	>1 gene, recessive	19
	L. perenne (B)	Paraquat	High SOD, GR	Unknown	20
	Geranium carolinianum (B)	None stated	No information	Quantitative-recessive	21
	Oryza sativa (C)	None stated	No information	1 gene, recessive	22
	Triticum aestivum (C)	None stated	Inducible SOD	Unknown	23
	Glycine max (max)	Some to ozone	No information	Quant. intermediate to recessive	24
Ozone	*Nicotiana tabacum* (C)	Some to paraquat[20]	Paraquat R has high SOD, AP, GR[20,25]	Quantitative	26
	Populus tremuloides (B)	None stated		Clone selection	27
	Phaseolus vulgaris (C)	None stated	No information	Few major recessive genes	28,29, 30
	G. max (C)	None stated	No information	Few major recessive genes	30

Herbicides/Xenobiotics

Cercosporin	*O. sativa* (C)	*Cercospora* sp.	High carotene (?)	Unknown	31
δ-Aminolevulinic acid	*P. vulgaris* (C)	None stated	High SOD	Unknown	32
Acifluorfen	*Ceratopteris richardii* (M)	Paraquat	No information	Unknown	33
	Solanum ptycanthum (M)	Paraquat, diquat	No information	Unknown	34
Paraquat	*L. perenne* (M)	SO_2[20]	High SOD, GR[20,35]	Quantitative[36]	
	S. americana (B)	None stated	Suppressed by SOD inhibitor	Unknown	37
	Conyza bonariensis (B)	Atrazine, acifluorfen, SO_2[20] Rose-bengal,[38] photoinhibition[39]	High SOD, AP, GR[40,41]	1 dominant nuclear	38
	C. canadensis (B)	Not acifluorfen	No information	No information	42
	C. canadensis (B)	None stated	High SOD, AP, DHAP, GR	No information	cf. 43
	C. philadelphicus (B)	None stated	No information	1 dominant nuclear	44
	C. sumatrensis (B)	None stated	No information	Unknown	cf. 43
	Youngia japonica (B)	None stated	No information	Unknown	cf. 43
	Poa annua (B)	None stated	No information	Unknown	cf. 45
	Epilobium ciliatum (B)	None stated	No information	Unknown	46
	Ceratopteris richardii (M)	None stated	Not SOD[47]	2 recessive genes	48
	Arctotheca calendula (B)	Diquat	No information	No information	49
	Hordeum glaucum (B)	None stated	Phoenix regrowth[50]	1 incomplete dominant	51
	H. leporinum (B)	None stated	Phoenix regrowth[50]	No information	
	Lycopersicon esculentum (M)	None stated	No information	Dominant gene(s)	52
	N. tabacum (M)	SO_2[53] Cercosporin[54] Acifluorfen[55] $HgCl_2$, fusaric acid[56]	High SOD[54,57] High ascorbate	No information	

TABLE 1 (continued)
Intraspecific Genetic Variations in Tolerance to Photooxidative Stresses

Primary stress selection	Species[a]	Cross-resistances	Physiology/biochemistry[b]	Genetics of resistance	Ref.
		Herbicides/Xenobiotics			
	N. tabacum (T)	None stated	High MnSOD	1 transgenic SOD gene	58
	N. tabacum (T)	None stated	High MnSOD	1 transgenic SOD gene	58
	N. tabacum (T)	Hyperbaric O_2[59]	High GR	1 transgenic GR gene	60
					59
		Not ozone[60]			61
	S. tuberosum (T)	None stated	High Cu,Zn SOD	1 transgenic SOD gene	62

[a] M, isolated mutant; B, naturally occurring biotype or strain; C, cultivar or inbred; T, transgenic plants.
[b] Pqt, paraquat; Acf, acifluorfen; SO_2, sulfur dioxide; H_2O_2, hydrogen peroxide; SOD, superoxide dismutase; GR, glutathione reductase; GSH, reduced glutathione.
[c] All environmental responses required light and oxygen.

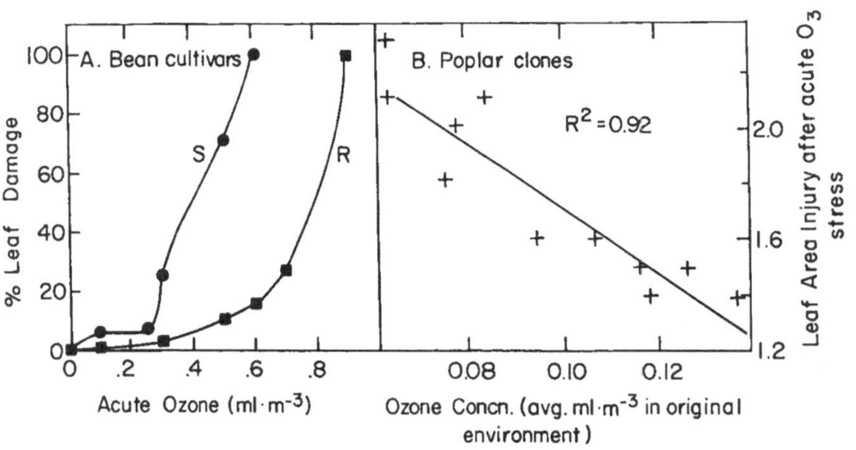

FIGURE 2. Differential susceptibility to ozone stress in varieties and clones. (A) Transient 1 h ozone stress of different *Phaseolus vulgaris* cultivars. The more susceptible cultivar is Bush Blue Lake and the more resistant (R) cultivar is Astro. (B) Poplar clones from areas with different levels of ozone pollution (bottom axis) were tested for tolerance to acute ozone. (Redrawn from Lee, E. H., Kramer, G. F., Rowland, R. A., and Agrawal, M., *Agric. Ecosys. Envir.*, 38, 99, 1992; Berrang, P., Karnosky, D. F., and Bennett, J. P., *Can. J. For. Res.*, 21, 1091, 1991.)

parent.[24] Various reports of ozone tolerance within species are described in Table 1.

2. Sulfur Dioxide Tolerance

Interspecific differences in resistance to sulfur dioxide stress have been shown in a large number of species (Table 1). Few precise genetic data have been reported on the modes of inheritance. Various sulfur dioxide-resistant ecotypes of *Geranium carolinianum* evolved a modicum of tolerance to sulfur dioxide near pollution sources. They were used to evaluate the genetic mechanisms controlling tolerance. Tolerance was inherited in a quantitative manner on the nuclear genome, with a narrow sense heriditability of near half.[21] Pretreatments of susceptible plants with certain chemicals can confer a modicum of sulfur dioxide tolerance,[30] suggesting that the tolerance is constitutive in the more resistant varieties, but resistance can be somewhat induced in the susceptible varieties. It is possible that such compounds indirectly induce higher levels of glutathione and/or glutathione pathway enzymes, as some mammalian P_{450} inhibitors have been shown to have similar properties of coinduction.[66] Indeed, there is evidence that biotypes with high constitutive levels of the oxidant detoxification pathway,[3] or that can be induced to higher levels,[18] are more resistant to SO_2 stress.[20] The most ozone-tolerant variety of the 21 soybean genotypes tested was also tolerant to sulfur dioxide, but many other varieties were tolerant to one or the other, or to neither gaseous pullutant.[24] Many biochemical pathways must be involved.

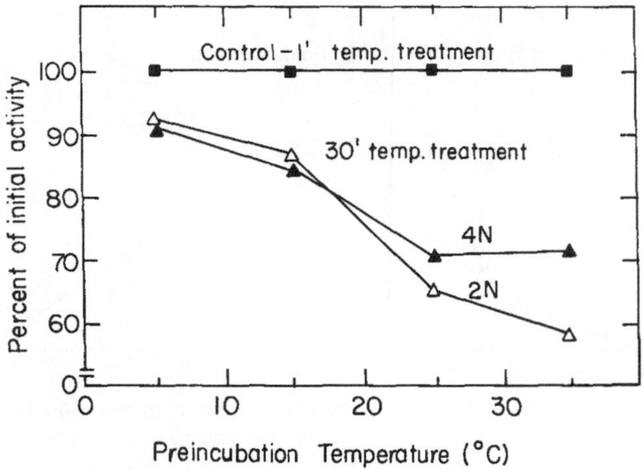

FIGURE 3. Increase in thermal stability of glutathione reductase in tetraploid vs. nearly isogenic diploid alfalfa. (Modified from Kidambi, S. P., Mahan, J. R., and Matches, A. G., *Plant Physiol.*, 92, 363, 1990.)

B. HEAT, DROUGHT, AND COLD TOLERANCE

Drought stress tolerance can be conferred by a multitude of mechanisms: those that control osmoticum, as well as those that temporarily protect against the photooxidants produced when drought-stress closes stomates. The osmotolerance protection against drought and saline environments by betaine derivatives has recently been reviewed.[67]

Drought stress has been correlated with induced levels of glutathione-reductase in cotton[68] and maize,[69] as well as ascorbate peroxidase and dehydroascorbate reductase in barley,[70] but without a discussion of the genetics. Constitutive drought stress in maize inbreds has been correlated with resistance to acute paraquat treatment along with constitutively elevated levels of superoxide dismutase and glutathione reductase,[17] the only two enzymes of the pathway studied. Unfortunately, genetic studies have not been performed to ascertain how these resistances are inherited. This is especially unfortunate, as some drought- and paraquat-sensitive inbreds were elevated in one, but not the other, of the measured enzymes, suggesting the possibility that the "right" crosses could confer higher tolerance.

Tetraploid alfalfa (lucerne) is better adapted to warm temperatures than the diploid, including near-isogenic doubled diploids.[71] This was related to a slight increase of thermostability of one glutathione reductase isozyme (130 kDa, in a one-dimensional activity gel) (Figure 3). This glutathione reductase was slightly more stable than other isozymes and had a much broader thermal kinetic window, where the K_m was doubled between 15 and 35°C compared to the diploid. The diploid window was only between 15 and 20°C.[73] The genetic mechanisms controlling this increase in enzyme thermal stability and heat tolerance are, as yet, obscure.

Chilling sensitivity/resistance has been correlated to different superoxide dismutase isozymes in two strains of the alga *Chlorella ellipsoidea*. The chilling-sensitive strain only contained an iron-containing isozyme, while the resistant strain also had high levels of a manganese-containing isozyme.[16] There is plasticity here as well; the level of resistance of the sensitive type can be enhanced by paraquat pretreatment. No information was presented stating whether paraquat pretreatment can further increase resistance in the resistant type or whether this resistant strain had cross-resistance to other oxidant stresses.

C. XENOBIOTICS THAT PRODUCE OXIDANTS

Chemical warfare among organisms is quite common. Many plant pathogens produce toxins that burn necrotic holes, assisting in penetration, or that cook their prey before devouring them. Such toxins include photodynamic pigments such as the cercosporins produced by many *Cercospora* species. Farmers have similarly used many herbicides to cause photooxidative death of unwanted weed species (see Chapter 9 and 14 of this volume). There was no need for surprise that weeds could evolve genetic resistance to these herbicides; they had millenia to evolve armor against pathogen-induced photooxidative stresses, as well as those stresses produced by the environment. Four major groups of herbicides result in the production of photooxidants: (1) photosystem II electron transport inhibitors (triazines, phenylureas, etc.), (2) photosystem I inhibitors (paraquat), (3) stimulators of protoporphyrin IX biosynthesis (e.g., nitrodiphenyl ethers such as acifluorfen, and (4) phytoene desaturase inhibitors that prevent carotene biosynthesis (e.g., norflurazon). Resistance has evolved to all four groups, but in the last group, resistance has only evolved in cyanobacteria.[74]

There are many evolutionary mechanisms that can confer resistance to xenobiotic-producing oxidants, and the different mechanisms can and sometimes must work in concert to attain resistance to the levels of xenobiotics found in the environment. The methodologies that a plant can evolve include:

1. Reduced rate of uptake or total blockage from uptake
2. Reduced rate of active or passive transport to the active site
3. Sequestration of the active xenobiotic away from its target (Sequestration has been misused by some to include localization of irreversibly inactivated metabolites.)
4. Metabolic inactivation of xenobiotics
5. Mutation in the target site of the xenobiotic, such that it can no longer bind and cause damage
6. Defenses that limit xenobiotic production by a pathogen; i.e., the many known direct responses to pathogenic organisms
7. Resistance to the photooxidative stress produced by the xenobiotic.

Much of the discussion below is devoted to the last mechanism, as it is the one directly related to this volume. Still, when resistances evolve, the other mechanisms must be evaluated and found to be either the causal mechanism or, in some cases, contributory to the mechanism. Despite the length of the list of possible mechanisms, there may still be other resistance mechanisms.

There are cases of resistance that are still hard to explain using the above list, e.g., "phoenix" paraquat resistance, which manifests itself as an evolved ability of seemingly herbicide-killed plants to regenerate and resume growth from the ashes of the paraquat-dessicated plants. This was seen with the dicot *Arctotheca calendula*[49] and the grass *Hordeum leporinum.*[50] Indeed, the same genus can evolve more than one mechanism of tolerance: *H. leporinum* and *H. glaucum* have evolved seemingly different mechanisms of paraquat tolerance,[50] and neither have direct resistance to photooxidation.[75] Similarly, one selected biotype of a fern with recessively inherited paraquat resistance had no cross-resistance to other photooxidative stresses.[48] Another mutant of the same fern had cross-resistance to both paraquat and acifluorfen,[33] suggesting a more general photooxidative resistance mechanism in this second biotype.

More than 50 species have evolved resistance to photosystem II-inhibiting triazine herbicides.[76] The vast majority are resistant due to a single amino acid transposition in the target site.[77] One species has evolved a biotype with elevated levels of a glutathione transferase that directly conjugates the herbicide, rendering it nontoxic,[78] while another species has evolved a biotype with higher levels of an *N*-dealkylase that oxidizes it to a nontoxic product.[79] No species seem to have directly evolved resistance to the photooxidants produced under selection pressure by atrazine, which is surprising, as these herbicides require 5 to 7 d to show extensive photooxidative damage to leaves. Still, one species that evolved a 100-fold resistance to paraquat had a fewfold increased resistance to atrazine.[20]

Many of the species that evolved resistance to one photooxidative stress had (as seen above) cross-tolerances to other photooxidative stresses. The modes of generating the photooxidative stresses are very different, and the chemical structures of the xenobiotics are different. This would suggest (but hardly prove) that such species have an increased capacity to cope with photooxidants. At times this has been correlated with elevated levels of carotenoids, ascorbate, and/or α-tocopherol (see Section II.D), but in many cases this has been correlated with elevated levels of enzymes related to dismutation of superoxide coupled with ascorbate and glutathione regeneration (Table 1).

The greatest amount of correlative data between genetics and enzyme levels are within the genus *Conyza≡Erigeron*. Six species within this synonymous genus have evolved resistance to paraquat in Japan (4 spp.), Hungary (1 sp.), and Egypt (1 sp.). The application of paraquat immediately stops photosynthesis in both the resistant and sensitive *Conyza* biotypes of the two species where this has been measured.[42,80] This indicates that resistance is not due to evolution of a barrier between the leaf cuticle and the target in the chloroplast, or to sequestration before reaching the chloroplast. Photosynthesis recovers in both resistant *Conyza* biotypes within 3 to 4 h, suggesting that the herbicide is either sequestered and/or detoxified during that period. The actual level of resistance (compared to the wild type) would then be related to the ability of the plant to dissipate the active compound vs. the ability of the herbicide to generate oxidants. Thus, a plant with a photooxidation protective mechanism need not be cross-tolerant to all photooxidant-generating xenobiotics, as had been incorrectly assumed.[38] Large amounts of NADPH are required to continually recycle ascorbate and glutathione to their reduced forms, and if a plant cannot be rid of the xenobiotic, it eventually will be damaged. Two of the *Conyza* species (where measured) have elevated levels of enzymes of the oxidant-detoxifying pathway.[41,43] Resistance could be abolished with chelators of copper and/or zinc, both in *Conyza* sp.[81] and in *Solanum* sp.,[37] and natural tolerance in pea was similarly lowered by a chelator.[82] Such data suggest involvement in resistance of a Cu,Zn superoxide dismutase and/or a copper-containing ascorbate peroxidase.

Resistance to paraquat was dominantly inherited on a single gene, in the two *Conyza* spp. where measured.[3,44] The levels of the enzymes measured were high in the F_1 population, suggesting a dominant control. Analysis of the F_2 generation required a plant-by-plant analysis of segregation of resistance with higher enzyme levels, which could only be done immunologically.[3] There was a cosegregation of high levels of enzyme with resistance. These results strongly suggest that there is relatedness between enzyme levels and resistance, under a pleiotropic control.

1. Cross-Tolerance Between Xenobiotic and Other Oxidative Stress Resistances

Superficially, one would expect that a plant possessing high levels of any mechanism that quenches active oxygen species should be tolerant of all photooxidant stresses. As different stresses, e.g., drought and SO_2,[83] amplify each other, one can assume that they overcome shared protective mechanisms. Conversely, an induction-enhanced ascorbate peroxidase activity provides cross-tolerance to peroxide, ozone, and paraquat.[84] Such a view of cross-tolerance may be oversimplified. Many active oxygen species must be quenched close to their sources of generation. Their exceedingly short lifetimes mean that they cannot, in most cases, diffuse long distances, even at the microscopic level of the cell. For example, singlet oxygen can diffuse only 100 nm, ten times the thickness of a membrane.[85] Additionally, it is statistically likely that the closest substrate to an active oxygen species is the one that will be oxidized. For this reason, there is considerable localization of the active oxygen-degrading enzymes and quenching chemicals in the different cellular compartments, as described in previous chapters.

It is therefore expected that there will be cross-resistance to photooxidant stresses localized to the same compartments. The chloroplast is a major source of photooxidants, even prior to stresses, and higher levels of the enzymatic pathway of oxidant detoxification in the chloroplast confers many cross-tolerances. This is true when the enzymes are induced, as by drought stress giving cross-tolerance to paraquat,[68] as well as when they are constitutive (Table 1).

A paraquat-resistant biotype of *Conyza* that evolved in the field was found to be also tolerant to photoinhibition, as measured by the effect of high light on photosynthetic CO_2 fixation, as well as measured photoacoustically as energy storage and quantum yield reductions.[39] As expected of an enzymatic process, it could not be measured as oxygen evolution at low temperature.[75] It was also not possible to measure the inherited resistance to photoinhibition by chlorophyll fluorescence,[75,86] for possible reasons discussed elsewhere.[86] As there are many ways to get some photooxidant-stress resistances, one should not expect cross-resistances in all cases. Indeed, of a group of ozone-tolerant tobacco varieties, only one had cross-tolerance to paraquat. That particular variety also possessed constitutively elevated levels of the oxidant-detoxifying enzymes.[20]

Many other inherited cross-tolerances are summarized in Table 1. There are many other cases where cross-tolerance was not checked, as well as many cases where they were sought, but not found. Only a few of the latter cases were published, and some are also summarized in Table 1.

D. ANTIOXIDANT CHEMICALS IN PLANT CELLS

There are many chemicals that can reduce oxidants in cells and could, in theory, confer a modicum of tolerance. Even when the levels of these compounds are high in some biotypes or species, compared with other biotypes, the levels are still very low in comparison with the potential of continual production of photooxidants if they will be used up stoichiometrically. Thus, reductants can possibly allay only an acute, but transient, "attack" by photooxidants if the oxidant quenchers are at high levels. Their continual effectivity would be a function of a genetic capacity to either rapidly synthesize such compounds *de novo* or to rapidly recycle them. Conversely, high levels of reductants can be useful if the plant also has the metabolic capacity to recycle them, or methods of sequestration to rid itself of the cause of photooxidation, or to block plant entry or translocation of most of the oxidant. For these reasons, the presence of antioxidant reductants must be considered as part of a larger stress-tolerance mechanism, and rarely as useful by themselves.

1. Ascorbate and Tocopherol

We have not seen reports on differences in genetic capacity to produce these oxidant quenchers within species. Still, an interesting correlation has

TABLE 2
Interspecific Relationship Between High Levels of Both Ascorbate and α-Tocopherol with Tolerance to Photooxidative Stress

	Level of		Stress tolerance	
Species	Ascorbate (mg/kg)	α-Tocopherol (mg/kg)	Damage (% inhibition)	Ethane evolution (nmols/kg/h)
Ipomea purpurea	20	100	67	225
Chenopodium album	580	120	25	99
Medicago sativa	1430	100	12	11
Amaranthus retroflexus	**5040**	100	49	212
Fagopyrum esculentum	**5370**	280	41	99
Datura stramonium	1140	**830**	59	285
Abutilon theophrasti	920	**500**	31	177
Cassia obtusifolia	**8610**	**600**	**3**	**0**
Sinapis alba	**4690**	**500**	**2**	**0**

Note: Stress was measured 6 d after spraying 10- to 15-d-old seedlings with 1 kg/ha oxyfluorfen. Damage was measured as inhibition of the dry weight increment (vs. controls). Bold numbers indicate high level of compound.

Collated and calculated from data in various tables in Finckh, B. F. and Kunert, K. J., *J. Agric. Food Chem.*, 33, 574, 1985.

been made among species.[87] Species with elevated levels of just one of these antioxidants were not capable of withstanding the effects of acifluorfen, a compound causing the metabolic generation of high levels of the photodynamic compound protoporphyrin IX. Only those species elevated in both had a modicum of tolerance to this chemical (Table 2). As the measurements of ascorbate and α-tocopherol were made only after the photooxidative stress, it is not clear whether the high levels of both that are correlated with tolerance were constitutive or induced.

2. Carotene

Carotenoids are not typical antioxidants, as they do not need an enzymatic reduction to be recycled. They chemically dissipate the energy they absorb through perturbations of their chemical structure and transform the energy to heat. In contrast, only a minute part of the photochemical energy adsorbed by chlorophyll is fluoresced when electron transport is blocked; some is dissipated as heat and a proportion produces radicals or causes irreversible bleaching of the chlorophyll. The presence of carotenoids can alleviate much damage to chlorophyll. This becomes obvious when carotenoid biosynthesis is blocked; the chlorophyll is oxidized at field-light intensity.

Dunaliella bardawil makes far more carotene in light than in dark, and the amount synthesized in the light is a function of light intensity. The use of inhibitors such as norflurazon to prevent carotene biosynthesis rendered

Dunaliella more susceptible to light damage.[88] Mutants producing high β-carotene levels (up to three times normal) in low light[13] are also resistant to photoinhibition when the mutant cells are transferred to high light.[88] This is not due to direct quenching of radicals by carotene, as the resistance to photoinhibition is only to high fluences of blue light. The cells are still photoinhibited by red light. Thus, the tolerance is due to a light-filtering effect of the specifically blue-adsorbing carotenoids and not to direct quenching of radicals produced by the photosynthetic apparatus. The excess carotene is not found in the chloroplasts; it is in special bodies along the cell periphery. Still, this is a photooxidant protective mechanism: many other plant pigments would have photodynamically produced active oxygen species under such high-light conditions, while the carotenes can dissipate the energy because of the special resonating strength of the isoprene double bonds.

Carotenes have been indirectly linked to cercosporin tolerance in rice cultivars; suppression of carotenoid biosynthesis rendered a cercosporin-resistant cultivar sensitive to the photodynamic fungal toxin.[31]

3. Flavonoids and Melanins

Whereas these types of chemicals are often implicated in protection against photooxidative stress (see, e.g., References 89 and 90), we have not found genetic data correlating quantities with resistance.

III. INHERITANCE OF ENZYMES OF PHOTOOXIDANT DEGRADATION PATHWAYS

A. SUPEROXIDE DISMUTASE

1. Inducible

The capacity to induce superoxide dismutase is found in the genetic complement of some species, but not in others. For example, drought stress, mimicked by osmotically stressing plants with polyethylene glycol, induced elevated levels of superoxide dismutase in maize, yet had no effect in *Calotropis procera.*[91] Even within wheat and related genera, the levels of this enzyme in drought-stressed plants decreased (*Aegilops squarrosa* and *Triticum monococcum*) or increased (*T. aestivum* and *T. persicum*),[91] allowing no broad generalizations. Low levels of sulfur dioxide induced superoxide dismutase to a much higher level in a sulfur dioxide-tolerant wheat variety than in a sensitive variety.[23] Unfortunately, no analysis of inheritance was made, even though only one isozyme was greatly elevated.

2. Constitutive

There has been only one extensive study on the inheritance of one of the loci coding for superoxide dismutase. Two codominant alleles code for a difference in electrophoretic mobility of one superoxide dismutase in soybean.[92] This locus was inherited independently, i.e., was not linked to another superoxide dismutase locus coding for a different isozyme, and was also independent of 11 other loci coding for other enzymes. Unfortunately, none of the other loci investigated were related to photooxidant stress tolerance, nor were any physiological parameters correlated with the isozymes of the various loci.

There can be large intraspecific differences in the constitutive levels of this enzyme. There might have been natural selection for oxidant tolerance, resulting in higher constitutive levels of this enzyme year round. Clones of Scots pine *(Pinus sylvestris)* from highly polluted north Finnish forests had double the constitutive level of superoxide dismutase than a clone from a less polluted area, when both were grown near each other in a less polluted area.[93]

High superoxide dismutase levels can also be negatively correlated with a genotypic response to stress, reminding us of the complexities of each stress system. Nematode-sensitive varieties of tomato have high superoxide dismutase levels and resistant varieties have lower levels.[94] The authors interpret this, after viewing tissue damage, as the need for the superoxide to generate hypersensitive tissue damage that blocks the nematodes' path. If the superoxide is enzymatically detoxified, the nematodes continue eating the plants.

Bean cultivars with high levels of superoxide dismutase had a much higher resistance to δ-aminolevulinic acid than sensitive cultivars.[32] Both δ-aminolevulinic acid and acifluorfen cause accumulation of protoporphyrin IX, a natural photodynamic agent. Other bean enzymes were measured, but the different varieties had the same levels of carotenes.[32] No information was presented on the genetic differences between the varieties.

Tobacco lines were selected with paraquat for resistance. The isolated resistant lines had high constitutive levels of superoxide dismutase compared to the wild type.[57] The callus enzyme was not immunologically reactive with the leaf enzyme.[57] Plants regenerated from resistant callus were resistant to paraquat, cercosporin, SO_2,[54,57] and acifluorfen,[55] as well as to other fungal toxins and diseases.[56] They also had higher constitutive levels of ascorbate, and oxidant treatment induced higher levels of glutathione reductase and ascorbate peroxidase.[55] No information was presented on how this is inherited. Plants transformed with higher levels of this enzyme are discussed in a later section.

B. INDUCIBLE ASCORBATE PEROXIDASE

Ascorbate peroxidase can be induced by photooxidative stress and by ethylene.[84] One novel stress that was used experimentally was the depletion of carotenoids from seedlings by treatment with the phytoene desaturase inhibitor norflurazon, and then illuminating the plants. Light was necessary, but not much (7 W m^{-2}) was sufficient for inducing a near-doubling of ascorbate peroxidase activity.[95] Interestingly, a cytoplasmic, not the plastid, isozyme was induced.

C. GLUTATHIONE REDUCTASE

1. Constitutive

Glutathione reductase activity is significantly higher in ozone-tolerant *N. tabacum* cv. Bel B than in the susceptible Bel W3.[25] Still, plants transformed with a glutathione reductase gene spliced to a strong 35S promoter had two to four times more glutathione reductase (but in the cytoplasm). These plants had paraquat tolerance, but no real dose-response curves were presented.[60]

2. Inducible

Two studies have shown that chronic oxidative stress can induce higher levels of this enzyme, conferring tolerance to further stress. Low levels of ozone (70 ppb) induced an increase in glutathione reductase activity in spinach leaves, with a concomitant increase in immunologically determined enzyme protein.[96] Acifluorfen-induced oxidant stress of bean leaves induced high levels of glutathione reductase, along with total glutathione, ascorbate, and galactonolactone oxidase, an enzyme-producing ascorbate.[97] Similarly, drought stress induced higher levels of this key enzyme in maize, conferring cross-tolerance to paraquat.[68]

IV. MOLECULAR BIOLOGY OF PHOTOOXIDATIVE STRESS ENZYMES

Molecular-genetic tools are useful for the study of the metabolic pathways that lead to scavenging of toxic oxygen species, just as they are useful for the study of other metabolic processes. Commonly, the first step is to identify the mRNA (messenger RNA) that is transcribed from the gene under study. This is performed, in practice, by isolation of all mRNAs from a tissue where the specifically sought mRNA should be relatively abundant. By reverse transcription, these mRNAs are converted into cDNAs (complementary DNAs) that are included in a cDNA library. The latter library is then screened by one of several procedures to isolate the specific cDNA of interest. The cDNA probes that code for specific isozymes provide a means to evaluate the transcript levels in specific plant organs and tissues during development and in response to external chemical and physical stimuli. The nucleotide sequences of these cDNAs can also be utilized to derive the amino acid sequences of the respective enzymes and thus can lead to the production of antibodies against these enzymes.

Southern blot hybridization of the DNA from the organisms, with the appropriate cDNA, is useful to estimate the number of genes or "gene copies" that code for the specific enzyme. The cDNAs can be cloned into transformation vectors, with a variety of promoters, terminators, and selective markers, to generate transgenic plants that have varying expression levels for the respective enzyme in different tissues or at different times during development or seasons. In addition, the cDNAs or fragments of them can serve to isolate the genomic sequences that code for the investigated enzymes in species other than the one used to isolate the cDNA clone. The latter sequences provide a way to study the control of gene activation imposed by *cis*-regulatory motifs, elements that either induce or suppress the expression of the coding sequence (in the promoter and the terminator), or by regulatory proteins coded by controlling *trans*-genes. For enzymes encoded in the nuclear DNA but located in organelles (such as chloroplasts and mitochondria), the cDNA is used to isolate DNA for decoding the amino acid sequences of the transit or prepeptides that allow them to be transported through the organelle membrane. The trafficking of the mature enzymes into the respective organellar sites can thus be studied.

The molecular biology of several enzymes involved in photooxidative stress-defense, such as superoxide dismutase, ascorbate peroxidase, and glutathione reductase, has been elucidated in the last 5 years (see other chapters in this volume). Most of the available information is based on the isolation of cDNAs; only a few reports are available on the genomic coding sequences, or elucidate the sequence based on the amino acid sequence. The cDNAs may also be utilized to produce transgenic plants that overexpress such enzymes. All this work is in its infancy.

A. Cu,Zn SUPEROXIDE DISMUTASE

Amino acid sequences of proteins are usually derived from the respective cDNAs. The opposite happened with plant Cu,Zn superoxide dismutase. The first amino acid sequences of plant superoxide dismutases were reported before the respective cDNAs were isolated. These reports were published 17 years after Fridovich and associates brought this enzyme to the limelight of biochemical research. Three laboratories published the amino acid sequences of Cu,Zn superoxide dismutases in the same year from the cytosol and chloroplasts of tomato plants,[98] the cytosol of cabbage,[99] and the chloroplasts of spinach.[100] Recently, the 40 amino-terminal amino acids of the cytosol Cu,Zn superoxide dismutase of soybean were also sequenced.[101] The first three sequences were based on purified protein extracted from vast quantities of plant material (e.g., 10 to 25 kg from "local-market" sources, which may explain their choice of plant material). These and further studies[102,103] indicated that plant Cu,Zn superoxide dismutases share similar amino acid sequences. Also, the homology of chloroplast Cu,Zn superoxide dismutases from various plant species is greater than the respective homology among the cytosol Cu,Zn superoxide dismutases from the same species, suggesting that the latter isozymes diverged very early in evolution.

The amino acid sequences were instrumental in generating synthetic nucleotide probes for screening cDNA libraries used for biochemical and physiological studies. Thus, the cDNA for the chloroplast Cu,Zn superoxide dismutase of tomato was isolated from a library, using a synthetic oligonucleotide

	1
Tomato	M A A H S I F T T T S T T N S F L Y P I - S S S S S - P N
Petunia	M A A H T I F T T T S T T N S F L F P I A S S N T N S A P S
Pea	M A - - S Q - T L V S P S P S - - - P L - - - - - S S - H S
	31
Tomato	I N - S S F L G V S L N V N A K F G Q S L T L Y A V T T P K
Petunia	L S - S S F H G V S L K V K S K P T Q S L T L S S V T S P K
Pea	L L R T S F S G V S - - V K L A P - Q F S T L A T S N F - K
	61 1
Tomato	P L T V F A A T K K A V A V L ...
Petunia	P F I V F A A T K K A V A V L ...
Pea	P L T V V A A T K K A V A V L ...
	↑

FIGURE 4. The deduced amino acid sequences of transit peptides of chloroplast superoxide dismutases from pea,[105] petunia,[106] and tomato[104] are aligned. The mature protein begins with the sequence ATKK, as denoted by the arrow.

probe.[104] This probe was constructed on the basis of the amino acid sequence of the chloroplast Cu,Zn superoxide dismutase of spinach and of other species. This tomato cDNA, as well as the cDNAs of chloroplast Cu,Zn superoxide dismutases from pea[105] and petunia,[106] provided interesting information. While the deduced amino acid sequences of the mature enzyme were almost identical, the amino acid sequences of the transit peptides that facilitate their movement into the chloroplast from the cytosol varied considerably (Figure 4).

The yet unresolved question is whether trafficking into the chloroplasts requires species-specific, different transit peptides or whether the transit peptide for the chloroplast Cu,Zn superoxide dismutase of a given species will also target this enzyme into the chloroplasts of other species. Moreover, it is also not known if the C-terminal sequence of the chloroplast Cu,Zn superoxide dismutase has a role in crossing the chloroplast membrane. Trafficking experiments with chloroplast Cu,Zn superoxide dismutase translated *in vitro* from appropriate chimeric cDNAs should furnish this information.

Scandalios and associates[107] used the NH_2 terminus amino acid sequence of maize to generate a synthetic nucleotide probe to isolate and sequence a cDNA clone that codes for the cytosol Cu,Zn superoxide dismutase (SOD-2) from maize. Subsequently, the cDNAs of cytosol Cu,Zn superoxide dismutases from various other plant species were isolated and sequenced (see, e.g., References 104 and 108 to 110). The coding sequences for the chloroplast and the cytosol Cu,Zn superoxide dismutases of tomato were each assigned to single locations on the chromosome map (in chromosomes 11 and 1, respectively).[111] This and previous Southern blot hybridizations[104] suggested that there is only one gene each for the two Cu,Zn superoxide dismutases of tomato. Conversely, two chloroplast Cu,Zn superoxide dismutase cDNAs were reported in peas.[112] They did not differ in their coding sequences, but did have different 3′ untranslated regions. Two cytosol cDNAs were found in maize that have identical coding regions (except for one codon),[113] but both the 3′ and the 5′ untranslated sequences varied considerably. It would be of interest to determine if these untranslated mRNA sequences have regulatory roles. Do they control the rate of translation in different plant organs or/and under different environmental conditions, as recently reported for a different plant mRNA?[114]

Valuable information was obtained with cDNA probes on the change in Cu,Zn superoxide dismutase transcript abundance during plant development as well as upon exposure to external chemical and physical stimuli using cDNA probes.[104,115] The cDNAs for the chloroplast and the cytosol Cu,Zn superoxide dismutases were used in northern blot hybridizations to evaluate the abundance of its mRNA transcripts in tomato. The transcript of each gene was developmentally regulated, and the two genes were differentially expressed in specific plant organs. Thus, the transcript of the cytosol Cu,Zn superoxide dismutase gene was abundant in roots where the expression of the gene for the plastid enzyme was very low or not detectable. The transcripts for the chloroplast Cu,Zn superoxide dismutase appeared very early in leaf and fruit development, but decreased drastically with leaf expansion and fruit maturation (Figure 5).

The transcripts for the cytosol Cu,Zn superoxide dismutase were much more stable in these organs. Illumination of young tomato seedlings strongly enhanced the mRNA transcript level of the chloroplast Cu,Zn superoxide dismutase, and to a lesser extent the amount of transcript of the cytosol Cu,Zn superoxide dismutase. The increase in enzymatic activities of these two Cu,Zn superoxide dismutases, as revealed by activity gels, occurred 4 or more h after the increase of the respective transcript levels.[104,115] The enzymatic activities then diminished only gradually and not abruptly. The levels of both transcripts increased in response to sublethal doses of paraquat as well as to mechanical wounding. These levels also increased in tomato plants treated with ethephon after it released ethylene in plant tissue. Conversely, prolonged sublethal drought stress dramatically increased the transcript level of the cytosol Cu,Zn superoxide dismutase (and to a lesser extent the enzyme activity).[115]

The level of the transcript of the cytosol Cu,Zn superoxide dismutase of mature *N. plumbaginifolia* plants was strongly affected by light: 3 d of darkness reduced the level of this transcript. Subsequent exposure to light increased the transcript to higher levels than before the dark period.[116] These data confirmed earlier results[115,117] that showed that the gene for the cytosol Cu,Zn superoxide dismutase is not induced by light stress in very young seedlings. This parallels previous studies showing that the transcript level of the cytosol Cu,Zn superoxide dismutase is strongly enhanced in plants sprayed with

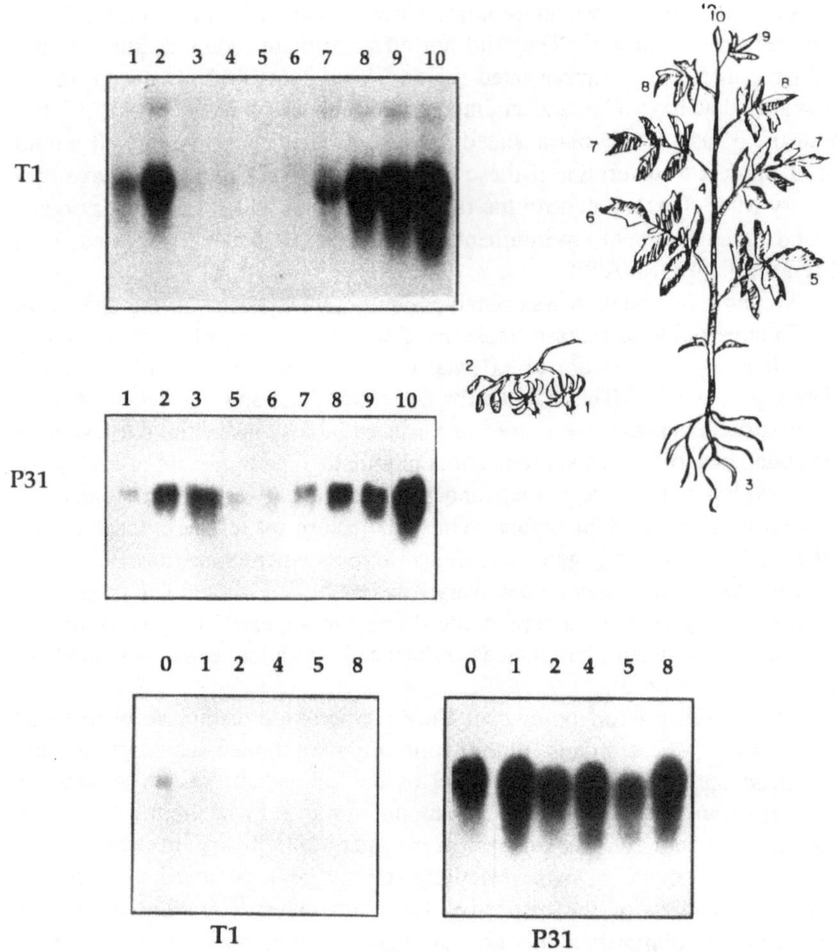

FIGURE 5. Levels of the transcripts for the chloroplast (T1) and the cytosol Cu,Zn superoxide (P31) dismutase in tomato plant organs. (A) Differential expression of the T1 and P31 transcripts in tomato plant organs. Total RNA samples (20 μg) from different tomato plant organs were separated by electrophoresis on formaldehyde gels, blotted to nitrocellulose, and hybridized to radioactive RNA probes from T1 (chloroplast Cu,Zn SOD)- and P31 (cytosolic Cu,Zn SOD)-derived plasmids. Lane designation: 1, flower; 2, flower bud; 3, root; 4, stem; 5, senescing (7th from apex) leaf; 6, mature (6th) leaf; 7, expanded (4th and 5th) leaves; 8, young (2nd and 3rd) leaves; 9, youngest (first unfolded) leaf; 10, shoot tip. (B) Transcript levels during fruit ripening. Samples of 20 μg total RNA from tomato fruit pericarp were subjected to northern blot analysis with the P31 and T1 RNA probes. Fruits were sampled starting with: 0, the "mature green" stage, 1 d before the "breaker" stage; 1, breaker, first pink color appears; 2, 1 d after the breaker (light orange color); 4, 2 d later, light red; next day, dark red, firm flesh; 8, 7 d after breaker, dark red and soft (overripe stage). (From Perl-Treves, R. and Galun, E., *Plant Mol. Biol.*, 17, 745, 1991. With permission of Kluwer Academic Publishers.)

paraquat.[115,117] The enhancement was much more dramatic in light, where paraquat causes more photooxidative damage. The increase is rather specific for the superoxide dismutase transcript. Probing the abundance of mRNAs of two other genes (*rbc*S and the β-1,3-glucanase gene) after paraquat treatment indicated that paraquat did not change the transcript levels of these other genes.

There is yet only some information on the genomic sequences that code for Cu,Zn superoxide dismutase in plants. The genomic sequence coding for the cytosol Cu,Zn superoxide dismutase was isolated from a genomic library of rice.[118] The eight exon-coding regions are split by seven noncoding introns. One of the introns precedes the first exon and is very long (572 bp). The other introns are much shorter. A genomic clone for the chloroplast Cu,Zn superoxide dismutase isolated from a tomato library was almost fully sequenced.[119] It consists of eight exons separated by seven introns. Two of these introns are extensive; the intron located downstream of the first exon (that consists of 276 nucleotides) spans about 1100 bp and the intron that is upstream of the last exon (that consists of 54 nucleotides) consists of over 2000 bp. This last intron is not yet fully sequenced. The other five introns are relatively short (less than 200 bp). The coding sequences of the exons are identical to the nucleotide sequence found in the cDNA for the chloroplast Cu,Zn superoxide dismutase of tomato.[104] Further studies should reveal the role of *cis*-regulatory sequences that modulate transcription in response to internal signals (e.g., organ, developmental stage) and to photooxidative stresses in this genomic sequence as well as in other Cu,Zn superoxide dismutase genes of plants.

One would assume that nature would evoke the most efficient superoxide dismutases. This may not be the case *vis a vis* the efficiency of the reaction. Site-directed mutagenesis of a Cu,Zn superoxide dismutase led to an enzyme with improved kinetic parameters.[120] Still, one has to ascertain its properties after the modified gene is reinserted into an organism. It may be more labile than the native enzyme, etc. If, indeed, it is better, such a gene may offer more efficient photooxidative stress tolerance than the wild-type genes presently used.

B. Mn SUPEROXIDE DISMUTASE

The Mn superoxide dismutase is constructed from four identical subunits. The enzyme is cyanide insensitive and peroxide resistant. It has structural homologies with Fe superoxide dismutase and has similar metal binding properties, rendering it distinctly different from the Cu,Zn superoxide dismutases. However, Mn superoxide dismutase, Cu,Zn superoxide dismutase, and Fe superoxide dismutase provide the same function: the dismutation of superoxide to oxygen and peroxide.[121,122] An amino acid consensus sequence derived from Mn superoxide dismutases of maize was used to synthesize an oligonucleotide probe.[123] The latter oligonucleotide was then used to isolate a cDNA for the maize Mn superoxide dismutase from a cDNA library. The deduced amino acid sequence consisted of 235 residues and had a calculated molecular mass of 25,525. The deduced amino acid sequence contained a prepeptide sequence of 31 amino acids. A cDNA for a Mn superoxide dismutase was also isolated from *N. plumbaginifolia.*[125] In both cases, these presequences are important for targetting the preprotein into mitochondria.[124,125] The amino acid sequence of the Mn superoxide dismutase of *N. plumbaginifolia* was deduced from the nucleotide sequence of the previously isolated cDNA.[126]

C. Fe SUPEROXIDE DISMUTASE

The isolation of cDNAs for Fe superoxide dismutase was only recently accomplished.[127] The isolation of these cDNAs from *N. plumbaginifolia* and *Arabidopsis thaliana* was achieved by a different strategy from that used to isolate plant Cu,Zn superoxide dismutases. A shotgun-cloning electroporation procedure was used to introduce a large number of random, plasmid-based cDNA clones coding for Fe superoxide dismutase into a superoxide dismutase-deficient *Escherichia coli* mutant. The appropriate cDNAs were then isolated from those *E. coli* colonies that could grow under highly aerobic selective conditions. This rather tedious isolation of the Fe superoxide dismutase cDNA was required[127] because there was no cDNA available for this enzyme from other eukaryotes. There was no information on the amino acid sequence of any other plant Fe superoxide dismutases. This study, as well as previous information (see, e.g., Reference 128) strongly suggest that plant Fe superoxide dismutase is located only in chloroplasts.

Fe superoxide dismutase has only been found in some angiosperm families (e.g., Cruciferae, Solanaceae, and Leguminoseae) and is absent from most others. The Fe superoxide dismutase cDNA was then used for RNA gel blot analyses to study transcript abundance in stressed *N. plumbaginifolia.*[116] More than 95% of the mRNA for Fe superoxide dismutase disappeared when light-grown plants were placed in the dark for 3 d, but all reappeared upon exposure to light for 17 h. Also, exposure of etiolated seedlings to white light vastly increased the transcript level of Fe superoxide dismutase mRNA, but this increase was gradual and lagged behind the increase of mRNA for other nonstress-related genes such as *rbc*S that codes for the small subunit of ribulose-1,5-bisphosphate carboxylase-oxygenase (RuBisCo). Generally, the transcription of Fe superoxide dismutase genes is under more stringent photocontrol than the transcription of Mn superoxide dismutase genes, or the cytosol Cu,Zn superoxide dismutase genes.[116] The transcription of chloroplast Cu,Zn superoxide dismutase was not analyzed in this study.

Many genes coding for chloroplast-located proteins are under red-light control via phytochrome, but not the Fe superoxide dismutase. The photosystem II-inhibiting herbicide diuron stimulated the production of the Fe superoxide dismutase. This suggests that the resulting oxidative stress, rather than light per se, increased the level of the Fe superoxide dismutase transcript.

A combination of chilling and light increased the abundance of the Fe superoxide dismutase transcript, but not the transcripts for Mn superoxide dismutase or cytosol Cu,Zn superoxide dismutase. Paraquat (50 μM) treatment of light-grown plants vastly increased the level of Fe superoxide dismutase transcripts, but paraquat had no detectable effect in the dark. Only the transcript level of the cytosol[116,117] and chloroplast[115] Cu,Zn superoxide dismutases were increased in dark-grown plants sprayed with paraquat.

Thus a versatile arsenal for superoxide dismutation is available with the four types of plant genes: cytosol and chloroplast Cu,Zn superoxide dismutases, mitochondrial Mn superoxide dismutase, and chloroplast Fe superoxide dismutase. Each is differentially activated by external effectors. Will an enhanced level of one type of superoxide dismutase compensate for low levels of others? An answer to this question might come from experiments with transgenic plants having enhanced expression of specific superoxide dismutase-coding genes. (For a detailed discussion of this, see Chapter 12 of this volume).

D. ASCORBATE PEROXIDASE AND GLUTATHIONE REDUCTASE

Ascorbate peroxidase has been found in all higher plants tested, but is only present in about half of the cyanobacteria,[129] suggesting that it evolved very long ago in some cyanobacteria. Molecular genetic studies with plant ascorbate peroxidase have only recently begun (see Chapter 13 of this volume). It is to be expected that these plant ascorbate peroxidase cDNAs will be utilized in the future to study the changes in the expression of this gene in response to external affectors. The available cDNAs should also be instrumental in the isolation of their genomic sequences coding for the various plant ascorbate peroxidases. Then the available regulatory sequences (i.e., untranslated upstream and downstream nucleotide sequences) should be useful for further studies on controlling the activity of ascorbate peroxidase and its involvement in photooxidative stress protection. Similarly, the characterization of plant glutathione reductase cDNAs is in progress (see Chapter 13 of this volume). As discussed below, bacterial genes were used to obtain transgenic plants that expressed this gene.

V. TRANSGENIC PLANTS EXPRESSING ALIEN SUPEROXIDE DISMUTASE GENES

In our own laboratory, transgenic potato plants were produced with cDNAs coding for either chloroplast or cytosol Cu,Zn superoxide dismutases.[62] The cDNAs were fused downstream of a strong promoter to give a high level of expression. Transgenic potato shoots overexpressing either the chloroplast or the cytosol Cu,Zn superoxide dismutase, as well as shoots from untransformed plants, were treated with 100 μM paraquat. Several plants transformed with

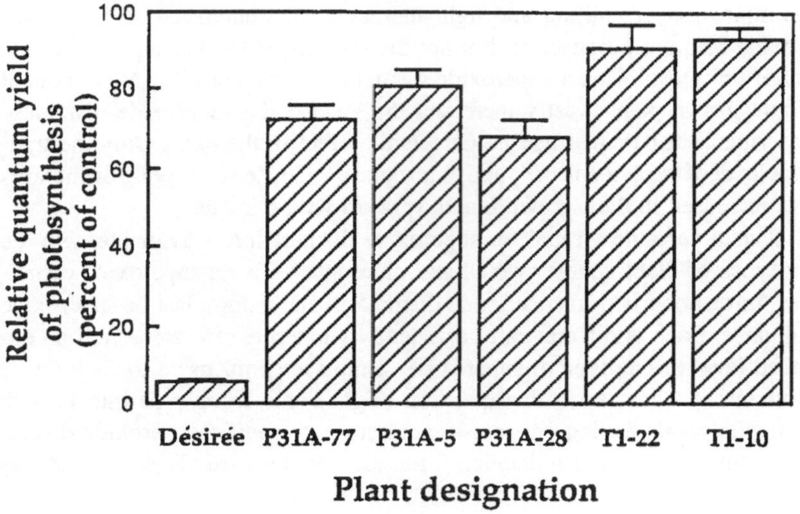

FIGURE 6. Relative quantum yield of photosynthetic oxygen evolution in the presence of paraquat. *In vitro* shoots were infiltrated for 10 min with three 10 μM of paraquat, followed by a 4-h incubation in the same solution — incubation in the light (70 $\mu E\ m^{-2}\ s^{-1}$). The relative values are expressed as a percentage of the control quantum yield (i.e., in the absence of paraquat for each plant). The first two fully expanded leaves from *in vitro* transformed shoots of the same developmental stage were used for all determinations. Vertical bars represent the standard error of the mean for triplicate determinations. (From Perl, A., Perl-Treves, R., Galli, S., Aviv, D., Shalgi, E., Malkin, S., and Galun, E., *Theor. Appl. Genet.*, 85, 568, 1993. With permission of Springer Verlag.)

either the cytosol or the chloroplast Cu,Zn superoxide dismutase withstood this paraquat treatment, while all control plants succumbed. Photosynthesis was less affected by paraquat in the transgenic plants. Potato roots grown *in vitro* were highly sensitive to paraquat (Figure 6). The transgenic potato roots expressing the cytosol Cu,Zn superoxide dismutase gene from tomato were much more resistant to paraquat than were roots from control plants.

There are several ways to explain differences in the reaction to paraquat by the transgenic tobacco and tomato (containing the petunia chloroplast Cu,Zn superoxide dismutase gene vs. the effects on transformed potato).[130,131] The promoter used to express the petunia chloroplast Cu,Zn superoxide dismutase gene required a few hours of light before expression could occur. The 35S CaMV promoter used in potato induces "constitutive" expression of the alien gene. Conversely, the level of superoxide dismutase in the chloroplasts of tobacco and tomato may not be the bottleneck of defense against oxidative damage. Indeed, transgenic *Nicotiana* plants overexpressing either the chloroplast or the cytosol Cu,Zn superoxide dismutase gene from tomato are apparently not more resistant to paraquat than control plants.[130,132] The bottleneck may be in the levels of the enzymes that cope with the peroxide produced from the superoxide. If they are low in tomato and tobacco, the increased level of superoxide dismutase would be of little help. Kinetic studies on the levels of the enzymes that scavenge peroxide in tomato, tobacco, and potato under nonstress and oxygen stress conditions should clarify the picture.

VI. CONCLUDING REMARKS

There is a basic difference between the classical genetic and the molecular genetic studies. The classical genetic studies utilized mutant biotypes that were resistant to one or more photooxidant stress. In no case could such resistance be traced to a modification in a structural gene or even to modification in a controlling element of a known gene. Indeed, from some of the genetic studies, it seems that many stress tolerance pathways are simultaneously initiated, and not necessarily just those for photooxidant stress (e.g., the paraquat-resistant tobacco that has cross-tolerance to mercurials, etc.).[56] Conversely, the vast majority of the molecular genetic studies assayed the effects of transformations with single genes, usually unmodified in their enzymatic properties, having different levels of expression and different subcellular localizations.

This single-gene approach of the molecular biologist can help elucidate the role of each gene product in a manner not yet used for such research. Genes can be inserted in inverse constructs, giving rise to antisense RNA. This causes, for unclear reasons, a suppression of production of the gene product in the resultant transformants.[133] The actual level of suppression varies from transformant to transformant (see, e.g., Reference 134). It would then be possible to assay both the loss of photooxidative stress tolerance and the level of the enzyme in question. It will also be possible to assay if the decrease in the level of the enzyme in question was compensated for by an increase in isoenzymes controlled by related genes. Additionally, one can ascertain whether the levels of other enzymes of the photooxidative pathway are decreased, along with the antisense-suppressed enzyme. It will be necessary to have such antisense constructs made from compatible plant genes. The bacterial genes used to enhance production of pathway enzymes probably do not have sufficient sequence homology for this approach. This approach cannot be paralleled by classical selection for the mutant having less of an enzyme, because of possible pleiotropy.

The molecular approach may well have many long-term practical advantages. The constitutive or induced levels of a plethora of stress tolerance pathways, so common in the genetic studies, must have a yield penalty (a decrease in competitive fitness), both in ideal growing conditions and when only a photooxidative stress is applied, because too many genes are activated. If other stress tolerance pathways are unneeded at such times, they are costly.

The molecular approach has the advantage, when fine tuned, of activating one or a few needed enzymes to confer stress tolerance, and only when needed. This could have far less cost in fitness than the field selection of tolerant mutants. The fine tuning of the molecular approach will require a far better understanding of the ideal balances between the different enzymatic and subcellular aspects of the photooxidative stress pathways. The molecular biologists have so far worked largely with blinkers, usually dealing with only one enzyme of the system at a time, ignoring others. They were lucky when that enzyme was the limiting factor in the system, unlucky when the gene was introduced into an inadequate background. So far, the genetic engineers have mainly used promoters that give high constitutive expression of the enzymes. This means that a substantial part of cellular resources go for their synthesis. Ultimately, light- or stress-inducible, growth-stage, or tissue-specific promoters will be used for greater economy. Eventually, the molecular approach has the potential of being far more effective when the optimal balances among the physiological components of the photooxidative stress tolerance pathways are understood. Until then, we are blind scientists groping different parts of an elephant.

ACKNOWLEDGMENTS

The authors deeply acknowledge the assistance of many colleagues for providing insights and unpublished data incorporated into this review. The authors' research in this area is supported by the Minerva Foundation, Munich, Germany, and by Grant No. U.S. 2043-01 from BARD, the United States-Israel Binational Agricultural Research and Development Fund. J. G. has the Gilbert de Botton-endowed chair of plant sciences and E. G. has the Irene and David Schwartz-endowed chair of plant genetics.

REFERENCES

1. **Sharon, A. and Gressel, J.,** Elicitation of a flavonoid phytoalexin accumulation in *Cassia obtusifolia* by a mycoherbicide: estimation by $AlCl_3$-spectrofluorimetry, *Pestic. Biochem. Physiol.*, 41, 142, 1991.
2. **Bradshaw, A. D. and Hardwick, K.,** Evolution and stress-genotypic and phenotypic components, *Biol. J. Linnean Soc.*, 37, 137, 1989.
3. **Shaaltiel, Y., Chua, N.-H, Gepstein, S., and Gressel, J.,** Dominant pleiotropy controls enzymes co-segregating with paraquat resistance in *Conyza bonariensis*, *Theor. Appl. Genet.*, 75, 850, 1988.
4. **Cakmak, I. and Marschner, H.,** Magnesium deficiency and high light intensity enhance activities of superoxide dismutase, ascorbate peroxidase, and glutathione reductase in bean leaves, *Plant Physiol.*, 98, 1222, 1992.
5. **Clark, C. W. and Harvell, C. D.,** Inducible defenses and the allocation of resources: a minimal model, *Am. Nat.*, 139, 521, 1992.
6. **Anderson, J. V., Chevone, B. I., and Hess, J. L.,** Seasonal variation in the antioxidant system of eastern white pine needles, *Plant Physiol.*, 98, 501, 1992.
7. **Gillham, D. J. and Dodge, A. D.,** Chloroplast superoxide and hydrogen peroxide scavenging systems from pea leaves: seasonal variations, *Plant Sci.*, 50, 105, 1987.

8. **Vaughan, D., DeKock, P. C., and Ord, B. G.,** The nature and localization of superoxide dismutase in fronds of *Lemna gibba* L. and the effect of copper and zinc deficiency on its activity, *Physiol. Plant.*, 54, 253, 1982.
9. **Cakmak, I. and Marschner, H.,** Enhanced superoxide radical production in roots of zinc-deficient plants, *Exp. Bot.*, 39, 207, 1988.
10. **Baron-Ayala, M. and Sandmann, G.,** Activities of Cu-containing proteins in Cu-depleted pea leaves, *Physiol. Plant.*, 72, 801, 1988.
11. **De Vos, C. H. R., Vonk, M. J., Vooijs, R., and Schat, H.,** Glutathione depletion due to copper-induced phytochelatin synthesis causes oxidative stress in *Silene cucubalus*, *Plant Physiol.*, 98, 853, 1992.
12. **Schat, H. and Ten Bookum, W. M.,** Genetic control of copper tolerance in *Silene vulgaris*, *Heredity*, 68, 219, 1992.
13. **Shaish, A.,** Production and selection of high β-carotene mutants of *Dunaliella bardawil* (Chlorophyta), *J. Phycol.*, 27, 652, 1991.
14. **Ben-Amotz, A., Shaish, A., and Avron, M.,** Mode of action of massively accumulated β-carotene of *Dunaliella bardawil* in protecting the alga against damage by excess irradiation, *Plant Physiol.*, 91, 1040, 1989.
15. **Steinitz, Y., Mazor, Z., and Shilo, M.,** A mutant of the cyanobacterium *Plectonema boryanum* resistant to photooxidation, *Plant Sci. Lett.*, 16, 327, 1979.
16. **Clare, D. A., Rabinowitch, H. D., and Fridovich, I.,** Superoxide dismutase and chilling injury in *Chlorella ellipsoidea*, *Arch. Bioch. Biophys.*, 231, 158, 1984.
17. **Malan, C., Greyling, M. M., and Gressel, J.,** Correlation between CuZn superoxide dismutase and glutathione reductase, and environmental and xenobiotic stress tolerance in maize inbreds, *Plant Sci.*, 69, 157, 1990.
18. **Madamanchi, N. R. and Alscher, R. G.,** Metabolic bases for differences in sensitivity of two pea cultivars to sulfur dioxide, *Plant Physiol.*, 97, 88, 1991.
19. **Wilson, G. B. and Bell, J. N. B.,** Studies on the tolerance to sulphur dioxide of grass populations in polluted areas. VI. The genetic nature of tolerance in *Lolium perenne* L., *New Phytol.*, 116, 313, 1990.
20. **Shaaltiel, Y., Glazer, A., Bocion, P. F., and Gressel, J.,** Cross tolerance to herbicidal and environmental oxidants of plant biotypes tolerant to paraquat, sulfur dioxide and ozone, *Pestic. Biochem. Physiol.*, 31, 13, 1988.
21. **Taylor, G. E.,** Genetic analysis of ecotypic differentiation within an annual plant species, *Geranium carolinianum* L. in response to SO_2, *Bot. Gaz.*, 139, 362, 1978.
22. **Omura, T., Satoh, H., Fujinuma, Y., and Aiga, I.,** Inheritance of sensitivity to SO_2 in rice *Oryza sativa*, *Res. Rep. Natl. Inst. Envir. Studies*, 11, 263, 1980; as cited in Wilson, G. B. and Bell, J. N. B., Studies on the tolerance to sulphur dioxide of grass populations in polluted areas. VI. The genetic nature of tolerance in *Lolium perenne* L., *New Phytol.*, 116, 313, 1990.
23. **Ranieri, A., Durante, M., Volterrani, A., Lorenzini, G., and Soldatini, G. F.,** Effects of low SO_2 levels on superoxide dismutase and peroxidase isoenzymes in two different wheat cultivars, *Biochem. Physiol. Pflanzen*, 188, 67, 1992.
24. **Miller, V. L., Howell, R. K., and Caldwell, B. E.,** Relative sensitivity of soybean genotypes to ozone and sulfur dioxide, *J. Envir. Qual.*, 3, 35, 1974.
25. **Tanaka, K., Machida, T., and Sugimoto, T.,** Ozone tolerance and glutathione reductase in tobacco cultivars, *Agric. Biol. Chem.*, 54, 1061, 1990.
26. **Heggestad, H. E.,** Origin of Bel-W3, Bel-C and Bel-B tobacco varieties and their use as indicators of ozone, *Environ. Pollut.*, 74, 264, 1991.
27. **Berrang, P., Karnosky, D. F., and Bennett, J. P.,** Natural selection for ozone tolerance in *Populus tremuloides:* an evaluation of nationwide trends, *Can. J. For. Res.*, 21, 1091, 1991.
28. **Butler, L. K., Tibbits, T. W., and Bliss, F. A.,** Inheritance of resistance to ozone in *Phaseolus vulgaris* L., *J. Am. Soc. Hortic. Sci.*, 104, 211, 1979.

29. **Mersie, W., Mebrahu, T., and Rangappa, M.,** Evaluation of bean introductions for ozone sensitivity, *HortScience,* 25, 1581, 1990.
30. **Lee, E. H., Kramer, G. F., Rowland, R. A., and Agrawal, M.,** Antioxidants and growth regulators counter the effects of O_3 and SO_2 in crop plants, *Agric. Ecosys. Envir.*, 38, 99, 1992.
31. **Batchvarova, R. B., Reddy, V. S., and Bennett, J.,** Cellular resistance in rice to Cercosporin, a toxin of *Cercospora, Phytopathology,* 82, 642, 1992.
32. **Askira, Y., Rubin, B., and Rabinowitch, H. D.,** Differential response to the herbicidal activity of δ-aminolevulinic acid in plants with high and low SOD activity, *Free Rad. Res. Comm.*, 12, 837, 1990.
33. **Hickok, L. G., Warne, T. R., and Slocum, M. K.,** *Ceratopteris richardii:* applications for experimental plant biology, *Am. J. Bot.*, 74, 1304, 1987.
34. **Yu, C. Y. and Masiunas, J. B.,** Characterization of acifluorfen tolerance in selected somaclones of eastern black nightshade (*Solanum ptycanthum* Dun.), *Weed Sci. Soc. Am. Abstr.*, 32, 201, 1992.
35. **Harvey, B. M. R. and Harper, D. B.,** Tolerance to bipyridylium herbicides, in *Herbicide Resistance in Plants,* Le Baron, H. M. and Gressel, J., Eds., John Wiley & Sons, New York, 215, 1982.
36. **Faulkner, J. S.,** Breeding herbicide-tolerant crop cultivars by conventional methods, in *Herbicide Resistance in Plants,* Le Baron, H. M. and Gressel, J., Eds., John Wiley & Sons, New York, 235, 1982.
37. **Bewick, T. A., Stall, W. M., Kostewicz, S. R., and Smith, K.,** Alternatives for control of paraquat-tolerant american black nightshade *(Solanum americana), Weed Tech.*, 5, 61, 1991.
38. **Vaughn, K. C., Vaughan, M. A., and Camilleri, P.,** Lack of cross-resistance of paraquat-resistant hairy fleabane *(Conyza bonariensis)* to other toxic oxygen generators indicates enzymatic protection is not the resistance mechanism, *Weed Sci.*, 37, 5, 1989.
39. **Jansen, M. A. K., Shaaltiel, Y., Kazzes, D., Canaani, O., Malkin, S., and Gressel, J.,** Increased tolerance to photoinhibitory light in paraquat-resistant *Conyza bonariensis* measured by photoacoustic spectroscopy and $^{14}CO_2$-fixation, *Plant Physiol.*, 91, 1174, 1989.
40. **Youngman, R. J. and Dodge, A. D.,** On the mechanism of paraquat resistance in *Conyza* sp., in *Proc. 5th Int. Cong. Photosynthesis,* Akoyunoglou, G., Ed., Balaban International Science, Philadelphia, 537, 1981.
41. **Shaaltiel, Y. and Gressel, J.,** Multienzyme oxygen radical detoxifying system correlated with paraquat resistance in *Conyza bonariensis, Pestic. Biochem. Physiol.*, 26, 22, 1986.
42. **Varadi, G., Lehoczki, E., Szigeti, Z., and Pölös, E.,** Short-term paraquat effects on paraquat-resistant horseweed, *Conyza canadensis* Cronq., in Proc. 9th Australian Weeds Conf., Adelaide, South Australia, 1990, 257.
43. **Matsunaka, S. and Ito, K.,** Paraquat resistance in Japan, in *Herbicide Resistance in Weeds and Crops,* Caseley, J. C., Cussans, G. W., and Atkin, R. K., Eds., Butterworths, Oxford, 1991, 77.
44. **Itoh, K.,** Paraquat resistance in *Erigeron philadelphicus* L., *Jpn. Agric. Res. Q.*, 22, 85, 1988.
45. **Gressel, J., Ammon, H. U., Fogelfors, H., Gasquez, J., Kay, Q. O. N., and Kees, H.,** Discovery and distribution of herbicide-resistant weeds outside North America, in *Herbicide Resistance in Plants,* LeBaron, H. M. and Gressel, J., Eds., John Wiley & Sons, New York, 1982, 31.
46. **Bulcke, R., Verstraete, F., Van Himme, M., and Stryckers, J.,** Biology and control of *Epilobium ciliatum* Rafin., in *Weed Control on Vine and Soft Fruits,* Cavalloro, R. and Robinson, D. W., Eds., A.A. Balkema, Amsterdam, 1987, 57.
47. **Carroll, E. W., Schwarz, O. J., and Hickok, L. G.,** Biochemical studies of paraquat-tolerant mutants of the fern *Ceratopteris richardii, Plant Physiol.*, 87, 651, 1988.

48. **Hickok, L. G. and Schwarz, O. J.,** Paraquat tolerant mutants in *Ceratopteris:* genetic characterization and reselection for enhanced tolerance, *Plant Sci.*, 47, 753, 1986.
49. **Powles, S. B., Tucker, E. S., and Morgan, T. R.,** A capeweed *(Arctotheca calendula)* biotype in Australia resistant to bipyridyl herbicides, *Weed Sci.*, 37, 60, 1989.
50. **Preston, C. J., Holtum, J. A. M., and Powles, S. B.,** On the mechanism of resistance to paraquat in *Hordeum glaucum* and *H. leporinum.* Delayed inhibition of photosynthetic activity following paraquat application, *Plant Phys.*, 100, 630, 1992.
51. **Islam, A. K. M. R. and Powles, S. B.,** Inheritance of resistance to paraquat in barley grass *Hordeum glaucum* Steud., *Weed Res.*, 28, 393, 1988.
52. **Thomas, B. R. and Pratt, D.,** Isolation of paraquat-tolerant mutants from tomato cell cultures, *Theor. Appl. Genet.*, 63, 169, 1982.
53. **Tanaka, K., Furusawa, I., Kondo, N., and Tanaka, K.,** SO_2 tolerance of tobacco plants regenerated from paraquat-tolerant callus, *Plant Cell Physiol.*, 29, 743, 1988.
54. **Furusawa, I. and Mizuguchi, A.,** Superoxide resistant plants regenerated from protoplasts, in 5th Int. Congr. Plant Pathology, Kyoto, Japan, 1988.
55. **Gullner, G., Komives, T., and Kiraly, Z.,** Enhanced inducibility of antioxidant systems in *Nicotiana tabacum* biotype results in acifluorfen resistance, *Z. Naturforsch. Teil C,* 46, 875, 1991.
56. **Barna, B., Adam, A. L., and Kiraly, Z.,** Juvenility and resistance of a superoxide tolerant plant to diseases and other stresses, submitted.
57. **Furusawa, I., Tanaka, K., Thanuton, P., and Mizuguchi, A.,** Paraquat resistant tobacco calluses with enhanced superoxide dismutase activity, *Plant Cell Physiol.*, 25, 1247, 1984.
58. **Bowler, C., Slooten, L., Vandenbranden, S., De Rycke, R., Botterman, J., Sybesma, C., Van Montagu, M., and Inze, D.,** Manganese superoxide dismutase can reduce cellular damage mediated by oxygen radicals in transgenic plants, *EMBO J.*, 10, 1723, 1991.
59. **Huckett, B. I., Wittaker, A., Roberts, M. A., Kunert, K. Y., Watt, M. P., Amory, A. M., and Cresswell, C. F.,** Use of transformed tobacco plants containing the *E. coli* glutathione oxido-reductase gene to investigate the role of antioxidant enzymes during oxygen toxicity, submitted.
60. **Aono, M., Kubo, A., Saji, H., Natori, T., Tanaka, K., and Kondo, N.,** Resistance to active oxygen toxicity of transgenic *Nicotiana tabacum* that expresses the gene for glutathione reductase from *Escherichia coli, Plant Cell Physiol.*, 32, 691, 1991.
61. **Foyer, C., Lelandais, M., Galap, C., and Kunert, K. J.,** Effects of elevated cytosolic glutathione reductase activity on the cellular glutathione pool and photosynthesis in leaves under normal and stress conditions, *Plant Physiol.*, 97, 863, 1991.
62. **Perl, A., Perl-Treves, R., Galili, S., Aviv, D., Shalgi, E., Malkin, S., and Galun, E.,** Enhanced oxidative-stress defense in transgenic potato expressing tomato Cu,Zn superoxide dismutases, *Theor. Appl. Genet.*, 85, 568, 1993.
63. **Aycock, M. K.,** Combining ability estimates for weather fleck in *Nicotiana tabacum, Crop Sci.*, 12, 672, 1972.
64. **Sung, C. H., Chen, H. H., Street, O. E., and Yang, Y. L.,** Gene effects for tolerance to weather fleck in tobacco, *Taiwan Agric. Q.*, 7, 173, 1971.
65. **Taylor, G. S.,** Ozone injury on Bel-W3 tobacco controlled by at least two genes, *Phytopathology,* 58, 1069, 1968.
66. **Prochaska, H. J. and Talalay, P.,** Regulatory mechanisms of monofunctional and bifunctional anticarcinogenic enzyme inducers in murine liver, *Cancer Res.*, 48, 4776, 1988.
67. **McCue, K. F. and Hanson, A. D.,** Drought and salt tolerance: towards understanding and application, *Tibtech,* 8, 358, 1990.
68. **Burke, J. J., Gamble, P. E., Hatfield, J. L., and Quisenberry, J. E.,** Plant morphological and biochemical responses to field water deficits. I. Responses of glutathione reductase activity and paraquat sensitivity, *Plant Physiol.*, 79, 415, 1985.

69. **Mahan, J. R. and Burke, J. M.,** Purification and characterization of glutathione reductase from corn mesophyll chloroplasts, *Physiol. Plant.*, 71, 352, 1987.
70. **Smirnoff, N. and Colombe, S. V.,** Drought influences the activity of enzymes of the chloroplast hydrogen peroxide scavenging system, *J. Exp. Bot.*, 39, 1097, 1988.
71. **Arbi, N., Smith, D., and Bingham, E. T.,** Dry matter and morphological responses to temperatures of alfalfa strains with differing ploidy levels, *Agron. J.*, 71, 573, 1979.
72. **Kidambi, S. P., Mahan, J. R., and Matches, A. G.,** Purification and thermal dependence of glutathione reductase from two forage legume species, *Plant Physiol.*, 92, 363, 1990.
73. **Kidambi, S. P., Mahan, J. R., Matches, A. G., and Bingham, E. T.,** Thermal dependence of glutathione reductase in one tetraploid and its near-isogenic diploid alfalfa, *Euphytica*, 52, 165, 1991.
74. **Chamovitz, D., Pecker, I., Sandmann, G., Böger, P., and Hirschberg, J.,** Cloning a gene coding for norflurazon resistance in Cyanobacteria, *Z. Naturforsch. Teil C*, 45, 482, 1990.
75. **Preston, C., Holtum, J. A. M., and Powles, S. B.,** Resistance to the herbicide paraquat and increased tolerance to photoinhibition are not correlated in several weed species, *Plant Physiol.*, 96, 314, 1991.
76. **LeBaron, H. M.,** Distribution and seriousness of herbicide-resistant weed infestations worldwide, in *Herbicide Resistance in Weeds and Crops*, Caseley, J. C., Cussans, G. W., and Atkin, R. K., Eds., Butterworths, Oxford, 1991, 27.
77. **Trebst, A.,** The molecular basis of resistance to photosystem II herbicides, in *Herbicide Resistance in Weeds and Crops*, Caseley, J. C., Cussans, G. W., and Atkin, R. K., Eds., Butterworths Oxford, 1991, 145.
78. **Anderson, M. P. and Gronwald, J. W.,** Atrazine resistance in a velvetleaf *(Abutilon theophrasti)* biotype due to enhanced glutathione S-transferase activity, *Plant Physiol.*, 96, 104, 1991.
79. **Burnet, M. W. M., Lovcys, B. R., Holtum, J. A. M., and Powles, S. B.,** Increased detoxification is a mechanism of simazine resistance, *Pestic. Biochem. Physiol.*, 46, 207, 1993.
80. **Shaaltiel, Y. and Gressel, J.,** Kinetic analysis of resistance to paraquat in *Conyza*, *Plant Physiol.*, 85, 869, 1987.
81. **Gressel, J. and Shaaltiel, Y.,** Biorational herbicide synergists, in *Biotechnology for Crop Protection*, Hedin, P. A., Menn, J. J., and Hollingworth, R. M., Eds., ACS Symp. Series, American Chemical Society, Washington, D.C., 1988, 4.
82. **Kunert, K. J. and Dodge, A. D.,** Herbicide induced radical damage and antioxidative systems, in *Target Sites of Herbicide Action*, Böger, P. and Sandmann, G., Eds., CRC Press, Boca Raton, FL 1989, 50.
83. **Pierre, M. and Queiroz, O.,** Air pollution by SO_2 amplifies the effects of water stress on enzymes and total soluble proteins of spruce needles, *Physiol. Plant.*, 73, 412, 1988.
84. **Mehlhorn, H.,** Ethylene-promoted ascorbate peroxidase activity protects plants against hydrogen peroxide, ozone and paraquat, *Plant Cell Environ.*, 13, 971, 1990.
85. **Moan, J., Pettersen, E. O., and Christensen, T.,** The mechanism of photodynamic inactivation of human cells *in vitro* in the presence of haematoporphyrin, *Br. J. Cancer*, 39, 398, 1979.
86. **Amsellem, Z., Jansen, M. A. K., Driesenaar, A. R. J., and Gressel, J.,** Developmental variability of photooxidative stress tolerance in paraquat resistant *Conyza*, paraquat resistance, photoinhibition, superoxide dismutase and glutathione reductase cross correlated, *Plant Physiol.*, in press..
87. **Finckh, B. F. and Kunert, K. J.,** Vitamins C and E, an antioxidative system against herbicide-induced lipid peroxidation in higher plants, *J. Agric. Food Chem.*, 33, 574, 1985.

88. **Ben-Amotz, A., Gressel, J., and Avron, M.,** Massive accumulation of phytoene induced by norflurazon in *Dunaliella bardawil* (Chlorophyceae) prevents recovery from photoinhibition, *J. Phycol.*, 23, 176, 1987.
89. **Yuting, C., Ronglian, Z., Zhongjian, J., and Yong, J.,** Flavonoids as superoxide scavengers and antioxidants, *Free Rad. Biol. Med.*, 9, 19, 1990.
90. **Sichel, G., Corsaro, C., Scalia, M., Di Bilio, A. J., and Bonomo, R. P.,** *In vitro* scavenger activity of some flavonoids and melanins against superoxide, *Free Rad. Biol. Med.*, 11, 1, 1991.
91. **Badiani, M., De Biasi, M. G., Colognola, M., and Artemi, F.,** Catalase, peroxidase and superoxide dismutase activities in seedlings submitted to increasing water deficit, *Agrochimica*, 34, 90, 1990.
92. **Griffin, J. D. and Palmer, R. G.,** Genetic studies with two superoxide dismutase loci in soybean, *Crop Sci.*, 29, 4, 968.
93. **Huttunen, S. and Heiska, E.,** Superoxide dismutase (SOD) activity in Scots pine (*Pinus sylvestris* L.) and Norway spruce (*Picea abies* L. Karst.) needles in northern Finland, *Eur. J. Forensic Pathol.*, 18, 343, 1988.
94. **Zacheo, G. and Bleve-Zacheo, T.,** Involvement of superoxide dismutases and superoxide radicals in the susceptibility and resistance of tomato plants to *Meloidogyne incognita* attack, *Physiol. Mol. Plant Pathol.*, 32, 313, 1988.
95. **Thomsen, B., Drumm-Herrel, H., and Mohr, H.,** Control of the appearance of ascorbate peroxidase (EC 1.11.1.11) in mustard seedling cotyledons by phytochrome and photooxidative treatments, *Planta*, 186, 600, 1992.
96. **Tanaka, K., Saji, H., and Kondo, N.,** Immunological properties of spinach glutathione reductase and inductive biosynthesis of the enzyme with ozone, *Plant Cell Physiol.*, 29, 637, 1988.
97. **Schmidt, A. and Kunert, K. J.,** Lipid peroxidation in higher plants, the role of glutathione reductase, *Plant Physiol.*, 82, 700, 1986.
98. **Kwiatowski, J. and Kaniuga, Z.,** Isolation and characterization of cytosolic and chloroplast isoenzymes of Cu,Zn-superoxide dismutase from tomato leaves and their relationships to other Cu,Zn-superoxide dismutases, *Biochim. Biophys. Acta*, 874, 99, 1986.
99. **Steffens, G. J., Michelson, A. M., Ötting, F., Puget, K., Strassburger, W., and Flohe, L.,** Primary structure of Cu,Zn superoxide dismutase of *Brassica oleracea* proves homology with corresponding enzymes of animals, fungi and prokaryotes, *Biol. Chem. Hoppe-Seyler*, 367, 1007, 1986.
100. **Kitagawa, Y., Tsunasawa, S., Tanaka, N., Katsubi, Y., Sakiyama, F., and Asada, K.,** Amino acid sequence of copper, zinc-superoxide dismutase from spinach leaves, *J. Biochem.*, 99, 1289, 1986.
101. **Chongpraditnum, P., Mori, S., and Chino, M.,** Excess copper induces a cytosolic Cu,Zn-superoxide dismutase in soybean root, *Plant Cell Physiol.*, 33, 239, 1992.
102. **Kanematsu, S. and Asada, K.,** Characteristic amino acid sequences of chloroplast and cytosol isozymes of Cu,Zn-superoxide dismutase in spinach, rice and horsetail, *Plant Cell Physiol.*, 31, 99, 1990.
103. **Kanematsu, S. and Asada, K.,** Cu,Zn-superoxide dismutases from the fern *Equisetum arvense* and the green alga *Spirogyra* sp.: occurrence of chloroplast and cytosol types of the enzyme, *Plant Cell Physiol.*, 30, 717, 1989.
104. **Perl-Treves, R., Nacmias, B., Aviv, D., Zeelon, E. P., and Galun, E.,** Isolation of two cDNA clones from tomato containing two different superoxide dismutase sequences, *Plant Mol. Biol.*, 11, 609, 1988.
105. **Scioli, J. R. and Zilinskas, B. A.,** Cloning and characterization of a cDNA encoding the chloroplastic copper zinc-superoxide dismutase from pea, *Proc. Natl. Acad. Sci. U.S.A.*, 85, 7661, 1988.
106. **Tepperman, J., Katayama, C., and Dunsmuir, P.,** Cloning and nucleotide sequence of a petunia gene encoding a chloroplast-localized superoxide dismutase, *Plant Mol. Biol.*, 11, 871, 1988.

107. **Cannon, R. E., White, J. A., and Scandalios, J. G.,** Cloning of cDNA for maize superoxide dismutase 2 (SOD[2]), *Proc. Natl. Acad. Sci. U.S.A.*, 84, 179, 1987.
108. **White, D. A. and Zilinskas, B. A.,** Nucleotide sequence of a complementary DNA encoding pea cytosolic copper/zinc superoxide dismutase, *Plant Physiol.*, 96, 1391, 1991.
109. **Hindges, R. and Slusarenko, A.,** cDNA and derived aminoacid sequence of a cytosolic Cu,Zn superoxide dismutase from *Arabidopsis thaliana, Plant Mol. Biol.*, 18, 123, 1992.
110. **Sakamoto, A., Ohsuga, H., and Tanaka, K.,** Nucleotide sequences of two cDNA clones encoding different Cu,Zn-superoxide dismutases expressed in developing rice seed (*Oryza sativa* L.), *Plant Mol. Biol.*, 19, 323, 1992.
111. **Perl-Treves, R., Abu-Abied, M., Magal, N., Galun, E., and Zamir, D.,** Genetic mapping of tomato cDNA clones encoding the chloroplastic and the cytosolic isozymes of superoxide dismutase, *Biochem. Genet.*, 28, 543, 1990.
112. **Isin, S. H., Burke, J. J., and Allen, R. D.,** Sequence divergence of pea Cu/Zn superoxide dismutase II cDNAs, *Plant Mol. Biol.*, 15, 789, 1990.
113. **Cannon, R. E. and Scandalios, J. G.,** Two cDNAs encode two nearly identical Cu/Zn superoxide dismutase proteins in maize, *Mol. Gen. Genet.*, 219, 1, 1989.
114. **Dietrich, R. A., Radke, S. E., and Harada, J. J.,** Downstream DNA sequences are required to activate a gene expressed in the root cortex of embryos and seedlings, *Plant Cell*, 4, 1371, 1992.
115. **Perl-Treves, R. and Galun, E.,** The tomato Cu,Zn superoxide dismutase genes are developmentally regulated and respond to light and stress, *Plant Mol. Biol.*, 17, 745, 1991.
116. **Tsang, E. W. T., Bowler, C., Herouart, D., Van Camp., W., Villarroel, R., Genetello, C., Van Montagu, M., and Inze, D.,** Differential regulation of superoxide dismutases in plants exposed to environmental stress, *Plant Cell*, 3, 783, 1991.
117. **Perl-Treves, R.,** Molecular Cloning and Expression of Cu,Zn Superoxide Dismutases in Tomato, Ph.D. thesis, Weizmann Institute of Science, Rehovot, Israel, 1990.
118. **Sakamoto, A., Okamura, T., Ohsuga, H., and Tanaka, K.,** Genomic structure of the gene for copper/zinc-superoxide dismutase in rice, *FEBS Lett.*, 301, 185, 1992.
119. **Magal, N., Perl-Treves, R., and Galun, E.,** unpublished results, 1992.
120. **Getzoff, E. D., Cabelli, D. E., Fisher, C. L., Parge, H. E., Viezzoli, M. S., Banci, L., and Hallewell, R. A.,** Faster superoxide dismutase mutants designed by electrostatic guidance, *Nature (London)*, 358, 347, 1992.
121. **Bannister, J. V., Bannister, W. H., and Rotillo, G.,** Aspects of the structure, function and applications of superoxide dismutase, *CRC Crit. Rev. Biochem.*, 22, 111, 1987.
122. **Beyer, W., Imlay, J., and Fridovich, I.,** Superoxide dismutase, *Prog. Nucleic Acid Res. Mol. Biol.*, 40, 221, 1991.
123. **White, J. A. and Scandalios, J. G.,** Isolation and characterization of a cDNA for mitochondrial manganese superoxide dismutase (SOD-3) of maize and its relation to other manganese superoxide dismutases, *Biochim. Biophys. Acta*, 951, 61, 1988.
124. **White, J. A. and Scandalios, J. G.,** Deletion analysis of the maize mitochondrial superoxide dismutase transit peptide, *Proc. Natl. Acad. Sci. U.S.A.*, 86, 3534, 1989.
125. **Bowler, C., Alliotte, T., Van der Bulcke, M., Bauw, G., Vandekerckhove, J., Van Montagu, M., and Inze, D.,** A plant manganese superoxide dismutase is efficiently imported and correctly processed by yeast mitochondria, *Proc. Natl. Acad. Sci. U.S.A.*, 86, 3237, 1989.
126. **Bowler, C., Alliotte, T., De Loose, M., Van Montagu, M., and Inze, D.,** The induction of manganese superoxide dismutase in response to stress in *Nicotiana plumbaginifolia, EMBO J.*, 8, 38, 1989.
127. **Van Camp, R., Bowler, C., Villarroel, R., Tsang, E. W. T., Van Montagu, M., and Inze, D.,** Characterization of iron superoxide dismutase cDNAs from plants obtained by genetic complementation in *Escherichia coli, Proc. Natl. Acad. Sci. U.S.A.*, 87, 9903, 1990.

128. **Kwiatowski, J., Safianowska, A., and Kaniuga, Z.,** Isolation and characterization of an iron-containing superoxide dismutase from tomato leaves, *Lycopersicon esculentum, Eur. J. Biochem.*, 146, 459, 1985.
129. **Miyake, C., Michihata, F., and Asada, K.,** Scavenging of hydrogen peroxide in prokaryotic and eukaryotic algae: acquisition of ascorbate peroxidase during the evolution of cyanobacteria, *Plant Cell Physiol.*, 32, 33, 1991.
130. **Tepperman, J. M. and Dunsmuir, P.,** Transformed plants with elevated levels of chloroplastic superoxide dismutase are not more resistant to superoxide toxicity, *Plant Mol. Biol.*, 14, 501, 1990.
131. **Pitcher, L. H., Brennan, E., Hurley, A., Dunsmuir, P., Tepperman, J. M., and Zilinskas, B.,** Overproduction of petunia chloroplastic copper/zinc superoxide dismutase does not confer ozone tolerance in transgenic tobacco, *Plant Physiol.*, 97, 452, 1991.
132. **Aviv, D., Perl Treves, R., and Galun, E.,** unpublished observations, 1992.
133. **Bird, C. R. and Ray, J. A.,** Manipulation of plant gene expression by antisense RNA, *Biotechnol. Genet. Eng. Rev.*, 9, 207, 1991.
134. **Höfgen, R. and Willmitzer, L.,** Transgenic potato plants depleted for the major tuber protein patatin via expression of antisense RNA, *Plant Sci.*, 87, 45, 1992.

Chapter 11

REGULATION AND PROPERTIES OF PLANT CATALASES

John G. Scandalios

TABLE OF CONTENTS

0-8493-5443-9/94/$0.00+$.50

I. INTRODUCTION

Aerobic organisms gain significant energetic advantages by using molecular oxygen as the terminal oxidant in respiration. However, the presence of molecular oxygen in the cell also carries potential physiological cost. Although O_2 is itself relatively harmless, it has the potential to be partially reduced and form toxic "activated" oxygen species. These partially reduced oxygen species are produced through programmed as well as aberrant metabolic processes and as a consequence of numerous environmental stresses. Activated forms of oxygen include the superoxide radical ($O_2^{\cdot-}$), the hydroxyl radical ($^{\cdot}OH$), and hydrogen peroxide (H_2O_2). All of these can react with certain biomolecules, altering or inactivating their biochemical activity. The combined biological effect of these toxic oxygen species on organisms is termed "oxidative stress". To minimize the damaging effects of activated oxygen, organisms have evolved various enzymatic and nonenzymatic mechanisms that can reduce oxidative stress by detoxifying harmful oxygen species. Among these defenses is the antioxidant enzyme catalase that converts H_2O_2 to oxygen and water.

A. STEPWISE REDUCTION OF O_2

During respiration, molecular oxygen accepts four electrons to produce two molecules of H_2O. However, because of spin restrictions, O_2 cannot accept four electrons at once, but, rather, accepts them one at a time.[1] Thus, during the reduction of O_2, stable intermediates are formed in a stepwise fashion. The consecutive univalent reductions of oxygen with reaction potentials are:

$$O_2 \xrightarrow[\ ]{e^-\ (-0.33V)} O_2^{\bar{\cdot}} \xrightarrow[2H^+]{e^-\ (+0.94V)} H_2O_2 \xrightarrow[H^+]{e^-\ (+0.46V)} H_2O + {}^{\cdot}OH \xrightarrow[H^+]{e^-\ (+2.18V)} H_2O$$

$$O_2 \xrightarrow{+4e^-\ (+0.82V)} H_2O$$

Molecular oxygen can accept one electron to form the superoxide radical. This first step is thermodynamically unfavorable and accounts for the low reactivity of O_2 itself. The superoxide radical is unique among the active oxygen species in that the thermodynamics are favorable for it to act as both an oxidant and a reductant. The superoxide radical will accept an electron and two protons to form hydrogen peroxide. Hydrogen peroxide, unlike the oxygen radicals, can readily diffuse across biological membranes; hence, it can cause oxidative stress far from the site of formation. It will accept an electron and proton to form a molecule of water and the hydroxyl radical. The hydroxyl radical is the most reactive of the oxygen species. It is an extremely potent oxidant and reacts with organic molecules at near-diffusion rates.

B. HYDROGEN PEROXIDE TOXICITY

H_2O_2 itself is a relatively stable oxidant and not highly reactive. Although some biomolecules (e.g., proteins) are directly sensitive to H_2O_2 (as will be discussed later), H_2O_2 has its primary cytotoxic effects through the formation of (${}^{\cdot}OH$).[2,3] In the Fenton reaction, H_2O_2 can form ${}^{\cdot}OH$ in the presence of a suitable transition metal.[4,5]

$$M^{n+} + H_2O_2 \rightarrow M^{n+1} + {}^{\cdot}OH + OH^-$$

In vivo, the transition metal believed to be available for promoting the formation of ${}^{\cdot}OH$ is iron, although copper may also play a minor role.[6] In the Haber-Weiss formulation of the reaction, $O_2^{\bar{\cdot}}$ acts as a reducing agent to recycle the iron.[5]

$$Fe^{3+} + O_2^{\bar{\cdot}} \rightarrow Fe^{2+} + O_2$$

There is a wide variety of sources of $O_2^{\bar{\cdot}}$ in the cell for the reduction of iron. These include radiation,[7] photosynthesis,[8] mitochondrial respiration,[9] herbicides,[10-12] and air pollution.[13,14]

As previously stated, ${}^{\cdot}OH$ is a highly reactive oxidant and will react indiscriminantly with biomolecules at near-diffusion rates; therefore, ${}^{\cdot}OH$

tends to oxidize molecules very near its site of formation. In this respect, it has been termed a "site-specific oxidant".[15] H_2O_2 diffuses throughout the cell, but the other component of the Fenton reaction, iron, is sequestered in the cell in metalloproteins and vacuoles. Biomolecules that are associated with metals *in vivo* are, therefore, often sensitive to H_2O_2-derived $^{\cdot}OH$.

Probably the most sensitive molecular target of H_2O_2-generated $^{\cdot}OH$ is DNA.[16-18] DNA is closely associated with a number of metalloproteins. Additionally, transition metals such as iron or copper may form salts along the phosphate backbone of DNA.[3,15] These metals are likely catalysts for the Fenton reaction-generated $^{\cdot}OH$.[19] *Escherichia coli* and mammalian cells subjected to H_2O_2 can suffer base loss, base modification, strand breakage, and chromosomal abnormalities.[17,20-22] The physiological consequences of DNA damage are, therefore, severe and can result in mutations and gene inactivation. Several studies suggest that DNA damage is the major cytotoxic effect of exogenously added H_2O_2.[16,17] In mammalian cells, H_2O_2-derived DNA damage leads to the activation of polyribose synthetase, an early enzyme in the proposed cascade of "programmed cell death".[22] Additionally, even under nonstressful conditions, oxidative damage on the DNA molecule can be significant. It has been estimated that the DNA of the average mammalian cell takes up to 3000 "oxidative hits" per day.[23]

Other targets of H_2O_2 toxicity are fatty acids and the bilayer lipid membranes.[24] Here again, although H_2O_2 is relatively unreactive with lipids, $^{\cdot}OH$ is very damaging, resulting in lipid peroxidation.[25] The $^{\cdot}OH$ attacks the lipid directly by extraction of a hydrogen atom, producing a carbon radical (Lipid$^{\cdot}$).

$$\text{Lipid} + {}^{\cdot}OH \rightarrow H_2O + \text{Lipid}^{\cdot}$$

The lipid carbon radical can then act as a catalyst in a progressive lipid peroxidation where lipids are converted to lipid peroxides. Lipid peroxides alter membrane quality and can interact and inactivate membrane proteins. Furthermore, lipid peroxides can spontaneously fragment, generating a variety of toxic compounds, including hydrocarbons and aldehydes.[26]

Many proteins are known to be sensitive to physiologically relevant levels of H_2O_2 (about 100 n*M*). Metalloproteins as a class seem to be sensitive to H_2O_2. These include heme proteins[27,28] and superoxide dismutase.[29,30] H_2O_2 reacts with these proteins, causing them to release the metal ion and lose their biological activity. Several Calvin cycle enzymes also appear to be especially sensitive to H_2O_2. These include fructose biphosphatase, sedoheptulose biphosphatase, glyceraldehyde-3-phosphate dehydrogenase, and phosphoribulokinase.[31-33] H_2O_2 appears to act here by oxidizing critical thiols in the enzymes, thereby inactivating them.

C. INTRACELLULAR PRODUCTION OF HYDROGEN PEROXIDE

H_2O_2 is produced in significant quantities in various subcellular organelles. Each organelle also has potential targets for H_2O_2 oxidative stress as well as mechanisms for eliminating H_2O_2. Further, H_2O_2 can readily diffuse through intra- and intercellular membranes, allowing for the interaction of organelles or even cell types.

1. Peroxisomes

In eukaryotic cells, much of the H_2O_2 produced is concentrated in specialized organelles called peroxisomes. They are widely distributed throughout eukaryotes, and in higher eukaryotes, they are found in every tissue examined. In peroxisomes, H_2O_2 is produced by oxidases, which are involved in the catabolic oxidation of a variety of biomolecules. Oxidases remove two electrons from substrates by way of a flavodoxin intermediate, ultimately transferring the electrons from O_2 to form H_2O_2.

Peroxisomes contain oxidases for a variety of general metabolic pathways.[37] The enzyme content of peroxisomes varies significantly with the metabolic needs of the specific cell type and developmental stage. In plants, peroxisomes associated with specific metabolic functions have been termed "specialized peroxisomes". These include the glyoxysomes found in tissues of oil-storing seeds during germination.[38] They are associated with the glyoxylate cycle and the utilization of fat reserves in oil-storing seeds. During germination, fatty acids are converted directly into carbohydrates in the glyoxylate cycle. Thus, seedlings are provided with a carbon source until the plant is photosynthetically competent. The fatty acids metabolized in the glyoxysomes are converted into succinate. An early enzyme in the glyoxylate pathway is acetylCoA oxidase, which generates H_2O_2. Ultimately, two H_2O_2 molecules are produced for every succinate molecule generated. The succinate can then be used as a carbon source or metabolized by mitochondria into cellular energy. In addition to the aforementioned enzymatic generators of H_2O_2, all peroxisomes contain the H_2O_2-consuming enzyme catalase (E.C. 1.11.1.6; H_2O_2:H_2O_2 oxidoreductase).

2. Chloroplasts

Illuminated chloroplasts generate copious amounts of H_2O_2. It is estimated that approximately 10 to 20% of the electrons that flow through photosystem I reduce O_2 to H_2O_2.[33] As electrons flow through the electron chains of photosystem I, they readily "leak" and produce $O_2^{\overline{\cdot}}$. The superoxide radical is then reduced to form H_2O_2. Herbicides (e.g., atrazine) that block electron transport or act as alternative electron acceptors and increase the flow of electrons to oxygen (i.e., paraquat) significantly increase the production of superoxide radicals and thereby H_2O_2.[33,39] It is hypothesized that the generation of activated oxygen species contributes to the toxic effects of these herbicides.

In the chloroplast, $O_2^{\overline{\cdot}}$ is rapidly converted to H_2O_2 through the action of superoxide dismutase.[40,41] The chloroplast isozyme of superoxide dismutase (SOD) converts $O_2^{\overline{\cdot}}$ into H_2O_2 and water.

$$2O_2^{\overline{\cdot}} + 2H^+ \rightarrow H_2O_2 + H_2O$$

There are also a number of nonenzymatic mechanisms that convert $O_2^{\overline{\cdot}}$ into H_2O_2.[33] For example, $O_2^{\overline{\cdot}}$ can react with two NAD(P)H molecules to produce H_2O_2.[42] Several electron donors can act in this manner, including ascorbate[43] and glutathione.[44] Additionally, electrons directly from the electron transport chains of the photosystems can reduce $O_2^{\overline{\cdot}}$ to H_2O_2.[45]

Chloroplasts scavenge and eliminate H_2O_2 via the ascorbate-glutathione cycle.[39,46] In this cycle, ascorbate peroxidase reduces H_2O_2 to form water and dehydroascorbate. Ascorbate reductase then utilizes glutathione to reduce the dehydroascorbate and thereby regenerate the ascorbate. The glutathione is then regenerated by glutathione reductase, utilizing reducing equivalents from NAD(P)H. NAD(P)H is not a limiting metabolite in chloroplasts under conditions where excess H_2O_2 is generated, namely, light.

Chloroplasts may have excess capacity to degrade H_2O_2 and may play a role in reducing H_2O_2 generated in other subcellular compartments. Evidence of this capacity comes from studies with isolated chloroplasts, which can rapidly eliminate H_2O_2 from a 5 m*M* H_2O_2 solution within a few minutes.[46] At this rate, illuminated chloroplasts might be able to scavenge all the H_2O_2 synthesized in the cell.

3. Mitochondria

In mitochondria, the production of H_2O_2 is associated with respiration. The rate of peroxide generated depends directly on the metabolic state of the mitochondria.[47] The greater the flow of electrons through the electron chain, the greater the H_2O_2 production. Under physiological conditions, it is estimated that 2% of the O_2 utilized by mitochondria is generated into H_2O_2.[48] Inhibitors of respiration that block electron flow through the electron transport chain can increase H_2O_2 production.[49] Inhibitors that block the entry of electrons onto the transport chain decrease H_2O_2 production.[50]

The superoxide radical is believed to be the precursor to H_2O_2 in the mitochondria, as it is in the chloroplast.[36] The electron-rich form of ubiquinone, ubisemiquinone, reduces O_2 to form $O_2^{\overline{\cdot}}$.[52] $O_2^{\overline{\cdot}}$, being charged, is not membrane permeable and accumulates in the mitochondria. A mitochondrial-specific isozyme of superoxide dismutase (SOD) degrades $O_2^{\overline{\cdot}}$ into H_2O_2 and O_2.[40,53]

In mammalian systems, some of the internally generated H_2O_2 may be partially metabolized by glutathione peroxidase.[54,55] There is some evidence that a significant portion of mitochondrially generated H_2O_2 diffuses out of the mitochondria and is scavenged by other systems. For instance, almost all conditions that raise mitochondrial H_2O_2 production also cause an increased flux of H_2O_2 through peroxisomal catalase.[34,51,56]

As discussed below, in maize, the CAT-3 catalase isozyme appears to be associated with the mitochondria.[57] To the extent that CAT-3 instead of glutathione peroxidase degrades mitochondrially produced H_2O_2, reducing equivalents are conserved, giving maize more efficient respiration.

4. Other Cellular Components

The cytosol, the endoplasmic reticulum, and the nucleus have all been shown to evolve small quantities of H_2O_2 *in vitro*.[35] Presently, little is known about the actual quantities generated *in vivo* or the metabolic processes that generate it. It is generally accepted that catalase is found in the cytosol and could effectively remove cytosolic H_2O_2. Additionally, H_2O_2 generated in these subcellular components may diffuse into other organelles with an excess capacity to degrade H_2O_2 (e.g., peroxisomes).

II. CATALASE

A. CHARACTERIZATION

Catalase (H_2O_2:H_2O_2 oxidoreductase, EC 1.11.1.6; CAT) is a tetrameric, heme-containing enzyme that is found in all aerobic organisms. Because of its wide distribution and capacity to rapidly degrade hydrogen peroxide, it has been proposed by many investigators that it plays an important role in systems which have evolved to allow organisms to live in aerobic environments.

Catalase is one of the most active catalysts produced by nature. It decomposes hydrogen peroxide at an extremely rapid rate, corresponding to a catalytic center activity of about 10^7 min^{-1}. Depending on the concentration of H_2O_2, it exerts a dual function.[58] At low concentrations ($<10^{-6}$ *M*) of H_2O_2, it acts peroxidatically, where a variety of hydrogen donors (e.g., ethanol, ascorbic acid) can be oxidized in the following manner.

$$RH_2 + H_2O_2 \rightarrow R + 2H_2O$$

At high concentrations of substrate, catalase decomposes toxic hydrogen peroxide at an extremely rapid rate, using the "catalatic" reaction in which H_2O_2 acts as both an acceptor and a donor of hydrogen molecules.

$$2H_2O_2 \rightarrow 2H_2O + O_2$$

Spectrophotometric and kinetic evidence suggest that catalase uses a two-step mechanism in both the peroxidatic and catalatic reactions.[58,59] In the first step, the heme iron of catalase interacts with hydrogen peroxide to form an oxygen-rich iron peroxide.

$$\text{CAT-Fe-OH} + H_2O_2 \rightarrow \text{CAT-Fe-OOH} + H_2O$$

This intermediate iron peroxide (CAT-Fe-OOH) is referred to as compound I. Compound I can be detected *in vitro* and *in vivo,* because it alters the spectrophotometric properties of the catalase heme. In fact, because of the special kinetic properties of catalase, compound I is utilized as an indirect indicator of H_2O_2 concentrations *in vivo*.[34]

At low H_2O_2 concentrations, compound I can be reduced by hydrogen donors.

$$\text{CAT-Fe-OOH} + RH_2 \rightarrow \text{CAT-Fe-OH} + H_2O + O_2 \text{ (Peroxidatic Rx)}$$

At high concentrations of H_2O_2, compound I reacts with a second hydrogen peroxide to produce water and molecular O_2.

$$\text{CAT-Fe-OOH} + H_2O_2 \rightarrow \text{CAT-Fe-OH} + H_2O + O_2 \text{ (Catalatic Rx)}$$

Catalase is unique among H_2O_2-degrading enzymes in that it degrades H_2O_2 without consuming cellular reducing equivalents. Hence, catalase provides the cell with a very energy-efficient mechanism to remove hydrogen peroxide. Therefore, when cells are stressed for energy and are rapidly generating H_2O_2 through "emergency" catabolic processes, H_2O_2 is degraded by catalase in an energy-efficient manner. This should result in a net gain of reducing equivalents and, therefore, cellular energy. It has been proposed that catalase may be uniquely suited to regulate the homeostasis of H_2O_2 in the cell.[35] In the catalatic mode, catalase has a very high apparent Michaelis constant and, therefore, is not easily saturated with substrate. Thus, the enzyme activity increases linearly over a wide range of H_2O_2 concentrations and, thereby, maintains a controlled intracellular H_2O_2 concentration. There is circumstantial evidence that catalase may function in this manner in mammalian systems. Organs with high concentrations of catalase (i.e., liver and kidney) have low levels of endogenous H_2O_2,[51] and organs with low concentrations of catalase (i.e., lung and heart) have high endogenous levels of H_2O_2.[60] Further, if catalase activity is inhibited, H_2O_2 concentrations rise in the liver.[35]

There are scant data on the relationship of subcellular pools of H_2O_2, but because H_2O_2 is soluble through membranes, it is generally thought that it diffuses readily from the site of synthesis to other cellular components. It is known that the membranes of the peroxisomes are very permeable to H_2O_2, and it is believed that this organelle may function as a scavenger of H_2O_2 produced in other subcellular compartments.[61] It has, for example, been shown that under conditions where H_2O_2 production is increased in the mitochondria, H_2O_2 metabolism by peroxisomal catalase increases.[51]

B. HISTORICAL

Catalase action in plant and animal tissues was first observed in 1818 by Thénard, who noted that such tissues readily degraded hydrogen peroxide, a substance he had also discovered some years earlier.[62] Loew (1901)[63] first established that the degradation of H_2O_2 in tissues was due to the effect of an individual, separable enzyme, which he named "catalase". Warburg (1923)[64] suggested that catalase is an iron-containing enzyme because it is inhibited by cyanide. Evidence for its hematin prosthetic group was presented by Zeile and Hellström (1930).[65] Catalase was first purified and crystallized from beef liver, and its identity made clear, by Sumner and Dounce (1937).[66] The earliest genetic studies on catalase were reported by the Russian biologist Koltzoff (1927),[67] who demonstrated that blood catalase levels in several animal species are inherited and segregate according to Mendelian rules.

Catalase has been found in all plants examined[68] and most thoroughly studied biochemically, genetically, and molecularly in the agronomically important species *Zea mays* L.[69] That catalases can exist in multiple molecular forms or isozymes in any organism was first demonstrated by Scandalios[70] with the maize catalases and has since been found to be the rule rather than the exception, as originally perceived. A detailed discussion of the state of knowledge of plant catalases follows, with special emphasis on the maize catalases.

III. THE MAIZE CATALASE (CAT) GENE-ENZYME SYSTEM

In maize, catalase exists as three distinct isozymes (CAT-1, CAT-2, and CAT-3) encoded by the respective, unlinked nuclear genes *Cat1*, *Cat2*, and *Cat3*.[57,68,70,71] Each of the *Cat* genes exhibits temporal and spatial specificity in its expression,[72] and each responds variably to different environmental signals.[73-76] In addition, the catalase isozymes exhibit cell[77] and organelle[78,79] specificities. Both overexpression and null mutants have been identified and characterized,[80-84] and have been of great utility in the resolution of this system in maize.

All three maize catalase isozyme subunits have a molecular weight of approximately 60 kDa and are structurally similar to catalases found in other organisms.[85] The differential spatial and temporal expression of these genes has been characterized in maize, and two temporal regulatory loci, *Car1* and *Car2*, have been genetically defined.[80,86] CAT-1 is the only catalase isozyme expressed in mature pollen, the milky endosperm, aleurone, and the scutellum during early kernel development.[87-89] During early sporophytic development, levels of CAT-1 in the scutellum decline, whereas levels of CAT-2 increase, with the CAT-2 developmental profile paralleling that of the glyoxysomes (peroxisomes), the primary intracellular location of these two isozymes.[78] CAT-1 and CAT-3 are the only catalase isozymes present in etiolated leaves and in the coleoptile of the germinating maize seedling. Upon exposure to light, there is rapid accumulation of CAT-2 in leaves due to both increased transcript accumulation[90] and increased translation of the *Cat2* message.[76] In mature green leaves of maize, CAT-2 is localized in the peroxisomes of bundle-sheath cells, whereas CAT-1 and CAT-3 are found in mesophyll cells.[77]

The CAT-3 isozyme is quite different from the other catalases biochemically. In maize, tobacco, and barley, the CAT-3 (i.e., the isozyme specifically immunoreactive with the maize anti-CAT- 3 monospecific Ab) isozyme has enhanced peroxidatic activity (70-, 30-, and 28-fold, respectively, over "typical" catalase, i.e., CAT-2). Catalase can catalyze either the direct dismutation of H_2O_2 into H_2O and O_2 (catalatic mode) or use H_2O_2 to oxidize substrates such as methanol, ethanol, formaldehyde, formate, or nitrite (peroxidatic mode).[91] The ratio of these is usually calculated as $R_{p/c}$ = (mU peroxidatic/U catalatic)[85] × 10. In maize bundle-sheath, the peroxisomal catalase (CAT-2) has high catalatic but low peroxidatic activity ($R_{p/c}$ = 0.25). In contrast, CAT-3, which is expressed in leaf mesophyll and is not peroxisomal (coisolates with mitochondria), has high peroxidatic but low catalatic activity ($R_{p/c}$ = 17.6).[72,91] Finally, each catalase isozyme exhibits varying degrees of sensitivity to inhibitors such as cyanide (KCN), azide (NaN_3), and aminotriazol (AT). CAT-3 proved the least sensitive of the three maize catalases to all inhibitors tested.[69] This again suggests that CAT-3 might have evolved to function under specific conditions or in a specific metabolic role distinct from that of the other catalases. For example, because CAT-3 is cyanide insensitive, one might hypothesize a role or function under conditions favoring cyanide-resistant respiration (i.e., alternate oxidase respiration).[92] Table 1 summarizes some of the properties of the maize catalases.

Although biochemically distinct, the three catalase isozymes are capable of interacting *in vitro* to form heterotetramers.[70,71] The CAT-1 and CAT-2 isozymes also form the expected heterotetramers when expressed together *in vivo* (e.g., in the scutellum of germinating embryos) (Figure 1). However, the CAT-3 isozyme is found only as a homotetramer *in vivo,* even when its temporal and tissue specificity of expression coincide with those of the CAT-1 and CAT-2 isozymes.[57,82] This further implies that CAT-3 is compartmentalized separately from the other catalases in the cell. Because all three catalases are encoded by nuclear genes, this suggests specific targeting of the CAT-3 protein. The CAT-1 and CAT-2 isozymes have been shown to be cytosolic or, where glyoxysomes/peroxisomes are present, peroxisomal. In contrast, CAT-3 is not found in these cell fractions in leaf, but coisolates exclusively with the mitochondrial cell fraction.[57,82] Also, as discussed below, recent findings concerning the novel light regulation of *Cat3* transcription further reinforce the inference of a special metabolic role for CAT-3.

A. CATALASE PROCESSING AND COMPARTMENTALIZATION

As mentioned above, CAT-1 and CAT-2 are associated with the cytosol and microbodies (peroxisomes and glyoxysomes); CAT-3 is associated with the mitochondria.[57,78] CAT-1 and CAT-2 are expressed in the scutellum at a time when glyoxysome biogenesis peaks in this tissue,[78,93] whereas CAT-3 is predominant in pericarp (a maternally derived tissue), leaf mesophyll, and stem.

TABLE 1
Summary of Biochemical, Genetic, and Molecular Properties of the Maize Catalase Gene-Enzyme System

	CAT-1	CAT-2	CAT-3
Subunit Number	4	4	4
Native M_r	240,000	240,000	240,000
Specific Activity (Units/mg protein)	561	2343	688
K_m	0.143*M*	0.040*M*	0.062*M*
pH Optima	7–9	7–9	7–9
pI	5.86	5.01	5.20
Turnover rate[a]	47 h	22 h	22–24 h
Thermostability[b]	12.5 min	Stable	18 min
Aminotriazole inhibition[c]	93.1%	98.2%	31.7%
Azide inhibition	39.1%	34.3%	6.7%
Cyanide inhibition	86.3%	73.4%	48.4%
Maize catalase inhibitor	36.3%	36.0%	17.3%
Subcellular localization	Peroxisomes/ cytosol	Peroxisomes/ cytosol	Coisolates w/mi- tochondria
Gene locus	*Cat1*	*Cat2*	*Cat3*
Chromosomal location	5S	1S	1L
Number of alleles	6	3	3
Transcript size[d]	2000 nt	1800 nt	1800 nt
Percent nt similarity[e]	—	67%	63%
Percent aa similarity	—	77%	67%
Light regulation	NO	POS	Circadian
ABA regulation	POS	NEG	NO

[a] Turnover rate, the time required to reach 50% saturation (50% of enzyme molecules are fully labeled) under our density-labeling conditions.
[b] Thermostability was measured at 45°C in 100 m*M* Tris-HCl (pH 7.8).
[c] % Inhibition by 10^{-2} *M* aminotriazole/10^{-7} *M* NaN_3/10^{-4} *M* KCN.
[d] Transcript size determined by RNA blot analysis.
[e] Percent similarity was determined by pairwise comparisons of the nucleic acid or predicted amino acid sequences of cDNA clones to the sequence of the *Cat1* cDNA sequence.

In vitro translation of *Cat2* mRNA using the rabbit reticulocyte lysate system revealed that the M_r of the *in vitro*-synthesized CAT-2 protein is 56 kDa compared to 54 kDa for the purified CAT-2 protein.[94] Two-dimensional PAGE of the *in vitro*- and *in vivo*-labeled CAT-2 protein and Western gel analysis indicated that the CAT-2 protein is processed from a precursor to a lower M_r form in the scutellum (Figure 2). The size reduction of 2 kDa corresponds to approximately 19 amino acids, being within the range estimated for signal peptides.[95] However, CAT-1, which is also associated with glyoxysomes, and CAT-3 are synthesized in the same size both *in vivo* and *in vitro*. Thus, the processing of CAT-2 may not be directly related to its microbody importation. In pumpkin, catalase is synthesized as a precursor of 59 kDa, which is larger than the mature form (55 kDa). Both precursor and mature forms are present in peroxisomes.[96]

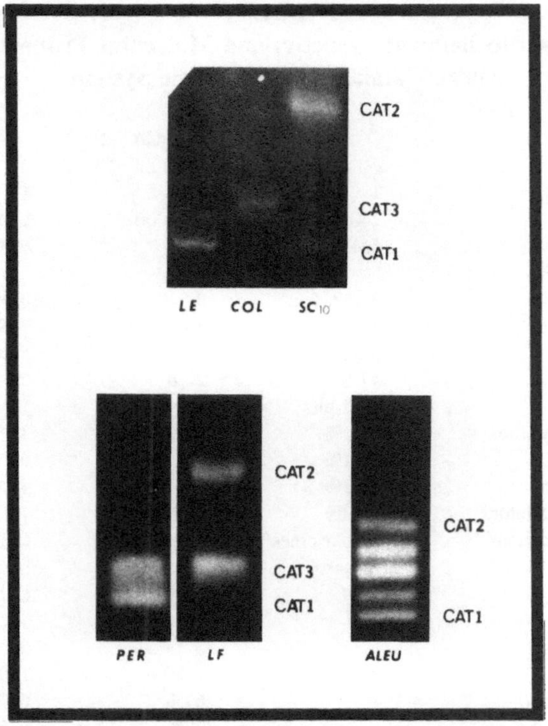

FIGURE 1. Upper panel: Zymogram showing tissue-specific expression of CAT-1 in immature milky endosperm, CAT-3 in coleoptile, and CAT-2 in scutellum at day 10 postgermination. Lower left panel: Shows lack of subunit interaction between CAT-3 when coexpressed in the same tissue with CAT-1 (pericarp, PER) or with CAT-2 (green leaf, LF). Lower right panel: Shows subunit interaction between CAT-1 and CAT-2 to form intergenic heterotetramers when they are coexpressed in the same tissue (aleurone, ALEU). Migration is anodal.

The genes coding for glyoxysomal enzymes in maize are noncoordinately regulated. Glyoxysome biogenesis and acquisition of glyoxysomal enzymes are independent of the level of catalase protein present.[97] Mutations affecting the levels of CAT-2[81] or isocitrate lyase[98] are specific and independent of other glyoxysomal proteins. It is not yet clear as to the extent this independent regulation of glyoxysomal enzymes may be true for microbodies of other fatty seeds or for other cell compartments. It has been demonstrated that the development and acquisition of enzyme activities within the glyoxysomes of cotton cotyledons during maturation occurs sequentially.[99]

It has been reported that the specific uptake of some enzymes into mammalian peroxisomes is directed by peroxisomal targeting sequences which are located at the extreme C terminus of such polypeptides.[100] The terminal tripeptide SKL (Ser-Lys-Leu-COOH) of firefly luciferase has been shown to be necessary and sufficient for directing cytosolic proteins to peroxisomes.[101] However, none of the catalases investigated have the SKL tripeptide at their

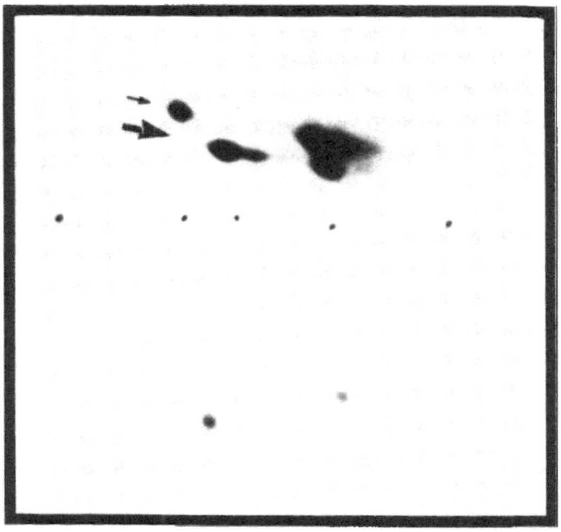

FIGURE 2. *In vivo*- vs. *in vitro*-labeled CAT-2 immunoprecipitates. Polysomes from 4-d-old W64A scutella were translated in the wheat germ system, and CAT-2 products (top arrow) were isolated by immunoprecipitation of the products with CAT-2 antibodies. Scutella from 7-d-old R6-67 were labeled *in vivo* by partial submersion in buffer containing [^{35}S]methionine. CAT-2 was isolated from scutellar homogenates by immunoprecipitation (bottom arrow). The two immunoprecipitates were electrophoresed together on 2-D gels. (Reprinted with permission from Skadsen, R. W. and Scandalios, J. G., *Biochemistry*, 25, 2027, 1986. Copyright 1986 American Chemical Society.)

extreme C terminus.[79,102] Human catalase has a variant of the tripeptide (Ser-His-Leu) within the last 27 amino acid residues.[100] None of the maize catalases have the SKL tripeptide motif at the extreme C terminus. However, CAT-1 and CAT-2 have a variant tripeptide, Ser-Arg-Leu (SRL), nine amino acid residues from the C terminus; this motif is not present in CAT-3. It is clear to date that the topogenic signal mechanism for protein entry into peroxisomes differs from those employed in mitochondria.[79,103] Although the SKL-type motifs are clearly essential for targeting and entry of some peroxisomal proteins, their universality remains ambiguous. Deletion mutation analyses, currently underway with the maize catalases, will allow us to identify catalase sequences relevant to its subcellular destination.

B. DEVELOPMENTAL EXPRESSION

Each of the maize catalases exhibits a unique and complex pattern of developmental expression throughout the maize life cycle.[72] In this respect, this system has served as a model to study the differential expression of genes in higher eukaryotes.[75] CAT-3 is detected in the pericarp (a maternal tissue) during ovule development, and during very early postpollination kernel development, but declines rapidly shortly after pollination. CAT-1 is also detected, but at extremely low levels, in the pericarp. By days 12 to 15 postpollination, CAT-1 becomes the predominant form, and is in fact the only catalase detected in the kernel at this time. CAT-1 is the only catalase detected in the milky endosperm of developing kernels. The aleurone, a specialized kernel tissue derived from endosperm cells, expresses CAT-1 and CAT-2 during the latter stages of kernel development. At days 9 to 20 postpollination, the scutellum expresses only the CAT-1 isozyme. However, at later stages, as the seed matures, the *Cat2* gene is induced to moderate levels until the seed desiccates and enters developmental arrest. In the scutellum isolated from dry "dormant" seed, CAT-1 appears to be the only isozyme expressed. Upon imbibition and germination, CAT-1 is initially the only form expressed in the scutellum, but as the embryo grows, *Cat2* is rapidly induced and by day 5 postimbibition, CAT-2 becomes the predominant isozyme. By day 10 postimbibition, catalase expression has shifted entirely to *Cat2*, and no *Cat1* expression can be detected at the protein level. In the coleoptile, CAT-3 is the only isozyme detected on zymograms. CAT-1 and CAT-3 are detected in dark-grown leaf tissue. CAT-2 is only expressed with greening of the leaf tissue in the light. The root has very low levels of catalase and is difficult to analyze. However, in young roots, CAT-1 and CAT-3 can be detected. As roots mature and grow, catalase activity decreases and rapidly drops below assay sensitivity. Only CAT-3 is detected in the stem of the mature plant, and only CAT-1 is detected in the pollen after anthesis.[89]

The overall developmental program of catalase gene expression presented herein has been corroborated by zymogram analyses, activity profiles, immunochemical measurements, and transcript levels for each of the isozymes.[75,88,90]

C. EXPRESSION IN EARLY SPOROPHYTIC DEVELOPMENT

Catalase activity in the scutellum increases rapidly after germination, peaking at day 4 postgermination and thereafter declining to moderate levels. A time-course zymogram analysis reflects the temporal shift in expression from *Cat1* to *Cat2* during this time period.[87] This shift in expression of the two *Cat* genes was shown to be largely due to changes in the rates of synthesis and degradation of the CAT-1 and CAT-2 isozymes during this period.[104]

That this shift in expression of the two *Cat* genes in the scutellum is under genetic control has been verified by use of several maize lines shown to exhibit altered patterns of catalase expression during the same developmental period.[80,82,86] For example, line R6-67 exhibits a continuous increase in catalase activity after day 4 of postimbibition compared to the usual decline seen in most lines (e.g., W64A). The continuous increase in catalase activity is concurrent with an increase in CAT-2 protein synthesis and accumulation. By genetic analysis, a genetic regulatory element, *Car1*, was identified and shown to be responsible for the overexpression of the *Cat2* gene in R6-67.[80] *Car1* maps 37 map units from the *Cat2* structural gene on the short arm of chromosome 1. The product of *Car1* and the mechanism by which it increases

CATALASE DEVELOPMENTAL TIME-COURSE MUTANTS (RELATIVE TO "WILD TYPE "(+/+) PROFILE)

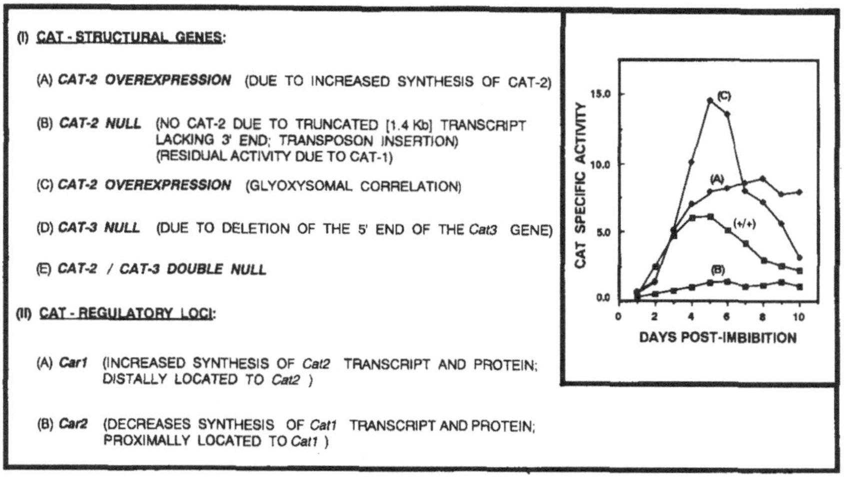

FIGURE 3. A listing of maize developmental time-course mutants of catalase. Activity profiles for some of these mutants, relative to the wild-type pattern (+/+), are shown on the inset to the right.

Cat2 expression are unclear. However, R6-67 has significantly more *Cat2* mRNA than do the normal lines, and it is suggested that *Car1* may encode a *trans*-acting factor affecting the transcription of *Cat2*.[105]

A second regulatory element, *Car2*, was identified and shown to function in decreasing the rate of CAT-1 synthesis in the scutellum during early sporophytic development in line TX303.[86] The decline in CAT-1 synthesis is associated with a decline in *Cat1* mRNA. *Car2* acts independently of *Car1* and is either closely linked or contiguous with the *Cat1* structural gene. A number of CAT developmental time-course mutants have been identified (Figure 3).

D. SIGNAL TRANSDUCTION AND *CAT2* GENE EXPRESSION IN THE SCUTELLUM

Catalase expression in the scutellum is apparently regulated by "signals" from tissues that spatially interact with the scutellum. Histoimmunological assays have shown that the postgermination induction of *Cat2* occurs as a consequence of a temporal and spatial gradient. CAT-2 is first detected in cells that interact with the aleurone. As development proceeds, CAT-2 accumulates in a spatial gradient (Figure 4), occurring first in cells nearest the aleurone, then in cells further and further away from the aleurone, until all scutellar cells express the CAT-2 protein.[106] This suggested that the aleurone may be releasing a diffusible molecular signal triggering the induction of *Cat2* in the scutellum.

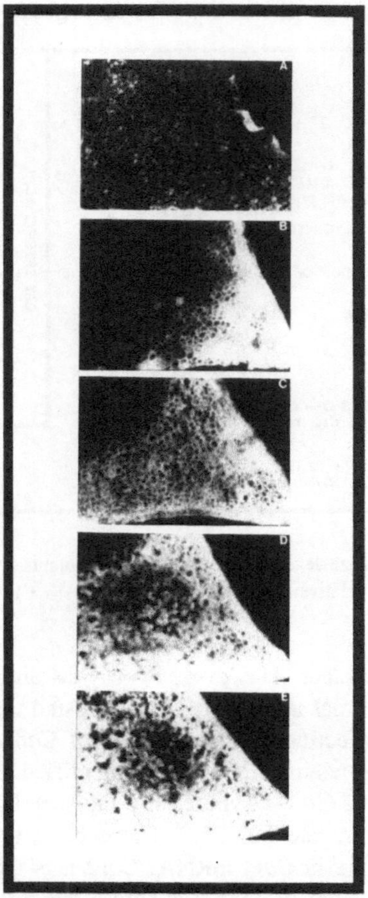

FIGURE 4. Immunofluorescent localization of CAT-2 in scutellar cells of W64A at different days of postgerminative development. An indirect immunofluorescent labeling technique was used in which thin sections of W64A scutella were incubated with anti-CAT-2 IgG, followed by conjugation with FITC-labeled goat antirabbit IgG. A-E: E 0-, 2-, 4-, 6-, and 8-d-old scutella, respectively. Note the progressive expression of the *Cat2* gene in scutellar cells from the outer to the inner perimeter with time. (From Tsaftaris, A. S. and Scandalios, J. G., *Proc. Natl. Acad. Sci. U.S.A.*, 83, 5549, 1986. With permission.)

It was also demonstrated[107] that the embryonic axis exerts a specific effect on the accumulation of glyoxysomal proteins, including catalase. Upon excising the embryonic axis from the scutellum prior to imbibition, the developmental accumulation of all glyoxysomal proteins is drastically reduced. In contrast, the developmental patterns of nonglyoxysomal proteins are unaffected; this suggested that an axis-specific factor modulates the level of expression of the glyoxysomal proteins, including catalase (Figure 5). The exact nature of this ''factor'' is yet unclear. However, it is apparent that possibly

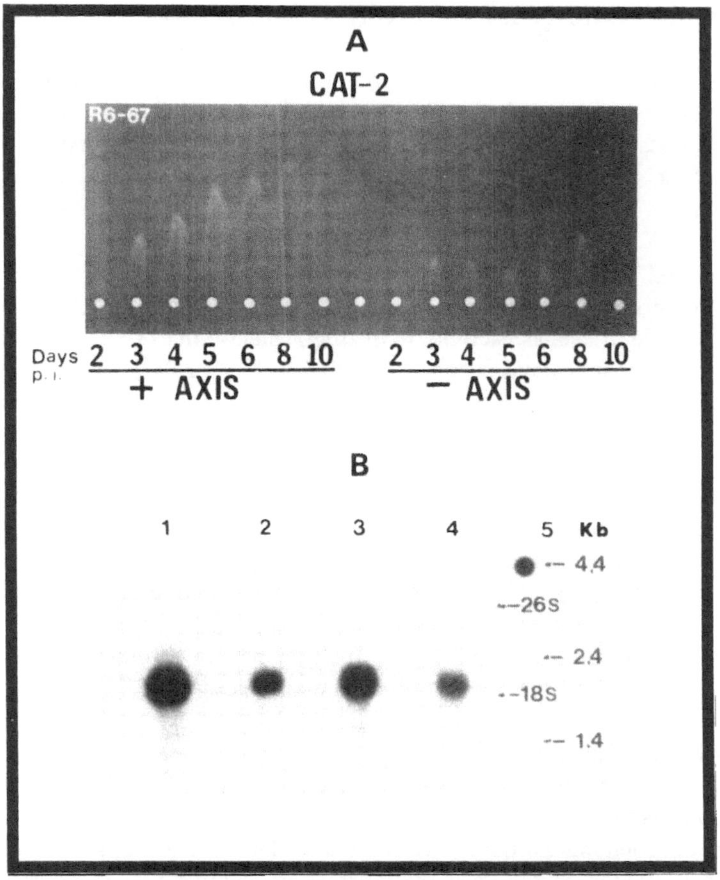

FIGURE 5. (A) Rocket immunoelectrophoresis showing alteration of genetically programmed CAT-2 expression after embryo axis removal. The inbred maize line R6-67 expresses progressively high levels of CAT-2 protein over the first 10 d after imbibition and germination. If the axis is removed prior to imbibition, the level of CAT-2 expression is drastically diminished during the same developmental period. Equal amounts of homogenate representing 0.24 mg of scutellar tissue were used in each lane. CAT-2 monospecific antibody was used. (B) RNA blot analysis of the effect of embryonic axis removal on *Cat2* RNA transcript levels. Blots were probed with a *Cat2* gene-specific sequence. Lane 1: mRNA from scutella with intact axes; lane 2: mRNA from scutella with axes removed prior to imbibition; lanes 3 and 4: total RNA from which mRNAs in lanes 1 and 2 (respectively) were prepared; lane 5: RNA size standards (in kilobases). 10 μg total RNA was used in lanes 3 and 4, and 0.5 μg mRNA was used in lanes 1 and 2. The positions of the 26S and 18S ribosomal RNAs are indicated. (From Skadsen, R. W. and Scandalios, J. G., *Dev. Genet.*, 10, 1, 1989. With permission.)

two molecular signals may be involved in regulating the expression of *Cat2* in the scutellum during early sporophytic development; one emanating from the aleurone may act to activate the gene, and the other emanating from the embryonic axis may modulate the level of expression of *Cat2*. Both aspects are being investigated.

E. ENVIRONMENTAL PERTURBATIONS OF THE DEVELOPMENTAL PROGRAM FOR CATALASE EXPRESSION

The temporal and spatial expression of the catalase structural genes is not only influenced by genetic and developmental factors, but is also responsive to exogenously applied environmental signals, both chemical and physical. Some of the signals to which the catalase genes in maize respond include the following.

1. Elevated Temperatures

Chronic elevated temperature has a differential effect on the developmental profile of catalase activity in the high catalase lines (e.g., R6-67) and lines with standard catalase expression (W64A).[108] At 25°C, R6-67 has much higher levels of catalase activity during the later stages of scutellar development (after day 5 postgermination). At 40°C, W64A catalase activity is slightly depressed, but R6-67 is depressed severalfold such that its developmental profile then resembles that of W64A. The catalase decrease in R6-67 at the higher temperature is associated with a decrease in immunologically detectable CAT-2 protein. It is speculated that the decrease of CAT-2 in R6-67 may be due to a decrease in the effective concentration of *Car1* factor. At the elevated temperatures, recent data from RNA blot analyses show a gradual decline in *Cat2* mRNA during the first 10 d of germination, whereas the level of *Cat1* mRNA present remains constant.

2. Hydrogen Peroxide

Catalase activity in the postgermination scutellum can be increased by incubating isolated scutella on solid agar containing H_2O_2.[72] Incubation in agar media containing 0.01% H_2O_2 more than doubles catalase activity in day 4 postgermination scutellum without affecting seedling growth or development. Lower amounts of H_2O_2 (below 0.004%) had no effect, and higher levels caused tissue damage in young seedlings. The increase in catalase activity in response to H_2O_2 was due to an increase in CAT-2 protein, as detected by immunoelectrophoresis.

3. Fungal Toxin

Many fungi of the genus *Cercospora* produce a light-induced, photoactivated polyketide toxin (cercosporin). In the presence of light and a suitable reducing substrate, an excited form of the cercosporin molecule (triplet state) reacts with molecular oxygen to form a superoxide radical. Dose-response studies showed that changes in total catalase activity, as well as individual isozyme levels, remained constant in all toxin treatments.[109]

Assays of 7- to 10-d total leaf protein extracts showed an approximately 25% increase in total CAT activity in plus-cercosporin vs. minus-cercosporin-treated leaves (Figure 6A). Rocket immunoelectrophoresis, however, showed an approximately 20 to 30% decrease in CAT-2 protein levels. This decrease

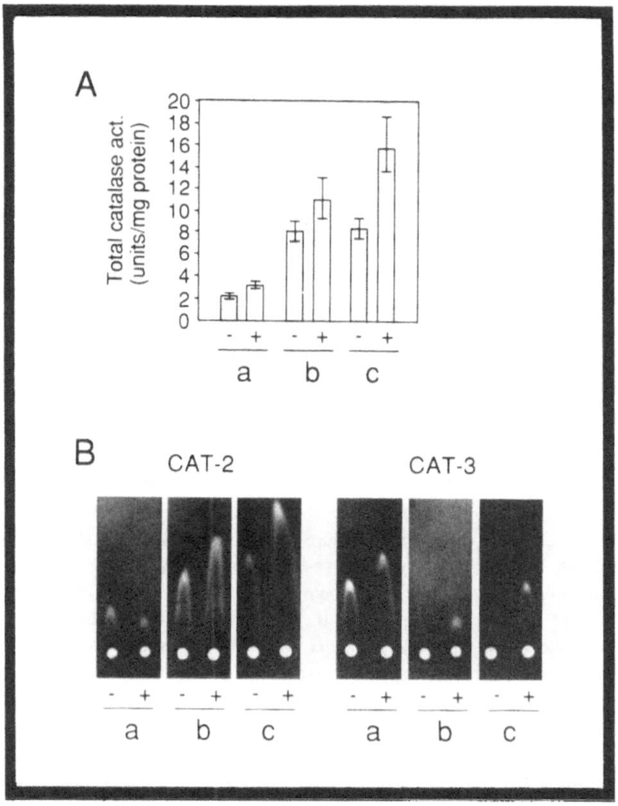

FIGURE 6. Changes in catalase activity and isozyme level in leaves and scutella of maize treated with *Cercospora* culture filtrate. Sterile, light-grown filtrate (plus-cercosporin) or dark-grown culture filtrate (minus-cercosporin) was applied to the surface of 7- to 10-d-old maize leaves (a). Plants were then grown for 24 h in constant light. Similarly, whole maize kernels (b) were imbibed in either light-grown culture filtrate (plus-cercosporin) or dark-grown culture filtrate (minus-cercosporin) and then incubated in the light for 3 d on germination pads soaked with fresh light- or dark-grown *Cercospora* culture filtrates. Finally, developing embryos (c) were excised from seed and placed on organ culture medium prepared with either light-grown culture filtrate (plus-cercosporin) or dark-grown culture filtrate (minus-cercosporin). Embryos were incubated in the light for 24 h. After treatment, whole leaves (a) and scutella isolated from whole seed (b) and excised embryos (c) were assayed for changes in total catalase activity (A). The mean and range for three independent replicates are shown. Equal protein from each treatment was also analyzed for accumulation of CAT-2 and CAT-3 isozyme proteins by rocket immunoelectrophoresis (B). Representative results are shown. (From Williamson, J. D. and Scandalios, J. G., *Plant J.*, 2, 351, 1992. With permission of Blackwell Scientific Publications Ltd.)

is apparently at least partially offset by a corresponding increase in the more abundant CAT-3 isozyme (Figure 6B). Nuclease S-1 analyses of leaf total RNA showed a comparable increase in steady-state levels of *Cat3* RNA in the presence of cercosporin, compared to a slight decrease in *Cat2* RNA accumulation (Figure 7).

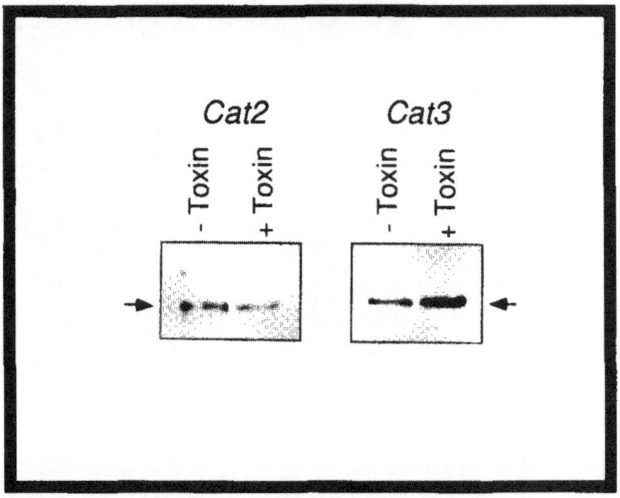

FIGURE 7. The effect of *Cercospora* culture filtrate on the steady-state levels of *Cat2* and *Cat3* catalase RNA in developing maize leaves. Leaves of germinating maize seedlings 7 to 10 d postimbibition were treated with light-grown culture filtrate (+ toxin) or dark-grown culture filtrate (− toxin) of *C. kikuchii*. Plants were then grown for 24 h in constant light. Total RNA was isolated, and the steady-state level of *Cat2* and *Cat3* RNAs were determined by nuclease S-1 protection analysis. Duplicate assays were performed, and representative results are presented. (From Williamson, J. D. and Scandalios, J. G., *Plant J.*, 2, 351, 1992. With permission of Blackwell Scientific Publications Ltd.)

We observed little or no change in total CAT activity in day 3 postimbibition (PI) embryo axes (not shown). In scutella (embryo minus axis) from day 3 PI seed, however, an approximately 50 to 60% increase in total CAT activity was observed. Rocket immunoelectrophoresis of corresponding scutellar protein extracts showed an increase in CAT-2 and CAT-3 isozyme levels of sufficient magnitude to account for the majority of this increased activity.

4. Hormones

Embryos removed from developing maize seed at days 18 to 21 postpollination will germinate precociously when placed on a simple sucrose/minimal salts growth medium, but when incubated on growth medium supplemented with the phytohormone abscisic acid, (ABA, 10^{-4} *M*), embryos do not germinate but continue to mature. Zymogram analyses of total extracted proteins from the scutella of these embryos showed that CAT-1 and CAT-3 proteins (but not CAT-2) accumulated in embryos incubated in the presence of ABA. In contrast, excised embryos incubated on growth media alone accumulated increased levels of the CAT-2 isozyme, whereas CAT-1 levels appeared to decrease. These observations suggested that expression (or accumulation) of the CAT-2 isozyme is downregulated by ABA in the scutellum of developing maize embryos. Conversely, CAT-1 and CAT-3 appeared to be at least candidates for positive ABA regulation.[110]

In light of these results, we proceeded to analyze changes in the accumulation of the three catalase messages in response to ABA. Initial dose response and time-course induction experiments suggested that increased *Cat1* RNA accumulation is a dose-dependent, direct response to ABA, with *Cat1* transcript levels reaching a maximum less than 2 h after exogenous application of ABA. Conversely, *Cat2* transcript failed to accumulate to detectable levels at any exogenous ABA concentration above 10^{-7} *M*. The results for *Cat3* were less definitive. To clarify what changes in *Cat3* accumulation, if any, could be attributed to exogenous application of ABA, we performed parallel time-course studies in both the presence and absence of ABA. Excised embryos were incubated for 0, 4, and 48 h at 25°C in the dark on growth media or growth medium supplemented with ABA. At the conclusion of these treatments, scutella were dissected from the embryo, and the levels of the various catalase transcripts determined by Northern analysis of total extracted RNA. *Cat1* RNA levels increased substantially in the presence of ABA, but not in its absence. Conversely, *Cat2* increased in the absence of ABA, but failed to accumulate in its presence. Surprisingly, levels of *Cat3* transcript decreased and then increased in both the presence and absence of ABA. The relevant point for these studies is that *Cat3* does not appear to respond to exogenously applied ABA, whereas *Cat1* is upregulated and *Cat2* is downregulated.[110]

These observations are consistent with sequence analyses of the corresponding catalase genomic clones and primer extension experiments.[111,112] *Cat1* was found to contain the ABA response element (CACGTGGA) reported for wheat, rice, and cotton,[113] whereas this consensus was not present in the *Cat3* 5′ genomic regions. In addition, the reported *Vp1* protein binding sequence (the Sph box) is not present in either of these catalase genomic sequences.

5. Light Regulation

In addition to exhibiting differential temporal and spatial patterns of expression, the three maize catalase genes respond differentially to light in developing maize leaves. CAT-1 protein and *Cat1* message accumulate at low levels in the mesophyll of maize leaves in a light-independent manner throughout development.[90,114] In contrast, both *Cat2* and *Cat3* respond to light, although in quite different ways. The *Cat2* gene appears to be positively regulated by light. The effect of light on CAT-2 is summarized in Table 2. CAT-2 protein and *Cat2* mRNA accumulate in the bundle-sheath cells of shoots grown either in a light/dark regime or in constant light,[77] but are not detected in young leaves grown in constant dark. The CAT-2 isozyme is dramatically induced in leaves when dark-grown seedlings are exposed to light.[76] The positive light response of the *Cat2* gene is not phytochrome mediated. When total poly(A)$^+$ RNA (mRNA), polysomes, or isolated polysomal mRNA from light- and dark-grown leaves was translated *in vitro,* CAT-2 protein was detected only among the light-grown leaf products.[76] RNA

TABLE 2
Summary of the Responses of the CAT-2 Catalase to Light

Property	Etiolated	Greening
CAT-2 enzyme activity	−	+
Detectable antigen	−	+
Labeled *in vivo*	−	+
mRNA present	+	+
mRNA translatable *in vitro*	−	+
mRNA on polysomes	+	+
Polysomes translatable *in vitro*	−	+
Isolated polysomal mRNA translatable *in vitro*	−	+

blot analyses using a gene-specific *Cat2* cDNA probe showed that *Cat2* mRNA was present in approximately equal quantities in total mRNA and polysomal mRNA derived from light- and dark-grown maize leaves.[76] Furthermore, *Cat2* mRNA was equally distributed in identical high molecular weight fractions in polysomes from light- and dark-grown leaves, indicating that the *Cat2* mRNA is not sequestered in ribonucleoprotein particles in dark-grown leaves.[76] Thus, the control of *Cat2* expression in maize leaves in response to light appears to involve a unique form of translational inhibition in dark-grown leaves, preventing translation of the isolated *Cat2* mRNA. The mRNA is rendered translatable only after the leaves are exposed to white light. There is evidence that, like maize *Cat2*, the mRNA for chlorophyll *a/b* binding protein is present in dark-grown peas[115] and *Lemma gibba,*[116] and may be activated by a translational control mechanism in the light.

It is tempting to speculate that a physiological role should attend this CAT-2-specific light response. A plausible explanation is that the cell reserves CAT-2 synthesis until it is needed to destroy H_2O_2 generated during photorespiration.[117] Photorespiration in maize, a C_4 plant, occurs to a lesser degree as compared to C_3 plants. C_3 plants often form large catalase crystals in their peroxisomes,[118] which may reflect the degree to which catalase is vital. This point is underscored in a catalase underexpression mutant of barley that reportedly does not survive under photorespiratory conditions.[119] However, the existence of normally growing *Cat2* and *Cat3* null mutants in maize[84,120] suggests that other mechanisms may accommodate the C_4 level of photorespiration and that each of the maize catalases may play a unique metabolic role.

In contrast, the *Cat3* gene appears to be light regulated in a much more complex manner. Accumulation of *Cat3* gene products is mesophyll specific in the leaves of light- and dark-grown maize seedlings. We recently described a novel circadian regulation of *Cat3* gene transcription in developing maize leaves (Figure 8A). The changing steady-state levels observed for *Cat3* mRNA were reflected in corresponding changes in the rate of gene transcription (Figure 8B). It was also established that this phenomenon is not phytochrome

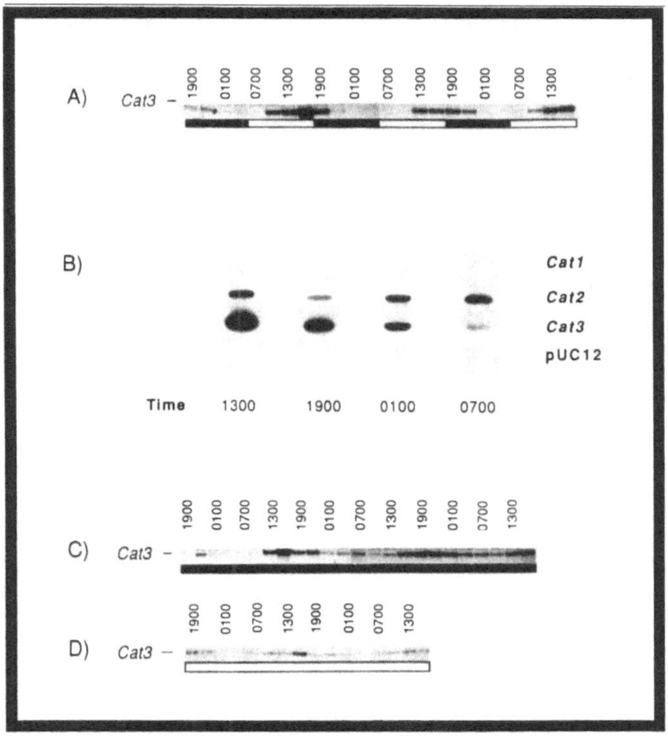

FIGURE 8. The diurnal variation in *Cat3* RNA accumulation in developing maize seedlings is regulated by a circadian rhythm. The relative steady-state levels of *Cat3* transcript in the total RNA of leaves of the normal inbred maize line W64A were determined by nuclease S-1 analysis. For one set of analyses (A), 6-d postimbibition W64A seedlings were grown under a 24-h photoperiod (12 h light/12 h dark) (LD), and leaves collected at 3-h intervals over a 48-h period. Samples were collected beginning at 7 p.m. (1900 hours), or 1 h into the dark period. (B) Rate of *Cat3* gene transcription in the leaves of maize seedlings grown under a 12-h light/12-h dark photoperiod. The relative rates of transcription of the *Cat1*, *Cat2*, and *Cat3* genes in the leaves of maize harvested 6 h into the light period (1300), 1 h into the dark period (1900), 7 h into the dark period (0100), or 1 h into the following light period (0700) were determined by nuclear transcription run-off analysis. Run-off transcripts did not bind to plasmid DNA (pUC12) at detectable levels. In addition, plants were grown as described in (A) and then transferred to either (C) continuous dark or (D) continuous light. Samples were collected as in (A). Note that the diurnal rhythm exhibited in an alternating light/dark regime continued after plants were transferred into constant light or constant dark. This indicates that the diurnal variation is circadian in nature.

mediated and that chlorophyll- and carotenoid-deficient mutants and their wild-type siblings exhibit the normal diurnal pattern of *Cat3* RNA accumulation.[114,121] This indicates that photosynthetic pigments, allelic variation, and genetic background do not directly affect the temporal pattern of *Cat3* accumulation in leaves. Most circadian phenomena previously described in plants at the molecular level involve the regulation of chloroplast-associated gene products directly involved in photosynthesis.[122-124] The *Cat3* gene product is not associated with chloroplasts, nor is it regulated by phytochrome.[114,121] The transcriptionally mediated accumulation of *Cat3* message late in the photoperiod, together with its localization in the leaf mesophyll mitochondrial fraction, suggested that the CAT-3 protein might have a role in nonphotosynthetic reactions in the dark. We observed, however, that when normally pigmented plants are grown in either continuous light or continuous dark, the *Cat3* transcript is present at uniformly high levels throughout the 24-h sampling period (Figure 9). Because the *Cat3* gene is continually transcribed in the absence of a cyclic light regime, the normally observed diurnal variation of *Cat3* gene expression is presumably, therefore, the result of a circadian-regulated transcriptional repressor.[121] This novel type of repressor-mediated circadian regulation has, to our knowledge, not been previously reported. Most previously reported light-induced circadian rhythms in plants are phytochrome mediated, involve genes directly related to photosynthesis, and are superimposed on the simple, positive light induction of gene transcription, i.e., genes are on in constant light and off in constant dark. Our results, in contrast, suggest that expression of the *Cat3* gene has a "default" mode of expression in leaves where, in the absence of an alternating light/dark cycle, the *Cat3* gene is maximally expressed in leaf mesophyll throughout the 24-h sampling period. Under a normal diurnal light regime, however, a cyclic repression superimposed on top of this quasiconstitutive mode of expression in leaves yields the observed variation in *Cat3* gene transcription. Present models of circadian regulation in plants hypothesize the existence of a transcriptional enhancer that is the product of a cyclic regulatory gene or "clock" gene, as has been described for *Neurospora* and *Drosophila*.[125,126] In our case, this clock gene product could be a specific repressor of *Cat3* gene expression or might regulate the cyclic expression of a specific *Cat3* gene repressor. Therefore, the light regulation pathway of *Cat3* represents a completely different set of light regulatory mechanisms and thus presents an opportunity to study a completely new type of light signal transduction pathway.

The circadian rhythm in *Cat3* transcript levels is unique among the three maize catalases. In addition, light affects the accumulation of the three catalase transcripts in different ways,[76,114] suggesting that each isozyme may fulfill a different physiological role. The CAT-3 isozyme, which in leaves is expressed only in mesophyll cells,[77] has considerably more peroxidatic than catalatic activity[91] compared to CAT-2, which is localized in the bundle-sheath cells of maize leaves. Because in maize there is a differential localization of the O_2-evolving and CO_2-fixing reactions of photosynthesis in the mesophyll and bundle-sheath cells, respectively, the different catalases may indeed perform distinct metabolic functions.

Studies with other plant genes have fortified the assumption that light-responsive sequences are located in the 5′ flanking regions of structural

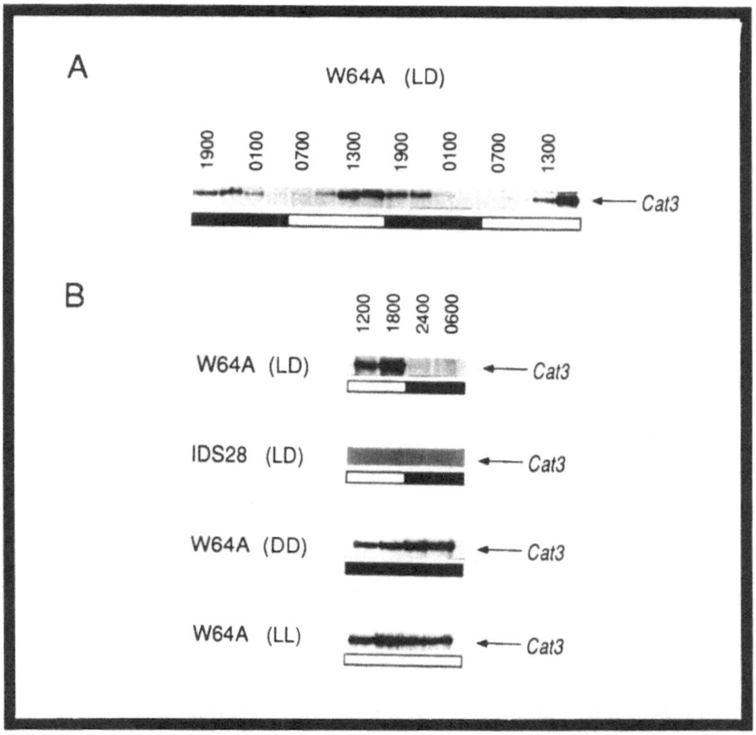

FIGURE 9. The effect of light regime on the accumulation of the *Cat3* catalase transcript in maize leaves. The steady-state levels of *Cat3* transcript in the total RNA population of leaves of the normal inbred maize line W64A and the *Cat3* null line IDS28 grown under three different light regimes was determined by nuclease S-1 protection analysis. For one set of analyses (A), 6-d postimbibition W64A seedlings were grown under a 24-h photoperiod (12 h light/12 h dark) and leaves collected at 3-h intervals over a 48-h period. Samples were collected beginning at 7 p.m. (1900 hours), or 1 h into the dark period. For all other analyses (B), samples were collected at four representative time points, spaced at 6-h intervals over the 24-h period. For these analyses, plants were grown from the time of imbibition in constant light (LL), constant dark (DD), or under a 24-h light/dark photoperiod (LD). Leaves of W64A grown under a 24-h light/dark regime exhibit the expected large diurnal variation in *Cat3* RNA levels. In contrast, plants grown in either constant light or constant dark from the time of imbibition accumulate constant, high steady-state levels of the *Cat3* transcript throughout the sampling period. As expected, *Cat3* RNA was not detected in leaves of the *Cat3* null line IDS28. (From Wadsworth, G. J. and Scandalios, J. G., *Genetics*, 125, 867, 1990. With permission.)

genes.[123,127] In fact, light-grown CAT-2 null plants (due to a deletion of the 3′ half of the *Cat2* transcript)[120] produce polyadenylated *Cat2* mRNA that is recruited onto leaf polysomes *in vivo*. This provides preliminary evidence that the light control of *Cat2* gene transcription lies at the 5′ flanking region of *Cat2* and *Cat3*. If so, *Cat2*-promoter constructs should not express CAT-2 protein in transformed dark-grown callus, whereas they will express CAT-2 in green callus; the opposite should hold for CAT-3. Such results would allow the manipulation of these antioxidant genes under different light regimens and should lead to a better understanding of their response to light and perhaps other environmental signals.

More recent studies of the circadian expression of the *Cat3* gene show that the steady-state level of *Cat3* mRNA is highest about 2 h before the light/dark transition (sunset) and lowest at about 2 h before the dark/light transition (sunrise) in both leaves and roots of plants grown under a 12-h light/12-h dark photoperiod.[128] Plants grown from imbibition in either constant light or constant dark show no variation in steady-state *Cat3* RNA levels. Thus, *Cat3* exhibits a circadian rhythm in roots as well as in leaves.

F. ISOLATION AND CHARACTERIZATION OF THE *Cat1*, *Cat2*, AND *Cat3* cDNA CLONES

To understand the molecular mechanisms underlying the temporal and spatial expression of the three catalases in maize and to understand more fully the biochemistry and specific physiological function of these isozymes, full-length cDNA clones for each of the maize catalase transcripts were isolated and characterized.[129-131] DNA sequence analyses confirmed that each cDNA encodes a unique catalase protein, and RNA blot analyses using gene-specific probes confirmed a clear correlation between the *Cat1*, *Cat2*, and *Cat3* transcripts and the tissue-specific expression of the CAT-1, CAT-2, and CAT-3 isozymes. The correlation between the presence of a catalase isozyme and its mRNA suggests that much of the tissue-specific regulation of the catalase gene family is under transcriptional control. However, because these analyses give the steady-state level of mRNA expression, and not the rate of gene transcription, differential stability of the catalase mRNAs in expressing and nonexpressing tissues cannot be ruled out.

The nucleotide sequences of *Cat2* and *Cat3* have a high G + C content of ≈61% overall and ≈66% in the coding region. This led to a remarkable codon bias in the gene: 93% of the time, G or C was found in the third position of the codon. In contrast, the *Cat1* cDNA has a G + C content of 51% and no discernible codon bias. In unicellular organisms, codon bias reflects the tRNA population and the level of specific gene expression.[132] In multicellular organisms, the determinants of codon usage are not clear, although the codon bias may reflect the local GC content of the genome or the tRNA population of a particular cell type. Thus, the codon bias of the *Cat2* and *Cat3* cDNAs might be involved in the regulation of the expression of these genes. In humans, the liver-specific glycogen phosphorylase transcript has a high bias for G or C in the third codon position, whereas the rabbit muscle-specific transcript does not.[133] Additionally, several other liver-specific transcripts have this same bias.

Hybridization of maize genomic DNA with the three gene-specific probes revealed a simple hybridization pattern for each of the catalase genes. Thus, each of the maize catalases is encoded by one, or a very few, genes. The

TABLE 3
Percent Similarity of Functional Domains of Various Catalases; SPOT, Sweet Potato Catalase

	CAT-1	CAT-2	CAT-3	SPOT	RAT
N-Terminal (a.a. 1–75; involved in subunit-subunit interactions)					
CAT-1	—	82	77	77	43
CAT-2		—	72	77	43
CAT-3			—	74	43
SPOT				—	42
β-Barrel (a.a. 76–327; involved in heme binding)					
CAT-1	—	76	74	76	49
CAT-2		—	75	73	50
CAT-3			—	75	46
SPOT				—	49
Wrapping (a.a. 328–441; contains catalytically essential α-helix 9)					
CAT-1	—	74	68	68	40
CAT-2		—	72	68	39
CAT-3			—	65	38
SPOT				—	38
C-Terminal (a.a. 442–C terminus; forms hydrophobic channel leading to active site)					
CAT-1	—	74	55	65	12
CAT-2		—	.47	68	12
CAT-3			—	55	12
SPOT				—	12

Note: The percent similarity was derived from identical amino acids in the aligned sequences.

dissimilarity of the restriction patterns indicates that the three genes are not closely linked. Both of these conclusions are in agreement with genetic data that each isozyme is encoded on a single, unlinked gene.[75,134]

A comparison of the similarity of the functional domains of plant and mammalian catalases shows that, although catalytically important regions of the proteins have a high degree of similarity, other domains of the proteins have diverged (Table 3). Bovine liver catalase has four functional domains: (1) an *N*-terminal domain involved in subunit-subunit interactions, (2) a β-barrel structure involved in heme binding, (3) a wrapping domain that contains the catalytically essential α-helix 9, and (4) a C-terminal domain which forms a hydrophobic channel leading to the active site and is near the bound pyridine nucleotide of mammalian catalase.[135,136] As would be expected, there is a high degree of sequence conservation in the functionally important β-barrel domain (Table 3; Figure 10, amino acids 76 to 327). The subunit-subunit interaction domain retains a high degree of sequence similarity between CAT-1, CAT-2, and CAT-3 and sweet potato catalase, whereas the conservation between these proteins and mammalian catalase is weaker in this region (Table 1, amino acids 1 to 75). Because CAT-1 and CAT-2 are known to form heterotetramers *in vivo,* and CAT-3 will form heterotetramers with CAT-1 and CAT-2, although only *in vitro,* this domain is functionally equivalent among maize catalases.[70,75] The fact that no heterotetramers are formed between CAT-3 and the other maize catalases *in vivo,* in spite of such a high degree of homology in the subunit-subunit interaction domain (Figure 11), further underscores the distinct intracellular location of CAT-3 in maize. The wrapping domain is slightly less conserved than the β-barrel domain (Table 3, amino acids 328 to 441). However, the catalytically important α-helix is highly conserved (Figure 10, amino acids 344 to 371). The C-terminal domain is not conserved between mammals and plants (Table 3; Figure 10, amino acids 442 to C termini). Among the plant catalases compared, CAT-1 and sweet potato catalase have a high degree of similarity in this domain, whereas CAT-2 and CAT-3 are less similar to CAT-1 and sweet potato catalase, and to one another. This sequence divergence of the C-terminal domain may be related to the differing ratios of catalatic/peroxidatic activity observed among plant catalases[85,137] and suggests different functions for this domain among the maize catalases. In most domains, the sequence similarity is greater between CAT-1 and sweet potato catalase, perhaps indicating analogous functions for these two enzymes. Catalase cDNAs have also been recently isolated from other plant species. These include sweet potato,[136] cotton,[137] pea,[138] and castor bean.[139]

G. MOLECULAR BASES FOR THE *Cat2* AND *Cat3* NULL MUTATIONS

As mentioned above, catalase activity and protein null mutations were discovered during a mutant screening program.[81,84] Because such mutants would be invaluable in deciphering the distinct physiological roles of each catalase isozyme, it was essential to also determine the molecular nature of such mutants, utilizing the available gene-specific cDNA probes.

RNA blot analyses of 5-d-old, scutellar poly(A)$^+$ RNA from the *Cat2* null lines showed that the *Cat2* transcript was ≈1400 nt compared to the 1850 nt *Cat2* transcript in normal maize lines.[83] Nuclease S1 protection assays further showed that the shortened transcript in the *Cat2* null lines was missing ≈100 bases; thus, the reason for the absence of normal *Cat2* mRNA in the null lines.

```
                10        20        30        40        50        60        70        80        90
CAT-1  -----------MDPYKHRPSSGSNSSFWTTNSGAPVWNNNSALTVGQRGPILLEDYHLIEKLAQFDRERIPERVVHARGASAKGFFEVTHDVSHLTCAD
CAT-2  -----------MDPYKHRPSSAFNAPYWTTNSGAPVWNMDSSLTVGARGPILLEDYHC-EKLANFDRERIPERVVHARGASAKGFFEVTHDITHLTCAD
CAT-3  ---------MTMDPTKFRPSSSHDTTVTTTNAGAPVWNDNEALTVGPRGPILLEDYHLIEKVAHFARERIPERVVHARGASAKGFFECTHDVTSLTCAD
SPOT   -----------MDPSKYRPSSSFNTPFCTTNSGAPVWNNTCALTVGSRGPILLEDYHLVEKIQNFTRERIPERVVHARGASAKGFFEVTHDITHLTCAD
RAT    MADSRDPASDQMKQWKEQRAPQ-KPDVLTTGGGNPIGDKLNIMTAGPRGPLLVQDVVFTDEMAHFDRERIPERVVHAKGAGAFGYFEVTHDITRYSKAK

       100       110       120       130       140       150       160       170       180       190
CAT-1  FLRAPGVQTPVIVRFSTVVHERGSPETLRDPRGFAVKFYTREGNFDLVGNNMPVFFIRDGMKFPDMVHAFKPNPKTNLQENWRIVDFFSHHPESLHMFT
CAT-2  FLRAPGVQTPVIVRFSTVIHERGSPETLRDPRGFAVKFYTREGNWDLVGNNFPVFFIRDGIKFPDMVHALKPNPRTHIQDNWRILDFFSHHPESLHMFS
CAT-3  FLRAPGVRTPVIVRFSTVIHERGSPETIRDPRGFAVKFYTREGNWDLLGNNFPVFFIRDGIKFPDVIHAFKPNPRSHVQEYWRVFDFLSHLPESLHTFF
SPOT   FLRAPGVQTPLIVRFSTVIHERGSPETYRDPRGFAVKMYTRGGNWDLVGNNFPVFFIRDGTQFPDVIHAFKPNPKSHIQENWRILDYLSHLPESLNTFA
RAT    VFEHIGKRTPIAVRFSTVAGESGSADTVRDPRGFAVKFYTEDGNWDLVGNNTPIFFIRDAMLFPSFIHSQKRNPQTHLKDPDMVWDFWSLCPESLHQVT

        200       210       220       230       240       250       260       270       280       290
CAT-1  FLFDDVGIPLNYRHMEGFGVNTYTLINRDGKPHLVKFHWKPTCGVKCLLDNEAVT-VGGTCHSHATKDLYDSIAAGNYPEWKLYIQTIDLDHEDKFDFD
CAT-2  FLFDDVGIPADYRHMDGSGVHTYTLVSRAGTVTYVKFHWRPTCGVRSLMDDEAVR-CGA-NHSHATKDLTDAIAAGNFPEWTLYIQTMDPEMEDRLD-D
CAT-3  FLFDDVGVPSDYRHMEGFGVNTYTFVSAAGKAQYVKFHWKPTCGVRCILTDEEAALVGGRNHSHATQDLYDSIAAGSFPEWTLYVQVMDPDTEEQYDFD
SPOT   WFYDDVGIPTDYRHMEGFGVHTFTMINKEGKANYVKFHWKPTCGVKCLLEEEAIR-IGGENHSHATQDLYESIAAGNYPEWKLYIQVMDPDHEDRFDFD
RAT    FLFSDRGIPDGHRHMNGYGSHTFKLVNANGEAVYCKFHYKTDQGIKNLPVEEAGR-LAQEDPDYGLRDLFNAIASGNYPSWTFYIQVMTFKEAETFPFN

         300       310       320       330       340       350       360       370       380       390
CAT-1  FDPLDVTKTWPEDIIPLQPVGRMVLNKNVDNFFAENEQIAFCPAIIVPGIHYSDDKLLQTRIFSYADTQRHRLGPNYLMLPVNAPKCAHHNNHH-DGFM
CAT-2  LDPLDVTKTWPEDTFPLQPVGRLVLNRNIDNFFAENEQLAFCPGLIVPGIYYSDDKLLQTRIFSYSDTQRHRLGPNYLLLPANAPKCAHHNNHY-DGSM
CAT-3  FDPLDDTKTWPEDLLPLRPVGRLVLDRNVDNFFNENEQLAFGPGLVVPGIYYSDDKMLQCRVFAYADTQRYRLGPNYLMLPVNAPRCAHHNNHY-DGAM
SPOT   FDPLDTTKIWPEELIPLQPVGRMVLNKNIDNFFAENEMLAMDPAHIVPPIYFSDDKMLQARVFAYADTHRHRLGPNYMLLPVNAPKCAHHNNAY-DGYM
RAT    FNPFDLTKVWPHKDYPLIPVGKLVLNRNPANYFAEVEQMAFDPSNMPPGIEPSPDKMLQGRLFAYPDTHRHRLGPNYLQIPVNCPYRARVANYQRDGPM

          400       410       420       430       440       450       460       470       480       490
CAT-1  NFMHRDEEV-NYFPSRFD-PARNAEKVPIPPRVLTGCREKCIIQKENNFKQAGERYRSFDPARQDRFIQRWVDALTDPRVTHEHRTIWISYWSQCDAAL
CAT-2  NFMHRHEEV-DYFPSRYD-AVRNAPRYPIPTAHIAGRREKTVISKENNFKQPGERYRAMDPARQERFITRWVDALSDPRLTHEIRTIWLSNWSQADRSL
CAT-3  NFMHRDEEV-DYYPSRHAPLRQAAPPTPLPPRPVAGRREKATIRKPNDFKQPGERYRSWDADRQDRFVRRFADSLGHPKVSQELRSIWIDLLAKCDASL
SPOT   NFVHRDEEV-DYFPSKFD-NTRNAERFPTPLRIVTGQRDKCVIEKENNFKQPGDRYRSWAPDRQDRFINRWVKALSEPRVTHEIRSTWISYLTQADRSL
RAT    CMHDNQGGAPNYYPNSFS-APEQQGSALEHHSQCSADVKRFNSANEDNVTQVRTFYTKVLNEEEKRLCENIANHLKDAQLFIQRKA--VKMFTDVHPDY

           500
CAT-1  GQKLASRLNLKPSM
CAT-2  GQKLASRLSAKPSM
CAT-3  GMKIATRLNMKANM
SPOT   GQKVASRLNIRPTM
RAT    GVQALLDQYNSQKPKNAIHTYVQAGSHIAAKGKANL
```

FIGURE 10. Comparison of the deduced amino acid sequences for the maize catalase isozymes. The sequences are given as standard one-letter symbols and are numbered from the beginning of the rat catalase sequence, including spaces. The shaded areas delineate conservation of an amino acid in at least four of the five proteins. The sequences for CAT-1, CAT-2, and CAT-3, sweet potato catalase, and rat catalase were deduced from cDNA sequences.

HOMOLOGY WITHIN THE SUBUNIT-SUBUNIT INTERACTION SITE OF THE MAIZE AND OTHER CATALASES

```
                   10        20        30
CAT-1   -----------MDPTKHRPSSGSNSSFWTTNSGAPVWNN
CAT-2   -----------MDPYKHRPSSAFNAPYWTTNSGAPVWNN
CAT-3   ---------MTMDPTKFRPSSSHDTTVTTTNAGAPVWND
SPOT    -----------MDPSKYRPSSSFNTPFCTTNSGAPVWNN
RAT     MSDSRDPASDQMKQWKEQRAPQ-RPDVLTTGGGNPIGDK

          40        50        60        70
CAT-1   NSALTVGQRGPILLEDYHLIEKLAQFDRERIPERVVHA
CAT-2   DSSLTVGARGPILLEDYHC-EKLANFDRERIPERVVHA
CAT-3   NEALTVGPRGPILLEDYHLIEKVAHFDRERIPERVVHA
SPOT    TCALTVGSRGPILLEDYHLVEKIQNFTRERIPERVVHA
RAT     LNIMTAGPRGPLLVQDVVFTDEMAHFDRERIPERVVHA
```

FIGURE 11. Comparison of the deduced amino acid sequences within the subunit-subunit interaction (N-terminal) domain of the various catalases. In spite of the strong homology, CAT-3 subunits (unlike CAT-1 and CAT-2) do not interact *in vivo* to form heterotetramers with either CAT-1 or CAT-2 (see text).

The molecular bases for the shorter *Cat2* transcript in maize lines null for the CAT-2 catalase isozyme was further investigated using cDNA libraries and genomic DNA blots. Sequence comparison of partial *Cat2* cDNAs obtained from two CAT-2 null lines and a wild-type CAT-2 encoding cDNA showed that the *Cat2* null transcripts diverged from the wild-type transcript, but remained homologous to each other (Figure 12). Genomic DNA blots indicated that the missing portion of the transcript is present in the genomes of lines null for the CAT-2 isozyme. Differences in the hybridization patterns of normal and null lines were revealed when genomic DNA blots were probed with the full-length *Cat2* cDNA, a *Cat2* gene-specific probe, and a "null sequence" probe. Together, the DNA blotting results suggest that a rearrangement of the *Cat2* gene has occurred in the CAT-2 null lines. The available data suggest that this CAT-2 null mutation in maize is due to a DNA insertion into the *Cat2* gene.[120]

Maize line IDS28 was identified as being homozygous for a *Cat3* null allele. Catalase protein (CAT-3) monospecific antibodies could not detect CAT-3 in extracts of any tissues of IDS28, which in any normal maize line possess CAT-3. RNA blot analysis indicated that IDS28 does not accumulate *Cat3* transcript in any tissues (Figure 13). Additionally, the zymogram and RNA blot analyses showed that the CAT-3 null mutation did not affect the expression of the *Cat1* or *Cat2* genes at the mRNA or protein level. Genomic DNA blots revealed significant structural alterations in the *Cat3* gene in IDS28. The results suggest that the molecular bases for the *Cat3* null phenotype in IDS28 is most likely due to a deletion in the 5′ end of the *Cat3* gene.[84] To investigate if the lack of *Cat3* transcript accumulation in IDS28 is due to a lack of transcription of the *Cat3* gene, run-off transcription assays

```
         994    1000           1010           1020             1030           1040           1050
R6-67    TTC GCG GAG AAC GAG CAG CTG GCC TTC TGC CCG GGC CTC ATC GTC CCT GGT ATC TAC TAC
A16      *** *** *** *** *** *** *** *** *** *** *** *** *** *** *** **C **G *** *** ***
A340     *** *** *** *** *** *** *** *** *** *** *** *** *** *** *** **C **G *** *** ***

         1054   1060           1070           1080             1090           1100           1110
R6-67    TCC GAC GAC AAG CTG CTG CAG ACC AGG ATC TTC TCC TAC TCC GAC ACG CAG CGC CAC CGC
A16      *** *** *** *** *** *** *** *** *** *** *** *** *** *TG TTA G*A ACA *CT GGG AA*
A340     *** *** *** *** *** *** *** *** *** *** *** *** *** *TG TTA G*A ACA *CT GGG AA*
                                                     Homology ends here ↑

         1114   1120           1130           1140             1150           1160
R6-67    CTC GGC CCC AAC TAC CTG CTG CTA CCG GCC AAC GCG CCC AAG TGC GCA CAC CAC .....
A16      GGG *AT *GG
A340     GGG *AT *GG *GA GGA AGA GCT *AG AG* AGA *TG *
```

FIGURE 12. Partial sequence comparison of the R6-67, A16, and A340 *Cat2* cDNAs. Numbers indicate the nucleotide position in the R6-67 *Cat2* cDNA. Homologous bases between the R6-67 *Cat2* cDNA and the *Cat2* null cDNAs (A16 and A340) are denoted by an asterisk (*). Homology ends at base 1093 of the R6-67 *Cat2* cDNA. Note that the A16 and A340 sequences are completely identical through the end of the A16 *Cat2* cDNA. (From Abler, and Scandalios, J. G., *Theor. Appl. Genet.*, 81, 635, 1991. With permission.)

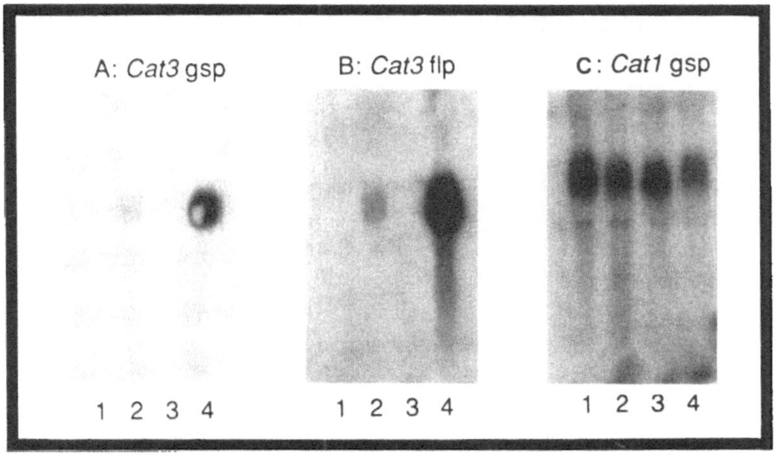

FIGURE 13. RNA blot analysis of *Cat1* and *Cat3* transcripts in W64A and IDS28. Total RNA (10 μg) from the immature kernel (9 d postpollination) of IDS28 (lane 1), W64A (lane 2), dark-grown epicotyls (8 d postimbibition) of IDS28 (lane 3), and W64A (lane 4) were examined by RNA blot analysis. RNA was hybridized to the *Cat3* gene-specific probe (gsp) (A), the full-length *Cat3* cDNA probe (flp) (B), and the *Cat1* gene-specific probe (gsp) (C). (From Wadsworth, G. J. and Scandalios, J. G., *Genetics*, 125, 867, 1990. With permission.)

were performed. Nuclei were isolated from tissues that normally express CAT-3 and were incubated under conditions which allow for the *in vitro* synthesis of RNA transcripts. Radiolabeled transcripts were isolated from the nuclei and hybridized to DNA bound to nitrocellulose filters. Hybridization was detected by autoradiography. There was a strong hybridization signal to the ribosomal DNA (a positive control), indicating that the nuclei of both the normal (W64A) and *Cat3* null mutant (IDS28) lines exhibit high levels of ribosomal RNA transcription *in vitro*. The plasmid pUC12, a bacterial vector with no sequences homologous to maize genes, was used as a negative control. The *Cat3* cDNA probe hybridized with the transcript from the W64A nuclei. However, the IDS28 nuclei did not synthesize a transcript homologous to *Cat3* to be detected under these conditions. Thus, the *Cat3* gene is likely not transcribed in IDS28.

Although there are no major discernible physiological effects of the *Cat2* and *Cat3* null mutations, such mutants will aid in deciphering the role each isozyme plays in the plant's metabolism. Perhaps the lack of observable major phenotypic effects of these mutants is due to the fact that CAT-1 is still expressed and may compensate for the defect. With breeding experiments, we have created a double null line lacking both CAT-2 and CAT-3. Preliminary data suggest that this line may, in fact, exhibit a higher germination rate than the normal lines, but it still expresses CAT-1. Eliminating CAT-1 by antisense mutations might be more telling as to the role of catalases in the life of the plant. It is also likely that each of the catalases may play a distinctly important physiological role under environmental conditions, not yet tested, which increase oxidative stress. Such experiments are currently in progress.

H. ISOLATION AND CHARACTERIZATION OF *Cat* GENOMIC CLONES

To identify developmental and stress-response elements in the *Cat* genes of maize, the respective cDNA probes were used to isolate and characterize *Cat* genomic clones. Genomic clones for *Cat1* and *Cat3* have been isolated from a maize genomic library. These are the first and only plant catalase genes to be isolated and fully sequenced to date. The *Cat1* genomic DNA (7.8 kb) was restriction mapped and intron-exon boundaries determined on comparison with the *Cat1* cDNA sequence.[111] The *Cat1* gene is comprised of six introns (ranging in size from 110 to 937 bp) and seven exons, a 2.5-kb promoter, and 1.5-kb 3′ untranscribed region (Figure 14). The start of transcription was identified by primer extension. The promoter region was sequenced, and computer searches identified the CAAT box, an *Sp*I binding site, and an abscisic acid (ABA)-response element. This may be the underlying basis for the ABA response observed for CAT-1. The *Cat3* gene promoter (2.6 kb) has been sequenced, and the transcription site has been mapped 68 bases from the start of translation by primer extension.[112] At approximately 29 bases 5′ of the transcriptional start site is a TATA box, and three possible CAAT boxes are located about 150 bases upstream of the start of transcription. Two small introns are located in the coding region of the *Cat3* gene. The first intron is 84 bases long and occurs at the same position in *Cat1* and *Cat3*. The second intron is 92 bases long and does not correlate to any of the *Cat1* introns. Both introns are "typical" in that they are AT-rich and contain the 5′ GT and 3′ AG splice junctions. Unlike *Cat1*, there is no ABA consensus sequence in *Cat3*.

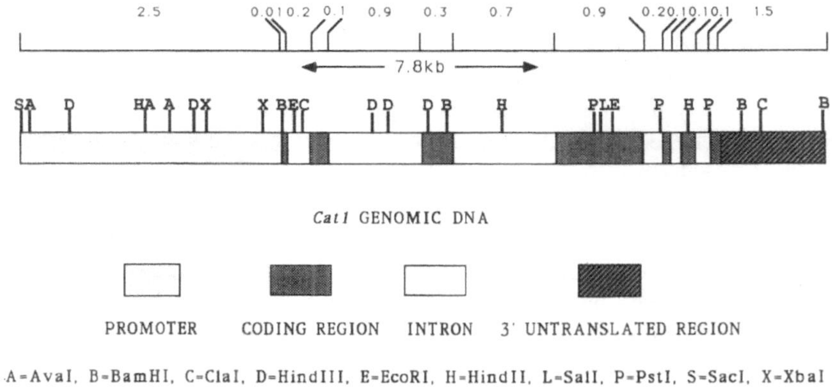

FIGURE 14. Restriction map of the maize *Cat1* gene. Introns, exons, and 5′ and 3′ noncoding regions are denoted by various degrees of shading. The 2.5-kb promoter region is to the left.

I. CHIMERIC *Cat* GENE CONSTRUCTS USING THE *uidA* (GUS) REPORTER GENE

To functionally define the promoter regions of the *Cat* genes, the putative *cis*-regions were placed in the pBI101 GUS vector, and various constructs were introduced into tobacco via *Agrobacterium*-mediated transformation in an attempt to express GUS under the control of the *Cat* promoters.[111] Chimeric *Cat1* promoter-GUS constructs have been used to transform tobacco plants using the binary vector pBI101.2. Preliminary data indicate that the *Cat1* promoter can drive the expression of GUS in the transgenic tobacco plants. Two transgenic tobacco lines have been established that contain 2.5 and 0.8 kb of the maize *Cat1* promoter sequence fused to the GUS coding sequence. Preliminary results have shown that GUS activity can be detected in young transgenic tobacco seedlings and seeds. The timing of *Cat1*-GUS expression in the transgenic tobacco seed is consistent with the *Cat1* expression pattern in maize and is also consistent with the expression of GUS in developing tobacco seeds driven by promoters of some other ABA-regulated genes. The ABA consensus sequence located within the *Cat1* promoter region may serve as a protein-binding site upon ABA induction of CAT-1 in maize. No GUS activity was found in tobacco plants containing "no promoter" or the *Cat1* promoter in antisense orientation. Deletion analysis is underway to determine the putative essential *cis*-acting elements for *Cat1* gene expression.

IV. CONCLUDING REMARKS

Catalase is an enzyme that is present in all aerobic cells and plays a critical role in protecting cell components from active oxygen species. Oxygen presents living organisms with a variety of physiological challenges collectively termed "oxidative stress". These challenges may be greater in plants than other eukaryotes, because plants both consume O_2 during respiration and generate O_2 during photosynthesis. Among the special problems incurred by aerobic organisms is the need to eliminate toxic H_2O_2 from the cell. Catalase, which is widely distributed in aerobic organisms and has efficient enzyme kinetics, is believed to play an important role in eliminating H_2O_2 from cells.

Plants, unlike animals, have multiple enzymatic forms of catalase. Nearly 30 years ago, Scandalios[70] demonstrated that the multiple isozymes of catalase were the products of separate, unlinked genes. The basis for the diversity of catalases in other species continues to be investigated, and indirect evidence suggests that multiple genes for catalase may also exist in other plants.[91,137-139] The existence of multiple molecular forms of catalase, their tissue specificities, and the specific changes they undergo during development and in response to various signals imply separate metabolic roles for each of the catalase isozymes, a subject of intense scrutiny at present.

The maize catalase gene-enzyme system discussed herein presents an attractive model for elucidating the specific biological role(s) of each isozyme. Maize catalases have been extensively characterized, and maize is the only plant system where the genetic basis of the different molecular forms is understood. The multiple forms of catalase in maize have been purified and biochemically characterized. Three different isozymic forms of catalase (CAT-1, CAT-2, and CAT-3) differ in several biochemical traits, including thermostability and sensitivity to inhibitors of catalase activity, and in their cellular and subcellular location. In addition, the maize catalase genes are some of the most highly regulated eukaryotic genes. The in-depth resolution of the maize and other plant catalases at the biochemical, developmental, genetic, and molecular levels will undoubtedly aid our understanding of their regulation, expression, and function. As new technologies are developed and applied to help unravel the underlying mechanisms by which these genes are regulated to respond to the various developmental, physiological, and environmental signals, it seems inevitable that their respective biological role(s) will begin to unfold.

The role of oxidative stress in effecting specific gene expression is quite intricate when one considers the fact that oxidative insult can affect specific organelles or the entire plant cell or tissue. The plant has defense responses (i.e., gene expression mechanisms) to accommodate either or both situations. How the plant responds to a specific organellar stress vs. a cellular or systemic stress, and what genetic elements may be involved in the defense response, can now be examined utilizing state-of-the-art molecular technologies. Additionally, the availability and use of *Cat* null and overexpression mutants, and the ability to transform alien genomes with the *Cat* genes (transgenics) are shedding new light on the functional role of each of these genes. The use of transgenic plants will aid in determining how the catalase genes are regulated and how they respond to specific signals. The resolution of *cis*-acting elements and *trans*-acting factors involved in *Cat* gene expression may help us understand the underlying mechanisms for the entire signal transduction pathway during oxidative stress. Such information is essential in any efforts to engineer higher plants for greater resistance to oxidative stress under hostile environments.

It seems appropriate to conclude, on the basis of the available data from various laboratories,[140] that catalases play significant roles in protecting living cells against the toxic and mutagenic effects of activated oxygen. Furthermore, the existence of multiple catalases in most plants, along with their distinct properties and complex regulation, imply that catalases may play a variety of important metabolic roles which await resolution.

ACKNOWLEDGMENTS

Research from the author's laboratory has been supported, in part, by grants from the U.S. Environmental Protection Agency, National Science Foundation, and National Institutes of Health to J.G.S. I thank past and present associates for their contributions to ongoing studies on catalase, especially J. D. Williamson, L. Guan, M. Abler, S. Ruzsa, and S. P. Kernodle. The expert typing of Suzanne Quick is greatly appreciated.

REFERENCES

1. **Halliwell, B. and Gutteridge, J. M. C.,** Oxygen-derived species and herbicide action, *Physiol. Plant.*, 15, 21, 1984.
2. **Fridovich, I.,** Superoxide dismutases, *Annu. Rev. Biochem.*, 44, 147, 1975.
3. **Imlay, J. A., Chin, S. M., and Linn, S.,** Toxic DNA damage by H_2O_2 through the Fenton reaction *in vivo* and *in vitro*, *Science*, 240, 640, 1988.
4. **Fenton, H. J. H.,** Oxidation of tartaric acid in presence of iron, *J. Chem. Soc.*, 65, 899, 1894.
5. **Haber, F. and Weiss, J.,** The catalytic decomposition of hydrogen peroxide by iron salts, *Proc. R. Soc. London Ser. A*, 147, 332, 1934.
6. **Halliwell, B. and Gutteridge, J. M. C.,** The importance of free radicals and catalytic metal ions in human diseases, *Mol. Aspects Med.*, 8, 89, 1985.
7. **Ward, J. F.,** Molecular mechanisms of radiation induced damage to nucleic acids, *Adv. Free Rad. Biol.*, 5, 182, 1975.
8. **Asada, K., Kiso, K., Yashikawa, K.,** Univalent reduction of molecular O_2 by spinach chloroplasts upon illumination, *J. Biol. Chem.*, 249, 2175, 1974.
9. **Turrens, J. and Boveris, A.,** Generation of superoxide anion by NADH-dehydrogenase of bovine heart mitochondria, *Biochem. J.*, 191, 421, 1980.
10. **Farrington, J. A., Ebert, M., Land, E. J., and Fletcher, K.,** Bipyridynium quaternary salts and related compounds, V. Pulse radiolysis studies of the reaction of paraquat with oxygen. Implications for the mode of action of bipyridyl herbicides, *Biochim. Biophys. Acta*, 314, 372, 1973.
11. **Harbor, J. R. and Bolton, J. R.,** Superoxide formation in spinach chloroplast, ESR detection by spin trapping, *Biochem. Biophys. Res. Commun.*, 64, 803, 1975.

12. **Halliwell, B.,** Oxygen derived species and herbicide action, *Physiol. Plant.,* 15, 21, 1984.
13. **Tanaka, K., Furusawa, I., Kondo, N., and Tanaka, K.,** SO_2 tolerance of tobacco plants regenerated from paraquat-tolerant callus, *Plant Cell Physiol.,* 29, 743, 1988.
14. **Grimes, H. D., Perkins, K. K., and Boss, W. F.,** Ozone degrades into hydroxyl radical under physiological conditions, *Plant Physiol.,* 72, 1016, 1983.
15. **Imlay, J. A. and Linn, S.,** DNA damage and oxygen radical toxicity, *Science,* 240, 1302, 1988.
16. **Halliwell, B.,** Oxidants and human disease: some new concepts, *FASEB J.,* 1, 358, 1987.
17. **Hoffmann, M. E., Mello-Filho, A. C., and Memghini, R.,** Correlation between cytotoxic effect of hydrogen peroxide and the yield of DNA strand breaks in cells of different species, *Biochim. Biophys. Acta,* 781, 234, 1984.
18. **Halliwell, B., Ed.,** *Oxygen Radicals and Tissue Injury,* Proc. Upjohn Symp., Bethesda, MD, 1988.
19. **Aruoma, O. I., Grootveld, M., and Halliwell, B.,** The role of iron in ascorbate dependent deoxyribose degradation, evidence consistent with site specific hydroxyl radical generation caused by iron ions bound to deoxyribose molecules, *J. Inorg. Biol.,* 29, 289, 1987.
20. **Parshad, R., Sanford, K. K., Jones, G. M., and Tarmone, R. E.,** Fluorescent light induced chromosomal damage and its prevention in mouse cells in culture, *Proc. Natl. Acad. Sci. U.S.A.,* 75, 1830, 1978.
21. **Larramendy, M., Mello-Filho, E. A., Leme, M., and Memeghini, R.,** Iron-mediated induction of sister chromatid exchange by H_2O_2 and superoxide anion, *Mutat. Res.,* 198, 57, 1983.
22. **Carson, D. A., Seto, S., Wasson, D. B., and Carrera, C. J.,** DNA strand breaks, NAD metabolism, and programmed cell death, *Exp. Cell Res.,* 164, 273, 1986.
23. **Adelman, R., Saul, R. L., and Ames, B.,** Oxidative damage to DNA: relation to species metabolic rate and lifespan, *Proc. Natl. Acad. Sci. U.S.A.,* 85, 2706, 1988.
24. **Halliwell, B. and Grootveld, M.,** The measurement of free radicals in humans, some thoughts for future experiments, *FEBS Lett.,* 1, 9, 1987.
25. **Wolff, S. P., Garner, A., and Dean, R. T.,** Free radicals, lipids, and protein degradation, *TIBS,* 11, 27, 1986.
26. **Shauenstein, E., Esterbauer, H., and Zollner, H.,** *Aldehydes in Biological Systems: Their Natural Occurrence and Biological Activities,* Pion Press, London, 1977.
27. **Puppo, A. and Halliwell, B.,** Formation of hydroxyl radicals from hydrogen peroxide in the presence of iron: is haemoglobin a biological Fenton reagent?, *Biochem. J.,* 249, 185, 1988.
28. **Puppo, A. and Halliwell, B.,** Generation of hydroxyl radicals by soybean nodule leg haemoglobin, *Planta,* 173, 405, 1988.
29. **Hodgson, E. K. and Fridovich, I.,** The interaction of bovine erythrocyte superoxide dismutase with H_2O_2: inactivation of the enzyme, *Biochemistry,* 14, 5294, 1975.
30. **Beyer, W. F. and Fridovich, I.,** Effect of H_2O_2 on the iron containing superoxide dismutase of *E. coli, Biochemistry,* 26, 1251, 1987.
31. **Ellyard, P. W. and Gibbs, M.,** Inhibition of photosynthesis by oxygen in isolated spinach chloroplasts, *Plant Physiol.,* 68, 65, 1969.
32. **Charles, S. A. and Halliwell, B.,** Light activation of fructose bisphosphate in isolated spinach chloroplasts and deactivation by H_2O_2, *Planta,* 151, 242, 1981.
33. **Asada, K. and Takahashi, M.,** Production and scavenging of active oxygen in photosynthesis, in *Photoinhibition,* Kyle, D. J., Osmond, C. B., and Arntzen, C. J., Eds., Elsevier, Amsterdam, 1987, 227.

34. **Oshino, N., Jamieson, D., Sugano, T., and Chance, B.,** Optical measurement of the catalase-hydrogen peroxide intermediate compound I in the level of anaesthetized rats and its implication to H_2O_2 production *in vivo, Biochem. J.*, 146, 67, 1975.
35. **Chance, B., Sies, H., and Boveris, A.,** Hydroperoxide metabolism in mammalian organs, *Physiol. Rev.*, 59, 527, 1979.
36. **Forman, H. J. and Boveris, A.,** Superoxide radical and H_2O_2 in mitochondria, in *Free Radicals in Biology*, Vol. 5, Pryor, W., Ed., Academic Press, New York, 1982, 69.
37. **Tolbert, N. E.,** Metabolic pathways in peroxisomes and glyoxysomes, *Annu. Rev. Biochem.*, 50, 133, 1981.
38. **Beevers, H.,** Glyoxysomes in higher plants, in *Peroxisomes and Glyoxysomes*, Kindl, H. and Lazarow, P. B., Eds., *Ann. N.Y. Acad. Sci.*, 386, 243, 1982.
39. **Halliwell, B.,** Oxidative damage, lipid peroxidation and antioxidant protection in chloroplast, *Chem. Phys. Lipids*, 44, 327, 1987.
40. **Baum, J. A. and Scandalios, J. G.,** Developmental expression and intracellular localization of superoxide dismutase in maize, *Differentiation*, 13, 133, 1979.
41. **Hayakawa, T., Kanematsu, S., and Asada, K.,** Occurrence of CuZn superoxide dismutase in the intra-thylakoid space of spinach chloroplast, *Plant Cell Physiol.*, 25, 883, 1984.
42. **Nadezhdin, D. and Dunford, H. B.,** Oxidation of nicotinamide adenine dinucleotide by hydroxyperoxyl radical, a flash photolysis study, *J. Phys. Chem.*, 83, 1957, 1979.
43. **Cabelli, D. E. and Bielski, B. H. J.,** Kinetics and mechanism for the oxidation of ascorbic acid/ascorbate by O_2^- radicals, A pulse radiolysis and stopped-flow photolysis study, *J. Phys. Chem.*, 87, 1809, 1983.
44. **Anderson, J. W., Foyer, C. H., and Walker, D. A.,** Light dependent reduction of H_2O_2 by intact spinach chloroplast, *Biochim. Biophys. Acta*, 724, 69, 1983.
45. **Allen, J. F.,** A two step mechanism for the photosynthetic reduction of oxygen by ferredoxin, *Biochem. Biophys. Res. Commun.*, 66, 36, 1975.
46. **Asada, K. and Nakano, Y.,** Spinach chloroplast scavenge H_2O_2 on illumination, *Plant Cell Physiol.*, 21, 1295, 1980.
47. **Loschen, G. A., Flohe, L., and Chance, B.,** Respiratory chain linked H_2O_2 production in pigeon heart mitochondria, *FEBS Lett.*, 18, 261, 1971.
48. **Chance, B., Boveris, A., Oshino, N., and Loschen, G.,** The nature of catalase intermediate and its biological function, in *Oxidases and Related Redox System*, King, T. E., Mason, H. S., and Morrison, M., Eds., University Park Press, Baltimore, MD., 1973, 350.
49. **Boveris, A. and Cadenas, E.,** Mitochondrial production of superoxide anions and its relationship to the antimycin insensitive respiration, *FEBS Lett.*, 54, 311, 1975.
50. **Boveris, A. and Chance, B.,** The mitochondrial generation of hydrogen peroxide: general properties and effect of hyperbaric oxygen, *Biochem. J.*, 134, 707, 1973.
51. **Oshino, N., Chance, B., Siess, H., and Bucher, T.,** The role of H_2O_2 generation in perfused rat liver, and the reaction of catalase compound I and hydrogen donor, *Biochim. Biophys. Acta*, 154, 117, 1973.
52. **Cadenas, E. and Boveris, A.,** Enhancement of hydrogen peroxide formation by protophores and ionophores in antimycin supplemented mitochondria, *Biochem., J.*, 188, 31, 1980.
53. **Fridovich, I.,** Superoxide dismutases, *Adv. Enzymol.*, 58, 61, 1986.
54. **Nicholls, P.,** Contributions of catalase and glutathione peroxidase to red cell peroxide removal, *Biochim. Biophys. Acta*, 279, 306, 1972.
55. **Starke, P. A. and Farber, J. I.,** Endogenous defenses against the cytotoxicity of H_2O_2 in culture rat hepatocytes, *J. Biol. Chem.*, 260, 86, 1985.
56. **Jones, D. P., Thor, H., Andersson, B., and Orrenius, S.,** Detoxification reactions in hepatocytes, role of glutathione peroxidase, formaldehyde peroxidase in reactions related to N-demethylation by cytochrome P-450 systems, *J. Biol. Chem.*, 253, 6031, 1978.

57. **Scandalios, J. G., Tong, W.-F., and Roupakias, D. G.,** *Cat3*, a third gene locus coding for a tissue-specific catalase in maize: genetics, intracellular location, and some biochemical properties, *Mol. Gen. Genet.*, 179, 33, 1980.
58. **Deisseroth, A. and Dounce, A.,** Catalase: physical and chemical properties, mechanism of catalysis and physiological role, *Physiol. Rev.*, 50, 319, 1970.
59. **Dounce, A. L.,** A proposed mechanism for the catalatic action of catalase, *J. Theor. Biol.*, 105, 553, 1983.
60. **Thayer, W. S.,** Adriamycin stimulated superoxide formation in submitochondrial particles, *Chem. Biol. Interact.*, 19, 250, 1977.
61. **Stadelman, E. J.,** Permeability of the plant cell, *Annu. Rev. Plant Physiol.*, 20, 585, 1969.
62. **Aebi, H. and Suter, H.,** Acatalasemia, *Adv. Hum. Genet.*, 2, 143, 1971.
63. **Loew, O.,** Catalase, new enzyme of general occurrence, with special reference to the tobacco plant, *USDA Rep.*, 68, 47, 1901.
64. **Warburg, O.,** Ueber die antikatalytische Wirkung der Blausäure, *Biochem. Z.*, 136, 266, 1923.
65. **Zeile, K. and Hellström, H.,** Ueber die aktive Gruppe der Leberkatalase, *Z. Physiol. Chem.*, 192, 171, 1930.
66. **Sumner, J. B. and Dounce, A. L.,** Catalase, in *Methods in Enzymology,* Vol. 2, Colowick, S. P. and Kaplan, N. O., Eds., Academic Press, NY, 1955, 775.
67. **Koltzoff, N. K.,** Catalase content in blood as an inherited trait, *J. Biol. Med. Exp.*, 5, 303, 1927.
68. **Scandalios, J. G.,** Genetic control of multiple molecular forms of catalase in maize, *Ann. N.Y. Acad. Sci.*, 151, 274, 1968.
69. **Scandalios, J. G.,** Response of plant antioxidant defense genes to environmental stress, *Adv. Genet.*, 28, 1, 1990.
70. **Scandalios, J. G.,** Subunit dissociation and recombination of catalase isozymes, *Proc. Natl. Acad. Sci. U.S.A.*, 53, 1035, 1965.
71. **Scandalios, J. G.,** Control of gene expression and enzyme differentiation, in *Physiological Genetics,* Scandalios, J. G., Ed., Academic Press, New York, 1979, 63.
72. **Scandalios, J. G., Tsaftaris, A. S., Chandlee, J. M., and Skadsen, R. W.,** Expression of the developmentally regulated catalase *(Cat)* genes in maize, *Dev. Genet.*, 4, 281, 1984.
73. **Matters, G. L. and Scandalios, J. G.,** Changes in plant gene expression during stress, *Dev. Genet.*, 7, 167, 1986.
74. **Matters, G. L. and Scandalios, J. G.,** Effect of the free radical-generating herbicide paraquat on the expression of the superoxide dismutase *(Sod)* genes in maize, *Biochim. Biophys. Acta,* 882, 29, 1986.
75. **Scandalios, J. G.,** The antioxidant enzyme genes *Cat* and *Sod* of maize: regulation, functional significance, and molecular biology, in *Isozymes: Current Topics in Biological and Medical Research,* Vol. 14, Rattazzi, M. C., Scandalios, J. G., and Whitt, G. S., Eds., Alan R. Liss, New York, 1987, 19.
76. **Skadsen, R. W. and Scandalios, J. G.,** Translational control of photo-induced expression of the *Cat2* catalase gene during leaf development in maize, *Proc. Natl. Acad. Sci. U.S.A.*, 84, 2785, 1987.
77. **Tsaftaris, A. S., Bosabalidis, A. M., and Scandalios, J. G.,** Cell-type-specific gene expression and acatalasemic peroxisomes in a null *Cat2* catalase mutant of maize, *Proc. Natl. Acad. Sci. U.S.A.*, 80, 4455, 1983.
78. **Scandalios, J. G.,** Subcellular localization of catalase variants coded by two genetic loci during maize development, *J. Hered.*, 65, 28, 1974.
79. **Scandalios, J. G.,** Targeting, import and processing of nuclear gene-encoded proteins into mitochondria and peroxisomes, *Progr. Clin. Biol. Res.*, 344, 515, 1990.

80. **Scandalios, J. G., Chang, D. Y., McMillin, D. E., Tsaftaris, A. S., and Moll, R.,** Genetic regulation of the catalase developmental program in maize scutellum: identification of a temporal regulatory gene, *Proc. Natl. Acad. Sci. U.S.A.*, 77, 5360, 1980.
81. **Tsaftaris, A. S. and Scandalios, J. G.,** Genetic and biochemical characterization of a *Cat2* catalase null mutant of *Zea mays, Mol. Gen. Genet.*, 181, 158, 1981.
82. **Chandlee, J. M. and Scandalios, J. G.,** Analysis of variants affecting the catalase developmental program in maize scutellum, *Theor. Appl. Genet.*, 69, 71, 1984.
83. **Bethards, L. A. and Scandalios, J. G.,** Molecular basis for the CAT-2 null phenotype in maize, *Genetics*, 118, 149, 1988.
84. **Wadsworth, G. J. and Scandalios, J. G.,** Molecular characterization of a catalase null allele at the *Cat3* locus in maize, *Genetics*, 125, 867, 1990.
85. **Chandlee, J. M., Tsaftaris, A. S., and Scandalios, J. G.,** Purification and partial characterization of three genetically defined catalases of maize, *Plant Sci. Lett.*, 29, 117, 1983.
86. **Chandlee, J. M. and Scandalios, J. G.,** Regulation of *Cat1* gene expression in the scutellum of maize during early sporophytic development, *Proc. Natl. Acad. Sci. U.S.A.*, 81, 4903, 1984.
87. **Scandalios, J. G.,** Multiple varieties of isozymes and their role in studies of gene regulation and expression during eukaryote development, in *Isozymes: Current Topics in Biological and Medical Research*, Vol. 9, Rattazzi, M. C., Scandalios, J. G., and Whitt, G. S., Eds., Alan R. Liss, New York, 1983, 1.
88. **Wadsworth, G. J. and Scandalios, J. G.,** Differential expression of the maize catalase genes during kernel development: the role of steady-state mRNA levels, *Dev. Genet.*, 10, 304, 1989.
89. **Acevedo, A. and Scandalios, J. G.,** Expression of the catalase and superoxide dismutase genes in mature pollen in maize, *Theor. Appl. Genet.*, 80, 705, 1990.
90. **Redinbaugh, M. G., Sabre, M., and Scandalios, J. G.,** The distribution of catalase activity, isozyme protein, and transcript in the tissues of the developing maize seedling, *Plant Physiol.*, 92, 375, 1990.
91. **Havir, A. E. and McHale, N. A.,** Enhanced-peroxidatic activity in specific catalase isozymes of tobacco, barley and maize, *Plant Physiol.*, 91, 812, 1989.
92. **Elthon, T. E. and McIntosh, L.,** Identification of the alternate terminal oxidase of higher plant mitochondria, *Proc. Natl. Acad. Sci. U.S.A.*, 82, 8399, 1987.
93. **Longo, C. P. and Longo, G. P.,** The development of glyoxysomes in peanut cotyledons and maize scutella, *Plant Physiol.*, 45, 249, 1970.
94. **Skadsen, R. W. and Scandalios, J. G.,** Evidence for processing of maize catalase-2 and purification of its messenger RNA aided by translation of antibody-bound polysomes, *Biochemistry*, 25, 2027, 1986.
95. **Watson, M. E.,** Compilation of published signal sequences, *Nucleic Acids Res.*, 12, 5145, 1984.
96. **Yamaguchi, J., Nishimura, M., and Akazawa, T.,** Purification and characterization of heme-containing low-activity form of catalase from greening pumpkin cotyledons, *Eur. J. Biochem.*, 159, 315, 1986.
97. **Huang, D., Scandalios, J. G., Skadsen, R. W., and Tsaftaris, A. S.,** Non-coordinate genetic alterations in the expression of catalase and other glyoxysomal enzymes during maize development, *J. Exp. Bot.*, 37, 1189, 1986.
98. **Tsaftaris, A. S. and Scandalios, J. G.,** Genetic analysis of isocitrate lyase enzyme activity levels in maize lines selected for high or low oil content, *J. Hered.*, 74, 70, 1983.
99. **Choinski, J. S. and Trelease, R. N.,** Control of enzyme activities in cotton cotyledons during maturation and germination, *Plant Physiol.*, 62, 141, 1978.
100. **Gould, S. J., Keller, G. A., and Subramani, S.,** Identification of peroxisomal targeting signals located at the carboxy terminus of four peroxisomal proteins, *J. Cell Biol.*, 107, 897, 1988.

101. **Gould, S. J., Keller, G. A., and Subramani, S.,** Identification of a peroxisomal targeting signal at the carboxy terminus of firefly luciferase, *J. Cell Biol.*, 105, 2923, 1987.
102. **Osumi, T. and Fujiki, Y.,** Topogenesis of peroxisomal proteins, *BioEssays*, 12, 217, 1990.
103. **White, J. A. and Scandalios, J. G.,** Molecular biology of intracellular protein trafficking, *Physiol. Plant.*, 74, 397, 1988.
104. **Quail, P. H. and Scandalios, J. G.,** Turnover of genetically defined catalase isozymes in maize, *Proc. Natl. Acad. Sci. U.S.A.*, 68, 1402, 1971.
105. **Kopczynski, C. C. and Scandalios, J. G.,** *Cat2* gene expression, developmental control of translatable *Cat2* mRNA levels in maize scutellum, *Mol. Gen. Genet.*, 203, 185, 1986.
106. **Tsaftaris, A. S. and Scandalios, J. G.,** Spatial pattern of catalase *(Cat2)* gene activation in scutella during post-germination development in maize, *Proc. Natl. Acad. Sci. U.S.A.*, 83, 5549, 1986.
107. **Skadsen, R. W. and Scandalios, J. G.,** Pretranslational control of the levels of glyoxysomal protein gene expression by the embryonic axis of maize, *Dev. Genet.*, 10, 1, 1989.
108. **Matters, G. L. and Scandalios, J. G.,** Effect of elevated temperature on catalase and superoxide dismutase during maize development, *Differentiation*, 30, 190, 1986.
109. **Williamson, J. D. and Scandalios, J. G.,** Differential response of maize catalases and superoxide dismutases to the photoactivated fungal toxin cercosporin, *Plant J.*, 2, 351, 1992.
110. **Williamson, J. D. and Scandalios, J. G.,** Differential developmental response of the maize catalases to abscisic acid: *Vp1* transcriptional activator is not required for ABA-regulated *Cat1* expression, *Proc. Natl. Acad. Sci. U.S.A.*, 89, 8842, 1992.
111. **Guan, L. and Scandalios, J. G.,** Characterization of the catalase antioxidant defense gene *Cat1* of maize and its developmentally regulated expression in transgenic tobacco, *Plant J.*, 3, 527, 1993.
112. **Abler, M. L. and Scandalios, J. G.,** Isolation and characterization of a genomic sequence encoding the maize *Cat3* catalase gene, *Plant Molec. Biol.*, in press.
113. **Guiltinam, M. J., Marcotte, W. R., and Quatrano, R. S.,** A plant leucine zipper protein that recognizes an abscisic acid response element, *Science*, 250, 267, 1990.
114. **Redinbaugh, M. G., Sabre, M., and Scandalios, J. G.,** Expression of the maize *Cat3* catalase gene is under the influence of a circadian rhythm, *Proc. Natl. Acad. Sci. U.S.A.*, 87, 6853, 1990.
115. **Giles, A. B., Grierson, D., and Smith, H.,** *In vitro* translation of mRNA from developing bean leaves, Evidence for the existence of stored mRNA and its light-induced mobilization onto polyribosomes, *Planta*, 136, 31, 1977.
116. **Slovin, J. P. and Tobin, E. M.,** Synthesis and turnover of the light-harvesting chlorophyll a/b-protein in *Lemma gibba* grown with intermittent red light: possible translational control, *Planta*, 154, 465, 1982.
117. **Zelitch, I.,** *Photosynthesis, Photorespiration, and Plant Productivity*, Academic Press, New York, 1971.
118. **Huang, A. H. C., Trelease, R. N., and Moore, T. S.,** *Plant Peroxisomes*, Academic Press, New York, 1983.
119. **Kendall, A. C., Keys, A. J., Turner, J. C., Lea, P. J., and Miflin, B. J.,** The isolation and characterization of a catalase-deficient mutant of barley, *Planta*, 159, 505, 1983.
120. **Abler, M. L. and Scandalios, J. G.,** The CAT-2 null phenotype in maize is likely due to a DNA insertion into the *Cat2* gene, *Theor. Appl. Genet.*, 81, 635, 1991.
121. **Acevedo, A., Williamson, J. D., and Scandalios, J. G.,** Photoregulation of the *Cat2* and *Cat3* catalase genes in pigmented and pigment-deficient maize: the circadian regulation of *Cat3* is superimposed on its quasi-constitutive expression in maize leaves, *Genetics*, 127, 601, 1991.

122. **Giuliano, G., Hoffman, N. E., Ko, K., Scolnik, P. A., and Cashmore, A. R.,** A light-entrained circadian clock controls transcription of several plant genes, *EMBO J.*, 7, 3635, 1988.
123. **Nagy, F., Kay, S. A., and Chua, N.-H.,** A circadian clock regulates transcription of the wheat *Cab-1* gene, *Genes Dev.*, 2, 376, 1988.
124. **Taylor, W. C.,** Transcriptional regulation by a circadian rhythm, *Plant Cell,* 1, 259, 1989.
125. **Hall, J. C. and Rosbash, M.,** Mutations and molecules influencing biological rhythms, *Annu. Rev. Neurosci.*, 11, 373, 1988.
126. **McClung, C. R., Fox, B. A., and Dunlap, J. C.,** The *Neurospora* clock gene *frequency* shares a sequence element with the *Drosophila* clock gene *period, Nature (London),* 339, 558, 1989.
127. **Herrera-Estrella, L. and Simpson, J.,** Influence of environmental factors on photosynthetic genes, *Adv. Genet.*, 28, 133, 1990.
128. **Abler, M. L. and Scandalios, J. G.,** unpublished data, 1993.
129. **Bethards, L. A., Skadsen, R. W., and Scandalios, J. G.,** Isolation and characterization of a cDNA clone for the *Cat2* gene in maize and its homology with other catalases, *Proc. Natl. Acad. Sci. U.S.A.*, 84, 6830, 1987.
130. **Redinbaugh, M. G., Wadsworth, G. J., and Scandalios, J. G.,** Characterization of catalase transcripts and their differential expression in maize, *Biochim. Biophys. Acta,* 951, 104, 1988.
131. **Guan, L., Ruzsa, S., Skadsen, R. W., and Scandalios, J. G.,** Comparison of the *Cat2* complementary DNA sequences of a normal catalase activity line (W64A) and a high catalase activity line (R6-67) of maize, *Plant Physiol.*, 96, 1379, 1991.
132. **Ikemura, T.,** Codon usage and tRNA content in unicellular and multicellular organisms, *Mol. Biol. Evol.*, 2, 13, 1985.
133. **Newgard, B., Nakano, K., Hwang, P., and Fletterick, R.,** Sequence analysis of the cDNA encoding human liver glycogen phosphorylase reveals tissue-specific codon usage, *Proc. Natl. Acad. Sci. U.S.A.*, 83, 8132, 1986.
134. **Roupakias, D. G., McMillin, D. E., and Scandalios, J. G.,** Chromosomal location of the catalase structural genes in *Zea mays,* using B-A translocations, *Theor. Appl. Genet.*, 58, 211, 1980.
135. **Murthy, M. R. N., Reid, T. J., III, Sicignano, A., Tanaka, N., and Rossman, M. G.,** Structure of beef liver catalase, *J. Mol. Biol.*, 152, 465, 1981.
136. **Sakajo, S., Nakamura, K., and Ashai, T.,** Molecular cloning and nucleotide sequence of full-length cDNA for sweet potato catalase mRNA, *Eur. J. Biochem.*, 165, 437, 1987.
137. **Ni, W., Turley, R., and Trelease, R.,** Characterization of a cDNA encoding cottonseed catalase, *Biochim. Biophys. Acta,* 1049, 219, 1990.
138. **Isin, S. H. and Allen, R. D.,** Isolation and characterization of a pea catalase cDNA, *Plant Mol. Biol.*, 17, 1263, 1991.
139. **Gonzales, E.,** The C-terminal domain of plant catalases, *Eur. J. Biochem.*, 199, 211, 1991.
140. **Scandalios, J. G.,** Ed., *Molecular Biology of Free Radical Scavenging Systems,* Cold Spring Harbor Laboratory Press, Cold Spring Harbor, NY, 1992.

Chapter 12

SUPEROXIDE DISMUTASES

Wim Van Camp, Marc Van Montagu, and Dirk Inzé

TABLE OF CONTENTS

0-8493-5443-9/94/$0.00+$.50

I. INTRODUCTION

Superoxide dismutases (SODs, EC 1.15.1.1) are metal-containing enzymes that catalyze the dismutation reaction of superoxide into oxygen and hydrogen peroxide:

$$2O_2^{\cdot -} + 2H^+ \rightarrow O_2 + H_2O_2 \quad (1)$$

Based on the metal cofactor, three types of SODs can be distinguished: a copper and zinc (Cu,Zn), a manganese (Mn), and an iron (Fe) form. Mn SOD and Fe SOD are structurally related, as illustrated by the observation that in some bacteria the metal cofactor can be exchanged without loss of activity.[1] Cu,Zn SOD is structurally unrelated to the other two SOD types and has been found in all higher eukaryotes studied to date except *Nuphar luteum.*[2] In prokaryotes, Cu,Zn SOD is generally absent except in a few cases.[3] Certain lower eukaryotes, including protozoa and most algae, possess Mn SOD and/or Fe SOD, but no Cu,Zn SOD, suggesting that the origin of Cu,Zn SOD is situated early in eukaryotic evolution, albeit not concurrent with the branching of eukaryotes and prokaryotes. The structure, properties, and identification of SODs have been extensively reviewed by Bannister et al.[4]

Plants usually contain a cytosolic Cu,Zn SOD, Mn SOD in the mitochondria, and Cu,Zn SOD and/or Fe SOD in the chloroplasts. Cu,Zn SOD appears to be present in the chloroplasts of most plants examined to date, whereas originally, Fe SOD had only been detected in three phylogenetically unrelated plant families.[2] It has been proposed that the peculiar distribution of Fe SOD in plants is the consequence of independent gene transfers from prokaryotes to plants or of metal replacement in a Mn SOD apoprotein. More detailed studies subsequently revealed the presence of Fe SOD in several other plant families,[5-11] and sequence analyses of plant and prokaryotic Fe SODs corroborate the endosymbiont theory of the Fe SOD origin in plants.[10]

In *Nicotiana plumbaginifolia,* plastidic Cu,Zn SOD and Fe SOD are expressed differentially, and they seem to be complementary in action.[138] It can be inferred that, depending on the plant species, tissue, stage of development, or environmental conditions, Cu,Zn SOD may be preferentially expressed to Fe SOD or vice versa. The former seems to occur most in plants, which may explain why Fe SOD has been greatly overlooked in the past. Therefore, former studies on SOD in chloroplasts should be reevaluated at this point.

In addition, SODs have been found in peroxisomes,[12,13] glyoxysomes,[14] and in the extracellular space,[15] but the corresponding genes have not been isolated. It is noteworthy that all three plant Fe SOD sequences identified to date contain a tripeptide (SRL for *N. plumbaginifolia* and soybean, and ARL for *Arabidopsis thaliana*) close to the carboxyl terminus, which has been shown to direct proteins to the peroxisomes.[16] This tripeptide is not conserved in the prokaryotic Fe SODs, suggesting that it is of no importance for Fe SOD activity. Yet, it is not clear whether this peroxisomal targeting sequence is functional in the Fe SODs, because the soybean Fe SOD was shown to contain a potential chloroplast transit sequence as well,[11] whereas in the two other cases, Fe SOD was mainly associated with chloroplasts.[10]

From these results, it becomes clear that many questions remain to be solved concerning the intracellular distribution of SODs. Until this matter is clarified, it will be difficult to correlate alterations in specific SOD activities with metabolic changes induced by oxidative stress in the different subcellular compartments.

Superoxide radicals are mainly produced by the univalent reduction of O_2 with electrons from the electron transport chains of chloroplasts (photosystem I), mitochondria (respiratory chain), and cytosol (cytochrome P_{450} in the endoplasmic reticulum)[17] and consequently, most of the SOD activity is localized in these compartments. In addition, superoxide can be formed by univalent oxidases, by some peroxidases, and by autooxidation of reduced compounds with a low reduction potential.[17] Production of superoxide has been reported in peroxisomes[18] and glyoxysomes,[19] but the molecular sources have not been identified yet.

The subcellular localization of radical production often coincides with the major site of metabolic activity because here electron leakage from electron transducing reactions will be the highest. Consistent with this view, chloroplast SODs are strongly induced by light in dark-adapted leaves or etiolated seedlings,[20,21] whereas Mn SOD activity is coinduced with cytochrome oxidase activity under conditions of enhanced mitochondrial respiration.[22] Despite the fact that SODs can easily be differentiated on the basis of mRNA level as well as activity level with an *in situ* staining technique on gel,[23] to date only a few studies have investigated the expression of the different SODs individually.[20,21,24] From these studies, it became clear that SODs are differentially regulated and respond in a specific way to environmental stresses. For example, in *N. plumbaginifolia,* cytosolic Cu,Zn SOD mRNA levels are affected most by heat shock, chilling, or paraquat treatment in the dark, whereas Fe SOD mRNA levels are induced most by chilling or paraquat treatment in the light.[21] In tomato, both chloroplastic and cytosolic Cu,Zn SOD respond at the level of transcription to paraquat in the light, but only the cytosolic isoform shows an induction after drought treatment.[20] In maize, both chloroplastic and cytosolic Cu,Zn SOD mRNA levels increase strongly after paraquat treatment in the dark, but only cytosolic Cu,Zn SOD responds to increased oxygen concentrations.[25,26] These variations in stress responses presumably reflect a diversity of mechanisms by which these environmental stresses may operate. Hence, detailed studies of SOD expression may provide insights into the cellular targets of a certain stress.

Little is known about the regulation of SODs. In bacteria, at least three regulons are involved in oxidative stress.[27] One of them is inducible by H_2O_2

via a transcriptional activator, named oxyR, which is only active in an oxidized form.[28] This suggests that some reactive oxidizing compound(s) generated during oxidative stress may induce this regulon. Similarly, reactive oxygen species have been implicated in the activation of the transcription factor NF-κB in mammalian systems.[29] At first sight, the use of radicals as signaling molecules seems to be the most economical strategy, as the defense response would only be induced when radicals are produced. However, eukaryotic systems are confronted with an additional problem with regard to the specificity of induction by reactive oxygen species. SODs are all nuclear encoded, but they still must react specifically to stresses in different subcellular compartments. Therefore, signal molecules, reporting oxidative stress to a transcriptional activator, must be specific for the subcellular sites of stress. This almost excludes the possibility that oxygen species themselves are the inducers, at least in the case of organellar SODs. At the moment, the nature of these signals is open to speculation, but it can be anticipated that the signal or signal pathway involved in reporting oxidative stress from the organelles to the nucleus must be compartment specific and must be able to pass the organellar membranes. The elucidation of these signal transduction pathways is not only one of the most exciting challenges for SOD biology, but in the future may turn out to be essential for the engineering of stress tolerance in plants.

II. SUPEROXIDE DISMUTASES AND PHOTOOXIDATIVE STRESS

A. HIGH LIGHT INTENSITIES

The primary symptom of light stress is photoinhibition of photosystem II.[30] It is well established that photosystem II inactivation results from the degradation of the D1 polypeptide, one of the two polypeptides that harbors most of the functionally important sites involved in photosystem II electron transduction reactions. However, the series of events leading to photosystem II inactivation and degradation are not well understood, and this has been the subject of much research activity and controversy in the past decade. Recent developments in this field are discussed in Chapter 2 of this volume. Current evidence indicates that superoxide radicals, presumably produced by photosystem I, may play a role in photosystem II photoinhibition. Barényi and Krause[31] showed that the addition of SOD and catalase to broken chloroplasts can protect photosystem II from inhibition. Similar results were obtained by Richter et al.,[32] who also demonstrated that even at very low O_2 concentrations, $O_2^{\cdot-}$ and H_2O_2 scavengers can provide protection against photoinhibition. This opposes former views that disfavor a role for toxic oxygen species in photosystem II inactivation based on the observation that photoinhibition can occur in anaerobic conditions. In addition, Sopory et al.[33] inferred from their studies with specific scavengers for toxic oxygen species that hydroxyl and superoxide radicals are involved in photoinhibition rather than singlet oxygen.

The underlying mechanisms of photoinhibition have gained interest in recent years as it became clear that photoinhibition may protect plants from irreversible photooxidative damage. Impairment of photochemistry by photoinhibition reduces the light-driven formation of highly toxic radicals and oxygen species and, hence, may serve as a drastic but very efficient strategy to counter photooxidative processes in an early stage. If true, protective functions should be attributed to molecules that promote photoinhibition, possibly including superoxide and hydrogen peroxide. This raises questions as to the benefit of enhancing the scavenging machinery of the plant (see Section III). At this time, the influence of increased SOD activities on photoinhibitory processes, generated solely by high-light treatment, is difficult to assess. Most studies investigating the role of SOD in stress tolerance apply light stress in combination with a second factor, mostly paraquat or chilling, with the advantage that photooxidative processes can be induced at lower light intensities. Moderate light intensities combined with paraquat have been shown to affect well-accepted parameters of photoinhibitory damage — CO_2-dependent oxygen evolution and photosystem II fluorescence — and, therefore, paraquat treatment is commonly used to assess the tolerance of plants to photooxidative processes. In one study with high light, Jansen et al.[34] reported that a paraquat-resistant biotype of *Conyza bonariensis* with elevated levels of SOD was more tolerant to photoinhibition; however, in a similar investigation, Preston et al.[35] did not observe any difference between the paraquat-resistant and sensitive biotype.

B. PARAQUAT

In plants, the herbicide paraquat (see Chapters 9 and 14 of this volume) is often used as a model system to study the effects of superoxide-mediated damage in the chloroplasts. Nevertheless, studies with paraquat must take into account that this herbicide not only affects photosynthetic electron transport, but also interferes with electron transducing reactions in the endoplasmic reticulum and the mitochondria.[17] Induction analysis of SOD transcript levels illustrates that paraquat treatment in the light not only affects chloroplastic SOD, but also cytosolic and mitochondrial SOD transcript levels, albeit to a lesser extent.[20,21] The cytosolic Cu,Zn SOD of *N. plumbaginifolia* was also induced at the level of transcription by paraquat treatment in the dark,[21] which indicates that some important sources of superoxide radicals, operating in the light as well as in the dark, are localized in the cytosol. Both chloroplastic and mitochondrial SODs were unaffected by paraquat treatment in the dark.[21] In the case of chloroplastic SOD, this may reflect the absence of electron transduction through the photosystems in the dark. Remarkably, and in contradiction to this hypothesis, both chloroplastic and cytosolic Cu,Zn SODs were induced by paraquat in the dark in maize,[25] suggesting that paraquat effects are species-specific and, in the case of chloroplasts, not strictly correlated with photosynthetic electron transport. A similar observation was made for the Mn SOD of *N. plumbaginifolia*. Basal levels of Mn SOD mRNA in *N. plumbaginifolia* are higher in the dark than in the light, possibly as a result of increased respiratory activity in the dark, but induction of Mn SOD expression by paraquat was only observed in the light.[21]

From these data, it can be inferred that the SOD response to paraquat in plants is not confined to chloroplastic SODs, but also includes mitochondrial and cytosolic SODs. The relative importance of the different SODs in the defense mechanism against paraquat will be governed by the physiological status of the cell, the plant species, and the conditions of paraquat administration, since these factors will influence the subcellular sites of paraquat toxicity. Hence, it can be anticipated that selection procedures to obtain mutant plant lines with high levels of SOD, using paraquat as a selecting agent, will yield different results, depending on the subcellular compartment on which the paraquat stress is imposed.

Several studies correlated enhanced activities of SOD, eventually in combination with increases in other scavenging enzymes, with tolerance to paraquat in photosynthetic organisms. In the green alga *Chlorella sorokiniana,* pretreatment with low doses of paraquat induces SOD activity and confers resistance to subsequent treatment with higher concentrations.[36] Paraquat-resistant lines of perennial rye grass,[37] *Conyza bonariensis,*[38] tobacco,[39] and maize[40] displayed higher SOD activities than sensitive lines. In tobacco and maize, these increases in SOD activity were accompanied by similar increases in glutathione reductase activity, whereas in rye grass induction was more pronounced for SOD than for glutathione reductase. However, in the paraquat-tolerant *C. bonariensis,* the activities of ascorbate peroxidase and glutathione reductase were much more elevated than SOD activity. Pea plants grown under different light conditions displayed different tolerances to paraquat which correlated with the increased activities of ascorbate peroxidase, dehydroascorbate reductase, and glutathione reductase, but not of SOD.[41] It can be concluded that paraquat tolerance in a plant often correlates with its scavenging capacities for oxyradicals, but that the increases in individual scavenging enzymes are very variable and species dependent. This probably reflects the fact that, depending on the species, different enzymes can be rate limiting in the scavenging machinery.

C. CHILLING

It is well established that the sensitivity of plants to high light intensities can be significantly enhanced when combined with chilling stress. At chilling temperatures, photoinhibition is manifested even at relatively moderate light intensities. Photoinhibition is usually associated with chilling sensitivity because it does not occur in plants which are essentially chilling tolerant.[42,43] Nevertheless, in chilling-sensitive plants, photoinhibition may constitute a rescue mechanism against photooxidative damage, as it prevents the formation of reactive species at the expense of excitation energy. This is supported by the observation that whereas during chilling, photosynthetic electron transport is higher in a chilling-sensitive tomato *(Lycopersicon esculentum)* than in the more tolerant one *(L. hirsutum)*, photosynthesis recovered at a faster rate and to a greater extent in *L. hirsutum* when plants were returned to a higher temperature.[44] Similarly, photoinhibition in spinach was fully reversible within 1 to 3 h.[45] Remarkably, in another study on tomato, photoinhibition was only reversible when the antioxidant levels were increased artificially by treating the plants with the triazole, uniconazole.[46] In the absence of uniconazole, substantial membrane damage was observed and recovery from photoinhibition was incomplete.

The involvement of oxygen in chilling-mediated photoinhibition was soon recognized, this in contrast to photoinhibition at higher temperatures.[30] Recently, a transient increase in H_2O_2 concentration, probably formed in a Mehler-type reaction, was observed during chilling in leaves of winter wheat.[47] More direct evidence for the production of superoxide radicals in response to chilling was presented by Walker et al.,[44] who demonstrated that the rate of superoxide production during chilling was lower in *L. hirsutum* than in the more chilling-sensitive *L. esculentum*. Several investigations illustrate the role of antioxidants and oxyradical-scavenging enzymes in the protection against chilling. In the chilling-sensitive cucumber, inactivation of photosynthesis by chilling was accompanied by a drop in antioxidant levels.[48] Cold-hardened spinach does not exhibit photoinhibition, in contrast to nonacclimated spinach, and this chilling tolerance, at least in part, was attributed to increased activities in SOD, ascorbate peroxidase, and monodehydroascorbate reductase, and a higher ascorbate and xanthophyll content.[45,49] In the green alga *Chlorella ellipsoidea*, a chilling-resistant strain contained more SOD than a chilling-sensitive one; chilling tolerance was inducible in the sensitive strain by pretreatment with paraquat, which resulted in a concomitant increase of SOD to levels comparable to the chilling-resistant strain.[50] In maize, SOD and ascorbate peroxidase are constitutively higher in the chilling-resistant *Zea diploperennis* than in the sensitive *Z. mays*, whereas glutathione reductase was induced in response to chilling only in the chilling-resistant maize.[51] In addition, exogenously added scavengers for superoxide and singlet oxygen decrease membrane damage due to chilling.[48,52] These results indicate that the inhibition of photosynthesis and the occurrence of photooxidative damage due to chilling stress in the light may largely depend on the oxyradical-scavenging capacities of the plant.

D. SULFUR DIOXIDE AND OZONE

SO_2 toxicity is mainly attributed to highly reactive intermediates such as the sulfur trioxide radical (HSO_3^-), the superoxide, and the hydroxyl radical, which are generated during the radical-initiated oxidation of SO_2.[53] Several studies on SO_2 toxicity revealed striking similarities between SO_2-mediated stress and photooxidative stress, as exerted by chilling in combination with high light (see above).

Illuminated chloroplasts can initiate the oxidation of SO_2 and, hence, may be a primary site of radical production during SO_2 treatment. This radical production can be inhibited by 3(3′,4′-dichlorophenyl)-1,1-dimethylurea (DCMU).[54] DCMU prevents the reoxidation of Q_A in the photosystem II reaction center and thereby lowers the amount of superoxide formed from photosynthetic electron transport.

Exposure to SO_2 results in lipid peroxidation and pigment destruction, but this occurs only in the light and can be inhibited by DCMU and by scavengers for $O_2^{\cdot -}$.[55,56] Hence, damage due to SO_2 treatment resembles photooxidative damage because both can be initiated by superoxide production in photosystem I. The sulfur trioxide radical is more important for the propagation of radical production than the superoxide radical, which could explain why SO_2 damage takes place even at relatively low light intensities.[57]

Inhibition of photosystem II has been observed in SO_2-treated plants before lipid peroxidation or pigment degradation. However, in contrast to photoinhibition, impairment of photosystem II activity by SO_2 is not light dependent[58,59] and, therefore, is believed to be due to a direct toxic effect of SO_2.

The induction of SOD in response to SO_2 stress is variable, depending on the species and SO_2 concentration. In response to SO_2 treatment, SOD was unaffected in maize,[26] and in spinach, SOD activity was reduced by 60%,[55] whereas in *C. sorokiniana*[60] and *Lemna* sp.,[61] SOD activity was enhanced twofold and three- to fourfold, respectively. Interestingly, induction of SOD by sulfite treatment in *C. sorokiniana* conferred resistance to paraquat.[60] Cross-resistances to SO_2 have been observed in paraquat-tolerant *Conyza bonariensis*,[38] maize,[40] perennial rye grass,[39] and in tobacco plants regenerated from paraquat-tolerant callus.[62] In poplar, young leaves containing more SOD activity are more resistant to SO_2 than old ones, and tolerance to SO_2 was inducible by pretreatment with low levels of SO_2.[63]

In summary, some striking similarities have been observed between SO_2 stress and photooxidative stresses regarding the subcellular targets of damage and the involvement of oxyradicals. This may explain why cross-tolerances to paraquat and SO_2 have been observed in several studies. However, the physiological relevance of the observed tolerances to SO_2 stress is not clear, as the SO_2 concentrations applied in most studies exceed by severalfold the maximal concentrations found in the atmosphere. In future studies, emphasis should be given to the effect of ambient SO_2 concentrations on plants, eventually in combination with other stresses, as SO_2 alone may not injure plants when applied in low doses.

Ozone can cause visible damage to plants at concentrations which correspond to peak episodes in Europe and North America. In particular, species from the Fabaceae and Solanaceae families are very sensitive to ozone.[64,65]

The mechanisms of ozone damage are not well understood. Ozone is a good oxidant, and it is generally believed that ozone injury is mediated by highly reactive free radicals formed by O_3 oxidation rather than by O_3 itself.[66,67]

As for SO_2, chloroplasts seem to be a primary target of ozone stress. Ozone effects a decrease in photosynthesis,[68] destruction of chlorophylls and carotenoids, alterations in ribulose-1,5-bisphosphate carboxylase-oxygenase activity,[69] the formation of 8-hydroxyguanine in chloroplast DNA,[70] and the vesiculation and swelling of thylakoids.[71]

The effect of ozone on the activity of oxyradical-scavenging enzymes is dose dependent. Fumigation with concentrations that correspond to maximal ambient levels (to 200 ppb) generally results in an induction of SOD activity,[15,72-75] ascorbate peroxidase,[76,77] and glutathione reductase.[76-78] Higher doses of ozone repressed the defense response against oxidative stress in two studies.[72,76]

Little is known about the expression of the individual SOD enzymes in response to ozone stress. In bean, only the cytosolic Cu,Zn SOD is induced after fumigation with 300 ppb ozone for 6 h, and this augmentation is strongly and significantly associated with increasing foliar injury.[79] Chloroplastic and mitochondrial SODs are unaffected by ozone treatment in bean. We observed a similar expression pattern in tobacco. Induction of the cytosolic Cu,Zn SOD in *N. tabacum* only occurred after 4 d of fumigation with ambient concentrations of ozone (0 ppb at night to maximal 120 ppb during the day), which is concurrent with the onset of visible leaf injury, whereas no changes in cytosolic Cu,Zn SOD levels were detected in *N. plumbaginifolia,* a species that did not display ozone damage under the same fumigation conditions.[138] In both *Nicotiana* species, Fe SOD levels decreased slowly but continuously during the ozonation experiment, whereas the Mn SOD was essentially unaffected.

Components of the defense system against pathogens such as chitinases, glucanases,[80] phenylalanine ammonia-lyase, chalcone synthase, and phytoalexins[81] are rapidly induced by ozone. The stimulation of the pathogenesis-related (PR) response by ozone is unexpected, as there is no obvious role for glucanases, chitinases, or phytoalexins in the defense against ozone. Therefore, the assumption can be made that ozone stress in its initial events resembles the hypersensitive reaction (HR) and consequently induces the PR response. Early events in the hypersensitive response[82] include the production of oxyradicals at the cellular membrane, increases in lipid peroxidation, and changes in membrane permeability, which are similar to those described for ozone stress. Apostol et al.[83] reported that H_2O_2 can function as a signal molecule for the activation of the HR response, which supports the view that any nonspecific oxidative event generating H_2O_2 at the cellular membrane can induce the PR response.

Several investigations aimed to identify the mechanisms which confer protection against ozone stress. Oxyradical scavengers attenuated pigment destruction and lipid degradation during O_3 fumigation of microsomal membranes[84] or leaf discs,[72] suggesting that oxyradicals do indeed mediate ozone damage. Pretreatment of wheat seedlings with the triazole S-3307 prevented membrane damage by ozone, and this protection was associated with the induction of lipid-soluble antioxidants.[85] Similarly, Lee and Bennett[86] attributed ethylene diurea (EDU)-mediated protection against ozone injury to increases in SOD activity, but these results were subsequently questioned.[79,87] Mehlhorn[88] reported increased resistance to ozone, paraquat, and H_2O_2 after preexposure of mung bean and pea to ethylene, and this tolerance was correlated to enhanced ascorbate peroxidase activity. However, considering the pleiotropic effects of ethylene, it cannot be ruled out that other mechanisms contributed to the observed tolerance.

The mechanisms of ozone tolerance have also been investigated by comparison of the biochemical properties of plant varieties which display different sensitivities to this air pollutant. In bean, ozone tolerance in the cultivar Goldcrop was not associated with stomatal behavior and higher levels of lipid-soluble antioxidants or SOD in comparison to the O_3-sensitive cultivar Seafarer.[89] Langebartels et al.[90] ascribed an important function to polyamines in the defense response to O_3 stress. Free and conjugated polyamine levels were rapidly induced in an O_3-tolerant tobacco cultivar, Bel B, in contrast to an O_3-sensitive one, Bel W3. Ozone injury in the sensitive variety was largely prevented by root application of polyamines.[91] The role of polyamines in protection against ozone is not elucidated, but possible functions include the antagonistic effect of polyamines on ethylene biosynthesis and the radical-scavenging properties of conjugated polyamines. Comparison of two O_3-tolerant and four O_3-sensitive spinach cultivars revealed no significant differences in scavenging enzymes.[76] However, both tolerant cultivars exhibited higher ascorbate and glutathione levels. Unfortunately, the influence of ozone treatment on the scavenging machinery in the different varieties was not investigated, and, therefore, differences in oxyradical-scavenging enzymes in response to ozone cannot be excluded. In tobacco, only one out of four ozone-tolerant cultivars contained higher SOD and glutathione reductase levels, and only this one cultivar displayed cross-tolerance to paraquat.[39] In contrast, an SO_2-tolerant tobacco line regenerated from paraquat-tolerant callus contained increased levels of SOD, but showed no cross-tolerance to ozone.[62] It should be mentioned, however, that the ozone concentrations used in the latter study were approximately twice the maximal ambient levels.

III. ENGINEERING SUPEROXIDE DISMUTASE IN TRANSGENIC PLANTS

The importance of SOD for aerobic organisms is clearly illustrated by the isolation of SOD-deficient mutants of *Escherichia coli,*[92] yeast,[93-95] and

Drosophila,[96] which are all hypersensitive to conditions of increased oxygen radical production. *SodASodB* double mutants of *E. coli* are unable to grow on minimal medium in aerobic conditions and display enhanced sensitivity to paraquat and hydrogen peroxide.[92] In addition, they have greatly enhanced mutation rates.[97,98] The SOD-deficient mutant can regain wild-type characteristics when complemented with SOD genes from homologous or heterologous systems[10,92,99-101] or by supplementing the 20 amino acids to the medium, the latter indicating that the amino acid pathways are especially sensitive to superoxide stress.[92] A Cu,Zn SOD-deficient *Saccharomyces cerevisiae* strain is hypersensitive to 100% oxygen, paraquat, and H_2O_2,[93] and shows increased mutation rates.[95] A yeast mutant deficient in the mitochondrial Mn SOD is growth inhibited in 100% oxygen[94] and it is unable to use a nonfermentable, two-carbon source in aerobic conditions.[102] Both yeast mutants have been complemented with heterologous SOD genes.[100,101,103] In *Drosophila*, Cu,Zn SOD deficiency has been associated with paraquat sensitivity, infertility, and reduction in life-span.[104]

For several reasons, SODs may form a primary target for engineering stress tolerance in plants. First, the dramatic effects of SOD deficiency illustrate the importance of SOD in the defense mechanism against oxidative stress. Second, in plants, damage arising from environmental stresses often appears to be caused by oxyradicals, and, hence, plants with an improved scavenging mechanism for active oxygen species may be of great agricultural value. Third, in many cases, enhanced SOD activity correlates with increased tolerance to one or several environmental stresses (see Table 1), suggesting that increased SOD levels are indeed beneficial for stress tolerance.

The advantage of elevated SOD levels is obvious: more efficient scavenging of the superoxide radical would result in less superoxide-mediated damage. Nonetheless, some disadvantages may accompany SOD overproduction. Elevated SOD levels may impair an adequate defense response against oxidative stress, as superoxide may function as a second messenger triggering the induction of certain defense mechanisms. In *E. coli*, about 40 proteins are induced by superoxide-generating agents.[105,106] Nine of these are positively regulated by the *soxR* (superoxide response) regulon, which is believed to be activated specifically by excess $O_2^{\cdot-}$.[107,108] Members of this regulon include an $O_2^{\cdot-}$-scavenging enzyme (Mn SOD), a DNA repair enzyme (endonuclease IV), and an NADPH-forming enzyme (glucose-6-phosphate dehydrogenase). In addition, an outer membrane protein (OmpF) that possibly is involved in the uptake of xenobiotics is negatively regulated by *soxR*. Hence, in *E. coli*, overproduction of SODs may impede other stress responses involved in the protection against oxidative damage. In eukaryotic systems, too, active oxygen species have been implied in signal transduction (activation of the transcription factor NF-κB in mammalian systems, induction of the hypersensitive response in plants), but we do not know to what extent superoxide radicals are involved in this process.

TABLE 1
Changes in Oxyradical-Scavenging Enzymes Induced by Environmental Stresses or Associated with Stress Resistance in Plants

Plant	Activity of normal (%)	Increase obtained by	(Cross) tolerance	Ref.
Tobacco	3000–5000% of chloroplastic Cu,Zn SOD	Engineering Cu,Zn SOD of petunia	No tolerance to paraquat	131
Tobacco	1500% of chloroplastic Cu,Zn SOD	Engineering Cu,Zn SOD of petunia	No tolerance to O_3	133
Tomato	200–2000% of chloroplastic Cu,Zn SOD	Engineering Cu,Zn SOD of petunia	No tolerance to chilling	131
Tobacco	160–300% of chloroplastic SOD	Engineering Mn SOD of *N. plumbaginifolia*	Paraquat	132
Tobacco	3000% of mitochondrial SOD	Engineering Mn SOD of *N. plumbaginifolia*	Little effect on paraquat tolerance	132
Tobacco	100–350% of GR	Engineering *gor* gene of *E. coli*	Paraquat, no tolerance to O_3	136
Tobacco	200–300% SOD, 75–100% APX, 125% GR	Screening for paraquat-resistant callus	SO_2, paraquat, no tolerance to O_3	62
Conyza bonariensis	170% SOD, 250–300% APX, 300–350% GR (chloroplast activities)	Screening for paraquat tolerance	SO_2,[39] paraquat, photoinhibitory light[34] [a]	38
Perennial rye grass	170% SOD, 120% GR	Screening for SO_2 tolerance	Paraquat, SO_2	39
Tobacco	160% SOD, 170% GR	Screening for O_3 tolerance	Paraquat, O_3	39
Maize	150% Cu,Zn SOD, 150–200% GR	Screening for drought tolerance, and paraquat tolerance	Paraquat; SO_2, drought	40

Tobacco callus	200–300% SOD, 100% APX, 100% CAT	Screening for paraquat tolerance	Paraquat	137
Maize	200% SOD, 200% APX, 150% GR	SOD and APX constitutively in chilling-tolerant species, GR induced by chilling in tolerant species	Chilling	51
Chlorella ellipsoidea	130% SOD	Screening for paraquat tolerance or by paraquat induction	Chilling, paraquat	50
C. sorokiniana	300% SOD	Induction by paraquat	Higher levels of paraquat	36
C. sorokiniana	200% SOD	Induction by SO_2	Paraquat	60
Pea	106% SOD, 170% APX, 157% DHAR, 157% GR (chloroplastic activities)	Induction by higher light	Paraquat	41
Poplar	440% SOD	Induction by SO_2	Higher levels of SO_2	63
Spinach	181% SOD, 131% APX, 136% MDAR, 100% DHAR (chloroplastic activities)	Induction by chilling	To photoinhibitory light at chilling temperature	49
Mung bean	500% APX	Induction by ethylene	O_3, H_2O_2, paraquat	88
Pea	170% APX	Induction by ethylene	O_3, H_2O_2, paraquat	88
Snap bean	190–240% SOD[b]	Treatment with EDU	O_3	86

[a] Preston et al.[35] observed no increased tolerance to photoinhibitory light.
[b] Chanway and Runeckles[87] and Pitcher et al.[79] observed no change in SOD activity after EDU treatment.

Note: GR, glutathione reductase; APX, ascorbate peroxidase; CAT, catalase; MDAR, monodehydroascorbate reductase; DHAR, dehydroascorbate reductase.

Second, under conditions of incipient oxidative stress, superoxide may inactivate components of radical-generating systems to prevent further radical formation. For example, the aconitase of *E. coli,* by its high sensitivity to superoxide, rapidly causes impairment of the tricarboxylic acid cycle and, hence, of electron flow to oxygen;[109] similarly, inactivation of the D1 protein in plants may prevent lethal superoxide production through photosynthesis (see above).

Moreover, by analogy to other systems (e.g., indoleamine dioxygenase in animals, nitropropane dioxygenase in yeast, and galactose oxidase in fungi), superoxide radicals are possibly involved in the catalytic mechanism of certain plant enzymes. Superoxide may also play a role in physiological processes such as cell wall biosynthesis, pathogen response, and fruit ripening.[17] As a consequence, SOD overproduction may interfere with some chemical and physiological functions of superoxide radicals in plants.

Nevertheless, the major drawback of SOD overproduction presumably concerns the fact that it increases the rate of H_2O_2 production and, hence, may generate conditions that favor $^{\bullet}OH$ formation via a Fenton-type reaction (see Chapter 2 of this volume). Some data indicate that cells can adapt to elevated SOD levels by enhancing H_2O_2-scavenging capacities, but this adaptation is most likely insufficient at high levels of SOD.

Several studies investigated the effects of SOD overproduction in *E. coli,* and with the exception of one report,[110] elevated SOD activity was associated with decreased tolerance to paraquat.[111-114] Scott et al.[112] reported that catalase, peroxidase, and glutathione reductase levels were not altered by SOD overproduction, whereas H_2O_2 production was increased, suggesting that the inability to adjust H_2O_2-scavenging capacities accounts for the enhanced sensitivity.

In *Drosophila* and mammalian systems, the consequences of SOD overproduction are controversial. A 50% increase in Cu,Zn SOD activity in *Drosophila* decreased paraquat resistance and had no effect on life-span, even though it increased resistance to ionizing radiation.[115,116] In contrast, transgenic *Drosophila* producing Cu,Zn SOD at levels about 35% higher than in control flies displayed enhanced tolerance to paraquat and a slight increase in mean life-span.[117] The authors never obtained levels of SOD production greater than 1.6-fold that normally found in these flies, and in some cases SOD overproduction seemed to cause death of the pupae during the process of hatching. It was suggested that high levels of SOD could be lethal because even in the case of a gene duplication for SOD in *Drosophila,* levels of expression twofold higher than normal were never observed. Hence, although moderate increases in SOD can be beneficial, high concentrations seem to be disadvantageous and may become lethal beyond a certain level.

A comparable situation may occur in mammalian cells. Low overproduction of SOD (1.5- to fivefold increase in total activity) conferred resistance to paraquat,[118-120] H_2O_2,[121] and hyperoxia,[122] whereas higher SOD levels decreased the tolerance to superoxide stress in cell cultures.[120,123] Only one study, in which the content of superoxide dismutases in Chinese hamster ovary cells had been augmented six- to 13-fold by scrape loading, reported an increase in tolerance associated with high levels of SOD.[124] However, the increment in activity may have been overestimated in this study because it was deduced from the amount of radioiodinated SOD that was taken up by the cells, without taking into account the eventual reductions in enzymic activity of the loaded SOD. The fact that low overproduction is beneficial whereas higher increases are detrimental seems paradoxical, but it could indicate that mammalian cells can compensate for up to fivefold increases in SOD by comparable inductions in H_2O_2 scavengers, whereas higher SOD levels unbalance the $O_2^{\bar{\cdot}}/H_2O_2$ ratio within the cell and hence enhance the toxicity of superoxide generators.

Such an adaptive response, involving glutathione peroxidase but not catalase, seems to exist in most mammalian cells. In Down's syndrome (DS) patients, Cu,Zn SOD activity is increased 1.5-fold due to the triplication of chromosome 21 or the chromosomal segment 21.q22. In DS erythrocytes, glutathione peroxidase activity is significantly increased along with that of Cu,Zn SOD,[125-127] but catalase levels are unaffected.[128] However, no compensation to increased SOD levels was observed in the brain of DS fetuses,[129] suggesting that the unbalanced increase in SOD levels contributes to some of the clinical symptoms associated with DS. Additional evidence for an adaptive mechanism comes from studies with transfected cells showing that overproduction of the human Cu,Zn SOD (1.5- to threefold) leads to corresponding increases in glutathione peroxidase (GPX) activity.[119,130] Remarkably, paraquat resistance correlated higher with GPX activity than with SOD activity, indicating that enhanced GPX activity or, more likely, enhanced oxyradical-scavenging activity with a constant ratio of SOD-to-GPX activity is of major importance for paraquat tolerance.[119] This may explain why in several studies no correlation was observed between paraquat tolerance and the magnitude of SOD overproduction.[118-120]

To date, only a few studies have investigated the effects of SOD overproduction in plants. Tepperman and Dunsmuir[131] obtained transgenic tobacco with 30- to 50-fold the normal chloroplastic Cu,Zn SOD activity. The actual increase in chloroplast activity is probably lower because tobacco contains Fe SOD in addition to Cu,Zn SOD in the chloroplasts.[132] This Fe SOD almost comigrates with cytosolic Cu,Zn SOD on nondenaturing polyacrylamide gels and, hence, escapes detection easily. Nevertheless, it is clear that the introduced SOD activity is severalfold higher than the endogenous chloroplastic activity. This high increment in SOD activity, however, did not result in any improved resistance against paraquat[131] or ozone.[133] In contrast, Bowler et al.[132] observed a significant, albeit not very pronounced, improvement in the protection against methyl viologen in old leaves from transgenic tobacco that display overproduction of Mn SOD to moderate levels (approximately threefold increase in chloroplastic SOD activity) in the chloroplasts. This effect was much less evident in young leaves of transgenic plants. These leaves contain less engineered activity (1.5-fold augmentation of chloroplastic SOD activity), and it was suggested that such small increases in SOD activity do not remove O_2^- completely, whereas they enhance H_2O_2 formation, thus creating favorable conditions for $^{\bullet}OH$ formation.

In the case of mitochondrial overproduction of Mn SOD, a 30-fold augmentation conferred enhanced tolerance to paraquat treatment in the dark, whereas a threefold increase had no positive effect.[132] Mitochondrial SOD overproduction did not accord significant protection against paraquat treatment in the light, which is consistent with the notion that the formation of superoxide radicals is biased toward the chloroplast when paraquat is administered in the light.

Recently, we tested the same transgenic lines that overproduce Mn SOD in chloroplasts or mitochondria for ozone tolerance. Transgenic tobacco overproducing Mn SOD in the chloroplasts displayed about fourfold less ozone injury than a control line[138] when exposed to ambient levels of ozone (0 ppb at night to 120 ppb during the day) for 7 d. A similar, but less pronounced, tendency was observed for mitochondrial overproduction, suggesting that ozone-induced radical formation also occurs in mitochondria, albeit to a lesser extent.

These conflicting data on SOD overproduction in plants could be explained in view of the results obtained in animal systems. As outlined above, evidence is accumulating that SOD overproduction in mammalian cell cultures is only beneficial when the H_2O_2-scavenging system can cope with the increased rate of H_2O_2 production that results from enhanced SOD activity. A similar situation may occur in plants, in which case the H_2O_2-scavenging capacity in every subcellular compartment will be determining for the level of SOD overproduction. Current data do not allow an estimation of the H_2O_2-scavenging capacities in plants, but the natural increases in SOD activities observed in stress-tolerant plants could be indicative of the increments we should aim for in transgenic plants.

Most studies on stress-tolerant plant varieties provide data on changes in total SOD activities only, and thus no accurate estimation of the increments in specific SOD activities can be made at this time. On average, total SOD activity was augmented about twofold in plants tolerant to oxidative stress, with a maximum of 4.4-fold in SO_2-tolerant poplar (Table 1). If we assume that SOD activities in mature leaves of plants are equally distributed between cytosol and chloroplasts, with only a minor part being present in the mitochondria, then a fourfold increase of SOD in chloroplasts could be sufficient to ensure improved stress tolerance. This corresponds quite well to the threefold increase in chloroplast activity that was shown to confer tolerance to paraquat and ozone in tobacco (see above).

This figure is probably quite different in mitochondria. Mitochondrial SOD abundance is very low in leaf mesophyll cells when compared to other cell types such as vascular cells, meristem cells, pollen,[134] and cells from suspension cultures;[22] as a consequence, only a small part of the H_2O_2-scavenging potential of mitochondria is utilized in leaf mesophyll cells under normal conditions. This may explain why a 30-fold augmentation of mitochondrial SOD levels conferred enhanced tolerance to paraquat treatment in the dark, whereas in this case a threefold increase had no positive effect.[132]

In summary, current evidence from both animal and plant systems indicates that the beneficial effects of SOD overproduction may be limited to a narrow concentration range that may vary, depending on the subcellular location, cell type, and species. Therefore, it is of particular importance that an extensive study on plant varieties resistant to oxidative stress is performed in order to establish the subcellular sites and the optimal levels for SOD engineering.

In addition, analysis of the effects of enhanced SOD levels on H_2O_2 concentration and H_2O_2-scavenging activities should clarify the need of engineering H_2O_2 scavengers in addition to SOD. There is little doubt that the introduction of a H_2O_2-removing enzyme in parallel with SOD in plants will reduce the risk of unbalancing the O_2^-/H_2O_2 ratio. However, H_2O_2 scavengers in plants are poorly characterized, and thus it is not clear which enzyme would be suited for this purpose. Depending on the subcellular compartment, ascorbate peroxidase or glutathione peroxidase seem to be the enzymes of choice, because their natural function is to cooperate with SOD. However, the fact that peroxidases require reduced substrates for activity, and hence are limited in activity by the endogenous mechanisms for the regeneration of these substrates, argues against the use of these enzymes. Alternatively, catalases could be targeted to different cell compartments, as they do not require substrates other than H_2O_2 for activity. On the other hand, catalases are presumably not functionally associated with SODs in plants, implying that their chemical properties are not adapted for this purpose. A possible disadvantage of catalase is its light sensitivity, and the fact that catalase activity generates oxygen, thus creating more favorable conditions for superoxide formation.[135] Hence, the best approach for obtaining tolerance to oxidative stress is to engineer regulatory genes that control the defense response against oxidative damage. In *Conyza bonariensis,* evidence for the existence of a regulatory gene controlling oxyradical-scavenging activities in chloroplasts has been presented.[38] The search for similar mutants in genetically well-characterized plants such as *Arabidopsis thaliana* and the identification of protein-binding DNA sequences with newly developed techniques in molecular biology (e.g., ligation-mediated PCR [LMPCR], *in vivo* footprinting) will ultimately lead to the isolation of such regulatory genes from plants.

ACKNOWLEDGMENTS

The authors wish to thank Dr. Luit Slooten for helpful discussions and Martine De Cock for help in preparing the manuscript. This research was supported by grants from the Services of the Prime Minister (IUAP 120CU192), ASLK-Kankerfonds, Vlaams Actieprogramma Biotechnologie, and International Atomic Energy Agency (#5285). W.V.C. is a Research Assistant of the National Fund for Scientific Research (Belgium), and D.I. is a Research Director of the "Institut National de la Recherche Agronomique" (France).

REFERENCES

1. **Amano, A., Shizukuishi, S., Tsunemitsu, A., Maekawa, K., and Tsunasawa, S.,** The primary structure of superoxide dismutase purified from anaerobically maintained *Bacteroides gingivalis, FEBS Lett.*, 272, 217, 1990.
2. **Bridges, S. M. and Salin, M. L.,** Distribution of iron-containing superoxide dismutase in vascular plants, *Plant Physiol.*, 68, 275, 1981.
3. **Steinman, H. M. and Ely, B.,** Copper-zinc superoxide dismutase of *Caulobacter crescentus:* cloning, sequencing, and mapping of the gene and periplasmic location of the enzyme, *J. Bacteriol.*, 172, 2901, 1990.
4. **Bannister, J. V., Bannister, W. H., and Rotilio, G.,** Aspects of the structure, function and applications of superoxide dismutase, *CRC Crit. Rev. Biochem.*, 22, 111, 1987.
5. **Kwiatowski, J. and Kaniuga, Z.,** Evidence for iron-containing superoxide dismutase in leaves of *Lycopersicon esculentum* and *Phaseolus vulgaris, Acta Physiol. Plant.*, 6, 197, 1984.
6. **Barón Ayala, M. and Sandmann, G.,** Activities of Cu-containing proteins in Cu-depleted pea leaves, *Physiol. Plant.*, 72, 801, 1988.
7. **Almansa, M. S., del Río, L. A., Alcaraz, C. F., and Sevilla, F.,** Isoenzyme pattern of superoxide dismutase in different varieties of citrus plants, *Physiol. Plant.*, 76, 563, 1989.
8. **Becana, M., Paris, F. J., Sandalio, L. M., and del Río, L. A.,** Isoenzymes of superoxide dismutase in nodules of *Phaseolus vulgaris* L., *Pisum sativum* L., and *Vigna unguiculata* (L.) Walp, *Plant Physiol.*, 90, 1286, 1989.
9. **Droillard, M.-J., Bureau, D., Paulin, A., and Daussant, J.,** Identification of different classes of superoxide dismutase in carnation petals, *Electrophoresis*, 10, 46, 1989.
10. **Van Camp, W., Bowler, C., Villarroel, R., Tsang, E. W. T., Van Montagu, M., and Inzé, D.,** Characterization of iron superoxide dismutase cDNAs from plants obtained by genetic complementation in *Escherichia coli, Proc. Natl. Acad. Sci. U.S.A.*, 87, 9903, 1990.
11. **Crowell, D. N. and Amasino, R. M.,** Nucleotide sequence of an iron superoxide dismutase complementary DNA from soybean, *Plant Physiol.*, 96, 1393, 1991.
12. **Sandalio, L. M. and Del Río, L. A.,** Intraorganellar distribution of superoxide dismutase in plant peroxisomes (glyoxysomes and leaf peroxisomes), *Plant Physiol.*, 88, 1215, 1988.
13. **Droillard, M.-J. and Paulin, A.,** Isozymes of superoxide dismutase in mitochondria and peroxisomes isolated from petals of carnation *(Dianthus caryophyllus)* during senescence, *Plant Physiol.*, 94, 1187, 1990.
14. **Bueno, P. and del Río, L. A.,** Purification and properties of glyoxysomal cuprozinc superoxide dismutase from watermelon cotyledons (*Citrullus vulgaris* Schrad), *Plant Physiol.*, 98, 331, 1992.
15. **Castillo, F. J., Miller, P. R., and Greppin, H.,** Extracellular biochemical markers of photochemical oxidant air pollution damage to Norway spruce, *Experientia*, 43, 111, 1987.

16. **Volokita, M.,** The carboxy-terminal end of glycolate oxidase directs a foreign protein into tobacco leaf peroxisomes, *Plant J.*, 1, 361, 1991.
17. **Halliwell, B. and Gutteridge, J. M. C.,** *Free Radicals in Biology and Medicine*, Clarendon Press, Oxford, 1989.
18. **del Río, L. A., Fernández, V. M., Rupérez, F. L., Sandalio, L. M., and Palma, J. M.,** NADH induces the generation of superoxide radicals in leaf peroxisomes, *Plant Physiol.*, 89, 728, 1989.
19. **Sandalio, L. M., Fernández, V. M., Rupérez, F. L., and Del Río, L. A.,** Superoxide free radicals are produced in glyoxysomes, *Plant Physiol.*, 87, 1, 1988.
20. **Perl-Treves, R. and Galun, E.,** The tomato Cu,Zn superoxide dismutase genes are developmentally regulated and respond to light and stress, *Plant Mol. Biol.*, 17, 745, 1991.
21. **Tsang, E. W. T., Bowler, C., Hérouart, D., Van Camp, W., Villarroel, R., Genetello, C., Van Montagu, M., and Inzé, D.,** Differential regulation of superoxide dismutase in plants exposed to environmental stress, *Plant Cell.*, 3, 783, 1991.
22. **Bowler, C., Alliotte, T., De Loose, M., Van Montagu, M., and Inzé, D.,** The induction of manganese superoxide dismutase in response to stress in *Nicotiana plumbaginifolia*, *EMBO J.*, 8, 31, 1989.
23. **Beauchamp, C. O. and Fridovich, I.,** Superoxide dismutase: improved assays and an assay applicable to acrylamide gel, *Anal. Biochem.*, 44, 276, 1971.
24. **Scandalios, J. G.,** Response of plant antioxidant defense genes to environmental stress, *Adv. Genet.*, 28, 1, 1990.
25. **Matters, G. L. and Scandalios, J. G.,** Effect of the free radical-generating herbicide paraquat on the expression of the superoxide dismutase *(Sod)* genes in maize, *Biochim. Biophys. Acta*, 882, 29, 1986.
26. **Matters, G. L. and Scandalios, J. G.,** Synthesis of isozymes of superoxide dismutase in maize leaves in response to O_3, SO_2 and elevated O_2, *J. Exp. Bot.*, 38, 842, 1987.
27. **Demple, B.,** Regulation of bacterial oxidative stress genes, *Annu. Rev. Genet.*, 25, 315, 1991.
28. **Storz, G., Tartaglia, L. A., and Ames, B. N.,** Transcriptional regulator of oxidative stress-inducible genes: direct activation by oxidation, *Science*, 248, 189, 1990.
29. **Schreck, R., Rieber, P., and Baeuerle, P. A.,** Reactive oxygen intermediates as apparently widely used messengers in the activation of the NF-κB transcription factor and HIV-1, *EMBO J.*, 10, 2247, 1991.
30. **Powles, S. B.,** Photoinhibition of photosynthesis induced by visible light, *Annu. Rev. Plant Physiol.*, 35, 15, 1984.
31. **Barényi, B. and Krause, G. H.,** Inhibition of photosynthetic reactions by light. A study with isolated spinach chloroplasts, *Planta*, 163, 218, 1985.
32. **Richter, M., Rühle, W., and Wild, A.,** Studies on the mechanism of photosystem II photoinhibition. II. The involvement of toxic oxygen species, *Photosynth. Res.*, 24, 237, 1990.
33. **Sopory, S. K., Greenberg, B. M., Mehta, R. A., Edelman, M., and Mattoo, A. K.,** Free radical scavengers inhibit light-dependent degradation of the 32 kDa photosystem II reaction center protein, *Z. Naturforsch. Teil C*, 45, 412, 1990.
34. **Jansen, M. A. K., Shaaltiel, Y., Kazzes, D., Canaani, O., Malkin, S., and Gressel, J.,** Increased tolerance to photoinhibitory light in paraquat-resistant *Conyza bonariensis* measured by photoacoustic spectroscopy and $^{14}CO_2$-fixation, *Plant Physiol.*, 91, 1174, 1989.
35. **Preston, C., Holtum, J. A. M., and Powles, S. B.,** Resistance to the herbicide paraquat and increased tolerance to photoinhibition are not correlated in several weed species, *Plant Physiol.*, 96, 314, 1991.
36. **Rabinowitch, H. D., Clare, D. A., Crapo, J. D., and Fridovich, I.,** Positive correlation between superoxide dismutase and resistance to paraquat toxicity in the green alga *Chlorella sorokiniana*, *Arch. Biochem. Biophys.*, 225, 640, 1983.

37. **Harper, D. B. and Harvey, B. M. R.,** Mechanism of paraquat tolerance in perennial ryegrass. II. Role of superoxide dismutase, catalase and peroxidase, *Plant, Cell Environ.*, 1, 211, 1978.
38. **Shaaltiel, Y., Chua, N.-H., Gepstein, S., and Gressel, J.,** Dominant pleiotropy controls enzymes co-segregating with paraquat resistance in *Conyza bonariensis, Theor. Appl. Genet.*, 75, 850, 1988.
39. **Shaaltiel, Y., Glazer, A., Bocion, P. F., and Gressel, J.,** Cross tolerance to herbicidal and environmental oxidants of plant biotypes tolerant to paraquat, sulfur dioxide, and ozone, *Pest. Biochem. Physiol.*, 31, 13, 1988.
40. **Malan, C., Greyling, M. M., and Gressel, J.,** Correlation between CuZn superoxide dismutase and glutathione reductase, and environmental and xenobiotic stress tolerance in maize inbreds, *Plant Sci.*, 69, 157, 1990.
41. **Gillham, D. J. and Dodge, A. D.,** Chloroplast superoxide and hydrogen peroxide scavenging systems from pea leaves: seasonal variations, *Plant Sci.*, 50, 105, 1987.
42. **Taylor, A. O. and Rowley, J. A.,** Plants under climatic stress. I. Low-temperature, high-light effects on photosynthesis, *Plant Physiol.*, 47, 713, 1971.
43. **Peeler, T. C. and Naylor, A. W.,** A comparison of the effects of chilling on leaf gas exchange in pea (*Pisum sativum* L.) and cucumber (*Cucumis sativus* L.), *Plant Physiol.*, 86, 143, 1988.
44. **Walker, M. A., McKersie, B. D., and Pauls, K. P.,** Effects of chilling on the biochemical and functional properties of thylakoid membranes, *Plant Physiol.*, 97, 663, 1991.
45. **Somersalo, S. and Krause, G. H.,** Photoinhibition at chilling temperature. Fluorescence characteristics of unhardened and cold-acclimated spinach leaves, *Planta*, 177, 409, 1989.
46. **Senaratna, T., Mackay, C. E., McKersie, B. D., and Fletcher, R. A.,** Uniconazole-induced chilling tolerance in tomato and its relationship to antioxidant content, *J. Plant Physiol.*, 133, 56, 1988.
47. **Okuda, T., Matsuda, Y., Yamanaka, A., and Sagisaka, S.,** Abrupt increase in the level of hydrogen peroxide in leaves of winter wheat is caused by cold treatment, *Plant Physiol.*, 97, 1265, 1991.
48. **Wise, R. R. and Naylor, A. W.,** Chilling-enhanced photooxidation. Evidence for the role of singlet oxygen and superoxide in the breakdown of pigments and endogenous antioxidants, *Plant Physiol.*, 83, 278, 1987.
49. **Schöner, S. and Krause, G. H.,** Protective systems against active oxygen species in spinach: response to cold acclimation in excess light, *Planta*, 180, 383, 1990.
50. **Clare, D. A., Rabinowitch, H. D., and Fridovich, I.,** Superoxide dismutase and chilling injury in *Chlorella ellipsoidea, Arch. Biochem. Biophys.*, 231, 158, 1984.
51. **Jahnke, L. S., Hull, M. R., and Long, S. P.,** Chilling stress and oxygen metabolizing enzymes in *Zea mays* and *Zea diploperennis, Plant Cell Environ.*, 14, 97, 1991.
52. **Michalski, W. P. and Kaniuga, Z.,** Photosynthetic apparatus of chilling-sensitive plants. X. Relationship between superoxide dismutase activity and photoperoxidation of chloroplast lipids, *Biochim. Biophys. Acta*, 637, 159, 1981.
53. **Asada, K. and Takahashi, M.,** Production and scavenging of active oxygen in photosynthesis, in *Photoinhibition*, Kyle, D. J., Osmond, C. B., and Arntzen, C. J., Eds., Elsevier, Amsterdam, 1987, 227.
54. **Asada, K. and Kiso, K.,** Initiation of aerobic oxidation of sulfite by illuminated spinach chloroplasts, *Eur. J. Biochem.*, 33, 253, 1973.
55. **Shimazaki, K.-I., Sakaki, T., Kondo, N., and Sugahara, K.,** Active oxygen participation in chlorophyll destruction and lipid peroxidation in SO_2-fumigated leaves of spinach, *Plant Cell Physiol.*, 21, 1193, 1980.
56. **Peiser, G. D., Lizada, M. C. C., and Yang, S. F.,** Sulfite-induced lipid peroxidation in chloroplasts as determined by ethane production, *Plant Physiol.*, 70, 994, 1982.

57. **Covello, P. S. and Thompson, J. E.,** Spin trapping evidence for formation of the sulfite radical anion during chloroplast-mediated oxidation of bisulfite ion, *Biochim. Biophys. Acta,* 843, 150, 1985.
58. **Shimazaki, K., Nakamachi, K., Kondo, N., and Sugahara, K.,** Sulfite inhibition of photosystem II in illuminated spinach leaves, *Plant Cell Physiol.,* 25, 337, 1984.
59. **Covello, P. S., Chang, A., Dumbroff, E. B., and Thompson, J. E.,** Inhibition of photosystem II precedes thylakoid membrane lipid peroxidation in bisulfite-treated leaves of *Phaseolus vulgaris, Plant Physiol.,* 90, 1492, 1989.
60. **Rabinowitch, H. D. and Fridovich, I.,** Growth of *Chlorella sorokiniana* in the presence of sulfite elevates cell content of superoxide dismutase and imparts resistance towards paraquat, *Planta,* 164, 524, 1985.
61. **Srivastave, A. and Tel-Or, E.,** Effect of some environmental pollutants on the superoxide dismutase activity in *Lemna, Free Rad. Res. Commun.,* 12-13, 601, 1991.
62. **Tanaka, K., Furusawa, I., Kondo, N., and Tanaka, K.,** SO_2 tolerance of tobacco plants regenerated from paraquat-tolerant callus, *Plant Cell Physiol.,* 29, 743, 1988.
63. **Tanaka, K. and Sugahara, K.,** Role of superoxide dismutase in defense against SO_2 toxicity and an increase in superoxide dismutase activity with SO_2 fumigation, *Plant Cell Physiol.,* 21, 601, 1980.
64. **Guderian, R.,** *Air Pollution by Photochemical Oxidants,* Springer-Verlag, Berlin, 1985.
65. **Heck, W. W., Taylor, O. C., and Tingey, D. T.,** *Assessment of Crop Loss from Air Pollutants,* Elsevier, London, 1988.
66. **Heath, R. L. and Castillo, F. J.,** Membrane disturbances in response to air pollutants, in *Air Pollution and Plant Metabolism,* Schulte-Hostede, S., Darrall, N. M., Blank, L. W., and Wellburn, A. R., Eds., Elsevier, London, 1988, 55.
67. **Mehlhorn, H., Tabner, B. J., and Wellburn, A. R.,** Electron spin resonance evidence for the formation of free radicals in plants exposed to ozone, *Physiol. Plant.,* 79, 377, 1990.
68. **Reich, P. B. and Amundson, R. G.,** Ambient levels of ozone reduce net photosynthesis in tree and crop species, *Science,* 230, 566, 1985.
69. **Lehnherr, B., Mächler, F., Grandjean, A., and Fuhrer, J.,** The regulation of photosynthesis in leaves of field-grown spring wheat (*Triticum aestivum* L., cv Albis) at different levels of ozone in ambient air, *Plant Physiol.,* 88, 1115, 1988.
70. **Floyd, R. A., West, M. S., Hogsett, W. E., and Tingey, D. T.,** Increased 8-hydroxyguanine content of chloroplast DNA from ozone-treated plants, *Plant Physiol.,* 91, 644, 1989.
71. **Athanassious, R.,** Ozone effects on radish (*Raphanus sativus* L. cv Cherry Belle): gradient of ultrastructural changes, *Z. Pflanzenphysiol.,* 97, 227, 1980.
72. **Sakaki, T., Kondo, N., and Sugahara, K.,** Breakdown of photosynthetic pigments and lipids in spinach leaves with ozone fumigation: role of active oxygens, *Physiol. Plant.,* 59, 28, 1983.
73. **Chanway, C. P. and Runeckles, V. C.,** The role of superoxide dismutase in the susceptibility of bean leaves to ozone injury, *Can. J. Bot.,* 62, 236, 1984.
74. **Tandy, N. E., Di Giulio, R. T., and Richardson, C. J.,** Assay and electrophoresis of superoxide dismutase from red spruce (*Picea rubens* Sarg.), loblolly pine (*Pinus taeda* L.), and Scotch pine (*Pinus sylvestris* L.). A method for biomonitoring, *Plant Physiol.,* 90, 742, 1989.
75. **Sen Gupta, A., Grene Alscher, R., and McCune, D.,** Response of photosynthesis and cellular antioxidants to ozone in *Populus* leaves, *Plant Physiol.,* 96, 650, 1991.
76. **Tanaka, K., Suda, Y., Kondo, N., and Sugahara, K.,** O_3 tolerance and the ascorbate-dependent H_2O_2 decomposing system in chloroplasts, *Plant Cell Physiol.,* 26, 1425, 1985.
77. **Mehlhorn, H., Cottam, D. A., Lucas, P. W., and Wellburn, A. R.,** Induction of ascorbate peroxidase and glutathione reductase activities by interactions of mixtures of air pollutants, *Free Rad. Res. Commun.,* 3, 193, 1987.

78. **Tanaka, K., Saji, H., and Kondo, N.,** Immunological properties of spinach glutathione reductase and inductive biosynthesis of the enzyme with ozone, *Plant Cell Physiol.*, 29, 637, 1988.
79. **Pitcher, L. H., Brennan, E., and Zilinskas, B. A.,** The antiozonant ethylenediurea does not act via superoxide dismutase induction in bean, *Plant Physiol.*, 99, 1388, 1992.
80. **Schraudner, M., Ernst, D., Langebartels, C., and Sandermann, H., Jr.,** Biochemical plant responses to ozone. III. Activation of the defense-related proteins β-1,3-glucanase and chitinase in tobacco leaves, *Plant Physiol.*, 99, 1321, 1992.
81. **Rosemann, D., Heller, W., and Sandermann, H., Jr.,** Biochemical plant responses to ozone. II. Induction of stilbene biosynthesis in Scots pine (*Pinus sylvestris* L.) seedlings, *Plant Physiol.*, 97, 1280, 1991.
82. **Dixon, R. A. and Lamb, C. J.,** Molecular communication in interactions between plants and microbial pathogens, *Annu. Rev. Plant Physiol. Plant Mol. Biol.*, 41, 339, 1990.
83. **Apostol, I., Heinstein, P. F., and Low, P. S.,** Rapid stimulation of an oxidative burst during elicitation of cultured plant cells. Role in defense and signal transduction, *Plant Physiol.*, 90, 109, 1989.
84. **Pauls, K. P. and Thompson, J. E.,** Effects of cytokinins and antioxidants on the susceptibility of membranes to ozone damage, *Plant Cell Physiol.*, 23, 821, 1982.
85. **Mackay, C. E., Seneratna, T., McKersie, B. D., and Fletcher, R. A.,** Ozone induced injury to cellular membranes in *Triticum aestivum* L. and protection by the triazole S-3307, *Plant Cell Physiol.*, 28, 1271, 1987.
86. **Lee, E. H. and Bennett, J. H.,** Superoxide dismutase. A possible protective enzyme against ozone injury in snap beans (*Phaseolus vulgaris* L.), *Plant Physiol.*, 69, 1444, 1982.
87. **Chanway, C. P. and Runeckles, V. C.,** Effect of ethylene diurea (EDU) on ozone tolerance and superoxide dismutase activity in bush bean, *Environ. Pollut.*, 35, 49, 1984.
88. **Mehlhorn, H.,** Ethylene-promoted ascorbate peroxidase activity protects plants against hydrogen peroxide, ozone and paraquat, *Plant Cell Environ.*, 13, 971, 1990.
89. **McKersie, B. D., Beversdorf, W. D., and Hucl, P.,** The relationship between ozone insensitivity, lipid-soluble antioxidants, and superoxide dismutase in *Phaseolus vulgaris*, *Can. J. Bot.*, 60, 2686, 1982.
90. **Langebartels, C., Kerner, K., Leonardi, S., Schraudner, M., Trost, M., Heller, W., and Sandermann, H., Jr.,** Biochemical plant responses to ozone. I. Differential induction of polyamine and ethylene biosynthesis in tobacco, *Plant Physiol.*, 95, 882, 1991.
91. **Bors, W., Langebartels, C., Michel, C., and Sandermann, H., Jr.,** Polyamines as radical scavengers and protectants against ozone damage, *Phytochemistry*, 28, 1589, 1989.
92. **Carlioz, A. and Touati, D.,** Isolation of superoxide dismutase mutants in *Escherichia coli:* is superoxide dismutase necessary for aerobic life?, *EMBO J.*, 5, 623, 1986.
93. **Biliński, T., Krawiec, Z., Liczmański, A., and Litwińska, J.,** Is hydroxyl radical generated by the Fenton reaction in vivo?, *Biochem. Biophys. Res. Commun.*, 130, 533, 1985.
94. **van Loon, A. P. G. M., Pesold-Hurt, B., and Schatz, G.,** A yeast mutant lacking mitochondrial manganese-superoxide dismutase is hypersensitive to oxygen, *Proc. Natl. Acad. Sci. U.S.A.*, 83, 3820, 1986.
95. **Butler Gralla, E. and Selverstone Valentine, J.,** Null mutants of *Saccharomyces cerevisiae* Cu,Zn superoxide dismutase: characterization and spontaneous mutation rates, *J. Bacteriol.*, 173, 5918, 1991.
96. **Campbell, S. D., Hilliker, A. J., and Phillips, J. P.,** Cytogenetic analysis of the *cSOD* microregion in *Drosophila melanogaster, Genetics*, 112, 205, 1986.
97. **Farr, S. B., D'Ari, R., and Touati, D.,** Oxygen-dependent mutagenesis in *Escherichia coli* lacking superoxide dismutase, *Proc. Natl. Acad. Sci. U.S.A.*, 83, 8268, 1986.
98. **Hoerter, J., Eisenstark, A., and Touati, D.,** Mutations by near-ultraviolet radiation in *Escherichia coli* strains lacking superoxide dismutase, *Mutat. Res.*, 215, 161, 1989.

99. **Natvig, D. O., Imlay, K., Touati, D., and Hallewell, R. A.,** Human copper-zinc superoxide dismutase complements superoxide dismutase-deficient *Escherichia coli* mutants, *J. Biol. Chem.*, 262, 14697, 1987.
100. **Bowler, C., Alliotte, T., Van den Bulcke, M., Bauw, G., Vandekerckhove, J., Van Montagu, M., and Inzé, D.,** A plant manganese superoxide dismutase is efficiently imported and correctly processed by yeast mitochondria, *Proc. Natl. Acad. Sci. U.S.A.*, 86, 3237, 1989.
101. **Bowler, C., Van Kaer, L., Van Camp, W., Van Montagu, M., Inzé, D., and Dhaese, P.,** Characterization of the manganese superoxide dismutase gene from *Bacillus stearothermophilus* and its ability to complement copper/zinc superoxide dismutase-deficiency in yeast, *J. Bacteriol.*, 172, 1539, 1990.
102. **Westerbeek-Marres, C. A. M., Moore, M. M., and Autor, A. P.,** Regulation of manganese superoxide dismutase in *Saccharomyces cerevisiae, Eur. J. Biochem.*, 174, 611, 1988.
103. **Zhu, D. and Scandalios, J. G.,** Expression of the maize Mn*Sod (Sod3)* gene in MnSOD-deficient yeast rescues the mutant yeast under oxidative stress, *Genetics*, 131, 803, 1992.
104. **Phillips, J. P., Campbell, S. D., Michaud, D., Charbonneau, M., and Hilliker, A. J.,** Null mutation of copper/zinc superoxide dismutase in *Drosophila* confers hypersensitivity to paraquat and reduced longevity, *Proc. Natl. Acad. Sci. U.S.A.*, 86, 2761, 1989.
105. **Greenberg, J. T. and Demple, B.,** A global response induced in *Escherichia coli* by redox-cycling agents overlaps with that induced by peroxide stress, *J. Bacteriol.*, 171, 3933, 1989.
106. **Walkup, L. K. B. and Kogoma, T.,** *Escherichia coli* proteins inducible by oxidative stress mediated by the superoxide radical, *J. Bacteriol.*, 171, 1476, 1989.
107. **Greenberg, J. T., Monach, P., Chou, J. H., Josephy, P. D., and Demple, B.,** Positive control of a global antioxidant defense regulon activated by superoxide-generating agents in *Escherichia coli, Proc. Natl. Acad. Sci. U.S.A.*, 87, 6181, 1990.
108. **Tsaneva, I. R. and Weiss, B.,** *soxR*, a locus governing a superoxide response regulon in *Escherichia coli* K-12, *J. Bacteriol.*, 172, 4197, 1990.
109. **Gardner, P. R. and Fridovich, I.,** Superoxide sensitivity of the *Escherichia coli* aconitase, *J. Biol. Chem.*, 266, 19328, 1991.
110. **Gruber, M. Y., Glick, B. R., and Thompson, J. E.,** Cloned manganese superoxide dismutase reduces oxidative stress in *Escherichia coli* and *Anacystis nidulans, Proc. Natl. Acad. Sci. U.S.A.*, 87, 2608, 1990.
111. **Bloch, C. A. and Ausubel, F. M.,** Paraquat-mediated selection for mutations in the manganese-superoxide dismutase gene *sodA*, *J. Bacteriol.*, 168, 795, 1986.
112. **Scott, M. D., Meshnick, S. R., and Eaton, J. W.,** Superoxide dismutase-rich bacteria. Paradoxical increase in oxidant toxicity, *J. Biol. Chem.*, 262, 3640, 1987.
113. **Touati, D.,** Molecular genetics of superoxide dismutases, *Free Rad. Biol. Med.*, 5, 393, 1988.
114. **Siwecki, G. and Brown, O. R.,** Overproduction of superoxide dismutase does not protect *Escherichia coli* from stringency-induced growth inhibition by 1 mM paraquat, *Biochem. Int.*, 20, 191, 1990.
115. **Staveley, B. E., Phillips, J. P., and Hilliker, A. J.,** Phenotypic consequences of copper-zinc superoxide dismutase overexpression in *Drosophila melanogaster, Genome*, 33, 867, 1990.
116. **Seto, N. O. L., Hayashi, S., and Tener, G. M.,** Overexpression of Cu-Zn superoxide dismutase in *Drosophila* does not affect life-span, *Proc. Natl. Acad. Sci. U.S.A.*, 87, 4270, 1990.
117. **Reveillaud, I., Niedzwiecki, A., Bensch, K. G., and Fleming, J. E.,** Expression of bovine superoxide dismutase in *Drosophila melanogaster* augments resistance to oxidative stress, *Mol. Cell. Biol.*, 11, 632, 1991.

118. **Krall, J., Bagley, A. C., Mullenbach, G. T., Hallewell, R. A., and Lynch, R. E.,** Superoxide mediates the toxicity of paraquat for cultured mammalian cells, *J. Biol. Chem.*, 263, 1910, 1988.
119. **Kelner, M. J. and Bagnell, R.,** Alteration of endogenous glutathione peroxidase, manganese superoxide dismutase, and glutathione transferase activity in cells transfected with a copper-zinc superoxide dismutase expression vector. Explanation for variations in paraquat resistance, *J. Biol. Chem.*, 265, 10872, 1990.
120. **Elroy-Stein, O., Bernstein, Y., and Groner, Y.,** Overproduction of human Cu/Zn-superoxide dismutase in transfected cells: extenuation of paraquat-mediated cytotoxicity and enhancement of lipid peroxidation, *EMBO J.*, 5, 615, 1986.
121. **Kyle, M. E., Nakae, D., Sakaida, I., Miccadei, S., and Farber, J. L.,** Endocytosis of superoxide dismutase is required in order for the enzyme to protect hepatocytes from the cytotoxicity of hydrogen peroxide, *J. Biol. Chem.*, 263, 3784, 1988.
122. **Freeman, B. A., Young, S. L., and Crapo, J. D.,** Liposome-mediated augmentation of superoxide dismutase in endothelial cells prevents oxygen injury, *J. Biol. Chem.*, 258, 12534, 1983.
123. **Scott, M. D., Eaton, J. W., Kuypers, F. A., Chiu, D. T.-Y., and Lubin, B. H.,** Enhancement of erythrocyte superoxide dismutase activity: effects on cellular oxidant defense, *Blood*, 74, 2542, 1989.
124. **Bagley, A. C., Krall, J., and Lynch, R. E.,** Superoxide mediates the toxicity of paraquat for Chinese hamster ovary cells, *Proc. Natl. Acad. Sci. U.S.A.*, 83, 3189, 1986.
125. **Sinet, P. M., Michelson, A. M., Bazin, A., Lejeune, J., and Jerome, H.,** Increase in glutathione peroxidase activity in erythrocytes from trisomy 21 subjects, *Biochem. Biophys. Res. Commun.*, 67, 910, 1975.
126. **Frischer, H., Cher, L. K., Ahmad, T., Justice, P., and Smith, G. F.,** Superoxide dismutase and glutathione peroxidase abnormalities in erythrocytes and lymphoid cells in Down's syndrome, *Prog. Clin. Biol. Res.*, 55, 269, 1981.
127. **Kedziora, J., Lukaszewicz, R., Koter, M., Bartosz, G., Pawlowska, B., and Aitkin, D.,** Red blood cell glutathione peroxidase in simple trisomy 21 and translocation 21/22, *Experientia*, 38, 543, 1982.
128. **Pantelakis, S. N., Karaklis, A. G., Alexion, D., Verdas, E., and Valaes, T.,** Red cell enzymes in trisomy 21, *Am. J. Hum. Genet.*, 22, 184, 1972.
129. **Brooksbank, B. W. L., Martinez, M., and Balazs, F.,** Altered composition of polyunsaturated fatty acyl groups in phosphoglycerides of Down's syndrome fetal brain, *J. Neurochem.*, 44, 869, 1985.
130. **Ceballos, I., Delabar, J. M., Nicole, A., Lynch, R. E., Hallewell, R. A., Kamoun, P., and Sinet, P. M.,** Expression of transfected human CuZn superoxide dismutase gene in mouse L cells and NS20Y neuroblastoma cells induces enhancement of glutathione peroxidase activity, *Biochem. Biophys. Acta*, 949, 58, 1988.
131. **Tepperman, J. M. and Dunsmuir, P.,** Transformed plants with elevated levels of chloroplastic SOD are not more resistant to superoxide toxicity, *Plant Mol. Biol.*, 14, 501, 1990.
132. **Bowler, C., Slooten, L. Vandenbranden, S., De Rycke, R., Botterman, J., Sybesma, C., Van Montagu, M., and Inzé, D.,** Manganese superoxide dismutase can reduce cellular damage mediated by oxygen radicals in transgenic plants, *EMBO J.*, 10, 1723, 1991.
133. **Pitcher, L. H., Brennan, E., Hurley, A., Dunsmuir, P., Tepperman, J. M., and Zilinskas, B. A.,** Overproduction of petunia chloroplastic copper/zinc superoxide dismutase does not confer ozone tolerance in transgenic tobacco, *Plant Physiol.*, 97, 452, 1991.
134. **Van Camp, W., Hérouart, D., Bowler, C., Van Montagu, M., and Inzé, D.,** Tissue-specific expression of a manganese superoxide dismutase gene from *Nicotiana plumbaginifolia*, in *Active Oxygen/Oxidative Stress and Plant Metabolism*, 6th Annu. Penn State Symp. Plant Physiology, Pell, E. and Steffen, K., Eds., American Society of Plant Physiologists, Rockville, 1991.

135. **Feierabend, J. and Engel, S.,** Photoinactivation of catalase *in vitro* and in leaves, *Arch. Biochem. Biophys.*, 251, 567, 1986.
136. **Aono, M., Kubo, A., Saji, H., Natori, T., Tanaka, K., and Kondo, N.,** Resistance to active oxygen toxicity of transgenic *Nicotiana tabacum* that expresses the gene for glutathione reductase from *Escherichia coli, Plant Cell Physiol.*, 32, 691, 1991.
137. **Furasawa, I., Tanaka, K., Thanutong, P., Mizuguchi, A., Yazaki, M., and Asada, K.,** Paraquat resistant tobacco calluses with enhanced superoxide dismutase activity, *Plant Cell Physiol.*, 25, 1247, 1984.
138. **Van Camp, W., Willekens, H., Bowler, C., Von Montagu, M., Inzé, D., Reupold-Popp, P., Sandermann, H., Jr., and Langebartels, C.,** Elevated levels of superoxide dismutase protect transgenic plants against ozone damage, submitted.

Chapter 13

GLUTATHIONE REDUCTASE AND ASCORBATE PEROXIDASE

Gary P. Creissen, E. Anne Edwards, and Philip M. Mullineaux

TABLE OF CONTENTS

0-8493-5443-9/94/$0.00 + $.50

I. INTRODUCTION

Oxidative damage in chloroplasts is minimized under "normal" conditions by the concerted activities of the superoxide dismutases[1,2] and the enzymes of the ascorbate-glutathione cycle (also referred to as the Halliwell-Asada pathway).[3,4] The ascorbate-glutathione cycle involves successive oxidations and reductions of ascorbate, glutathione, and NADPH by the enzymes ascorbate peroxidase, glutathione reductase, and dehydroascorbate reductase (Figure 1). Although much of the discussion of this pathway in the literature has focussed on its central role in the protection against oxidative damage in the chloroplast, the presence of ascorbate-glutathione cycle enzymes and their substrates in nonphotosynthetic tissues[5-7] clearly demonstrates that the pathway also functions in other cell compartments. To date, the most extensively studied enzymes of this pathway are ascorbate peroxidase and glutathione reductase (GR), but while these enzymes have been the subject of numerous biochemical and physiological studies, their precise roles and importance to the plant have yet to be clearly defined. A classical genetic approach to the role of GR has not been extensively explored (see Chapter 10 of this volume for a more detailed treatment of the subject), but may be worthy of more effort in the future.

Our interests are to arrive at a more complete understanding of the roles of these enzymes in oxidative stress by applying molecular genetic techniques in conjunction with a biochemical and physiological approach. The ability to engineer plants which express introduced genes at high levels provides an opportunity to manipulate the levels of these two enzymes *in vivo*. Such plants will be invaluable in assessing the benefits of increased glutathione reductase and ascorbate peroxidase activities. As the reader will discern, this work is still in its early stages.

In this chapter, we review the current understanding of the biochemical and physiological functions of glutathione reductase and ascorbate peroxidase under normal and stressed conditions, and describe the progress being made in cloning the genes encoding these enzymes and the prospects for genetic manipulation of the ascorbate-glutathione pathway.

II. ASCORBATE PEROXIDASE

Ascorbate functions as an excellent antioxidant in both the cytosol and the chloroplast, with a fundamental role in the removal of hydrogen peroxide via the ascorbate-glutathione cycle,[8-10] and it may form a redox link between the two compartments via an ascorbate-specific translocator.[11,12]

While most of the chloroplast ascorbate is stromal, a portion is also bound to the thylakoids.[13] This appears to be important in the regeneration of membrane-bound carotenoids and tocopherols, and particularly in the regeneration of α-tocopherol from its free radical (the α-chromanoxyl radical) produced

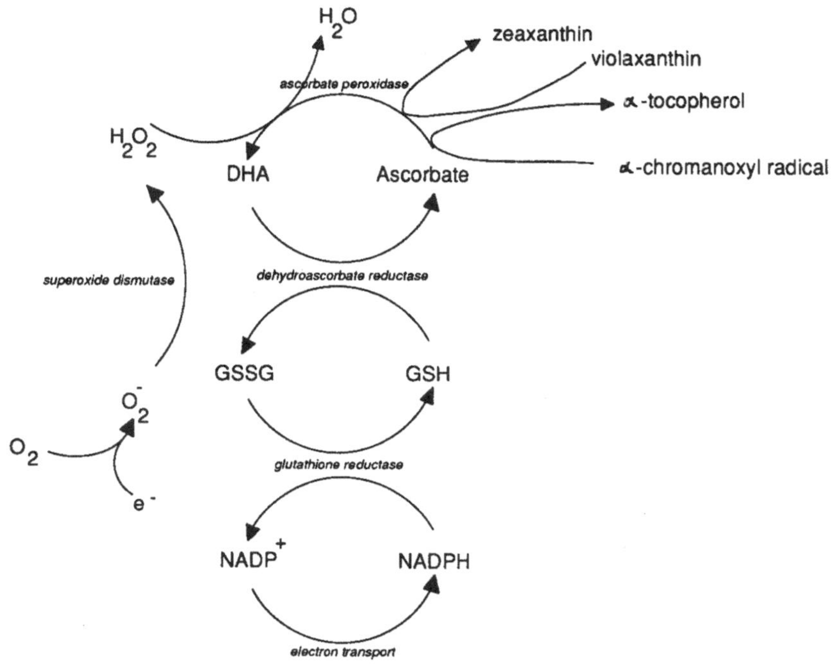

FIGURE 1. Ascorbate-glutathione cycle and associated reactions. DHA is dehydroascorbate.

via the chain termination action of vitamin E during lipid peroxidation[14] and of zeaxanthin from its diepoxide, violaxanthin[15,16] (see Figure 1). The xanthophyll cycle is considered to be the major photoprotective dissipation process in leaves,[16-18] and consequently the importance of ascorbate as well as other components of the ascorbate-glutathione cycle in this aspect of photoprotection is clear. For a detailed consideration of these pathways, see Chapters 1 and 4 of this volume.

The removal of H_2O_2 produced in the chloroplast as a result of dismutation of superoxide is essential to avoid inhibition of the Calvin cycle enzymes.[19,20] In other cell compartments, H_2O_2 degradation is largely performed by the action of catalases, (see Chapter 11 of this volume). Catalases are absent from chloroplasts, and the degradation of H_2O_2 in chloroplasts is achieved by the activity of ascorbate peroxidase.[21-23] Ascorbate peroxidases perform a function similar to that of glutathione peroxidases in animal cells. Glutathione peroxidases are not thought to be present in plant tissues, although recently a cDNA clone showing homology to a number of glutathione peroxidases was isolated from *Nicotiana sylvestris*.[24]

Like the extensively studied guaiacol peroxidases (e.g., horseradish peroxidase), ascorbate peroxidase is a heme-containing protein. However, ascorbate peroxidase(s) may be distinguished from the guaiacol peroxidases by a number of characteristics: ascorbate peroxidases are highly specific for

$$2\ \text{Ascorbate} + H_2O_2 \xrightarrow{\text{Ascorbate peroxidase}} 2\ \text{Monodehydroascorbate} + 2H_2O$$

FIGURE 2. The reaction catalyzed by ascorbate peroxidase.

ascorbate as the electron donor, whereas the guaiacol peroxidases can utilize a wide range of electron donors and, unlike ascorbate peroxidases, have a high specificity for guaiacol;[35] in addition, guaiacol peroxidases, unlike ascorbate peroxidases, are glycoproteins.[21] The reaction catalyzed by ascorbate peroxidase is shown in Figure 2.

Ascorbate peroxidase activity has been identified in a number of different tissues, and the enzyme has been purified from tea,[21] spinach,[25,26] and pea.[27] Two isozymes of ascorbate peroxidase have been identified in each of these species; one is localized in the chloroplasts and the other is thought to be a cytosolic form. In tea leaves, the molecular weights of the two isoforms were estimated by gel filtration at 57 kDa (cytosolic) and 34 kDa (chloroplastic). The chloroplastic isoform was purified to homogeneity and found to account for 60 to 80% of the ascorbate peroxidase activity.[21] The chloroplastic isozyme also had an apparent molecular weight of 34 kDa when resolved on SDS-PAGE, indicating that it is a monomer. Antiserum raised against the purified plastidial enzyme cross-reacted with both isoforms. In addition to differences in molecular weight, the two isoforms were found to differ in a number of other enzymatic and molecular properties, including specificity for electron donor, lability in the absence of ascorbate, affinities for H_2O_2, and ascorbate and pH optima. In contrast to the data for tea ascorbate peroxidase, the isozymes from spinach leaves were found to be identical in size.[38] However, the two spinach isoforms were immunologically distinguishable; antiserum raised against the purified putative chloroplastic isoform failed to cross-react with the other isoform.[26]

Partially purified ascorbate peroxidase from pea leaves resolved into two bands on SDS-PAGE of 30 and 50 kDa, both of which cross-reacted with anti-tea leaf antibodies.[12] However, antibodies raised against the cytosolic isoform from pea failed to cross-react with the plastidial isoform.[27] A summary of the properties of ascorbate peroxidases from different sources is presented in Table 1.

III. GLUTATHIONE REDUCTASE

Glutathione, γ-glutamyl-cysteinyl glycine (GSH), is the major low molecular weight thiol compound in most plants. Glutathione is not an essential metabolite in bacteria,[28] but it does appear to enhance tolerance to numerous xenobiotics and oxidative agents. However, glutathione is considered essential in plant and animal cells, although in some plants (e.g., some of the legumes),

TABLE 1
Properties of Ascorbate Peroxidases from Different Sources

Source	Mol wt (kDa)	Location	K_m (μM) Ascorbate	H_2O_2	pH_{opt}	Ref.
Spinach	30	—	300	30	7.2	25
Spinach	31	Cyt	—	—	—	26
	31	Cp	—	—	—	26
Tea	57	Cyt	400	80	6.0	21
	34	Cp	220	30	7.0	21
Pea	29.5	Cyt	—	—	5.8	27
	29.5	Cp	—	—	8.0	27
Pea	57	—	6500	>5	5.0	38
	57	—	2900	>5	6.0	38

Note: Cyt., cytosolic; Cp, chloroplast.

homoglutathione (γ-glutamyl-cysteinyl alanine) may partly or wholly replace glutathione.[29] In addition to its major role in sulfur transport, glutathione is involved in a number of cellular processes in plants, including acting as a protein disulfide reductant,[30] detoxifying xenobiotics (e.g., herbicides) by conjugation either spontaneously or by the activity of one of a number of glutathione-*S*-transferases, and regulating gene expression in response to environmental stress and pathogen attack.[31-34]

Cellular glutathione concentrations of about 0.1 m*M* have been reported,[35] but this figure can vary considerably between species and in different organs of the same plant. The concentrations of glutathione in different cell compartments can also differ, with the highest concentrations being recorded in the chloroplasts.[3,36,37] The high GSH content of chloroplasts is consistent with the accepted role for GSH in the recycling of ascorbate via the activities of the enzymes of the ascorbate-glutathione pathway.

Many of the important metabolic, regulatory, and antioxidative roles of glutathione result in its oxidation to glutathione disulfide (GSSG). For most of its functions, glutathione must be available in its reduced form (GSH). The reduction of oxidized glutathione to GSH is carried out by the flavoprotein oxidoreductase glutathione reductase (GR), in a NADPH-dependent reaction (Figure 3).

Glutathione reductase is very widely distributed in prokaryotes and eukaryotes and has been studied from a large number of organisms, ranging from heterotrophic and photosynthetic bacteria to plants and animals. The most extensively studied glutathione reductase is that from human erythrocytes. The primary amino acid sequence for human GR is known,[38] and X-ray crystallography of the purified protein has provided a detailed understanding of the three-dimensional structure of the homodimer, including elucidation of the binding sites for FAD, NADPH, and glutathione.[40-44] Detailed

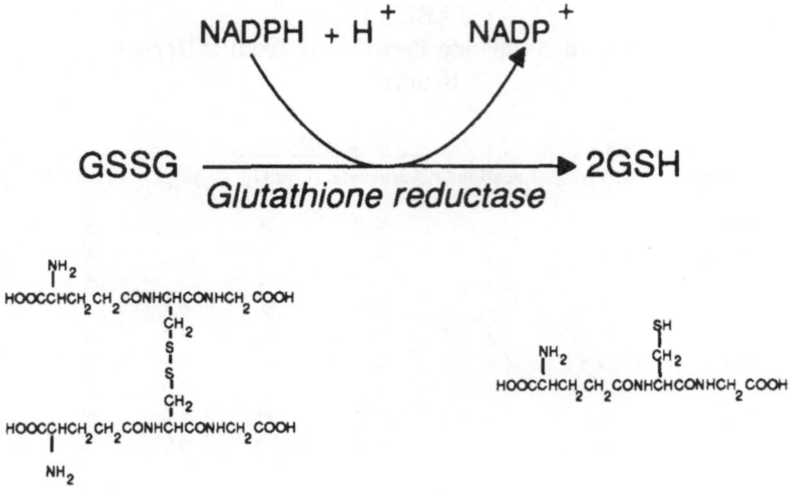

OXIDIZED GLUTATHIONE **REDUCED GLUTATHIONE**

FIGURE 3. The reaction catalyzed by glutathione reductase.

structural information is not available for GR from plant sources; however, the enzyme has been characterized from a wide range of higher plants.

A. HIGHER PLANT GLUTATHIONE REDUCTASES

Glutathione reductase has been studied in a wide range of plants, including both gymnosperms and angiosperms, and extensively purified from at least four species: maize,[45] pea,[37,46-48] spinach,[49] and pine.[50]

In animal cells, most of the GR activity is cytosolic, with a small proportion located in the mitochondrial matrix. In plants, most of the GR activity is present in chloroplasts.[36,48] Recently, Edwards et al.[37] showed that, in pea leaves, the chloroplasts and mitochondria contained 77 and 3% of the total GR activity, respectively, with the balance being cytosolic. Immunogold labeling of thin sections of pea leaves strongly supports the distribution of GR determined by cell fractionation (Figure 4).

Pea leaf GR can be resolved into a single band of M_r 55 kDa on SDS-PAGE; however, on 2-D gels, up to eight isoforms are detected with isoelectric points (pIs) ranging from 5.6 to 6.5 (Figure 5). On Western blots of organelle proteins, five isoforms (*pI* 5.6–6.3) were detected in the chloroplast fraction, while three isoforms (*pI* 6.3–6.5) were identified in the mitochondrial fraction. In addition, on native polyacrylamide gels, up to six bands of GR activity were detected (Figure 6). Western blotting of the eluted activity bands following separation by SDS-PAGE confirmed that the observed forms were GR, and failed to reveal differences in molecular weight (Figure 6). It was not possible to conclude from the biochemical data for the

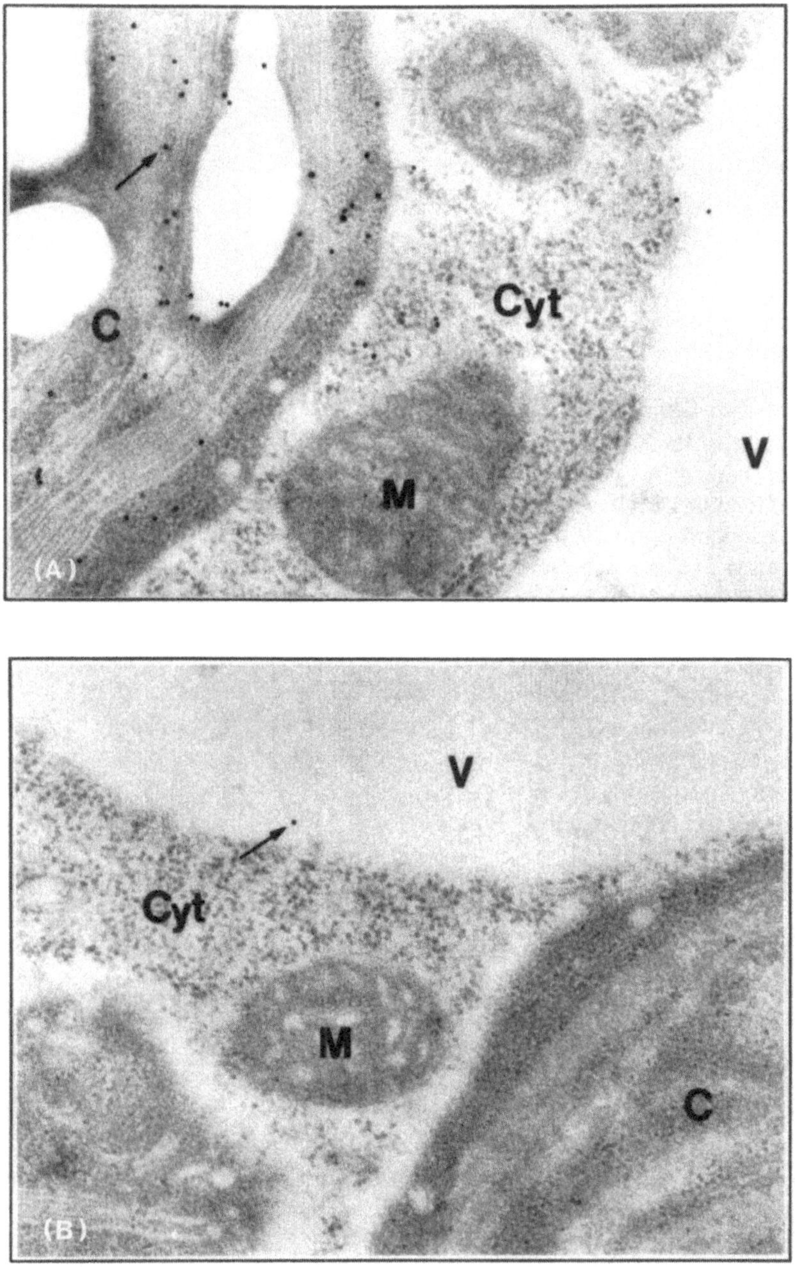

FIGURE 4. Immunogold localization of glutathione reductase in pea leaf sections. (A) Anti-pea leaf glutathione reductase; (B) null antiserum. (Courtesy of Chris Hylton.)

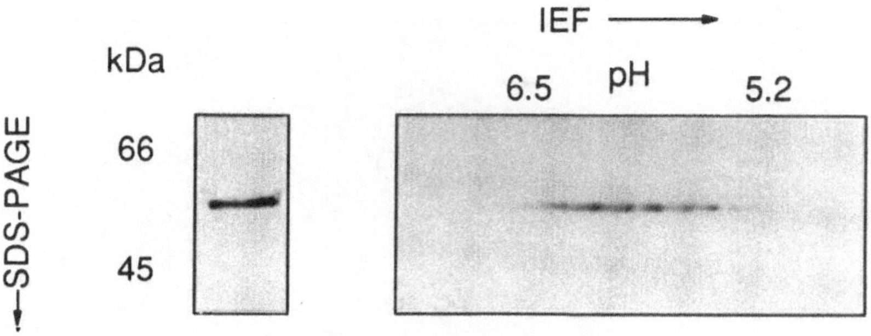

FIGURE 5. Western blot of one- and two-dimensional PAGE of purified pea leaf glutathione reductase.

pea GRs whether the different isoforms represented different gene products or arose as the result of posttranscriptional or posttranslational modification of a single gene product, although a number of possible posttranslational modifications such as changes in oxidation state, glycosylation, or phosphorylation appear to have been ruled out. Further evidence for multiple isoforms of GR has been found in other plants. Guy and Carter[51] identified GR isoforms in cold-hardened and nonhardened spinach leaves which differed in several characteristics, including thermal inactivation and electrophoretic mobility. Two isoforms of GR were also purified from eastern white pine. These were distinguishable on the basis of differences in molecular weight, K_ms for GSSG and NADPH and pH optima.[50] In tobacco leaves, at least six GR isoforms were detected on native polyacrylamide gels stained for GR activity.[52] A summary of physical and kinetic characteristics reported for purified GR is presented in Table 2.

B. GLUTATHIONE REDUCTASE, ASCORBATE PEROXIDASE, AND OXIDATIVE STRESS

The activities of some of the ascorbate-glutathione cycle enzymes have been reported to be influenced by various environmental factors known to increase oxyradical formation. These include exposure to gaseous pollutants or xenobiotics, extremes of temperature, pathogen attack, and senescence. In most instances, relatively small changes in enzyme activities have been reported. Low-level exposure to ozone (100 ppb) resulted in a threefold increase in ascorbate peroxidase activity in spinach,[54] and similar exposures (70 ppb) produced a doubling of GR activity.[55] A doubling in both GR and ascorbate peroxidase activities was reported in pea plants following fumigation with ozone or with mixtures of ozone, NO_2, and SO_2.[56] The sensitivity of pea plants to ozone was found to be determined by the formation of stress ethylene.[57] Furthermore, preexposure to ethylene resulted in increased resistance to ozone due to elevated ascorbate peroxidase activity.[58] In contrast to the earlier reports on the induction of GR by ozone fumigation, Madamanchi et al.[59] found no increase in GR activity in response to ozone fumigation in pea.

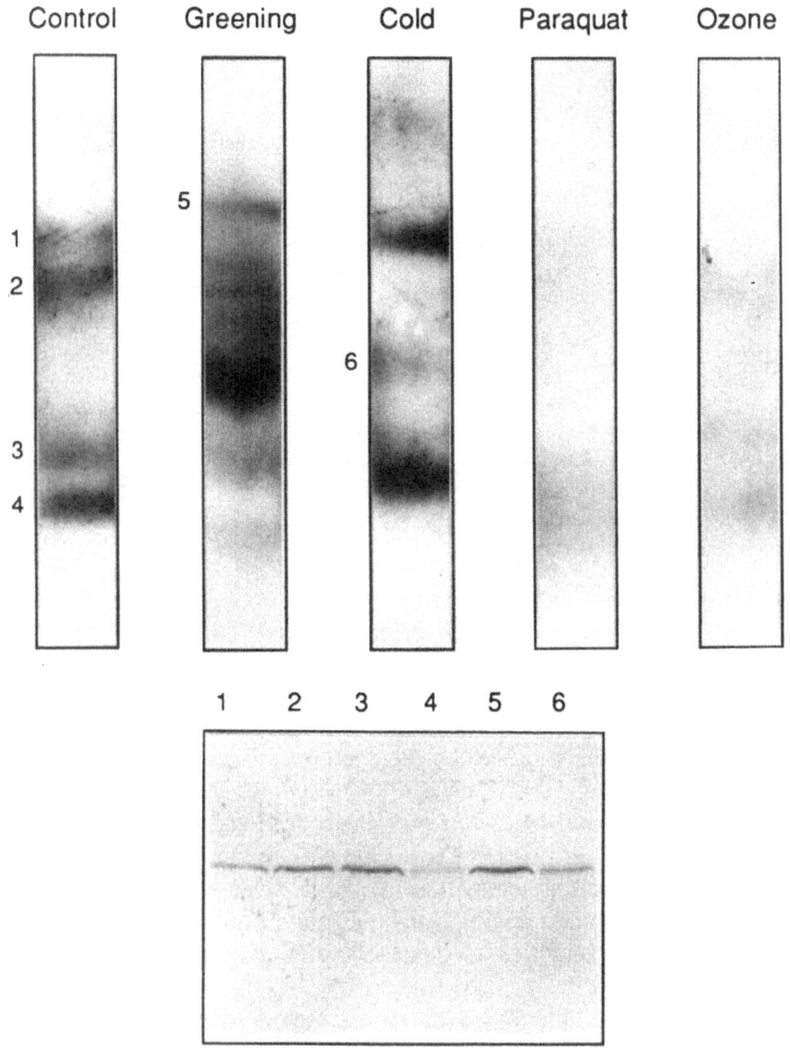

FIGURE 6. Upper panel: glutathione reductase activity staining of total pea leaf extracts resolved on 7.5% native PAGE. Lower panel: Western blot of GR isoforms (1-6 in upper panel) eluted from native gel and separated by SDS-PAGE.

The combination of high light with magnesium deficiency, which would be expected to cause severe photooxidative damage in the chloroplasts, resulted in approximately three- and sevenfold increases in ascorbate peroxidase and glutathione reductase activities, respectively, in French bean leaves.[60] In cold-acclimated spinach leaves, a 31% increase in ascorbate peroxidase activity was reported.[61] Small changes (up to twofold) in GR activity have been reported for other species in response to low-temperature stress (see

TABLE 2
Properties of Purified Glutathione Reductase

Source	Mol wt		K_m (μM)			
	Native	Subunit	GSSG	NADPH	Ph_{opt}	Ref.
E. coli	94–105	49	70	25	—	53
Human	106	53	65	8.5	6.8	39
Maize	190	65, 63, 35, 32	14	1.7	—	45
Spinach	145	72.5	196	2.8	8.5–9	49
Pea (cp)	60		62	3		47
Pea (cp)	156	41, 42	11	1.7	7.6	46
Pea (cp)	135	—	28	2.7	8.2	48
Pea (root)	135	—	10	2.3	7.7	48
Pea (cp)	—	55	27	2.8	—	37
Pea (mt)	—	55	12	2.4	—	37
Pine (cyt?)	103	53, 54	15.3	3.7	7.25–7.75	50
Pine (cp?)	95	53	39.8	8.8	7.25	50

Note: cp, chlorophat; mt, mitochondrion; ?, indicated probable location.

Reference 62), but the significance of these increases is unclear. More dramatic evidence for the induction of detoxifying enzymes under potentially oxidative conditions was reported for eastern white pine. Increased activities of glutathione reductase and ascorbate peroxidase (40- to 122-fold and 40- to 65-fold, respectively) were found in needles collected in the winter compared with those collected in the summer, and the magnitude of the increase in activities correlated with the severity of the winter season: warm winter temperatures were associated with lower levels of glutathione reductase and ascorbate peroxidase activities.[63] These changes are consistent with the requirement for increased protection against photooxidative injury under conditions of low temperature and high light. Examples of changes in GR and ascorbate peroxidase activities in response to a number of different stresses are presented in Table 3.

We have examined the effects of various factors likely to result in oxidative stress in peas. These included transfer of etiolated seedlings to the light, low temperature in combination with high light, paraquat treatment, and ozone fumigation. Increases in the specific activity of GR in the stressed plants were never found to be greater than 2.5-fold under any of the experimental conditions tested. On native PAGE, the relative amounts of the six isozymes visualized by activity staining varied between the different stress treatments (Figure 7). The appearance of different GR isoforms was accompanied by changes in the K_m for GSSG (Table 4).

The significance of these relatively small increases in GR activity is difficult to assess. While it is quite clear that increased GR activity results from a relative increase in the abundance of the enzyme, it should be borne in mind that under conditions of oxidative stress there are changes in the

TABLE 3
Effects of Various Stresses on Glutathione Reductase (GR) and Ascorbate Peroxidase (AP) Activities

Stress	Species	GR[a]	AP[a]	Ref.
Cold	*Pinus strobus*	122	65	63
Cold/high light	*Spinacia oleracea*	—	1.3	61
Cold acclimation	*Malus pumila*	1.5	2	64
Cold	*Zea mays*	1	1.3	65
Cold	*Z. diploperennis*	1.7	1.3	65
High light/low Mg^{2+}	*Phaseolus vulgaris*	7	5.3	60
High O_2	*Z. mays*	3	—	66
High O_2	*Gossypium herbaceum*	3	—	67
O_3	*S. oleracea*	2	2	55
O_3	*Pisum sativum*	1	—	59
$O_3/NO_2/SO_2$	*P. sativum*	2	2	56
Drought	*Tortula ruralis*	2	—	68

[a] Figures represent the maximum increase observed. The value of 1 indicates no change in activity.

abundance of many proteins. Coomassie-stained SDS-PAGE gels of total leaf protein extracts from stressed plants showed decreases in ribulose-1,5-bisphosphate carboxylase-oxygenase (RuBisCo) in all but one case (Figure 7), and consequently in these examples, increased GR activity may simply represent the relative stability of GR against *in vivo* degradation.

IV. MOLECULAR CLONING OF GLUTATHIONE REDUCTASE AND ASCORBATE PEROXIDASE cDNAs

A. GLUTATHIONE REDUCTASE

The gene encoding GR from *Escherichia coli* has been cloned and sequenced,[69] and more recently, cDNA clones have been obtained for the human enzyme, which enabled the amino acid sequence to be confirmed,[70] and for the enzyme from *Pseudomonas aeruginosa.*[71] Difficulties with purification and amino acid sequencing of the enzyme from higher plant sources have, until recently, prevented direct comparisons of the amino acid sequence of a plant glutathione reductase with those from other organisms.

As a first step in studying the molecular biology of glutathione reductase, we have isolated a full-length cDNA encoding the enzyme from a pea leaf λ-gt11 expression library by probing with an antiserum raised against pea leaf GR[72] and from a tobacco cDNA library by probing with a PCR product generated by using an oligonucleotide primer derived from the pea GR sequence.[81] The pea and tobacco cDNAs shared 78% homology at the nucleotide level and were nearly identical in the amino acid sequence of the mature GR protein (over 78% identity; over 86% similarity). Alignments of the deduced

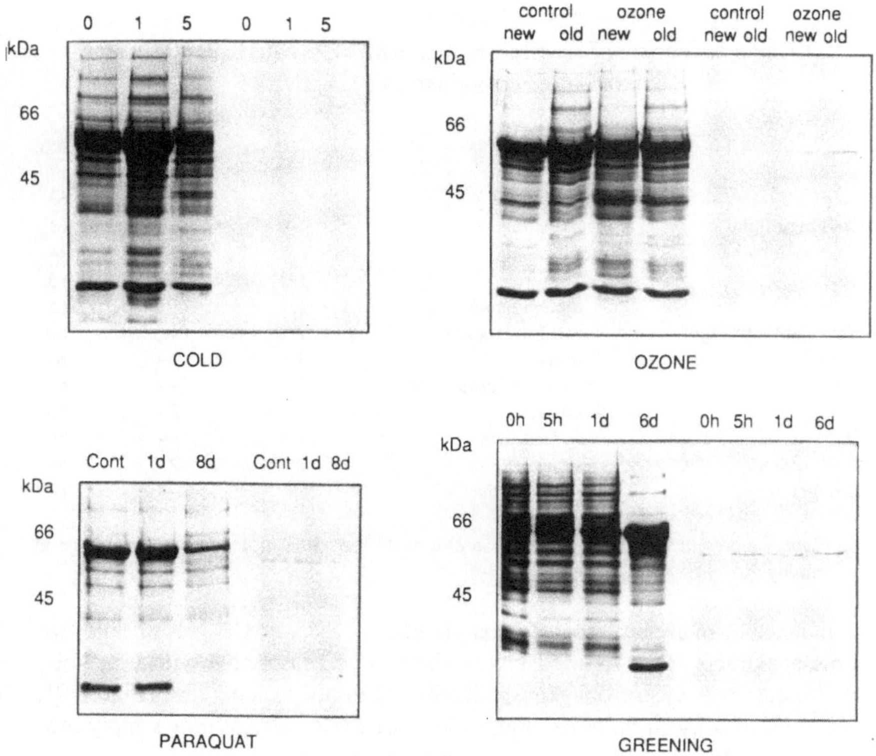

FIGURE 7. Coomassie-stained SDS-PAGE and corresponding Western blots of pea leaf protein extracts from plants grown under different stress conditions. Cold: plants were grown under control conditions (20°C light/15°C dark; 16 h/d) for 14 d before transfer to 4°C. Samples were taken after 1 and 5 months. Ozone: 1-week-old plants were fumigated with 200 nl/l ozone, 7 h/d, 18 d. Samples were collected from new growth as well as existing leaves. Paraquat: leaves of 2-week-old plants were painted with 150 μM paraquat daily for 8 d. Samples were harvested after 0, 1, and 8 d. Greening: seeds were germinated in the dark for 9 d before being transferred to control conditions. Samples were harvested after 0 h, 5 h, 1 d, and 6 d.

amino acid sequences of all the GRs for which data are available are shown in Figures 8 and 9. Overall, there was a high degree of amino acid sequence similarity between the different GRs (Table 5). The region responsible for forming the redox-active disulfide bridge is nearly identical in all cases. Additionally, the arginine residues required for binding of the 2′-phosphate group of NADPH[73] are conserved. The cysteine residue responsible for the intersubunit disulfide bond in the human enzyme is, however, absent in all other GRs, indicating that, unlike the human enzyme, the other GRs are unlikely to form covalently linked dimers.

Both plant GR sequences contained amino terminal extensions, which, based on their amino acid compositions,[74] are probably chloroplast transit peptides. In addition, the plant proteins have similar C-terminal extensions

TABLE 4
Apparent K_m Values for GSSG and Relative GR Activities in Total Leaf Protein Extracts from Oxidatively Stressed Pea Plants

Stress	$K_{m[GSSG]}$ (μM)	GR activity[a]
Control	23	100
Greening	15	100
Cold	4.5	200
Paraquat	9.5	175
Ozone	11	200

[a] GR activity was determined in triplicate and expressed as a percentage of the control levels.

that are rich in alanine, lysine, and serine, which suggests a common function for this domain, possibly functioning as an additional signal sequence.

Interestingly, although the anti-pea leaf GR antiserum recognizes all multiple isoforms of the pea enzyme, to date we have only been able to identify a single class of pea GR cDNA. The presence of GR in three cell compartments (chloroplast, cytosol, and mitochondrion)[37] suggests that there should be at least three GR genes within the pea genome, since targeting of the GR polypeptide subunits into organelles would require different transit peptide sequences. Mapping of the cDNA onto the pea genome identified a single GR locus. Even when Southern blots were probed and washed at low stringency, no additional hybridizing bands could be identified, indicating that no closely related sequences were present elsewhere in the genome, and consequently, unlike some other multicompartment enzymes, GR is not encoded by a multigene family. It is possible that more than one GR gene resides at the same locus in pea, and that the different genes differ only in the regions encoding the transit peptide sequences. Genomic mapping of the GR locus indicates that there are at most two very closely related genes.[82] We are currently analyzing genomic GR clones in order to clarify their genetic organization.

B. ASCORBATE PEROXIDASE

Amino-terminus peptide sequence data have been determined for putative chloroplastidic ascorbate peroxidases from tea[21] and spinach,[26] and for the presumed cytosolic ascorbate peroxidase from pea.[27] A cDNA encoding ascorbate peroxidase has been cloned and sequenced from a pea leaf expression library.[75] The derived amino acid sequence of the amino terminus matched that determined by amino acid sequencing of the purified enzyme and there was no evidence for the presence of a transit peptide, thus providing additional evidence that this cDNA encodes the cytosolic isoform. We have recently

```
  1 MNQAMATPLSLSCCSPTLTRST...LFFTKTFPFSRSFSTP........L
  1 .....ATSLSTPKLSTTLSSPTLHSLLYKHKFSL.LSLSNPIKPLHFNFL

 40 PLSTKTLISLSPPHRTF.AVRAESQNGADPARQYDFDLFTIGAGSGGVRA
 45 THSRSTPSSLSCTRRRFTAPRAESSNGADAPRHYDFDLFTIGAGSGGVRA

 89 SRFASNFGASSAVCELPFSTISSDTTGGVGGTCVIRGCVPKKLLVYASKF
 95 SRFASNFGASVAVCELPFSTISSDSTGGVGGTCVLRGCVPKKLLVYASKY

139 SHEFEESNGFGWRYDSEPKHDWSSLIANKNAELQRLTGIYKNTLKNAGVK
145 SHEFEESCGFGWNYDVEPRFDWSTLIANKNAELQRLTGIYKNILKNAGVT

189 LIEGRGKIVDAHTVDVDGKLYSAKHILVSVGGRPFIPDIPGKEYAIDSDA
195 LIEGRGKVVDPHTVDVDGKLYSAKNILISVGGRPFIPDIPGSEYAIDSDA

239 ALDLPSKPQKIAIVGGGYIALEFAGIFNGLKSEVHVFIRQKKVLRGFDEE
245 ALDLPTKPNKIAIVGGGYIALEFAGIFNGLKSEVHVFIRQKKVLRGFDEE

289 IRDFVAENMALRGIEFHTEESPVAITKAADGSLSLKTNKGTEEGFSHIMF
295 IRDFVGEQMSLRGIEFHTEESPQAIVKSADGSLSLKTSRGTVEGFSHIMF

339 ATGRSPNTKDLGLESVGVKVAKDGSIEVDEYSQTSVPSIWAIGDATNRVN
345 ATGRRPNTKNLGLETVGVKMTKNGAIEVDEYLVLS.PSIWAVGDVTDRIN

389 LTPVALMEGVALAKTLFQNEPTKPDYRAIPSAVFSQPPIGGVGLTEEQAA
394 LTPVALMEGGALAKTIFAHEPTKPDYRNVPAAVFSQPPIGQVGLMEEQAI

439 EQYGDIDVFTANFRPMKATLSGLPDRVFMKLIVSAETNVVLGLHMCGEDA
444 KEFGDVDVYTANFRPLKATISGLPDRVFMKLIVCAKTSKVLGLHMCGDDA

489 AEIAQGFAVGIKAGLTKADFDATVGIHPTAAEEFVTMRTPTRKVRKNQAS
494 PEIVQGFAIAVKAGLTKADFDATVGIHPTSAEEFVTMRTPTRKVRSS.PS

539 QGKSDSKAKAVAGS
543 EGKAEHDIKAAAGV
```

FIGURE 8. Amino acid alignments of pea (upper line) and tobacco glutathione reductases.

isolated ascorbate peroxidase cDNAs from a pea expression library by probing with an antiserum raised against plastidial ascorbate peroxidase from tea (generous gift of K. Asada), one of which has been sequenced and found to be identical to the published pea cytosolic ascorbate peroxidase sequence except for the presence of an additional three bases at the 5′ end. The only other ascorbate peroxidase clone published to date is from *Arabidopsis thaliana*.[76] The derived amino acid sequence of *A. thaliana* ascorbate peroxidase shows a high homology to the amino terminus of the spinach putative chloroplast enzyme, but is also very highly homologous (75% identity) to the pea cytosolic isoform, and shows no similarity to the amino terminal sequence of the presumed chloroplastidic enzyme from tea leaves[77] (Figure 10). These data clearly raise some questions regarding the differences between the different isoforms and their subcellular localization. The fact that immunolog-

```
      1                                                 50
Peagr APPGTERETL MNQAMATPLS LSCCSPTLTR STLFFTKTFP FSRSFSTPLP

      51                                                97
Peagr LSTKTLISLS PPHRTFAVRA ESQNGADPAR Q...YDFDLF TIGAGSGGVR
Psagr                                   MSFDFDLF VIGAGSGGVR
 Ecgr                                   MTKHYDYI AIGGGSGGIA
Humgr                  MACRQ EPQPQGPPPA AGAVASYDYL VIGGGSGGLA
                                                    .....
      98                                               147
Peagr ASRFASNFGA SSAVCELPFS TISSDTTGGV GGTCVIRGCV PKKLLVYASK
Psagr AARFAAGFGA RVAVAESRY. .........L GGTCVNVGCV PKKLLVYGAH
 Ecgr SINRAAMYGQ KCALIEAKE. .........L GGTCVNVGCV PKKVMWHAAQ
Humgr SARRAAELGA RAAVVESHK. .........L GGTCVNVGCV PKKVMWNTAV
                                     .......  ...
      148                                              196
Peagr FSHEFEE.SN GFGWRYDSEP KHDWSSLIAN KNAELQRLTG IYKNTLKNAG
Psagr FSEDFEQ.AR AYGWS.AGEA QFDWATLIGN KNREIQRLNG IYRNLLVNSG
 Ecgr IREAIHMYGP DYGFD.TTIN KFNWETLIAS RTAYIDRIHT SYENVLGKNN
Humgr HSEFMHD.HA DYGFP.SCEG KFNWRVIKEK RDAYVSRLNA IYQNNLTKSH
                  ^                     ....... ....

      197                                              241
Peagr VKLIEGRGKI VDA..HTVDV DGKLYSAKHI LVSVGGRPFI P...DIPGKE
Psagr VTLLEGHARL LDA..HSVEV DGQRFSAKHI LVATGGWPQV P...DIPGKE
 Ecgr VDVIKGFARF VDA..KTLEV NGETITADHI LIATGGRPSH P...DIPGVE
Humgr IEIIRGHAAF TSDPKPTIEV SGKKYTAPHI LIATGGMPST PHESQIPGAS

      242                                              291
Peagr YAIDSDAALD LPSKPQKIAI VGGGYIALEF AGIFNGLKSE VHVFIRQKKV
Psagr HAITSNEAFF LERLPRRVLV VGGGYIAVEF ASIFNGLGAE TTLLYRRDLF
 Ecgr YGIDSDGFFA LPALPERVAV VGAGYIAVEL AGVINGLGAK THLFVRKHAP
Humgr LGITSDGFFQ LEELPGRSVI VGAGYIAVEM AGILSALGSK TSLMIRHDKV
                                                      *
      292                                              341
Peagr LRGFDEEIRD FVAENMALRG IEFHTEESPV AITKAADGSL SLKTNKGTEE
Psagr LRGFDRSVRE HLRDELGKKG LDLQFNSDIA RIDKQADGSL AATLKDGRVL
 Ecgr LRSFDPMISE TLVEVMNAEG PQLHTNAIPK AVVKNTDGSL TLELEDGRSE
Humgr LRSFDSMIST NCTEELENAG VEVLKFSQVK E.VKKTLSGL EVSMVTAVPG
       *
               342                                     383
Peagr ........GF SHIMFATGRS PNTKDLGLES VGVKVAKDGS IEVDEYSQTS
Psagr ........EA DCVFYATGRR PMLDDLGLEN TAVKLTDKGF IAVDEHYQTS
 Ecgr ........TV DCLIWAIGRE PANDNINLEA AGVKTNEKGY IVVDKYQNTN
Humgr RLPVMTMIDV DCLLWAIGRV PNTKDLSLNK LGIQTDDKGH IIVDEFQNTN

      384                                              432
Peagr VPSIWAIGDA TNRVNLTPVA LMEGVALAKT LFQNEPTKP. DYRAIPSAVF
Psagr EPSILALGDV IGRVQLTPVA LAEGMAVARR LFKPEEYRPV DYKLIPTAVF
 Ecgr IEGIYAVGDN TGAVELTPVA VAAGRRLSER LFNNKPDEHL DYSNIPTVVF
Humgr VKGIYAVGDV CGKALLTPVA IAAGRKLAHR LFEYKEDSKL DYNNIPTVVF
                     .... ......
      433                                              480
Peagr SQPPIGGVGL TEEQAAEQYG D..IDVFTAN FRPMKATLSG LPDRVFMKLI
Psagr SLPNIGTVGL TEEEALS..A GHKVKIFESR FRPMKLTLTD DQEKTLMKLV
 Ecgr SHPPIGTVGL TEPQAREQYG DDQVKVYKSS FTAMYTAVTT HRQPCRMKLV
Humgr SHPPIGTVGL TEDEAIHKYG IENVKTYSTS FTPMYHAVTK RKTKCVMKMV
                                  ....
      481                                              530
Peagr VSAETNVVLG LHMCGEDAAE IAQGFAVGIK AGLTKADFDA TVGIHPTAAE
Psagr VDAHDDRVLG CHMVGAEAGE ILQGIAVAMK AGATKQAFDE TIGIHPTAAE
 Ecgr CVGSEEKIVG IHGIGFGMDE MLQGFAVALK MGATKKDFDN TVAIHPTAAE
Humgr CANKEEKVVG IHMQGLGCDE MLQGFAVAVK MGATKADFDN TVAIHPTSSE
                                                    ......
      531                            562
Peagr EFVTMRTPTR KVRKNQASQG KSDSKAKAVA GS
Psagr EFVTLRTPTR
 Ecgr EFVTMR
Humgr ELVTLR
      ....
```

FIGURE 9. Alignments of pea, *Pseudomonas aeruginosa*, *E. coli*, and human glutathione reductases. Regions of homology between all four peptides are shaded. The redox active disulfide bridge domain is boxed. Residues shown to be important in binding GSSG in the human enzyme are indicated by dots beneath the alignment. The conserved Arg residues required for NADPH binding are indicated by *. The Cys residue responsible for the formation of the intersubunit disulfide bridge in the human enzyme is indicated by ^.

TABLE 5
Amino Acid Homologies Between Glutathione Reductases from Different Sources

	N. tabacum	*Ps. aeruginosa*	*E. coli*	*H. sapiens*
P. sativum	90 (83)	70 (54)	62 (43)	59 (40)
N. tabacum	—	71 (52)	62 (42)	59 (40)
Ps. aeruginosa	—	—	64 (45)	65 (44)
E. coli	—	—	—	69 (55)

Note: Comparisons were made between the mature peptide sequences. Homologies are shown as the percentage similarity, with the percentage identity in parentheses.

```
    1
Pea GLTSSFLLLRFRESFAMGKSYPTVSPDYQKAIEKAKRKLRGFIAEKKCAP
Ata ...DFIILPRNKSLAKMTKNYPTVSEDYKKAVEKCRRKLRGLIAEKNCAP
Tea ....................FASDPDELKSAREDIKELLNTKVATPQQAQ
Spi...................GSYPTVHENYQKSIEK

    51
Pea LILRLAWHSAGTFDSKTKTGGPFGTIKHQAELAHGANNGLDIAVRL....
Ata IMVRLAWHSAGTFDCQSRTGGPFGTMRFDAEQAHGANSGIHIALRL....
Tea EV.........FYDTK...........HXAVRDHAANAGLVNALKWHIFW
        97
Pea ....LEPIKEQFPIVSYADFYQLAGVVAVEITGGPEVPFHP.GREDKPEP
Ata ....LDPIREQFPTISFADFHQLAGVVAVEVTGGPDIPFHP.GREDKPQP
Tea VDERVVPXKGSVP.IAFGGTEEPAAYGELVSIGGLNPSVNKYGRVDTSSP

       142
Pea ...PPEGRLPDATKGSDHLRDVFGKAMGLSVVQDIVALSGGHTIGAAHKE
Ata ...PPEGRLPDATKGCDHLRDVFAKQMGLS.DKDIVALSGAHTLGRCHKD
Tea DQXPEEGRLXDAA.................XNEIVALSGAHTLGRSRPE

    189
Pea RSGFEGPWTSNPLIFDNSYFTELLTGEKDGLLQLPSDKALLTDSVFRPLV
Ata RSGFEGAWTSNPLIFDNSYFKELLSGEKEGLLQLVSDKALLDDPVFRPLV
Tea RSGWGEPXT.... FDNSYFK.LLETIK......................

    239
Pea EKYAADED.....VFFADYAEAHLKLSELGFAEA
Ata EKYAADED.....AFFADYAEAHMKLSELGFAD.
Tea ..YAADQDEFFKWVILGHSEERNVKGSNFGWNGTTF
```

FIGURE 10. Alignment of the deduced amino acid sequences for ascorbate peroxidase from pea and *A. thaliana* with peptide sequences from tea and the N terminus of the spinach enzyme.

ically distinct cytosolic and plastidial isoforms of the enzyme exist in both pea and spinach implies that the proteins should differ markedly, and this certainly appears to be the case if one considers the relationship between the tea plastidial ascorbate peroxidase and the pea cytosolic isoform. However, it should also be remembered that the antiserum raised against the plastidial ascorbate peroxidase from tea also recognizes a cytosolic isoform in this species.[21] The evidence that the *A. thaliana* cDNA encodes the plastidial form of the enzyme rests on the fact that it was isolated using monoclonal antibodies raised against the spinach chloroplast isoform; however, this cDNA lacks a transit peptide sequence, which would be required for targeting the protein to the chloroplast.

By comparing the amino acid sequence of tea ascorbate peroxidase with those from spinach, pea, and *A. thaliana,* it is clear that there are two distinct classes of ascorbate peroxidases, one of which is probably only found in the chloroplast (represented by the tea isoform), while the other form is found in both chloroplast and cytosol. If this is the case, it may be postulated that there are at least three ascorbate peroxidase genes, two of which, as we have proposed for GR, are encoded by nearly identical genes which differ only in the presence or absence of plastidial transit peptide. This question will only be resolved when clones have been isolated which correspond to all of the compartment-specific isoforms from a single species. Nevertheless, the availability of cDNAs for ascorbate peroxidase will open the way to manipulating the levels of expression of the enzyme in transgenic plants, as has already been attempted for the other key enzyme of the ascorbate-glutathione cycle, glutathione reductase.

C. MANIPULATION OF GLUTATHIONE REDUCTASE EXPRESSION IN TRANSGENIC PLANTS

The first attempts to manipulate glutathione reductase activities by overexpressing an introduced glutathione reductase gene in transgenic tobacco plants were carried out with the *gor* gene isolated from *E. coli.* Tobacco plants transformed via *Agrobacterium tumefaciens* with the *gor* gene under the control of the cauliflower mosaic virus (CaMV) 35S promoter and nopaline synthase terminator sequences were found to have between 1 and 3.5 times the GR activity of the controls.[78] No differences were detected in sensitivity to ozone, even in the highest-expressing plants, but a slightly increased tolerance to paraquat (which is reduced by PSI and generates superoxide and H_2O_2 within the chloroplasts) was detected in the transgenics. The significance of this apparent increased tolerance is unclear, since, lacking a chloroplast leader sequence, the transgene would be expected to be expressed only in the cytosol, and consequently the levels of GR within the chloroplast would not be expected to change. A more detailed analysis of overexpression of the *gor* gene in transgenic tobacco was carried out by Foyer et al.[52] The GR activities in these plants ranged from two to ten times that of the control plants. The increased GR activity did not appear to have a significant effect on either the total glutathione content or the ratio of GSH to GSSG under normal or oxidatively stressed conditions; however, there was a decrease in the reduction state of the ascorbate pool in the control plants compared with that seen in the transgenics when exposed to oxidative stress (paraquat treatment in the light), which suggests that elevated cytosolic GR activities may contribute to the chloroplast ascorbate-glutathione cycle, possibly via active transport of ascorbate across the chloroplast envelope.

Increased GR activities in the cytosol of transgenic plants which express the *E. coli* enzyme have so far resulted in only slight increases in tolerance of oxidative stress conditions. We have attempted to obtain increased GR activities in transgenic tobacco plants by overexpression of the pea enzyme. We have generated 29 independent transgenic tobacco lines containing the pea GR cDNA under the control of the CaMV 35S promoter. Preliminary analyses of GR expression in these lines has confirmed that the introduced GR gene is transcriptionally active, yielding pea-GR mRNA levels in transgenic tobacco greater than fivefold higher than those of the native GR in pea. This increased level of GR transcript is reflected in a two- to fivefold increase in GR activity. We are currently screening these lines for individuals with high GR activities for further analysis of their responses to a variety of oxidative stresses.

V. SUMMARY AND FUTURE PROSPECTS

A large body of evidence has been accumulated over the last decade or so which demonstrates a clear role for glutathione reductase and ascorbate peroxidase in protecting plants against damage associated with oxyradical production under "normal" conditions. However, it is still difficult to assess the actual benefits to the plant of elevated levels of either of these enzymes under conditions of oxidative stress. The production of transgenic plants which overexpress the *E. coli* GR in the cytosol has yielded equivocal results in terms of elevated stress tolerance, and there is as yet insufficient evidence that cytosolic overexpression of the gene from a plant source will provide any greater protection.

By designing appropriate gene constructs, it will be possible to engineer plants which overexpress ascorbate-glutathione cycle enzymes in specific cell compartments (chloroplast and mitochondrion) and thus to target the sites at greatest risk of oxyradical production. It is known that the *E. coli* enzyme is rapidly inactivated in the presence of low (1 μM) NADPH concentrations.[79] The concentration of NADPH in the stroma in the light is about 300 μM,[80] and consequently the *E. coli* enzyme would be expected to be rapidly inactivated in the chloroplast, especially under conditions of high light (i.e., when the demand for GR activity would be greatest). Consequently, the greatest benefit may be achieved by overexpressing the plant enzyme.

The availability of cDNAs for ascorbate peroxidase now make it possible to overexpress both enzymes in the same background. A comparison of transgenic plants which express either GR or ascorbate peroxidase at significantly higher levels than control plants with transgenics which overexpress both enzymes will provide a valuable insight into the significance of these enzymes in the overall protection mechanism.

REFERENCES

1. **Bannister, J. V., Bannister, W. H., and Rotilio, G.,** Aspects of the structure, function and applications of superoxide dismutase, *CRC Crit. Rev. Biochem.*, 22, 111, 1987.
2. **Bowler, C., Van Montagu, M., and Inze, D.,** Superoxide dismutase and stress tolerance, *Annu. Rev. Plant Physiol. Plant Mol. Biol.*, 43, 83, 1992.
3. **Foyer, C. H. and Halliwell, B.,** The presence of glutathione and glutathione reductase in chloroplasts: a proposed role in ascorbic acid metabolism, *Planta*, 133, 21, 1976.
4. **Nakano, Y. and Asada, K.,** Spinach chloroplasts scavenge hydrogen peroxide on illumination, *Plant Cell Physiol.*, 21, 1295, 1980.
5. **Bielawski, W. and Joy, K. W.,** Reduced and oxidised glutathione and glutathione-reductase activity in tissues of *Pisum sativum*, *Planta*, 169, 267, 1986.
6. **Dalton, D. A., Russell, S. A., Hanus, F. J., Pascoe, G. A., and Evans H. J.,** Enzymatic reactions of ascorbate and glutathione that prevent peroxide damage in soybean root nodules, *Proc. Natl. Acad. Sci. U.S.A.*, 83, 3811, 1986.
7. **Klapheck, S., Zimmer, I., and Cosse, H.,** Scavenging of hydrogen peroxide in the endosperm of *Ricinus communis* by ascorbate peroxidase, *Plant Cell Physiol.*, 31, 1005, 1990.
8. **Loewus, F. A.,** Ascorbic acid and its metabolic products, in *The Biochemistry of Plants*, Preiss, J., Ed., Academic Press, New York, 1988, 85.
9. **Anderson, J. W., Foyer, C. H., and Walker, D. A.,** Light-dependent reduction of hydrogen peroxide by intact spinach chloroplasts, *Biochim. Biophys. Acta*, 724, 69, 1983.
10. **Law, M. Y., Charles, S. A., and Halliwell, B.,** Glutathione and ascorbic acid in spinach *(Spinacia oleracea)* chloroplasts. The effect of hydrogen peroxide and paraquat, *Biochem. J.*, 210, 899, 1988.
11. **Anderson, J. W., Foyer, C. H., and Walker, D. A.,** Light-dependent reduction of dehydroascorbate and uptake of exogenous ascorbate by spinach chloroplasts, *Planta*, 158, 442, 1983.
12. **Foyer, C. H., Lelandais, M., Edwards, E. A., and Mullineaux, P. M.,** The role of ascorbate in plants, interactions with photosynthesis, and regulatory significance, in *Plant Physiology*, Pell, E. and Steffen, K., Eds., 1991.
13. **Miyake, C. and Asada, K.,** Thylakoid-bound ascorbate peroxidase in spinach chloroplasts and photoreduction of its primary oxidation product monodehydroascorbate radicals in thylakoids, *Plant Cell Physiol.*, 33, 541, 1992.
14. **Fryer, M.,** The antioxidant effects of thylakoid Vitamin E (α-tocopherol), *Plant Cell Environ.*, 15, 381, 1992.
15. **Foyer, C., Furbank, R., Harbinson, J., and Horton, P.,** The mechanisms contributing to photosynthetic control of electron transport by carbon assimilation in leaves, *Photosynth. Res.*, 25, 83, 1990.
16. **Demmig-Adams, B. and Adams, W. W.,** Photoprotection and other responses of plants to high light stress, *Annu. Rev. Plant Physiol. Plant. Mol. Biol.*, 43, 599, 1992.
17. **Demmig-Adams, B., Winter, K., Kruger, A., and Czygan, F.-C.,** Light stress and photoprotection related to the carotenoid zeaxanthin in higher plants, in *Photosynthesis*, Briggs, W. R., Ed., Alan R. Liss, New York, 1989, 375.
18. **Noctor, G., Rees, D., Young, A., and Horton, P.,** The relationship between zeaxanthin, energy-dependent quenching of chlorophyll fluorescence, and trans-thylakoid pH gradient in isolated chloroplasts, *Biochim. Biophys. Acta*, 1057, 320, 1991.
19. **Kaiser, W. M.,** Reversible inhibition of the Calvin cycle and activation of the oxidative pentose phosphate cycle by hydrogen peroxide, *Planta*, 145, 377, 1979.
20. **Tanaka, K., Otsubo, T., and Kondo, N.,** Participation of hydrogen peroxide in the inactivation of Calvin-cycle SH enzymes in SO_2-fumigated spinach leaves, *Plant Cell. Physiol.*, 23, 1009, 1982.

21. **Chen, G.-X. and Asada, K.**, Ascorbate peroxidase in tea leaves: occurrence of two isozymes and the differences in their enzymatic and molecular properties, *Plant Cell Physiol.*, 30, 987, 1989.
22. **Asada, K.**, Ascorbate peroxidase — a hydrogen peroxide-scavenging enzyme in plants, *Physiol. Plant.*, 85, 235, 1992.
23. **Nakano, Y. and Asada, K.**, H_2O_2 is scavenged by an ascorbate-specific peroxidase in spinach chloroplasts, *Plant Cell Physiol.*, 22, 867, 1981.
24. **Criqui, M. C., Jamet, E., Parmentier, Y., Marbach, J., Durr, A., and Fleck, J.**, Isolation and characterisation of a plant cDNA showing homology to animal glutathione peroxidases, *Plant Mol. Biol.*, 18, 623, 1992.
25. **Nakano, Y. and Asada, Y.**, Purification of ascorbate peroxidase from spinach chloroplasts: its inactivation in ascorbate-depleted medium and reactivation by monodehydroascorbate radical, *Plant Cell Physiol.*, 28, 131, 1987.
26. **Tanaka, K., Takeuchi, E., Kubo, A., Sakaki, T., Haraguch, K., and Kawamura, Y.**, Two immunologically different isozymes of ascorbate peroxidase from spinach leaves, *Arch. Biochem. Biophys.*, 286, 371, 1991.
27. **Mittler, R. and Zilinskas, B. A.**, Purification and characterisation of pea cytosolic ascorbate peroxidase, *Plant Physiol.*, 97, 962, 1991.
28. **Fuchs, J. A. and Warner, H. R.**, Isolation of an *Escherichia coli* mutant deficient in glutathione synthesis, *J. Bacteriol.*, 124, 140, 1975.
29. **Klapheck, S.**, Homoglutathione: isolation, quantification and occurence in legumes, *Physiol. Plant.*, 74, 727, 1988.
30. **Ziegler, D. M.**, Role of reversible oxidation-reduction of enzyme thiol-disulphides in metabolic regulation, *Annu. Rev. Biochem.*, 54, 305, 1985.
31. **Dron, M., Clouse, S. D., Dixon, R. A., Lawton, M. A., and Lamb, C. J.**, Glutathione and fungal elicitor regulation of a plant defense gene promoter in electroporated protoplasts, *Proc. Natl. Acad. Sci. U.S.A.*, 85, 6738, 1988.
32. **Wingate, V. P. M., Lawton, M. A., and Lamb, C. J.**, Glutathione elicits a massive and selective induction of plant defense genes, *Plant Physiol.*, 87, 206, 1988.
33. **Edwards, R., Blount, J. W., and Dixon, R. A.**, Glutathione and elicitation of the phytoalexin response in legume cell cultures, *Planta,* 184, 403, 1991.
34. **Bradley, D. J., Klellbom, P., and Lamb, C. J.**, Elicitor and wound-induced oxidative cross-linking of a proline-rich plant cell wall protein: a novel, rapid defense response, *Cell,* 70, 21, 1992.
35. **Smith, I. K., Polle, A., and Rennenberg, H.**, Glutathione, in *Stress Responses in Plants: Adaptation and Acclimation Mechanisms,* Alscher, R. G. and Cumming, J. R., Eds., Wiley-Liss, NY, 1990, 201.
36. **Gillham, D. J. and Dodge, A. D.**, Hydrogen-peroxide-scavenging systems within pea chloroplasts, *Planta,* 167, 246, 1986.
37. **Edwards, E. A., Rawsthorne, S., and Mullineaux, P. M.**, Subcellular distribution of multiple isoforms of glutathione reductase in leaves of pea (*Pisum sativum* L.), *Planta,* 180, 278, 1990.
38. **Gerbling, K. P., Kelly, G. J., Fischer, K. H., and Latzko, E.**, Partial purification and properties of soluble ascorbate peroxidases from tea leaves, *J. Plant Physiol.*, 115, 59, 1984.
39. **Krauth-Siegel, R. L., Blattespiel, R., Saleh, M., Schiltz, E., Schirmer, R. H., and Untucht-Grau, R.**, Glutathione reductase from human erythrocytes, *Eur. J. Biochem.*, 121, 259, 1982.
40. **Thieme, R., Pai, E. F., Schirmer, R. H., and Schultz, G. E.**, Three-dimensional structure of glutathione reductase at 2 A resolution, *J. Mol. Biol.*, 152, 763, 1981.
41. **Schultz, G. E., Schirmer, R. H., and Pai, E. F.**, FAD-binding site of glutathione reductase, *J. Mol. Biol.*, 160, 287, 1982.

42. **Pai, E. F. and Schultz, G. E.,** The catalytic mechanism of glutathione reductase as derived from X-ray diffraction analyses of reaction intermediates, *J. Biol. Chem.*, 258, 1752, 1983.
43. **Karplus, P. A. and Schultz, G. E.,** Refined structure of glutathione reductase at 1.54A resolution, *J. Mol. Biol.*, 195, 701, 1987.
44. **Karplus, P.A., Pai, E. F., and Schultz, G. E.,** A crystallographic study of the glutathione binding site of glutathione reductase at 0.3-nm resolution, *Eur. J. Biochem.*, 178, 693, 1989.
45. **Mahan, J. R. and Burke, J. J.,** Purification and characterisation of glutathione reductase from corn mesophyll chloroplasts, *Physiol. Plant.*, 71, 352, 1987.
46. **Kalt-Torres, W., Burke, J. J., and Anderson, J. M.,** Chloroplast glutathione reductase: purification and properties, *Physiol. Plant.*, 61, 271, 1984.
47. **Connell, J. P. and Mullet, J. E.,** Pea chloroplast glutathione reductase: purification and properties, *Plant Physiol.*, 82, 351, 1986.
48. **Bielawski, W. and Joy, K. W.,** Properties of glutathione reductase from chloroplasts and roots of pea, *Phytochemistry*, 25, 2261, 1986.
49. **Halliwell, B. and Foyer, C. H.,** Properties and physiological function of a glutathione reductase purified from spinach leaves by affinity chromatography, *Planta*, 139, 9, 1978.
50. **Anderson, J. V., Hess, J. L., and Chevone, B. I.,** Purification, characterisation and immunological properties of two isoforms of glutathione reductase from eastern white pine needles, *Plant Physiol.*, 94, 1402, 1990.
51. **Guy, C. L. and Carter, J. V.,** Characterisation of partially purified glutathione reductase from cold-hardened and nonhardened spinach leaf tissue, *Cryobiology*, 21, 454, 1984.
52. **Foyer, C. H., Lelandais, M., Galap, C., and Kunert, K. J.,** Effects of elevated cytosolic glutathione reductase activity on the cellular glutathione pool and photosynthesis in leaves under normal and stress conditions, *Plant Physiol.*, 97, 863, 1991.
53. **Scrutton, N. S., Berry, A., and Perham, R. N.,** Purification and characterisation of glutathione reductase encoded by a cloned and overexpressed gene in *E. coli*, *Biochem. J.*, 245, 875, 1987.
54. **Tanaka, K., Suda, Y., Kondo, N., and Sugahara, K.,** O_3 tolerance and the ascorbate-dependent H_2O_2 decomposing system in chloroplasts, *Plant Cell Physiol.*, 26, 1425, 1985.
55. **Tanaka, K., Saji, H., and Kondo, N.,** Immunological properties of spinach glutathione reductase and inductive biosynthesis of the enzyme with ozone, *Plant Cell Physiol.*, 29, 637, 1988.
56. **Mehlhorn, H., Cottam, D. A., Lucas, P. W., and Wellburn, A. R.,** Induction of ascorbate peroxidase and glutathione reductase activities by interactions of mixtures of air pollutants, *Free Rad. Res. Commun.*, 3, 193, 1987.
57. **Mehlhorn, H. and Wellburn, A. R.,** Stress ethylene formation determines plant sensitivity to ozone, *Nature (London)* 327, 417, 1987.
58. **Mehlhorn, H.,** Ethylene-promoted ascorbate peroxidase activity protects plants against hydrogen peroxide, ozone and paraquat, *Plant Cell Environ.*, 13, 971, 1990.
59. **Madamanchi, N. R., Anderson, J. V., Alscher, R. G., Cramer, C. L., and Hess, J. L.,** Purification of multiple forms of glutathione reductase from pea (*Pisum sativum* L.) seedlings and enzyme levels in ozone-fumigated pea leaves, *Plant Physiol.*, 100, 138, 1992.
60. **Cakmak, I. and Marschner, H.,** Magnesium deficiency and high light intensity enhance activities of superoxide dismutase, ascorbate peroxidase and glutathione reductase in bean leaves, *Plant Physiol.*, 98, 1222, 1992.
61. **Schoner, S. and Krause, H.,** Protective systems against active oxygen species in spinach: response to cold acclimation in excess light, *Planta*, 180, 383, 1990.
62. **Smith, I. K., Vierheller, T. L., and Thorne, C. A.,** Properties and functions of glutathione reductase in plants, *Physiol. Plant.*, 77, 449, 1989.

63. **Anderson, J. V., Chevone, B. I., and Hess, J. L.**, Seasonal variation in the antioxidant system of eastern white pine needles, *Plant Physiol.*, 98, 501, 1992.
64. **Kuroda, H., Sagisaka, S., Asada, M., and Chiba, K.**, Peroxide-scavenging systems during cold-acclimation of apple callus in culture, *Plant Cell Physiol.*, 32, 635, 1991.
65. **Jahnke, L. S., Hull, M. R., and Long, S. P.**, Chilling stress and oxygen metabolising enzymes in *Zea mays* and *Zea diploperennis, Plant Cell Environ.*, 14, 97, 1991.
66. **Foster, J. G. and Hess, J. L.**, Oxygen effects on maize leaf superoxide dismutase and glutathione reductase, *Phytochemistry,* 21, 1527, 1982.
67. **Foster, J. G. and Hess, J. L.**, Responses of superoxide dismutase and glutathione reductase activities in cotton leaf tissue exposed to an atmosphere enriched in oxygen, *Plant Physiol.*, 66, 482, 1980.
68. **Dhindsa, R. S.**, Drought stress, enzymes of glutathione metabolism, oxidation injury, and protein synthesis in *Tortula ruralis, Plant Physiol.*, 95, 648, 1991.
69. **Greer, S. and Perham, R. N.**, Glutathione reductase from *Escherichia coli:* cloning and sequence analysis of the gene and relationship to other flavoprotein disulphide oxidoreductases, *Biochemistry,* 25, 2736, 1986.
70. **Tutic, M., Lu, X., Schirmer, R. H., and Werner, D.**, Cloning and sequencing of mammalian glutathione reductase cDNA, *Eur. J. Biochem.*, 188, 523, 1990.
71. **Perry, A. C. F., Bhriain, N. Ni., Brown, N. L., and Rouch, D. A.**, Molecular characterisation of the *gor* gene encoding glutathione reductase from *Pseudomonas aeruginosa:* determinants of substrate specificity among pyridine nucleotide-disulphide oxidoreductases, *Mol. Microbiol.*, 5, 163, 1991.
72. **Creissen, G., Edwards, E. A., Enard, C., Wellburn, A., and Mullineaux, P.**, Molecular characterisation of glutathione reductase cDNAs from pea (*Pisum sativum* L.), *Plant J.*, 2, 129, 1992.
73. **Scrutton, N. S., Berry, A., and Perham, R. N.**, Redesign of the coenzyme specificity of a dehydrogenase by protein engineering, *Nature (London),* 343, 38, 1990.
74. **von Heijne, G., Stepphuhn, J., and Herrmann, R. G.**, Domain structure of mitochondrial and chloroplast targeting peptides, *Eur. J. Biochem.*, 121, 259, 1989.
75. **Mittler, R. and Zilinskas, B. A.**, Molecular cloning and nucleotide sequence analysis of a cDNA encoding pea cytosolic ascorbate peroxidase, *FEBS Lett.*, 289, 257, 1991.
76. **Kubo, A., Saji, H., Tanaka, K., Tanaka, K., and Kondo, N.**, Cloning and sequencing of a cDNA encoding ascorbate peroxidase from *Arabidopsis thaliana, Plant Mol. Biol.*, 18, 691, 1992.
77. **Chen, G.-X., Sano, S., and Asada, K.**, The amino acid sequence of ascorbate peroxidase from tea has a high degree of homology to that of cytochrome c peroxidase from yeast, *Plant Cell Physiol.*, 33, 109, 1992.
78. **Aono, M., Kubo, A., Saji, H., Natori, T., Tanaka, K., and Kondo, N.**, Resistance to active oxygen toxicity of transgenic *Nicotiana tabacum* that expresses the gene for glutathione reductase from *Escherichia coli, Plant Cell Physiol.*, 32, 691, 1991.
79. **Arscott, D. A., Drake, D. M., and Williams, C. H., Jr.**, Inactivation-reactivation of two-electron reduced *Escherichia coli* glutathione reductase involving a dimer-monomer equilibrium, *Biochemistry,* 28, 3591, 1989.
80. **Heineke, D., Riens, B., Grosse, H., Hoferichter, P., Peter, U., Flugge, U.-I., and Heldt, H. W.**, Redox transfer across the inner chloroplast envelope membrane, *Plant Physiol.*, 95, 1131, 1991.
81. **Creissen, G. P., Edwards, E. A., and Mullineaux, P. M.**, unpublished data.
82. **Enard, C.**, unpublished data.

Chapter 14

TOLERANCE TO HERBICIDES AND AIR POLLUTANTS

Kiyoshi Tanaka

TABLE OF CONTENTS

0-8493-5443-9/94/$0.00 + $.50

I. INTRODUCTION

It is widely known that higher plants show different tolerance to environmental stresses. For example, radish plants show high sensitivity to O_3[1] and, therefore, are used as ozone-indicating plants. On the other hand, *Polygonum cuspidatum* has a high tolerance to SO_2 and is one of the dominant species of natural vegetation found at volcanoes.[2] It is conceivable that it may play a role as a natural air cleaner. The tolerance of higher plants to environmental stresses is controlled by several factors. For instance, the tolerance of higher plants to air pollutants such as O_3, SO_2, and NO_2 may be decided by the capacity for physiological detoxification, the sensitivity of stomatal closure, and structural rigidity. The tolerance of higher plants to herbicides seems to resemble that to air pollutants. It influences the tolerance of some weeds to paraquat, affecting how much of the herbicide they absorb into leaves.[3] Furthermore, plant tolerance to paraquat is controlled by the physiological detoxifying capacity.[4] To clarify the tolerance mechanism of higher plants to environmental stresses, it is very important to know how plants are damaged by environmental stresses. We have investigated the effects of air pollutants on higher plants and the mechanism of tolerance of higher plants to air pollutants. We have found a common mechanism of damage occurrence and tolerance of higher plants between the air pollutant, SO_2, and the herbicide, paraquat. We are now developing air-pollutant-tolerant plants which also show high resistance to some species of herbicides. In this chapter, we discuss our investigations on the relationship between active oxygen and air pollutant toxicity, and also show the similarities between the toxicities of air pollutants and herbicides.

II. EFFECTS OF AIR POLLUTANTS ON HIGHER PLANTS

Higher plants suffer severe damage from air pollutants. Typical injuries are bleaching of leaves and inhibition of photosynthesis. Most investigators have been interested in clarification of these phenomena. Within a few hours after plants have been fumigated with a high concentration of air pollutants such as O_3 and SO_2, the green color of the leaves gradually disappears.[5,6] Even when O_3 or SO_2 was removed after plants had been fumigated, the pigments of the fumigated leaves were destroyed, but only in the light. Leaf segments were collected from plants which had been fumigated with SO_2 or O_3 and were incubated in petri dishes under light. The pigments of leaf segments were gradually destroyed in the same manner as the pigments of leaves of whole plants fumigated with SO_2 or O_3 were decomposed. This light-dependent destruction of the pigments was inhibited by 3-(3′,4′-dichlorophenyl)-1,1-dimethylurea (DCMU). Since the destruction of pigments was induced only in the presence of oxygen, it was conceivable that active oxygen might participate in the destruction of pigments. To confirm this possibility, the effects of several active oxygen scavengers on the destruction of the pigments were examined. Ascorbate and tiron, which scavenge superoxide radicals, depressed the destruction of chlorophyll and carotenoid. The inactivation of superoxide dismutase (SOD) in air pollutant-fumigated leaves may be one of the causes for the accumulation of active oxygen during air pollutant fumigation.

When higher plants were fumigated with a low concentration of O_3 and SO_2, which gave no visible injury on plant leaves, the activities of enzymes which may protect against active oxygen increased (Figure 1).[7-9] This increase was accompanied by an increase in the amount of enzyme proteins.[9] Evidence shows several enzymes that scavenge active oxygen species were induced, in many organisms, under conditions in which active oxygen increased.[10-14] Therefore, the increase of the activities of active-oxygen-scavenging enzymes indicates that active oxygen, accumulated in O_3-or SO_2-fumigated leaves, induces the biosynthesis of active-oxygen-defending enzymes.

Of the physiological influences of air pollutants on higher plants, the inhibition of photosynthesis is one of the most conspicuous phenomena. Most investigators thought that the inhibition of photosynthesis by SO_2 might be caused by the competitive inhibition of ribulose-1,5-bisphosphate carboxylase (RuBPC) by SO_2 in chloroplasts. However, this hypothesis is merely based on the *in vitro* inhibition of isolated RuBPC by the addition of sulfite ion.[15] To clarify the mechanism of inhibition of photosynthesis with SO_2, the leaf segments were periodically sampled during SO_2 fumigation and coupled to the measurements of Calvin-Benson cycle enzymes.[16] Accompanying the inhibition of photosynthetic carbon fixation, of the ten enzymes measured, three thiol-regulated (−SH) enzymes — fructose-1,6-bisphosphatase (FBPase), ribulose-5-phosphate kinase (Ru5PK), and NADP-glyceraldehyde-3-phosphate dehydrogenase (NADP-GAPD) — were inactivated (Figure 2). Interestingly, RuBPC was not affected by SO_2 fumigation. When gaseous SO_2 is incorporated into chloroplasts, it is conceivable that a sulfite-mediated chain reaction may occur there and that active oxygen may accumulate.[17] H_2O_2 is a final and comparatively stable product of active oxygen and is a strong inhibitor of photosynthetic carbon fixation.[18,19] It was, therefore, anticipated that the accumulation of H_2O_2 in SO_2-fumigated leaves might cause a decrease of photosynthesis. The accumulation of hydrogen peroxide in SO_2-fumigated leaves was confirmed experimentally.[20] When isolated chloroplasts were incubated with sulfite ion in the light, H_2O_2 accumulated. When chloroplasts isolated from SO_2-fumigated leaves were illuminated, they produced H_2O_2, and at the same time, thiol-regulated enzymes such as FBPase, Ru5PK, and NADP-GAPD were specifically inactivated. Since it was confirmed that thiol-regulated enzymes were inactivated not by sulfite itself, but by H_2O_2, it may be that the inactivation of enzymes in SO_2-fumigated leaves was caused by H_2O_2. Of the three SH enzymes, FBPase has the lowest specific activity in

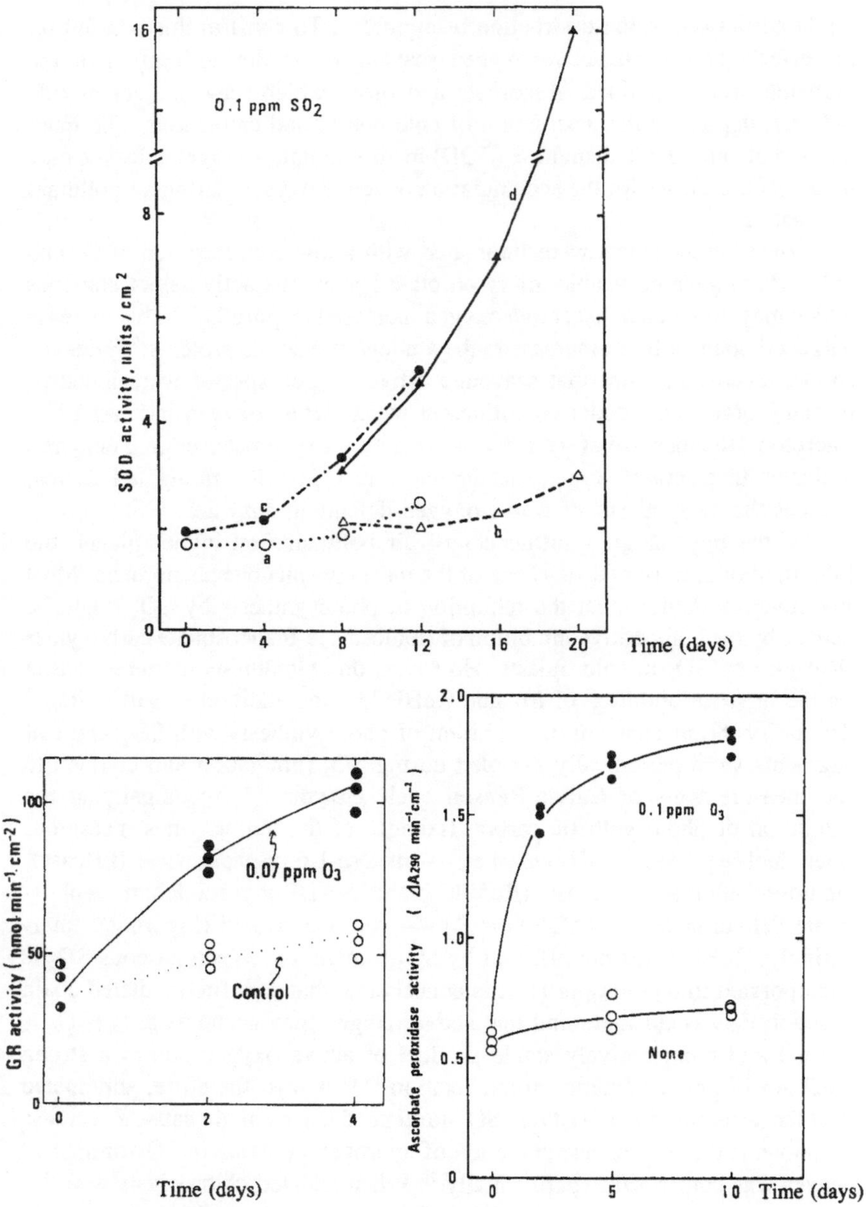

FIGURE 1. Effects of a low concentration of air pollutants on active-oxygen-scavenging enzymes in higher plants. Upper: effect of 0.1 ppm SO_2 on superoxide dismutase in poplar leaves. Closed circles and triangles show SO_2-fumigated plants. Open circles and triangles show control plants. Lower left: effects of O_3 on glutathione reductase; lower right: ascorbate peroxidase in spinach leaves. Closed circles show O_3-fumigated leaves and open circles show control leaves.

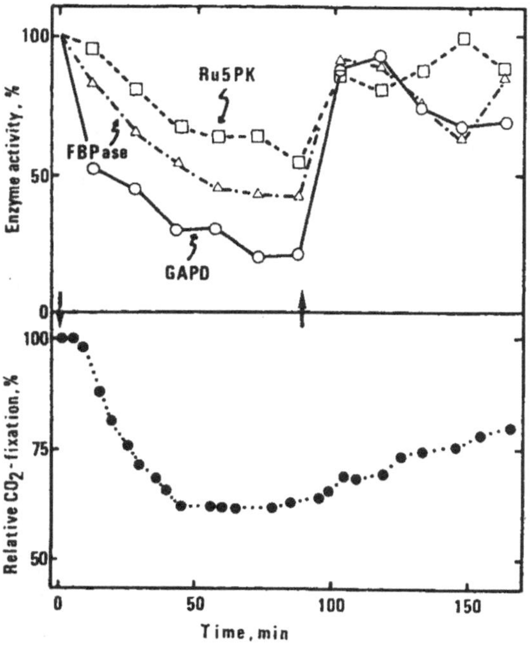

FIGURE 2. Relationship between photosynthesis and activities of chloroplast SH enzymes during a 2.0 ppm SO_2 fumigation. The light-dependent CO_2 uptake of spinach plants in a pot was measured using an assimilation chamber (lower) and concomitantly SH enzymes of leaves were periodically measured (upper). At the indicated times, SO_2 fumigation was started (↓) and stopped (↑).

chloroplasts, and it was found that the inhibition of CO_2 fixation caused by adding H_2O_2 to intact chloroplasts might be due to the decrease in FBPase activity.[21] When the substrate levels of the photosynthetic carbon cycle were examined in spinach leaves during SO_2 fumigation, an increase in fructose-1,6-bisphosphate (FBP) and a decrease in fructose-6-phosphate (F6P) were observed, while the level of ribulose-1,5-bisphosphate (RuBP) did not change and 3-phosphoglyceric acid (PGA) rapidly decreased (Figure 3).[22] These results indicate that the rate-determining enzyme in the inhibition of photosynthesis with SO_2 fumigation is FBPase. The reasons for H_2O_2 accumulation in SO_2-fumigated leaves may be as follows:

1. Active oxygen accumulates in chloroplasts by a sulfite-mediated chain reaction.
2. The decrease of photosynthetic carbon fixation stimulates the electron flow to O_2 by overreduction.
3. The scavenging system of active oxygen species is impaired by SO_2 fumigation.
4. The integrity of the thylakoid membrane is damaged and the electron flow to O_2 becomes more active.

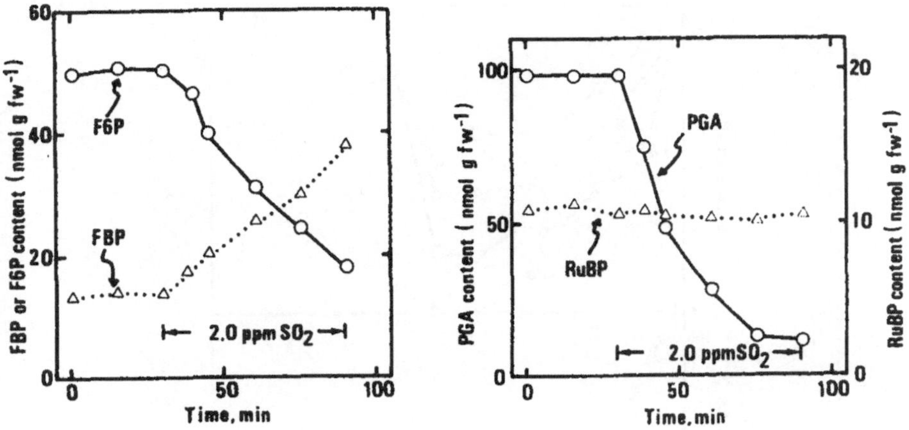

FIGURE 3. Effects of 2.0 ppm SO_2 on FBP, F6P, RuBP, and PGA levels in spinach leaves.

III. TOLERANCE TO AIR POLLUTANTS AND ACTIVE-OXYGEN-SCAVENGING SYSTEM IN HIGHER PLANTS

When higher plants were fumigated with a low concentration of SO_2 or O_3, several enzymes that scavenge active oxygen were induced. In poplar plants fumigated for several days with 0.1 ppm SO_2,[7] SOD activity increased severalfold. Since the increased SOD was sensitive to cyanide treatment, the induced enzyme was shown to be Cu,Zn SOD. The plants which had the increased SOD exhibited a greater tolerance to a high concentration of SO_2 than the control plants (Figure 4). This result indicated that plants elevated the tolerance to SO_2 toxicity by strengthening the active-oxygen-scavenging system. Similarly, ascorbate peroxidase (AP) and glutathione reductase (GR) activities increased a fewfold in 0.07 to 0.1 ppm O_3-fumigated spinach plants, accompanied by increased biosynthesis of the enzyme proteins.[9] The relationship between O_3 tolerance and inherent antioxidative substances in higher plants was investigated by several workers. Plant species and cultivars which contain more antioxidative substances generally have a greater tolerance to O_3. In our experiments, spinach cultivars with higher ascorbate and GSH content had a higher tolerance to O_3,[8] and the tobacco cultivar, Bel W3, with less GR content, was more sensitive to O_3.[23] Although the inconsistency between the active-oxygen-scavenging system and O_3 tolerance in higher plants remains, it seems that plants rich in antioxidative substances are apt to have high tolerance to O_3.

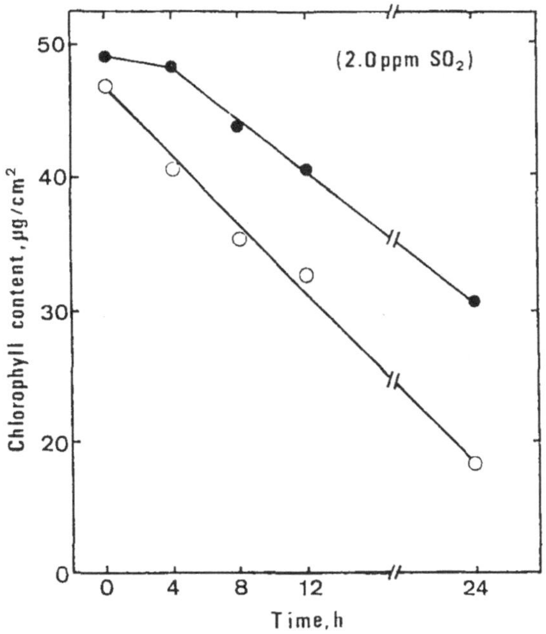

FIGURE 4. Effect of 2.0 ppm SO_2 fumigation on chlorophyll contents of poplar plants containing SOD previously increased by a 0.1 ppm SO_2 fumigation. Poplar plants at 20 d after 0.1 ppm SO_2 fumigation (●) and controls (○) were fumigated with 2.0 ppm SO_2.

IV. TOLERANCE OF PLANTS GENETICALLY ENRICHED IN ACTIVE-OXYGEN-SCAVENGING ENZYMES TO AIR POLLUTANTS AND PARAQUAT

Many workers have a great interest in determining whether high tolerance to air pollutants may be found in plants which are genetically enriched in active-oxygen-scavenging enzymes. There have been controversies surrounding the tolerance to active oxygen toxicity of transgenic plants with elevated levels of SOD.[24,25] When tobacco cultured cells were treated with paraquat, most of them died, but the remaining cells were recultured. The surviving cells were again treated with paraquat. The remaining cells had a greater paraquat tolerance and higher SOD content.[4,27] Whole tobacco plants were regenerated from paraquat-tolerant tobacco callus. The seedlings of these plants were paraquat treated, and paraquat-tolerant seedlings were selected in a manner similar to that of the cultured cells. The final tobacco plants had a greater tolerance to paraquat and a higher SOD content than the control plants.[27] Since SO_2 toxicity was thought to be caused by active oxygen, it

FIGURE 5. Effects of 2.0 ppm SO_2 on paraquat-tolerant (PQ) and control (Control) tobacco plants. The two plants were fumigated for 22 h with 2.0 ppm SO_2.

was anticipated that these paraquat-tolerant, SOD-rich plants would be SO_2 tolerant. As expected, these plants had a higher tolerance to SO_2 than the controls (Figure 5). Interestingly, these paraquat- and SO_2-tolerant plants had more absorption capacity for SO_2 than the controls. Therefore, the plants have the same property as the SO_2-purifying plants. In this experiment, the paraquat- and SO_2-tolerant plants had a higher level of SOD activity than the controls, but the activities of AP and GR in the tolerant plants were similar to those in the controls. It is, however, undeniable that paraquat treatment may induce scavengers against active oxygen in addition to the above three enzymes. *Escherichia coli,* in which SOD was enriched, did not show tolerance to paraquat.[28] Efficient detoxification of active oxygen species requires the removal of H_2O_2 in addition to the superoxide radical. Elevated SOD activity results in an increased rate of H_2O_2 production from superoxide radical dismutation. Since the capacity for removing H_2O_2 is not changed, H_2O_2 itself may then present a new threat. The production of very toxic hydroxyl radicals is greatly increased by the interaction of the superoxide radical and H_2O_2 in the Haber-Weiss reaction. It was recently reported that SOD itself is able to catalyze hydroxyl radical production directly from H_2O_2.[29]

The SOD gene has been introduced into transgenic plants in several laboratories. Although the sensitivities of these transgenic plants to active oxygen were examined, they did not always exhibit increased tolerance. The explanation is as follows. First, only some kinds of SOD isozymes participate in the detoxification of active oxygen. Second, the cellular localization of SOD is most important in the scavenging of active oxygen. It was recently found that a low-level overproduction of Mn SOD in chloroplasts could significantly reduce the amount of cellular damage which would normally occur upon paraquat treatment.[25] The authors proposed that the subcellular location is likely to play an important role in determining whether positive

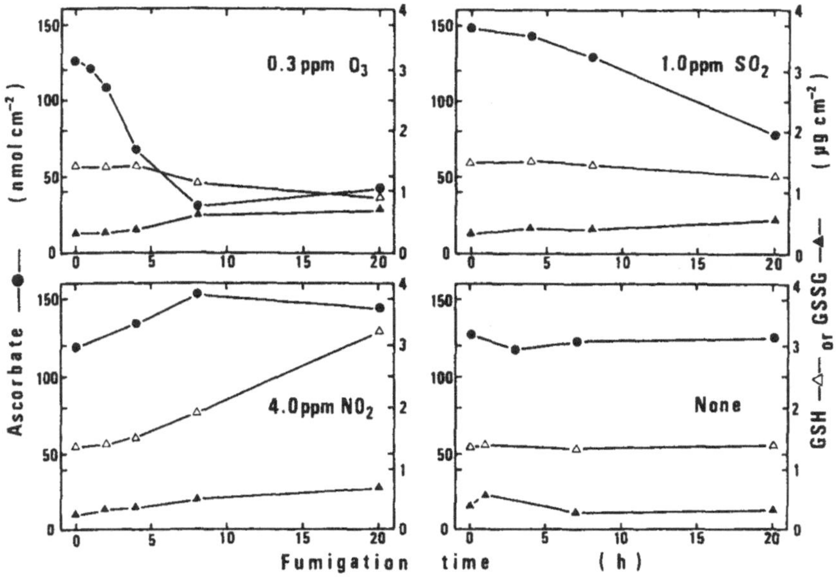

FIGURE 6. Effects of air pollutants on ascorbate and glutathione in spinach leaves. Spinach plants were fumigated with 0.3 ppm O_3, 4.0 ppm NO_2, or 1.0 ppm SO_2, and the levels of ascorbate and glutathione were determined.

or negative effects arise from SOD overproduction. Third, for higher plants to conquer active oxygen toxicity, not only SOD, but also the scavenging system of H_2O_2 is very important, as is the case in *E. coli*.

As described above, GR and AP activities in leaves increased in response to a low concentration of O_3. The following explanation may be given as to why the increases of these enzymes occurred. These enzymes are constituents of the chloroplast H_2O_2-decomposing system. Hence, it is conceivable that by increasing the activities of these enzymes, higher plants defend themselves from active oxygen toxicity. Alternatively, since ascorbate and GSH in higher plant leaves are oxidized by O_3 or SO_2, the increases in the activities of these enzymes as a result of fumigation may contribute to the regeneration of the reductants (Figure 6). The relationship between O_3 tolerance and the contents of active-oxygen-scavenging enzymes in nine cultivars of tobacco plants was examined.[23] One of them, Bel W3, had half the GR activity of the other cultivars and was very sensitive to O_3. There were no differences in SOD and AP in these cultivars. GR has a comparatively lower specific activity than other active-oxygen-scavenging enzymes. There is much evidence that GR activity increases with environmental stresses, which cause an increase in active oxygen.[30-32] GR may be a rate-limiting enzyme and a key enzyme in higher plants for the defense against active oxygen toxicity. Since the GR gene has already been obtained from *E. coli*,[33] a chimeric gene consisting of

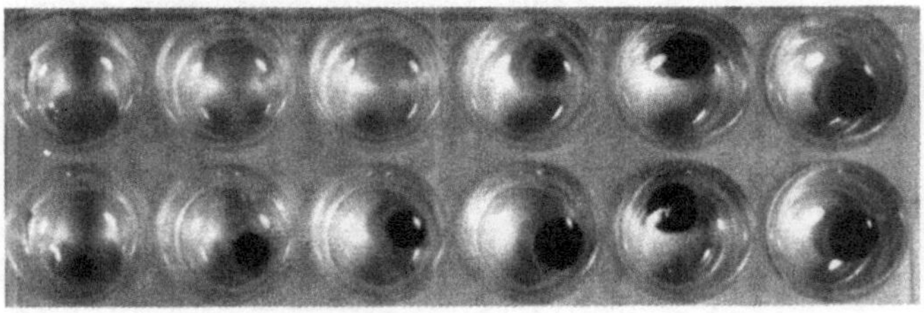

FIGURE 7. Effects of paraquat on leaf discs of transgenic tobacco plants with high GR activity. Leaf discs from transgenic (1415-32) and control (1901-32) plants were incubated on solutions that contained paraquat at the indicated concentrations. The activities of GR in the leaf extracts of 1415-32 and 1901-32 were 0.028 and 0.008 unit/mg protein, respectively.

the GR gene, the 35S promoter of cauliflower mosaic virus, and the terminator sequences of the gene for nopaline synthase was prepared and introduced into tobacco cells via a Ti plasmid vector. Expression of the bacterial gene in transformed plants and their descendants was confirmed by immunochemical analysis. GR activity in leaf extracts varied among transgenic plants, ranging from about 1.0 to 3.5 times the control levels. These transgenic plants exhibited higher tolerance to paraquat in terms of visible foliar damage than the controls (Figure 7).[34] The transgenic plants also showed SO_2 tolerance. Therefore, GR may play an important role in the detoxification of active oxygen in the cytoplasmic matrix of plant cells. Since the transgenic plants, however, did not display O_3 tolerance, there may be differences between SO_2 and O_3 toxicities. O_3 may impair another rate-limiting step of the active-oxygen-scavenging system concomitant with GR. By adding the transit peptide gene to the above chimeric gene, GR activity in chloroplasts was increased. However, regardless of whether GR activity was increased in the cytoplasm or in the chloroplasts, there were no differences in tolerance in transgenic plants to paraquat or SO_2. There is much evidence to show that the active-oxygen-scavenging system may occur in many organs and cellular compartments. Both types of chloroplast and cytoplasmic GR and AP are present in higher plants. In mustard seedlings, these types of GRs were different from each other in physicochemical properties such as the isoelectric point. Also, chloroplastic GR is photoregulated while the cytoplasmic type is not.[35] When leaf and root GRs were compared, the pH optimum of leaf (chloroplastic) GR was more alkaline than the one of root (cytoplasmic) GR. It is well-known that the pH of chloroplasts is more alkaline than the pH of cytoplasm. Although the role of cytoplasmic GR has not yet been clarified, the fact that tobacco plants, enriched in cytoplasmic GR, exhibited the same tolerance to SO_2 and paraquat as those enriched in chloroplastic GR may indicate the great importance of cytoplasmic GR in the active-oxygen-scavenging system. Chloroplasts have developed the active-oxygen-scavenging system because much production of active oxygen is unavoidable in chloroplasts in which light-dependent electron transport reactions occur. Since organelles, except for chloroplasts, seem to be very sensitive to active oxygen toxicity, it is conceivable that the production of active oxygen in such organelles beyond a limited concentration may cause greater toxicity than in the chloroplasts. An increase in the active-oxygen-defense system in the cytoplasm is more effective for the maintenance of plant life than that in chloroplasts. Unexpectedly, the transgenic plants in which GR activity was enriched did not show increased tolerance to O_3 compared to the controls. Tobacco plants regenerated from paraquat-tolerant cultured cells exhibited higher tolerance to SO_2, but not to O_3. The relationship between O_3 and active oxygen toxicity seems to be a difficult problem. Various genes of the enzymes participating in the active-oxygen-scavenging system have been isolated in many laboratories, and the methods for producing transgenic plants have become routine. By using this technique, environmental stress-tolerant plants will be developed soon. Concomitantly, clarification of the roles of enzymes in the active-oxygen-defense system will be done.

V. ACTIVE OXYGEN TOXICITY AND HERBICIDES

Plant tolerance to herbicides is intimately related to the mode of action of the herbicide. The latter can be roughly grouped into chloroplast-associated reactions and other reactions.[36] Among the herbicides producing chloroplast-associated reactions, the action mechanism of bipyridylium ions (paraquat, diquat, etc.) and inhibitors of carotenoid synthesis (aminotrole, norflurazon, acifluorfen, etc.) have been shown to be closely related to the formation of active oxygen. The bipyridyliums accept electrons from PSI and are reoxidized by molecular oxygen. Reactive superoxide anions result from reoxidation of the bipyridylium radicals. The inhibitors of carotenoid synthesis depress the dissipation of the absorbed excitation energy of chlorophyll to carotenoids and then the excess electrons divert to the oxygen molecule. It is a well-known fact that carotenoids themselves scavenge active oxygen.[37] Therefore, the primary reaction mechanism of these herbicides can be defined as the perturbation of plant metabolism by active oxygen.

When photosynthetic electron transport is inhibited by herbicides such as diuron and monuron, chlorosis occurs. The degradation of pigments in the presence of electron transport inhibitors is considered to result from photooxidations induced by the inability of chlorophyll to dissipate its absorbed excitation energy. The order of the sequential destruction of pigments is as follows: β-carotene, chlorophyll *a*, xanthophylls, and chlorophyll *b*. As the pigments decreased in concentration, lipid peroxidation, measured by the production of malondialdehyde, increased. Thus, since all herbicides cause as a final expression, phytotoxicity, the tolerance mechanism of higher plants to the herbicides seems to relate, more or less, to the active-oxygen-scavenging system.

We examined the relationship between the active-oxygen-scavenging system and the tolerance of green algae to cimetryne, a photosystem II electron transport inhibitor[38] that is frequently found in contaminated rivers in Japan. *Chlorella* exhibited a much stronger tolerance to cimetryne than other green algae. *Chlorella* has more active oxygen scavengers than other green algae. Although there are many kinds of herbicides, it is conceivable that active oxygen may appear, in some process, from the primary reaction to the last withering step of leaves with plant injuries.

REFERENCES

1. **Heck, W. W.,** Effects of photochemical oxidants on vegetation and certain microorganism, in Air Quality Criteria for Photochemical Oxidants, Publ. No. AP-63-6-I-6-23, National Air Pollution and Control Administration, 1970, chap.6.
2. **Natori, T. and Totsuka, T.,** Responses of dry weight growth under SO_2 stress in an SO_2 -tolerant plant, *Polygonum cuspidatum, Ecol. Res.*, 3, 1, 1988.
3. **Tanaka, Y., Chisaka, H., and Saka, H.,** Movement of paraquat in resistant and susceptible biotypes of *Erigeron philadelphicus* and *E. canadensis, Physiol. Plant.*, 66, 605, 1986.
4. **Furusawa, I., Tanaka, K., Thanutong, P., Mizuguchi, A., Yazaki, M., and Asada, K.,** Paraquat-resistant tobacco calluses with enhanced superoxide dismutase activity, *Plant Cell Physiol.*, 25, 1247, 1984.
5. **Simazaki, K., Sakaki, T., Kondo, N., and Sugahara, K.,** Active oxygen participation in chlorophyll destruction and lipid peroxidation in SO_2-fumigated leaves of spinach, *Plant Cell Physiol.*, 21, 1193, 1980.
6. **Sakaki, T., Kondo, N., and Sugahara, K.,** Breakdown of photosynthetic pigments and lipids in spinach leaves with ozone fumigation: role of active oxygens, *Physiol. Plant.*, 59, 28, 1983.
7. **Tanaka, K. and Sugahara, K.,** Role of superoxide dismutase in defense against SO_2 toxicity and an increase in superoxide dismutase activity with SO_2 fumigation, *Plant Cell Physiol.*, 21, 601, 1983.
8. **Tanaka, K., Suda, Y., Kondo, N., and Sugahara, K.,** O_3 tolerance and the ascorbate-dependent H_2O_2 decomposing system in chloroplasts, *Plant Cell Physiol.*, 26, 1425, 1985.
9. **Tanaka, K., Saji, H., and Kondo, N.,** Immunological properties of spinach glutathione reductase and inductive biosynthesis of the biosynthesis of the enzyme with ozone, *Plant Cell Physiol.*, 29, 637, 1988.
10. **Abeliovich, A., Kellenberg, D., and Shilo, M.,** Effects of photooxidative conditions on levels of superoxide dismutase in *Anacystis nidulans, Photochem. Photobiol.*, 19, 379, 1974.
11. **Fridovich, I.,** Biological aspects of superoxide radical and superoxide dismutases, in *Biochemical and Medical Aspects of Active Oxygen*, Hayaishi, G. and Asada, K., Eds., University of Tokyo Press, Japan. 1977, 171.

12. **Gregory, E. M. and Fridovich, I.,** Induction of superoxide dismutase by molecular oxygen, *J. Bacteriol.*, 114, 543, 1973.
13. **Gregory, E. M., Goscin, S. A., and Fridovich, I.,** Superoxide dismutase and oxygen toxicity in a eukaryote, *J. Bacteriol.*, 117, 456, 1974.
14. **Hassan, H. M. and Fridovich, I.,** Regulation of the synthesis of superoxide dismutase in *Escherichia coli:* induction by methyl viologen, *J. Biol. Chem.*, 252, 7667, 1977.
15. **Ziegler, I.,** The effects of SO_3^{--} on the activity of ribulose-1,5-diphosphate carboxylase in isolated spinach chloroplasts, *Planta*, 103, 155, 1972.
16. **Tanaka, K., Otsubo, T., and Kondo, N.,** Participation of hydrogen peroxide in the inactivation of Calvin-Benson cycle SH enzymes in SO_2-fumigated spinach leaves, *Plant Cell Physiol.*, 23, 1009, 1982.
17. **Asada, K. and Kiso, K.,** Initiation of aerobic oxidation of sulfite by illuminated chloroplasts, *Eur. J. Biochem.*, 23, 253, 1973.
18. **Kaiser, W. M.,** Reversible inhibition of the Calvin cycle and activation of the oxidative pentose phosphate cycle in isolated intact chloroplasts by hydrogen peroxide, *Planta*, 145, 377, 1979.
19. **Kaiser, W. M.,** The effect of hydrogen peroxide on CO_2 fixation of isolated intact chloroplasts, *Biochem. Biophys. Acta*, 440, 476, 1976.
20. **Tanaka, K., Kondo, N., and Sugahara, K.,** Accumulation of hydrogen peroxide in chloroplasts of SO_2-fumigated spinach leaves, *Plant Cell Physiol.*, 23, 999, 1982.
21. **Heldt, H. W., Chan, C. J., Lilley, R. McC., and Portis, A.,** The role of fructose and sedoheptulose bisphosphatase in the control of CO_2 fixation. Evidence for the effects of Mg^{2+} concentration, pH and H_2O_2, in *Proceedings of the Fourth International Congress on Photosynthesis*, Hall, D. O., Coombs, J., and Goodwin, T. W., Eds. Biochemical Society, London, 1978, 459.
22. **Tanaka, K., Mitsuhashi, H., Kondo, N., and Sugahara, K.,** Further evidence for inactivation of fructose-1,6-bisphosphatase at the beginning of SO_2 fumigation. Increase in fructose-1,6-bisphosphate and decrease in fructose-6-phosphate in SO_2-fumigated spinach leaves, *Plant Cell Physiol.*, 23, 1467, 1982.
23. **Tanaka, K., Machida, T., and Sugimoto, T.,** Ozone tolerance and glutathione reductase in tobacco cultivars, *Agric. Biol. Chem.*, 54, 1061, 1990.
24. **Tepperman, J. M. and Dunsmuir, P.,** Transformed plants with elevated levels of chloroplastic SOD are not more resistant to superoxide toxicity, *Plant Mol. Biol.*, 14, 501, 1990.
25. **Bowler, C., Slooter, L., Vandenbrander, S., Rycke, R. D., Botterman, J., Sybesma, C., Montagu, M. V., and Inze, D.,** Manganese superoxide dismutase can reduce cellular damage mediated by oxygen radicals in transgenic plants, *EMBO J.*, 10, 1723, 1991.
26. **Furusawa, I. and Mizuguchi, A.,** Production of herbicide resistant plants using cell culture, *Biol. Ind.*, 4, 11, 1987 (in Japanese).
27. **Tanaka, K., Furusawa, I., Kondo, N., and Tanaka, K.,** SO_2 tolerance of tobacco plants regenerated from paraquat-tolerant calluses, *Plant Cell Physiol.*, 29, 743, 1988.
28. **Scott, M. D., Meshnick, S. R., and Eaton, J. W.,** Superoxide dismutase rich bacteria. Paradoxical increase in oxidant toxicity, *J. Biol. Chem.*, 262, 3640, 1987.
29. **Yim, M. B., Chock, P. B., and Stadtman, E. R.,** Copper, zinc superoxide dismutase catalyses hydroxyl radical production from hydrogen peroxide, *Proc. Natl. Acad. Sci. U.S.A.*, 87, 5006, 1990.
30. **Foste, J. G. and Mess, J. L.,** Responses of superoxide dismutase and glutathione reductase activities in cotton leaf tissue exposed to an atmosphere enriched in oxygen, *Plant Physiol.*, 66, 482, 1980.
31. **Gamble, P. E. and Burke, J. J.,** Effect of water stress on the chloroplast antioxidant system. I. Alterations in glutathione reductase activity, *Plant Physiol.*, 76, 615, 1984.
32. **Schmidt, A. and Kunert, K. J.,** Lipid peroxidation in higher plants: the role of glutathione reductase, *Plant Physiol.*, 82, 700, 1986.

33. **Greer, S. and Perham, R. N.,** Glutathione reductase from *Escherichia coli:* cloning and sequence analysis of the gene and relationship to other flavoprotein disulfide oxidoreductases, *Biochemistry,* 25, 2736, 1986.
34. **Aono, M., Kubo, A., Saji, H., Natori, T., Tanaka, K., and Kondo, N.,** Resistance to active oxygen toxicity of transgenic *Nicotiana tabacum* that expresses the gene for glutathione reductase from *Escherichia coli, Plant Cell Physiol.,* 32, 691, 1991.
35. **Drumm-Herrel, H., Gerhausser, U., and Mohr, H.,** Differential regulation by phytochrome of the appearance of plastidic and cytoplasmic isoforms of glutathione reductase in mustard *(Sinapis alba L.)* cotyledons, *Planta,* 178, 103, 1989.
36. **Moreland, O. E.,** Mechanisms of action of herbicides, *Annu. Rev. Plant Physiol.,* 31, 597, 1980.
37. **Foote, C. S.,** in *Singlet Oxygen,* Wasserman, H. H. and Myrray, R. W., Eds., Academic Press, New York, 1979, 139.
38. **Matsumoto, H. and Ishizuka, K.,** Effects of simetryne on the photochemical reactions of isolated chloroplasts of rice and barnyard grass plants, *Weed Res. (Japan),* 27, 91, 1982.

INDEX

A

B

C

D

E

F

G

H

I

K

L

M

N

O

P

Q

R

S

T

U

V

W

X

Z